AF411219

Sustainable Environmental Solutions

Editors

Sergio Ferro
Marco Vocciante

MDPI • Basel • Beijing • Wuhan • Barcelona • Belgrade • Manchester • Tokyo • Cluj • Tianjin

Editors
Sergio Ferro
Ecas4 Australia Pty Ltd
Mile End South
South Australia
Australia

Marco Vocciante
Department of Chemistry and
Industrial Chemistry
University of Genoa
Genova
Italy

Editorial Office
MDPI
St. Alban-Anlage 66
4052 Basel, Switzerland

This is a reprint of articles from the Special Issue published online in the open access journal *Applied Sciences* (ISSN 2076-3417) (available at: www.mdpi.com/journal/applsci/special_issues/ Sustainable_Environmental_Solutions).

For citation purposes, cite each article independently as indicated on the article page online and as indicated below:

LastName, A.A.; LastName, B.B.; LastName, C.C. Article Title. *Journal Name* **Year**, *Volume Number*, Page Range.

ISBN 978-3-0365-1812-1 (Hbk)
ISBN 978-3-0365-1811-4 (PDF)

© 2021 by the authors. Articles in this book are Open Access and distributed under the Creative Commons Attribution (CC BY) license, which allows users to download, copy and build upon published articles, as long as the author and publisher are properly credited, which ensures maximum dissemination and a wider impact of our publications.
The book as a whole is distributed by MDPI under the terms and conditions of the Creative Commons license CC BY-NC-ND.

Contents

About the Editors

Sergio Ferro

Sergio Ferro was born in Italy in 1972 and obtained his Ph.D. in Chemical Sciences in 2002. His interests include electrochemical reactivity, material chemistry, surface science, and environmental chemistry, with particular attention to application aspects, such as the development of new electrode materials for industrial electrochemistry, the use of electrochemical methods for water disinfection, and the remediation of water and soils. In 2015, he left the University of Ferrara and accepted the position of technical manager at Ecas4 Australia, where he deals with the technical aspects related to the use of electrochemically activated solutions. Having worked with companies such as Industrie De Nora, Henkel, ENI, Saipem, HERA, and Electrolux, he has learned to combine commercial interests (and related intellectual property) with a decent scientific production. He authored about a hundred publications, including scientific articles and book chapters, and about ten patent applications.

Marco Vocciante

Marco Vocciante was born in Italy in 1987 and obtained his Ph.D. in Chemical Engineering in 2016; he is currently an Assistant Professor at the Department of Chemistry and Industrial Chemistry (DCCI), University of Genoa, Italy. His interests include process intensification and energy optimization, equipment fluid dynamics investigation and modeling, remediation technologies, raw materials recovery from urban and industrial waste, circular economy, management and automation of metallurgical processes, and nanoparticles synthesis by innovative and sustainable methods.

He is author of about 60 publications, including scientific articles in peer-reviewed ISI journals and book chapters. He is currently a member of the research team of the ReTeST project on the technical, economic, sustainability and risk analysis of innovative environmental technologies, in collaboration with ENI SpA and IRET-CNR, and the Erasmus+ coordinator for the interchange between DCCI and the Paderborn University, Germany.

Editorial

Sustainable Environmental Solutions

Sergio Ferro [1,*] and **Marco Vocciante** [2]

1   Ecas4 Australia Pty Ltd., Mile End South 5031, Australia
2   Department of Chemistry and Industrial Chemistry, University of Genova, 16146 Genova, Italy; marco.vocciante@unige.it
*   Correspondence: sergio@ecas4.com.au

Abstract: In recent decades, increasing attention has been paid to the sustainability of products and processes, including activities aimed at environmental protection, site reclamation or treatment of contaminated effluents, as well as the valorization of waste through the recovery of resources. Although implemented with 'noble intentions', these processes are often highly invasive, unsustainable and socially unacceptable, as they involve significant use of chemical products or energy. This Special Issue is aimed at collecting research activities focused on the development of new processes to replace the above-cited obsolete practices. Taking inspiration from real problems and the need to face real cases of contamination or prevent potentially harmful situations, the development and optimization of 'smart' solutions, i.e., sustainable not only from an environmental point of view but also economically, are discussed in order to encourage as much as possible their actual implementation.

Keywords: environmental pollution and remediation; hazardous waste management; circular economy; soil and water reclamation; nanomaterials; sustainable processes

check for **updates**

Citation: Ferro, S.; Vocciante, M. Sustainable Environmental Solutions. *Appl. Sci.* **2021**, *11*, 6868. https://doi.org/10.3390/app11156868

Received: 14 July 2021
Accepted: 23 July 2021
Published: 26 July 2021

Publisher's Note: MDPI stays neutral with regard to jurisdictional claims in published maps and institutional affiliations.

Copyright: © 2021 by the authors. Licensee MDPI, Basel, Switzerland. This article is an open access article distributed under the terms and conditions of the Creative Commons Attribution (CC BY) license (https://creativecommons.org/licenses/by/4.0/).

1. Introduction

The term 'sustainability' is generally used today when discussing possible improvements to problems such as excessive exploitation of natural resources, excessive use of energy, or the release of polluting by-products during manufacturing operations. Starting from the assumption (not always assured) that ecosystems will continue to operate and maintain the conditions that make it possible not to diminish the quality of life of today's modern societies, if a process or action causes little, less or no damage to the natural world, this process or action is considered 'sustainable'. The goal of a sustainable process/action is to maintain a balance between the exploitation of resources and the improvement of the quality of life of our modern societies and to increase the current and future potential to satisfy human needs and aspirations.

This Special Issue proposes 'sustainable environmental solutions' in relation to various activities, which have been divided into the following categories: sustainable remediation, sustainable development, and sustainable production. We hope that readers will be able to find some interesting answers or an incentive to contribute to Volume 2 of this editorial work.

2. Sustainable Remediation

Sustainable remediation is a modality of intervention in which the effects of the implementation of environmental restoration are taken into account through actions that minimize the environmental footprint, i.e., the demands in terms of energy (through the use of renewable energy), the use of materials and the production of waste (through the reuse and recycling of materials and waste), the use of water and/or the impact on water resources, the emission of air pollutants and greenhouse gas and the use of land and the impact on ecosystems.

The use of Luminescent Solar Concentrators (LSC) in combination with phytoremediation is an example of how energy savings can be exploited to achieve self-sufficiency of greenhouses [1].

The efficiency improvement of simple and economical processes such as solid–liquid separation (a key operation in wastewater treatment) can be promoted through the optimization of the hydrodynamic behavior of suspended particles and the rheology of sludge [2].

With regard to wastewater treatment, Pietrelli et al. [3] evaluated the potential application of chitosan, a low-cost and environmentally friendly adsorbent, in the treatment of compounds with highly toxic and carcinogenic effects on biological systems such as chromium ions, while Kim and Han [4] investigated on the use of hydrogen nanobubbles to improve the electrokinetic remediation of copper-contaminated soils. Heidarrezaei et al. [5] managed to isolate and characterize a new bacterium (*Lysinibacillus boronitolerans*) capable of breaking down trichloroacetic acid, a member of the class of halogenated organic compounds widely used as solvents, herbicides and pesticides, but unfortunately carcinogenic to humans and animals, and Kulikova et al. [6] proposed a cost-effective approach to the synthesis of a magnesium potassium phosphate matrix, which is promising for the solidification of radioactive waste on an industrial scale.

On the other hand, a sustainable approach to remediation can also start from the sustainable detection of the contaminants to be addressed. An example of this is the development of a cork-modified carbon paste electrode for the determination of Pb (II), which has proved to be a sensitive electrochemical sensor capable of meeting stringent environmental control requirements while being economical, simple and highly selective [7]. In some cases, it may also be useful to conduct an economic evaluation of the intervention through the contingent evaluation (CV) to verify if the reduction is also socially advantageous or if the willingness to pay (WTP) for the reduction is greater than the costs involved in the reduction [8].

3. Sustainable Development

Speaking of sustainable development means referring to practices that make it possible to satisfy the needs of the present without compromising the ability of future generations to satisfy theirs. It means thinking about the future by balancing environmental, social and economic considerations to pursue a better quality of life. The logical or necessary consequences for society, culture, economy and the environment are interconnected and must be considered as such.

In the line of pursuing better energy efficiency in human activities, which would result in more sustainable use of resources, techniques aimed at improving the energy performances of buildings are of paramount importance, with the construction sector responsible for almost 40% of both energy consumption and the release of pollutants into the atmosphere. Among these, green roofs are becoming increasingly popular due to their ability to reduce the (electrical) energy requirements for the (summer) climatization of buildings, thus also positively influencing the internal comfort levels for the occupants [9]. The transition towards a low-carbon path should also involve agritourism buildings through the issuing of community directives, laws in member states and technical rules, including evaluation tools to assess the environmental improvements resulting from energy efficiency interventions in buildings [10].

The main obstacles to sustainable consumption are the lack of adequate infrastructure and a lack of knowledge. Infrastructure barriers in some situations make sustainable consumer behavior impossible or inconvenient (who therefore prefer other types of consumption), or in some cases require additional expenditure of time and money, thus leading to a reduction in the practice of sustainable consumer behavior [11].

To achieve the goal of ecological sustainability, influencing factors that could reduce ecological consumption need to be explored, to provide guidance for evidence-based policymaking on reducing ecological consumption [12].

Finally, the concept of virtual water, as a new approach to addressing water shortage and safety issues, can help support sustainable development in water-scarce regions [13].

4. Sustainable Production

The production of products is always linked to the extraction and consumption of natural raw materials and the use of the land. During the production process, pollutants are released into the soil, air and water and along the entire supply chain. The goal of sustainable production is to guarantee the conservation of resources and the ability of the environment to regenerate; this can be achieved by relying on processes and systems that are: non-polluting; able to limit the consumption of energy and natural resources; economically sustainable; safe and healthy for workers, communities and consumers and socially and creatively rewarding for all workers.

The growing pressure to comply with legislation and to adopt environmental strategies due to environmental concerns has led to the development of new sustainable supply chains, where a new area for a production and reconditioning system has emerged. In this complex scenario, optimal decisions can no longer avoid simultaneously considering strategies on carbon emissions, carbon tax and compulsory emissions [14].

This also applies to the food industry, with the constant search for sustainable strategies to maximize the effectiveness of the approaches and minimize the number of processes required for the production of safe food [15], the amount of drinking water required [16] and the chemicals involved, which must be as environmentally friendly and cost-effective as possible [17].

To minimize the environmental impact of industrial production, the treatment of waste linked to production is decisive, which at least in the past has always suffered from scarce attention as it is perceived as not aimed at generating value and therefore profit. In this regard, it is also appropriate to evaluate investment projects for waste treatment and try to understand their impact on the development of environmental policies [18].

5. Future Advances in Sustainable Environmental Solutions

It is widely believed that reconciling economic and environmental interests is not possible because they are conflicting interests. Many wonder whether it is actually possible to meet people's needs for food, water and energy by doing more to protect nature. We think the answer is yes, but we need a path to get there, and we need to make it urgently. Changing course over the next 10 years requires global collaboration at levels likely to be comparable to those seen after the Second World War. Protecting nature and providing water, food and energy to a growing world do not necessarily need to be mutually exclusive interests; on the contrary, energy, water, air, health and ecosystem initiatives are needed that intelligently balance economic growth and resource conservation needs. Achieving a sustainable future will depend on our ability to ensure both thriving human communities and abundant and healthy natural ecosystems.

We look forward to reporting on further advances in Volume 2 of the Special Issue 'Sustainable Environmental Solutions', which will soon be available to receive contributions from authors from around the world.

Funding: This research received no external funding.

Conflicts of Interest: The authors declare no conflict of interest.

References

1. Pedron, F.; Grifoni, M.; Barbafieri, M.; Petruzzelli, G.; Franchi, E.; Samà, C.; Gila, L.; Zanardi, S.; Palmery, S.; Proto, A.; et al. New light on phytoremediation: The use of luminescent solar concentrators. *Appl. Sci.* **2021**, *11*, 1923. [CrossRef]
2. Trofa, M.; D'Avino, G. Sedimentation of fractal aggregates in shear-thinning fluids. *Appl. Sci.* **2020**, *10*, 3267. [CrossRef]
3. Pietrelli, L.; Francolini, I.; Piozzi, A.; Sighicelli, M.; Silvestro, I.; Vocciante, M. Chromium(III) removal from wastewater by chitosan flakes. *Appl. Sci.* **2020**, *10*, 1925. [CrossRef]
4. Kim, D.; Han, J. Remediation of copper contaminated soils using water containing hydrogen nanobubbles. *Appl. Sci.* **2020**, *10*, 2185. [CrossRef]

5.	Heidarrezaei, M.; Shokravi, H.; Huyop, F.; Rahimian Koloor, S.; Petrů, M. Isolation and characterization of a novel bacterium from the marine environment for trichloroacetic acid bioremediation. *Appl. Sci.* **2020**, *10*, 4593. [CrossRef]

6.	Kulikova, S.; Vinokurov, S.; Khamizov, R.; Vlasovskikh, N.; Belova, K.; Dzhenloda, R.; Konov, M.; Myasoedov, B. The use of MgO obtained from serpentinite in the synthesis of a magnesium potassium phosphate matrix for radioactive waste immobilization. *Appl. Sci.* **2021**, *11*, 220. [CrossRef]

7.	Silva, I.; de Araújo, D.; Vocciante, M.; Ferro, S.; Martínez-Huitle, C.; Dos Santos, E. Electrochemical determination of lead using a composite sensor obtained from low-cost green materials: Graphite/cork. *Appl. Sci.* **2021**, *11*, 2355. [CrossRef]

8.	Jin, S.; Kwon, Y.; Yoo, S. Economic valuation of reducing submerged marine debris in South Korea. *Appl. Sci.* **2020**, *10*, 6086. [CrossRef]

9.	Cirrincione, L.; La Gennusa, M.; Peri, G.; Rizzo, G.; Scaccianoce, G.; Sorrentino, G.; Aprile, S. Green roofs as effective tools for improving the indoor comfort levels of buildings—An application to a case study in Sicily. *Appl. Sci.* **2020**, *10*, 893. [CrossRef]

10.	Cirrincione, L.; La Gennusa, M.; Peri, G.; Rizzo, G.; Scaccianoce, G. Towards nearly zero energy and environmentally sustainable agritourisms: The effectiveness of the application of the European ecolabel brand. *Appl. Sci.* **2020**, *10*, 5741. [CrossRef]

11.	Ratner, S.; Lazanyuk, I.; Revinova, S.; Gomonov, K. Barriers of consumer behavior for the development of the circular economy: Empirical evidence from Russia. *Appl. Sci.* **2021**, *11*, 46. [CrossRef]

12.	Zhang, S.; Zhu, D.; Zhang, J.; Li, L. Which influencing factors could reduce ecological consumption? Evidence from 90 countries for the time period 1996–2015. *Appl. Sci.* **2020**, *10*, 678. [CrossRef]

13.	Wang, W.; Adamowski, J.; Liu, C.; Liu, Y.; Zhang, Y.; Wang, X.; Su, H.; Cao, J. The impact of virtual water on sustainable development in Gansu province. *Appl. Sci.* **2020**, *10*, 586. [CrossRef]

14.	Turki, S.; Sahraoui, S.; Sauvey, C.; Sauer, N. Optimal manufacturing-reconditioning decisions in a reverse logistic system under periodic mandatory carbon regulation. *Appl. Sci.* **2020**, *10*, 3534. [CrossRef]

15.	Tenzin, S.; Ferro, S.; Khan, S.; Deo, P.; Trott, D. Spray and aerosolised pH-neutral electrochemically activated solution reduces Salmonella Enteritidis and total bacterial load on egg surface. *Appl. Sci.* **2021**, *11*, 732. [CrossRef]

16.	Gámiz, J.; Grau, A.; Martínez, H.; Bolea, Y. Automated chlorine dosage in a simulated drinking water treatment plant: A real case study. *Appl. Sci.* **2020**, *10*, 4035. [CrossRef]

17.	Tenzin, S.; Ogunniyi, A.; Ferro, S.; Deo, P.; Trott, D. Effects of an eco-friendly sanitizing wash on spinach leaf bacterial community structure and diversity. *Appl. Sci.* **2020**, *10*, 2986. [CrossRef]

18.	Shvetsova, O.; Lee, J. Minimizing the environmental impact of industrial production: Evidence from South Korean waste treatment investment projects. *Appl. Sci.* **2020**, *10*, 3489. [CrossRef]

Article

New Light on Phytoremediation: The Use of Luminescent Solar Concentrators

Francesca Pedron [1], Martina Grifoni [1], Meri Barbafieri [1], Gianniantonio Petruzzelli [1], Elisabetta Franchi [2], Carmen Samà [3], Liliana Gila [3], Stefano Zanardi [3], Stefano Palmery [3], Antonio Proto [3] and Marco Vocciante [4,*]

[1] Institute of Research on Terrestrial Ecosystems, National Council of Research, 56124 Pisa, Italy; francesca.pedron@cnr.it (F.P.); martina.grifoni@cnr.it (M.G.); meri.barbafieri@cnr.it (M.B.); gianniantonio.petruzzelli@cnr.it (G.P.)

[2] Eni S.p.A., Renewable Energy & Environmental Laboratories, 20097 Milan, Italy; elisabetta.franchi@eni.com

[3] Eni S.p.A., Renewable Energy & Environmental R&D, 28100 Novara, Italy; carmen.sama@eni.com (C.S.); liliana.gila@eni.com (L.G.); stefano.zanardi@eni.com (S.Z.); stefano.palmery@eni.com (S.P.); antonio.alfonso.proto@eni.com (A.P.)

[4] Department of Chemistry and Industrial Chemistry, University of Genova, 16146 Genova, Italy

*   Correspondence: marco.vocciante@unige.it

Featured Application: Under Luminescent Solar Concentrators (LSC), plants used in a phytoremediation feasibility test appear to grow better than plants grown in conventional greenhouse. This result and the energy savings characteristics of LSC highlight the prospective of LSC to further contribute in developing green remediation strategies.

Abstract: The latest developments in photovoltaic studies focus on the best use of the solar spectrum through Luminescent Solar Concentrators (LSC). Due to their structural characteristics, LSC panels allow considerable energy savings. This significant saving can also be of great interest in the remediation of contaminated sites, which nowadays requires green interventions characterized by high environmental sustainability. This study reported the evaluation of LSC panels in phytoremediation feasibility tests. Three plant species were used at a microcosm scale on soil contaminated by arsenic and lead. The experiments were conducted by comparing plants grown under LSC panels doped with Lumogen Red F305 (BASF) with plants grown under polycarbonate panels used for greenhouse construction. The results showed a higher production of biomass by the plants grown under the LSC panels. The uptake of the two contaminants by plants was the same in both the growing conditions, thus resulting in an increased total accumulation (defined as metal concentration times produced biomass) in plants grown under LSC panels, indicating an overall higher phytoextraction efficiency. This seems to confirm the potential that LSCs have to be building-integrated on greenhouse roofs, canopies, and shelters to produce electricity while increasing plants productivity, thus reducing environmental pollution, and increasing sustainability.

Keywords: soil remediation; soil contamination; greenhouse; phytoextraction; mobilizing agents; photosynthetic efficiency; photovoltaics; luminescent dyes; energy savings; sustainability

Citation: Pedron, F.; Grifoni, M.; Barbafieri, M.; Petruzzelli, G.; Franchi, E.; Samà, C.; Gila, L.; Zanardi, S.; Palmery, S.; Proto, A.; et al. New Light on Phytoremediation: The Use of Luminescent Solar Concentrators. *Appl. Sci.* **2021**, *11*, 1923. https://doi.org/10.3390/app11041923

Academic Editor: Luisa F. Cabeza

Received: 19 January 2021
Accepted: 19 February 2021
Published: 22 February 2021

Publisher's Note: MDPI stays neutral with regard to jurisdictional claims in published maps and institutional affiliations.

Copyright: © 2021 by the authors. Licensee MDPI, Basel, Switzerland. This article is an open access article distributed under the terms and conditions of the Creative Commons Attribution (CC BY) license (https://creativecommons.org/licenses/by/4.0/).

1. Introduction

In recent years, research for photovoltaics development has been oriented towards the search for lower costs and higher conversion efficiencies. One of the fields of investigation concerns the optimal use of the solar spectrum by means of Luminescent Solar Concentrator (LSC). A typical LSC consists of a sheet of transparent material (generally polymeric matrix as PolyMethylMethAcrylate (PMMA)) where luminescent particles are homogeneously dispersed. The luminescent particles can selectively absorb the solar radiation and re-emit the energy absorbed at longer wavelengths, where PhotoVoltaic (PV) cells exhibit the highest efficiency. Therefore, the light is captured by the larger, planar surface of the slab, and the main part (~75%) of the converted radiation is waveguided in the slab's plane (thanks to the total internal reflection) and concentrated on the smaller PV cell area on the edges to produce electricity (Figure 1). This reduces the amount of Silicon cells needed to generate a particular amount of energy and the overall cost of the panel, since the waveguiding material is inexpensive.

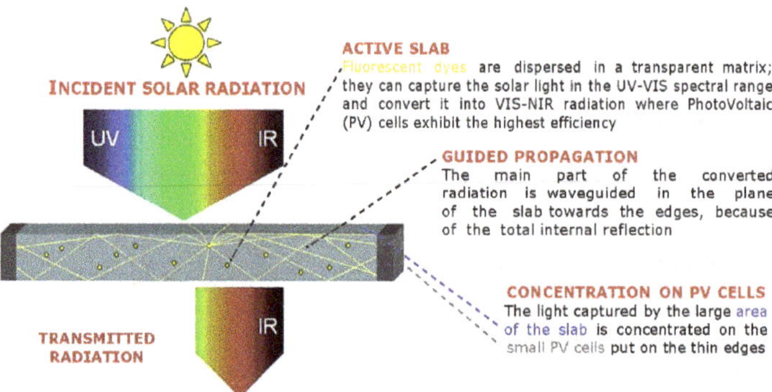

Figure 1. Diagram of incident photons and photons emitted by a dye molecule inside the Luminescent Solar Concentrator (LSC).

The hypothesis of converting the incident solar spectrum into monochromatic light in the LSCs was already proposed at the end of the 1970s, when LSCs based on organic dyes were first introduced as a low-cost technology to enhance the power conversion efficiency (PCE) of PV solar cells through improving their spectral response [1]. They can collect both diffuse and direct solar radiation, making them a suitable technology to be used in countries where diffuse solar radiation is dominant (more than 50%) such as in northern European countries [2]. Unlike traditional PV panels, LSCs are transparent and can be made with a wide variety of colors and shapes. Thanks to these characteristics, LSCs can be seen as potential structural energy components in the design of PV windows, skylights, and colored PV panels in building facades, but also of noise barriers, advertisement signs, shelters, agricultural covers, and so on, reducing environmental pollution caused by fossil fuels.

The considerable energy savings achievable by using LSC panels could also have an interesting application in the field of remediation of contaminated sites. Several different physicochemical and biological approaches have been suggested for the remediation of contaminated water [3–5] and soils [6,7], so that selecting a suitable technology is often a difficult yet crucial step for the successful reclamation of a contaminated site [8,9]. However, activities aimed at the remediation of contaminated sites or the treatment of effluents also have an environmental impact, since they make use of chemical products or processes, with consequent consumption of raw materials and energy. In many cases these aspects are not negligible, and could compromise the sustainability of the approach or even invalidate its beneficial aspects. This is the case with the Electro Kinetic Remediation Technology (EKRT) [10,11], which has proved particularly interesting and efficient in dealing with various types of contaminations allowing in situ interventions, and demonstrates a more ecological character, compared to other approaches [12], but suffers from some critical issues including high energy consumption [13] that require new solutions to confirm the technology as operationally valid.

In light of a growing demand for remediation interventions characterized by high environmental sustainability, there is a considerable request to promote those technologies with a reduced impact on the environment, among which a primary role is attributed to phytoremediation. This term includes a series of technologies based on the use of plants to remediate organic and inorganic contaminants in soil and other environmental matrices (sediments, water). The interest in these phytotechnologies has increased over time, given some significant advantages in their use compared to traditional remediation technologies: low cost, simplicity of operation, environmental benefits. Recent contributions based on Life Cycle Assessment (LCA) comparison of different technologies clearly show the major advantage of phytoremediation in environmental impact and ecological footprint with respect to consolidated technologies or excavation and landfill disposal [14].

Phytotechnologies fall entirely within the green remediation category [15], and given their minimal environmental impact, have been proposed as an effective nature-based solution (NBS), as they ensure environmental remediation in a sustainable and economically efficient way [16,17].

Phytoremediation includes several decontamination processes; the most used are:

- Phytoextraction: process of contaminants extraction (organic and inorganic) from the soil through roots and subsequent translocation and accumulation in plant tissues (roots and shoots);
- Phytodegradation: degradation of organic contaminants through plant biochemical processes by the increase in the microbial activity, which promotes the degradation of a contaminant in the soil;
- Rhizofiltration: decontamination of polluted water carried out by aquatic plants either floating or submerged which uptake and concentrate, by their roots, the contaminants, removing them from aqueous environments;
- Phytostabilization: containment or immobilization of contaminants in the rhizospheric region of plants by adsorption or precipitation preventing the leaching of dissolved contaminants and the aerial dispersion of contaminated soil particles;
- Phytovolatilization: adsorption of contaminants by the root system, followed by translocation into the shoots and release into the atmosphere through the leaves' transpiration process.

Among the various phytotechnologies in contaminated sites, phytoextraction is the most important for the removal of heavy metals. In the last twenty years, several innovative strategies have been developed to maximize the efficiency of the approach, which depends both on the biomass produced and the quantity of metal absorbed by the plants [18]. These strategies are aimed both to improve the performance of plants by using species with high biomass production and by increasing the bioavailability of contaminants through appropriate soil treatments with additives capable of increasing the concentration of metals in the liquid phase of the soil [19,20]. Among these innovative strategies, the possibility of relying on plant growth-promoting bacteria (PGPB) to improve the effectiveness of phytoextraction processes is of particular interest. These beneficial plant bacteria living in close association with roots have several positive effects on plant growth and develop PGPB can significantly contribute to increase plants metal uptake and, consequently, the efficiency and the rate of phytoextraction [20,21]. The increase in uptake is often further enhanced by simultaneous addition of metals mobilizing agents such as chelating compounds [22], to be selected and dosed appropriately in order not to penalize the sustainability of the approach, and possibly monitor with non-invasive approaches to avoid uncontrolled diffusion in the environment [23].

However, since phytoremediation is a highly site-specific approach, each innovation requires preliminary tests on an increasing scale to verify its effectiveness before full-scale applicability, including the setting up of dedicated greenhouses as a suitable environment for feasibility tests to optimally prepare full-scale interventions [24]. These increasing-scale trials mostly take place in greenhouses that are subject to considerable costs when they require the maintenance of optimal temperature conditions.

In a controlled environment, it is possible to better study the use of additives that increase the bioavailability of contaminants. The evaluation of the responses to plant stress and all those measures that can have significant positive effects on the efficiency of phytoremediation, particularly root-microorganism interactions, is often crucial to the success of the technology. Proper management of greenhouse conditions (light, temperature, humidity, and irrigation) can considerably improve phytoremediation feasibility tests' efficiency and speed. In semi-controlled and protected conditions, some obstacles to plant growth can be studied and overcome, such as the reduced biological activity of plants due to seasonality and the bioavailability dynamics of pollutants.

In particular, aim of this study was to provide the information necessary for evaluating the possibility of coupling Luminescent Solar Concentrators with phytoremediation in remediation procedures. Indeed, as the wave-guiding material is semi-transparent and wavelength selective, LSCs could also find promising applications in greenhouse roof panels: by selecting only the light that plants do not use for photosynthesis [25], it should be possible to produce electricity without penalizing plants growth, or possibly even increase agricultural productivity.

For this purpose, phytoextraction tests were carried out with three plant species (*Brassica juncea*, *Helianthus annuus* and *Lupinus albus*) to assess the effect of LSC panels on plant growth and the absorption capacity of contaminants by plants on a soil polluted by arsenic and lead. Experiments were carried out by comparing plants grown under LSC panels versus plants grown under polycarbonate panels. LSC panels doped with Lumogen Red F305 (BASF) as luminescent dye were used to be evaluated as optical filter and not as a PV device for electricity production. The parameters examined were those essential for a phytoremediation feasibility test:

- possibility of plant growth in contaminated soil;
- biomass production;
- uptake of contaminants by plants;
- total removal of pollutants by plants.

This is a highly innovative perspective since, to the best of our knowledge, there are no consolidated studies and results on the use of LSC in the remediation of contaminated sites using phytoremediation technologies, when, due to soil contamination, plants should grow under significant stress conditions.

2. Materials and Methods

2.1. Soil

The soil considered was collected from a former industrial site in Tuscany (Italy) contaminated by lead (Pb) and arsenic (As) arising from manufacturing activities of various chemicals. Soil samples were collected from the 0 to 20 cm layer, air-dried, and sieved through a 2 mm sieve before laboratory analysis. In Table 1, soil pH was determined in a soil/water ratio of 1:2.5 [26], cation exchange capacity (CEC) using barium chloride (pH = 8.1) [27], and texture by the pipette method [28]. Total nitrogen (N) was determined by the Kjeldahl method [29], available phosphorus (P) by extraction with sodium bicarbonate [30], and organic matter by wet combustion [31].

Table 1. Physical–chemical properties of As and Pb contaminated soil. Values represent the mean ($n = 3$) $\pm$ standard deviation.

pH (H_2O)	8.20 ± 0.1
CEC ($Cmol^+\ kg^{-1}$)	17.5 ± 0.3
Sand (%)	73.6 ± 0.3
Silt (%)	18.0 ± 0.2
Clay (%)	8.46 ± 0.2
Textural class (USDA)	Sandy Loam
Inorganic C ($g\ kg^{-1}$)	15.8 ± 1.00
Organic C ($g\ kg^{-1}$)	4.80 ± 0.20
Total N ($g\ kg^{-1}$)	0.30 ± 0.03
Available P ($mg\ kg^{-1}$)	7.60 ± 0.4
As ($mg\ kg^{-1}$)	878 ± 93.2
Pb ($mg\ kg^{-1}$)	572 ± 61.4

2.2. Chelating Agents

Plant uptake is mainly influenced by the bioavailable fractions rather than the total amount in the soil. For the highest efficiency in soil phytoextraction, an increased availability of soluble forms of the contaminants is required. Bioavailability depends on the soil characteristics that determine the release of Pb and As in the soil solution and plants ability to uptake and transfer the metals to their tissues.

A high cation exchange capacity (CEC) and alkaline pH reduce Pb mobility and bioavailability. Consequently, in soils contaminated by Pb, phytoextraction has many limitations, deriving from the behavior of the element in the soil environment.

To increase bioavailability, the uptake and translocation of metals, the addition of chelating agents has been extensively used in phytoextraction, with organic acids being particularly effective in increasing the solubility of metals [32,33].

For many years, chelating agents have been used to increase plants uptake of micronutrients from the soil. Chelating agents' action is mainly based on the release of metals from the soil–solid surfaces and the formation of stable metal complexes in soil solution available for plant uptake. In this experimental campaign, Ethylene Diaminete Traacetic Acid (EDTA) was selected being one of the most used chelant, which increases the uptake of several metals, Pb in particular [34]. Indeed, it was preferred to opt for a well-known and commonly used solution, with positive results in assisted phytoextraction processes, even very recent ones [35,36], in order to reduce uncertainty about this factor and focus more attention on the effect of using LSC panels.

2.3. Photosynthetic Process and Selection of Luminescent Dye

In the process of photosynthesis, plants absorb the solar radiation in the range 400–700 nanometers (this range is called Photosynthetically Active Radiation, PAR), primarily using the pigment Chlorophyll, the most abundant in the plant. In addition to chlorophyll, plants also use other pigments belonging to the carotenoid group.

The chlorophyll exists in two forms: chlorophyll *a* and chlorophyll *b*. Chlorophyll *a* shows a strong light absorption in the blue zone of the solar spectrum as well in the red zone; chlorophyll *b* absorbs mostly blue and orange light. In contrast, both absorb poorly the green and near-green light, which reflects the typical green color of the leaves. In the near-green zone, the absorption of carotenoids takes place (Figure 2).

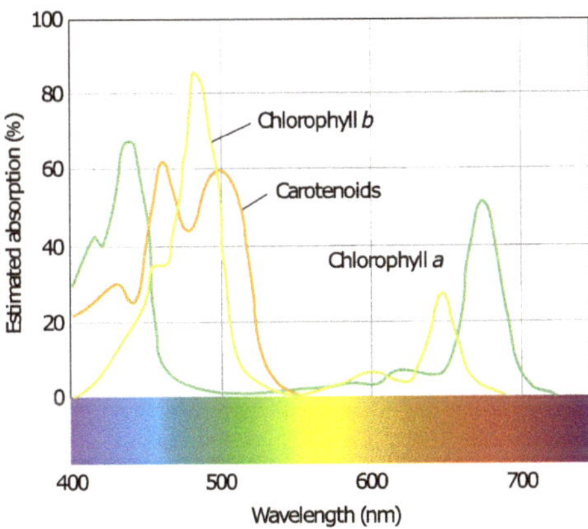

Figure 2. Absorption spectrum for chlorophylls and carotenoids (based on data from [37]).

The photosynthetic efficiency (i.e., the fraction of light energy converted into chemical energy during photosynthesis) depends on the wavelength of light. The red light (600–700 nm) is the most efficient; the efficiency increases when coupled to an equal far-red light (700–800 nm). The violet-blue light, even if less efficient (efficiency = 0.65–0.75) [38], is necessary for photosynthesis because it promotes the development of chloroplasts, where photosynthesis takes place [39]. A low percent of violet-blue light (~7%) is enough to ensure plants good health [40]. Outside the visible spectrum, the UV light (200–400 nm) damages the chloroplasts hindering the photosynthetic process [41].

Considering this, luminescent dye Lumogen F305 (perylene-based molecule by BASF) was chosen for its spectroscopic features, high quantum yield (100%) and good photostability (5% of degraded dye after 4600 h accelerated ageing). Indeed, the absorption spectrum of Lumogen is characterized by a strong band in the range 500–600 nm with maximum absorption at 576 nm and a weak band between 400 and 500 nm; the photoluminescence emission takes place between 600 and 750 nm with a maximum fluorescence peak at 615 nm (Figure 3).

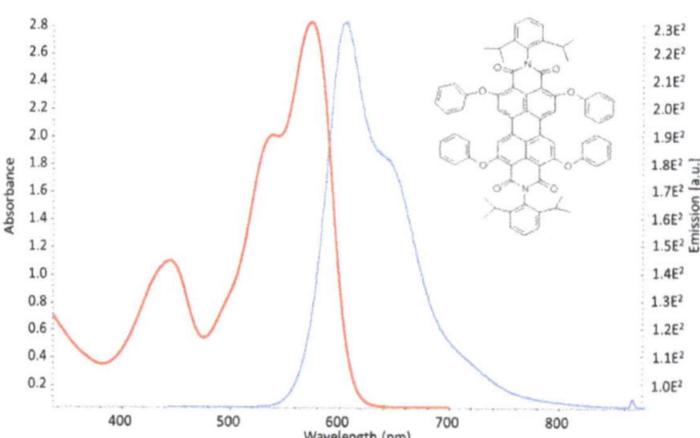

Figure 3. Absorption (red curve) and photoluminescent emission (blue curve) spectra of Lumogen F305.

Green light of solar spectrum, which is largely captured by Lumogen, is not absorbed by chlorophylls and carotenoids. Therefore, a LSC panel doped with Lumogen does not interfere with the photosynthetic process.

High intensity and correct distribution of transmitted sunlight are required to LSC plates for greenhouse application; these conditions allow high growth and good development of plants. Increased conversion of sunlight absorbed in electricity is also needed for making self-supporting greenhouses. In this respect, the amount of luminescent dye in the plate plays a key role.

Different dye concentrations were evaluated to determine the best one in terms of quantity and quality of transmitted light and electricity production. Transmitted light and its distribution in PAR range were measured from the transmittance UV–Vis spectra by using a Perkin Elmer UV–Vis–NIR Lambda 950 spectrometer. Every spectrum was firstly weighted for the solar spectrum by the AM 1.5 Reference Solar Spectrum. Then, the integrated area of the solar weighted transmission spectrum (in the range 400–700 nm) of the sample was perceptualized using the AM 1.5 as reference.

PAR attenuation (%) is the attenuation of the transmitted light (I_{TL}) with respect to the incident light (I_{IL}), as reported in Equation (1):

$$(I_{TL} - I_{IL})/I_{IL} * 100 \tag{1}$$

2.4. LSC Panel Design

A typical greenhouse hard cover is made of double-wall PolyCarbonate (PC) panels with sizes that range from 1.2 × 0.6 m to 3 × 2 m. In contrast with this, large LSC modules rarely exceed a short-side length of 50–60 cm. In fact, the performances of larger devices are limited by self-absorption of the fluorescent dye (multiple absorption-emission events that reduce the light transport efficiency, due to the overlap between the absorption and emission spectra, Figure 3).

A reasonable trade-off is a module with a fixed short side of 50–60 cm and a variable length of 0.5 to 2 m; this size permits to have modules with a large area but where the optical path inside the slab is not so long to limit the performance due to self-absorption [42].

In addition, proper structural characteristics as rigidity and thermal insulation are required in practical use as roofing. Hence, a suitable LSC device for greenhouses requires a dedicated design with the integration of additional materials to fulfill these specifications.

In the present work, LSC panels with dimensions of 50 × 50 × 0.6 cm and 100 × 100 × 0.6 cm were used, fabricated by Altuglas (Arkema Group) using an industrial method of "cell-casting polymerization" [43]. In particular, PMMA ShieldUp® (impact resistant) was used as transparent material. LSCs with ShieldUp® are a patented technology [44] resulting from the collaboration between Arkema Group and Eni S.p.A, specially developed to provide a polymer composition that is highly resistant to shocks, remains transparent regardless of temperature, and possesses greater flexibility, all while it absorbs and re-emits light. These characteristics make it suitable for greenhouse roofing.

Measures of electrical efficiency of a complete LSC device were also performed. Eight silicon cells IXYS SLMD142H01LE (dimensions 24.7 × 0.6 cm each and an active surface of 14.8 cm^2) were glued on the four edges of the slab, wired in series, and connected to a Keithley 2602A (3A DC, 10A Pulse) digital multimeter to record the power response. A 50 × 50 × 0.6 cm device was exposed directly to the sun and the current-voltage (I-V) curves were collected.

The corresponding power conversion efficiency (PCE) was estimated though the following formula:

$$\frac{V_{OC}\cdot J_{SC}\cdot FF}{P_{in}} \tag{2}$$

where in V_{oc} is the open-circuit voltage, J_{sc} is the short-circuit current density, P_{in} is the intensity of the light incident on the device (Global Normal Irradiance GNI = 1000 W/m^2), and FF (Fill Factor) is defined by the following ratio:

$$FF = \frac{V_{MPP}\cdot J_{MPP}}{V_{OC}\cdot J_{SC}} \tag{3}$$

with V_{MPP} and J_{MPP} defined as the voltage and current density, respectively, corresponding to the maximum power point.

Considering Equations (2) and (3), it results $PCE = P_{max}/P_{in} = V_{MPP} * J_{MPP}/P_{in}$.

The power production from the panel estimated in this way represents the peak value, obtained with a naked LSC slab. However, the effective power conversion efficiency is influenced by the final configuration of the device, so measures on the building-integrated panel are deemed necessary.

2.5. Experimental Design

The microcosm tests were carried out in two different conditions: inside a greenhouse (first phase) and outdoor (second phase). In the first phase (Figure 4), the plants were grown on a microcosm scale under a red LSC panel of 100 × 100 cm supplied by Eni, positioned at the height of about 50 cm to allow adequate plant growth. As a comparison, the same tests were set up outside the LSC panel. The first phase aimed to evaluate if the red LSC panel could hinder or reduce the biomass production of the selected species.

Figure 4. Picture of the experiments inside the greenhouse.

In the second phase (Figure 5), outside the greenhouse, tests were set up using small boxes made of polycarbonate and LSC panels, built and supplied by Eni with a size of 50 × 50 cm. Microcosms were placed inside the red LSC box, so that the plants were totally subject to the action of the LSC panels. The same number of microcosms for each species were placed in the polycarbonate box, in order to have a comparison at the same conditions. In this second phase, metals bioavailability has been increased by the addition of EDTA for some samples.

Figure 5. Picture of outdoor experiments. The boxes are made of transparent polycarbonate and red LSC panels. Note: pots filled with white stones at the base of the box structures have the only function of anchoring the panels to avoid possible adverse effects in the event of a mighty wind.

In both phases, the growth of plant species commonly used for phytoremediation, *Brassica juncea (B), Lupinus albus (L)*, and *Helianthus annuus (H)* [20,22] was considered to evaluate the biomass yield and the accumulation of the target metals (As and Pb).

2.5.1. Indoor Microcosm Tests

The phytoextraction test was carried out in 400 mL microcosms, i.e. pots filled with soil in which the selected species are grown. Pots were filled with 300 g of As and Pb contaminated soil.

Ten microcosms per species were prepared, for a total number of microcosms of 30, 15 grown under the LSC panel and 15 outside the panel. Sowing was carried out using 0.5 g of *Brassica juncea*, 6 seeds of *Helianthus annuus,* and 5 seeds of *Lupinus albus*. The experiment was organized in a randomized complete block design.

Microcosms were watered daily (at least twice a day) according to the needs of the plants. *B. juncea* was the plant species that needed the least water. On average 20 mL for *B. juncea* and 25 mL for *L. albus* and *H. annuus*, twice a day.

The whole experiment lasted 30 days. Plants were separated into roots and shoots. Vegetal samples were accurately washed with deionized water, and roots were further sonicated for 5 min with a Branson Sonifier 250 ultrasonic processor (Branson Ultrasonic Corporation, Danbury, CT, USA) to remove the soil particles possibly still present, and then rinsed with deionized water. Vegetal samples were dried up to constant weight in a ventilated oven at 40 °C and each dry weight (DW) was recorded.

2.5.2. Outdoor Microcosm Tests

The microcosm test was conducted outdoor under two boxes measuring $50 \times 50 \times 50$ cm, one in transparent polycarbonate and the other consisting of red LSC panels. The amount of soil per microcosm was 300 g. The selected plant species were the same as the first phase: *B. juncea* (B), *L. albus* (L) and *H. annuus* (H). Sowing was carried out using 0.5 g of brassica seeds, 6 sunflower seeds, and 5 lupine seeds. For each species, tests were conducted in triplicate both without and with the addition of EDTA. The trial lasted about 30 days. Irrigation was carried out based on the daily need of the plants. About 20 days after sowing, treatment with the 2 mM EDTA solution started. A total dose of 10 mL of EDTA solution was added for each treated microcosm, divided into 5 days, diluting the daily dose with water. At the end of both tests, the plants were harvested by separating the leaves and stems from the roots. The fresh weight of the developed biomass was measured and after careful washing, the plant samples were placed in an oven at about 45 °C to dry. The dry weight was then determined, and the samples were prepared to be analyzed and evaluate the amount of As and Pb accumulated in vegetable tissues.

2.6. Lead and Arsenic Analysis in Soil and Plants

Each plant sample (roots and shoots) was ground into fine particles (<1 mm) and digested according to US-EPA 3052 [45]. Total As (using a method for the generation of hydrides) and Pb concentration in soil and plants (aerial part and roots) were analyzed by ICP-OES (Varian AX Liberty. Varian Inc., Palo Alto, CA, USA).

2.7. Quality Assurance and Quality Control

QA/QC were performed by testing two standard solutions (0.5 and 2 mg L^{-1}) every 10 samples. Certified reference materials, CRM ERM e CC141 for soil and CRM ERM - CD281 for plants, were used. The limit of quantification (LOQ) for Pb and As were of 5 and 50 mg L^{-1}, respectively. The recovery of spiked samples ranged from 95% to 101% with a Relative Standard Deviation (RSD) of 1.88 of the mean for Pb and from 94 to 101% with a RSD of 1.91 of the mean for As.

2.8. Statistical Analysis

Statistical analysis was performed using STATISTICA version 6.0 (Statsoft, Inc., Tulsa, OK, USA). Effects of treatments were analyzed using one-way analysis of variance (ANOVA). Differences between means were compared and a post-hoc analysis of variance was performed using the Tukey's honestly significant difference test ($P < 0.05$).

3. Results

3.1. LSC Panels Properties and Performance

Tests with various concentrations of luminescent dye were performed in the range 40–160 ppm, in order to identify the right composition to both maximize the photosynthesis process and the energy production. LSC plate doped with 160 ppm of Lumogen F305 has been identified to have the right characteristics to be used in microcosm experiments. As reported in Table 2, it transmits a sufficient PAR light (about 30%), a low UV light (0.6%), and a high red and far-red light (90%); the blue light is also satisfactory (7.8%) [42].

Table 2. Photosynthetically Active Radiation (PAR) attenuation and distribution (%) of transmitted radiation at different wavelengths through PolyMethylMethAcrylate (PMMA) ShieldUp® doped with 160 ppm of Lumogen F305 luminescent dye.

PAR Attenuation % (400–700 nm)	UV % (300–400 nm)	Blue % (400–510 nm)	Green % (510–610 nm)	Red % (610–720 nm)	Far-Red % (720–800 nm)
−70.5	0.6	7.8	1.8	55.0	34.7

A power conversion efficiency (PCE) of 1.5% (corresponding to 15 W/m^2) was obtained for the naked LSC slab (0.5 × 0.5 m^2) using Equations (1) and (2). Starting from 1.5% that is the maximum value obtained, the PMMA ShieldUp® implementation in the final configuration (to achieve the necessary structural characteristics) determines a drop, albeit modest, of the effective power conversion efficiency.

The power produced is reasonably sufficient to fulfill all or part of greenhouse needs, as airflow and water pumping irrigation, based on the characteristics of the specific greenhouse. Indeed, greenhouse electrical consumption depends on several parameters (as the location, the season of the year, technological innovation level and so on). However, the large surfaces of the greenhouse roofs can be potentially entirely covered by LSC ShieldUp® devices to maximize energy production.

3.2. First phase: Indoor LSC Panel

The first phase aimed to evaluate if the red LSC panel could hinder or reduce the biomass production of the selected species. The obtained results for the fresh weight for the aerial part and roots are reported in Figure 6. In the following, B LSC, H LSC, and L LSC indicate the brassica, sunflower and lupine plants grown under the LSC panel. B, H, and L refer to species grown outside the LSC panel, also indicated with the label CT (control).

From data showed, it can be noticed a trend towards a higher production of fresh biomass for the aerial part of the plants grown under the LSC panel. On average, considering the mean values, this increase ranges from around 24% to 31%. In Figure 6, the fresh weight values obtained for the roots have also been reported, even if these data are not particularly significant on a microcosm scale due to the difficulty of their harvesting and the modest production. Additionally, in this case, a general trend towards higher (root) biomass production under the LSC panel can be seen, although the differences between the plants grown outside or below the LSC panel are not very large.

To compare the overall biomass production, it is possible to consider the whole set of microcosms and calculate the sum of the biomass produced of the five microcosms of each species. The results are reported in Figure 7.

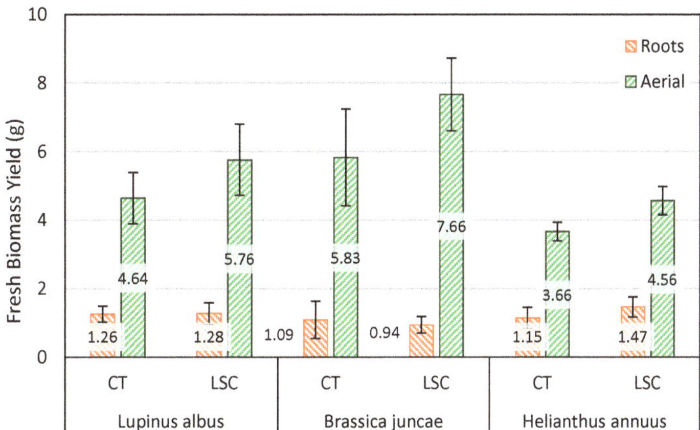

Figure 6. Fresh biomass yield (g) for aerial part and roots. Values are reported as mean ± standard deviation.

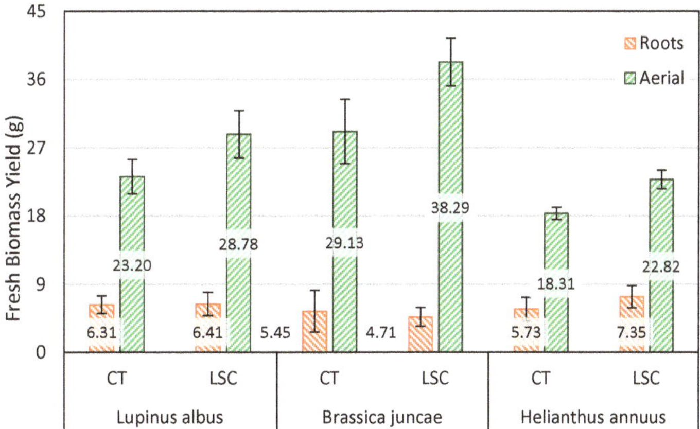

Figure 7. The yield of the biomass (g) of the five microcosms for each species. Data refer to the fresh weight of shoots and roots. Values are reported as mean ± standard deviation.

These results from fresh weights are also confirmed by the trend of the dry weight of the biomass, as shown in Table 3.

Table 3. Dry weight (mg) of the biomass of the aerial part and of the roots of plants. Values are reported as mean ± standard deviation (SD).

	Aerial Part			Roots		
	Mean	**SD**	**Diff. %**	**Mean**	**SD**	**Diff. %**
B	608	61.7	—	96.3	63.0	—
B LSC	761	77.3	+25.12	100	59.7	+4.34
H	399	47.1	—	81.8	31.6	—
H LSC	506	66.0	+27.05	104	26.5	+26.7
L	750	54.2	—	158	25.1	—
L LSC	961	96.5	+28.17	166	31.9	+5.61

Note: Diff% is the increased percentage of the mean values of biomass grown under the LSC panel compared to the plants grown outside the panel.

In this case, there are significant differences between the aerial parts of plants grown under the red panel and those grown outside the panel, with an increase of about 25%, 27%, and 28% of the mean values for *B. Juncea*, *H. Annuus*, and *L. Albus*, respectively.

Considering that energy saving is one of the strengths of "green remediation", the usefulness of LSC panels can be demonstrated if the plants cultivated grow and develop like those grown in traditional greenhouses. The results obtained seem to support this thesis. By absorbing mainly green light, red LSC panels maintain the blue spectral range necessary to activate photosynthesis. At the same time, the quantity and quality of light transmitted by the luminescent dye incorporated in these panels can improve the spectrum red fraction where the photosynthetic activity is highest [46–48]. Indeed, it should also be considered that these results were obtained with plants grown in contaminated soil, therefore under stress conditions.

3.3. Arsenic and Lead Uptake by Plants

An essential parameter to evaluate the feasibility of a phytoremediation intervention is the plant ability to absorb contaminants. The concentration of As absorbed by the plants is reported in Table 4.

Table 4. Concentration values (mg kg^{-1}) of As absorbed by the plants. Values are reported as mean ± standard deviation (SD).

	Aerial Part		Roots	
	Mean	**SD**	**Mean**	**SD**
B	93.8	25.9	1668	208
B LSC	96.9	31.0	1716	344
H	64.5	7.2	1306	335
H LSC	64.8	15.0	1183	320
L	70.5	25.2	377	50.2
L LSC	78.6	41.0	439	92.0

No difference was found between the amount absorbed by plants grown under the LSC panel and those grown outside the panel. The values obtained, which are the average over five replicates, are not significantly different from each other; thus, it appears that the LSC panel did not have a negative influence on As uptake by plants both in the aerial part and in the roots of plants.

The average values of Pb concentration were shown in Table 5. Further, in the case of Pb, the LSC panel did not affect the absorption of the metal. The mean Pb concentration values are not significantly different in plants grown under or outside the panel.

Table 5. Concentration values (mg kg^{-1}) of Pb absorbed by the plants. Values are reported as mean ± standard deviation (SD).

	Aerial Part		Roots	
	Mean	**SD**	**Mean**	**SD**
B	8.50	1.50	134	40.1
B LSC	10.0	1.25	112	34.2
H	6.32	1.06	109	40.5
H LSC	5.58	1.02	96.1	32.1
L	3.56	0.27	77.5	15.7
L LSC	3.50	0.25	98.1	38.8

3.4. Total Accumulation

The "total accumulation" (i.e., the total metal amount extracted by plants) was evaluated as product of metal concentration and aerial biomass [49]. This parameter provides an estimation of phytoextraction efficiency, since it includes both metal uptake and vegetal biomass production [20].

Data of total accumulation are reported in Figure 8.

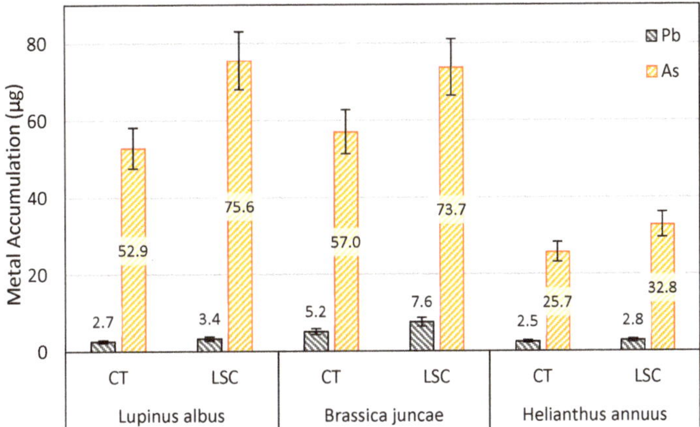

Figure 8. Total accumulation of As and Pb in the aerial part of plants grown under and outside the red panel. Values are reported as mean ± standard deviation.

The tests conducted in the greenhouse show positive effects on plants grown under the red LSC panel. On the contrary, since there are no differences in the absorption of contaminants, the increase in plant biomass grown under the red LSC panel also shows a beneficial effect on the total uptake values. On balance, the LSC panel could improve plant growth and development, with a consequent increase in the amount of metals removed from the contaminated soil.

3.5. Second Phase: Outdoor Comparison between LSC and Polycarbonate Boxes

As described above, the microcosm test was conducted outdoor using two boxes, one in transparent polycarbonate and the other consisting of red LSC panels.

In this case, the lighting conditions of the plants in the microcosms are significantly different from those of the first phase of the experiments. Indeed, there is no longer the shielding due to the greenhouse under which the first phase tests were conducted: the two boxes were placed outdoor, directly under the sunlight. In addition, for the LSC box all sides are made of red LSC panels, condition that should simulate on a small scale the effect of a hypothetical greenhouse made up exclusively of red LSC panels (in the comparison box all sides are made of polycarbonate).

In general, the plants grew well, even after adding EDTA. Additionally, in this case, the results obtained show that the production of fresh biomass, especially in the aerial part, is higher for plants grown under the red box (Figure 9).

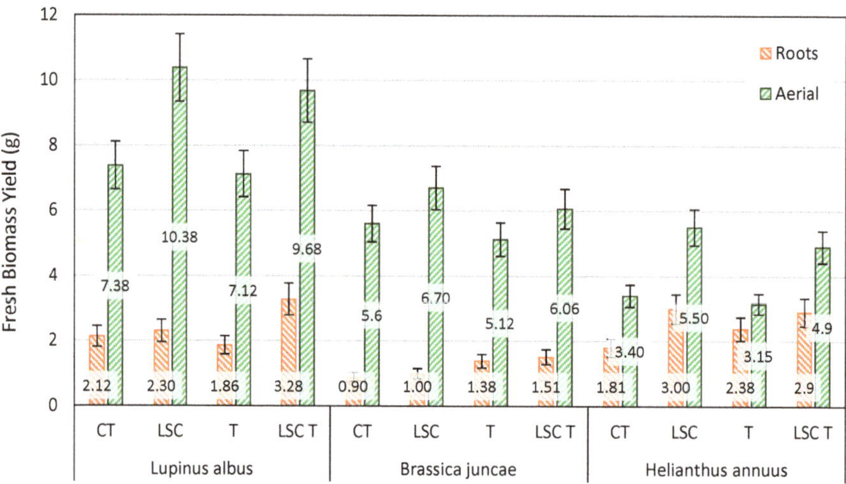

Figure 9. Fresh biomass production of shoots and roots. LSC the plants grown under the red panels and T the plants that have been treated with Ethylene Diamine Tetraacetic Acid (EDTA). Values are reported as mean $\pm$ standard deviation.

The trend of dry weight of the biomass of the plants is similar to that of the fresh weight; the data has been reported in Table 6.

Table 6. Dry weight (mg) of the biomass of the aerial part and of the roots of plants. Values are reported as mean $\pm$ standard deviation (SD).

	Aerial Part			Roots		
	Mean	SD	Diff. %	Mean	SD	Diff. %
B	749	65.1	—	197	82.7	—
B LSC	877	71.0	+17.1	230	73.0	+16.6
B T	712	69.8	—	242	74.1	—
B LSC T	959	96.9	+34.7	261	80.7	+8.0
H	489	97.3	—	224	26.7	—
H LSC	615	141	+25.8	228	38.2	+1.8
H T	484	97.2	—	239	29.4	—
H LSC T	560	117.0	+15.7	220	28.8	−8.1
L	1051	58.6	—	309	12.2	—
L LSC	1310	212	+24.7	308	83.0	−0.5
L T	1120	82.9	—	296	19.3	—
L LSC T	1220	117.1	+8.9	432	63.5	+45.9

Note: Diff% is the increased percentage of the mean values of biomass grown under the LSC panel compared to the plants grown outside the panel.

A picture of plants grown under the two different boxes before EDTA addition is reported in Figure 10. As an example, for each of the three plant species, two microcosms grown in the LSC box (on the left of the viewer) are compared with two microcosms grown in the polycarbonate box (on the right).

Figure 10. A picture of plants grown under the two different boxes, (**a**) *L. albus*, (**b**) *H. annuus*, and (**c**) *B. juncea*. On the left of the beholder, the plants grown under the LSC box.

It must be emphasized that the addition of EDTA (not reported in Figure 10) did not show a negative effect on biomass production. This result can be attributed to both the fractional addition of the chelating agent and the short growth period of the plants in the microcosm tests.

The concentration values of As and Pb in the aerial part and in the roots of the plants are shown in Table 7.

From the results obtained, it can be seen that the concentration of Pb and As is in general very similar for the plants grown under the red box and under the polycarbonate one. In some cases, the plants under the red box even showed a slightly improvement in the absorption of the contaminants. In general, the present experiment results showed that in all plants, EDTA addition increased Pb concentrations in shoots compared with the control.

The addition of EDTA increased the Pb content in the brassica plants in the aerial part by more than four times, without notable differences between the plants grown under the red box and those under the polycarbonate box.

A similar increase was also found in the aerial part of the sunflower, with an increase in the concentration of Pb of about 4.5 times in plants grown under the polycarbonate box and about 6 times in those grown under the red box. The most relevant effects of the addition of EDTA were found in the lupine plants with increases in Pb concentration of the aerial part of about 10 times the value found in the plants not treated with the chelating agent. The results were the same for both boxes.

The effect of EDTA was instead not very evident in the root system, where the difference in the concentration of Pb between treated and untreated microcosms was always minimal.

Lead is not easily transferred to above-ground plant biomass, since it is mainly stored in root cells [50,51]. In this experiment, the addition of EDTA has proved to be particularly effective for phytoextraction because it seems to have favored the translocation of the metal in the aerial part. It can be supposed that EDTA chelates Pb in the soil liquid phase then the soluble Pb–EDTA complex enters the roots and Pb is transported through the plant and accumulated in the aerial part [52].

Table 7. Mean concentration values (mg kg^{-1}) of As and Pb absorbed by the plants. Values are reported as mean $\pm$ standard deviation (SD).

	Aerial Part						Roots					
		As			Pb			As			Pb	
	Mean	SD	Diff. %	Mean	SD	Diff. %	Mean	SD	Diff. %	Mean	SD	Diff. %
B	42.2	1.6	—	4.7	0.4	—	2181	82.1	—	178	9.6	—
B LSC	45.2	4.2	+7.2	4.3	0.5	−7.8	2238	79.3	+2.6	232	8.4	+30.3
B T	40.2	2.6	—	21.1	2.1	—	2376	86.8	—	189	8.5	—
B LSC T	43.5	3.0	+8.2	18.7	2.0	−11.4	2401	83.3	+1.1	305	13.1	+61.4
H	62.3	3.8	—	3.5	0.5	—	1368	44.7	—	289	12.1	—
H LSC	69.0	2.6	+10.8	3.3	0.3	−6.7	1198	50.3	−12.4	355	13.2	+22.9
H T	71.3	3.5	—	15.1	1.8	—	1412	52.7	—	308	12.2	—
H LSC T	65.2	3.1	−8.6	19.5	2.2	+29.1	1430	56.0	+1.3	383	14.4	+24.4
L	27.1	2.8	—	2.2	0.3	—	199	11.5	—	152	9.5	—
L LSC	26.8	1.0	−1.2	2.0	0.2	−7.6	227	10.2	+14.1	154	10.2	+1.3
L T	32.1	2.3	—	21.2	2.5	—	232	11.9	—	132	8.5	—
L LSC T	31.7	2.1	−1.3	20.9	2.3	−1.4	241	11.8	+3.9	177	12.0	+34.1

The results did not show any adverse effect of EDTA on arsenic uptake. This can be ascribed to the action that EDTA carries out on iron oxides, partially solubilized by the complexing agent [53]. The mobility of arsenic in soil is greatly influenced by the presence of Fe-oxides where significant amounts of As are adsorbed [54]. Arsenate forms outer-sphere complexes by electrostatic coulombic interactions on all variable charge minerals [54]. It can be hypothesized that the disruptive effects of the oxides by EDTA release the arsenic that goes into soil solution becoming available for plants. As a matter of fact, in the specific contaminated soil, the plants can uptake both the contaminants even if not essential elements. There were no differences in the concentration of the two metals between the plants grown in the polycarbonate box and the LSC red box. After all, the EDTA metal complexes are poorly photodegradable in the soil, mostly when plants have grown and in the alkaline conditions of soil pH [55].

The total accumulation of the two metals for the three investigated plants is reported in Figure 11 for As and Pb.

This parameter, which, as previously stated, is essential in evaluating the efficiency of phytoremediation, shows a positive effect of the LSC panels, as it is generally higher for tests carried out in the red box. The results showed that plants development under the red box also appears visually higher than that under the polycarbonate box. Thus, the positive effects are more explicit in outdoor tests rather than in greenhouses, as in general the light conditions in the greenhouse tend to decrease due to the shielding effect of the walls of the structure.

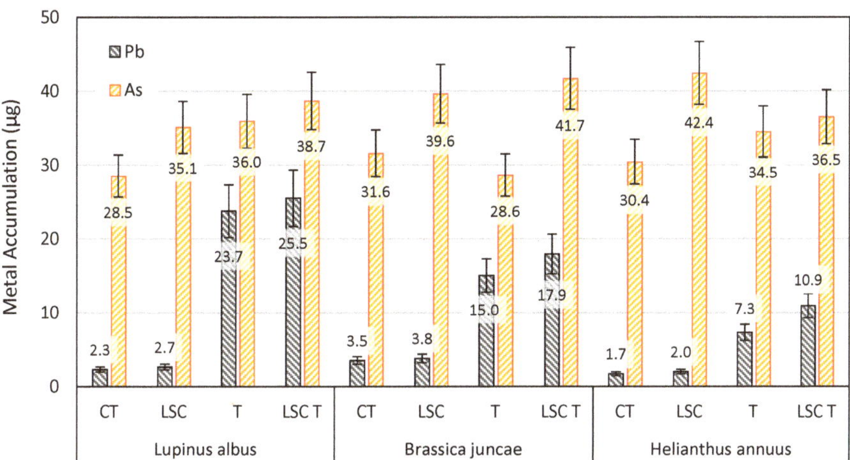

Figure 11. Total accumulation (µg) of Pb and As in the aerial part of the plants. Values are reported as mean ± standard deviation.

4. Conclusions

In a context of growing international attention linked to the need to resort to renewable energy sources, the possibility of using LSC panels for the growth of vegetables is becoming a path pursued with great interest. LSCs can collect both diffuse and direct solar radiation, making them a suitable technology to be used in countries where diffuse solar radiation is dominant such as in northern European countries (with more than 50% diffuse light). Research is still at an early stage regarding the beneficial effects of LSCs on the plant growth, which will require a better understanding of the potential impacts of this technology on growth across the huge diversity of vegetable species.

Nevertheless, the results of the present experimentation show that LSC panels in PMMA ShieldUp doped with 160 ppm Lumogen Red F305 have the right characteristics to be used for greenhouse application, since they do not penalize the growth of plants, but rather they contribute to enhance the photosynthetic efficiency: the fluorescent dye transmits a sufficient PAR light (about 30%), a low UV light (0.6%) and a high red and far-red light (90%), while maintaining a satisfactory blue light (7.8%), necessary to improve the photosynthetic efficiency.

Despite the preliminary nature of the conducted tests, which to the best of the authors knowledge are absolutely innovative as there are no similar experiences even at an international level, the results seem absolutely promising. Indeed, despite stressful conditions due to a high concentration of contaminants in the soil, the plants under the LSC panel grew well showing a higher uptake capacity with respect to plants grown in the traditional greenhouse in polycarbonate. At the same time, plants grown in the LSC greenhouse showed an interesting increase in fresh biomass production on microcosm scale. These results offer some ideas for the possible use of these materials in the field of remediation through phytoremediation, especially on a larger scale where both LSCs and phytoremediation techniques deemed to be further proved.

An on-site greenhouse capable of being energy self-sufficient allows feasibility tests to be carried out in optimal times, regardless of the climatic conditions (protection from high temperatures in summer and low temperatures in winter) even with the addition of chemical additives and biological (PGPR) to choose the best strategy for full-scale remediation activities. In this sense, the power produced by relying on LSC panels could be utilized to fulfill all or part of the greenhouse needs, as airflow and water pumping irrigation. Indeed, the large surfaces of the greenhouse roofs can be potentially entirely building-integrated with by LSC devices to maximize energy production.

About this, measures of electrical efficiency of a complete LSC device ($50 \times 50 \times 0.6$ cm^3) with 160 ppm Lumogen Red F305 concentration showed a power conversion efficiency (PCE) of 1.5%, equal to a maximum power production of 15 W/m^2.

Depending on the greenhouse size, location, year season and technological innovation degree, a use could also be envisaged for ex situ phytoremediation activities. For example, by digging the contaminated soil at a certain depth it may be possible to place it inside the greenhouse and program the growth of plants on the soil to be reclaimed in any climatic season, considerably reducing the remediation times especially in the presence of contamination by organic compounds.

On a full scale, LSC panels can also be used as canopies, placed at a certain height on the area to be reclaimed, using appropriate anchoring systems, allowing the LSC properties to be exploited. The height of the panels should be established based on the potentially used agricultural machinery, and sowing should also be arranged in such a way as to allow access and use of these means.

Author Contributions: Conceptualization, C.S., F.P., G.P., and E.F.; methodology, C.S. and G.P.; validation, F.P., G.P., M.B., and M.G; formal analysis, M.V.; investigation, A.P., S.P., S.Z., F.P., G.P., M.B., and M.G.; resources, F.P., G.P., C.S., and L.G.; data curation, M.V.; writing—original draft preparation, C.S., L.G., F.P., and G.P.; writing—review and editing, M.V. and E.F.; visualization, M.V.; supervision, M.V.; project administration, C.S.; funding acquisition, E.F. All authors have read and agreed to the published version of the manuscript.

Funding: This research was supported by Eni S.p.A, Research & Technological Innovation Department, San Donato Milanese (Italy) and fully funded by Syndial S.p.A. (now Eni Rewind S.p.A.), agreement number 3500047416.

Institutional Review Board Statement: Not applicable.

Informed Consent Statement: Not applicable.

Data Availability Statement: Data is contained within the present article.

Acknowledgments: The Authors thank Irene Rosellini, IRET-CNR, for technical assistance in Phytoremediation experiments.

Conflicts of Interest: The authors declare no conflict of interest.

References

1. Rafiee, M.; Chandra, S.; Ahmed, H.; McCormack, S.J. An Overview of Various Configurations of Luminescent Solar Concentrators for Photovoltaic Applications. *Opt. Mater.* **2019**, *91*, 212–227. [CrossRef]
2. Debije, M.G.; Verbunt, P.P.C. Thirty Years of Luminescent Solar Concentrator Research: Solar Energy for the Built Environment. *Adv. Energy Mater.* **2012**, *2*, 12–35. [CrossRef]
3. Farhadian, M.; Vachelard, C.; Duchez, D.; Larroche, C. In situ bioremediation of monoaromatic pollutants in groundwater: A review. *Bioresour. Technol.* **2008**, *99*, 5296–5308. [CrossRef]
4. Vocciante, M.; Finocchi, A.; De Folly D'Auris, A.; Conte, A.; Tonziello, J.; Pola, A.; Reverberi, A. Enhanced oil spill remediation by adsorption with interlinked multilayered graphene. *Materials* **2019**, *12*, 2231. [CrossRef] [PubMed]
5. Pietrelli, L.; Francolini, I.; Piozzi, A.; Sighicelli, M.; Vocciante, M. Chromium(III) removal from wastewater by chitosan flakes. *Appl. Sci.* **2020**, *10*, 1925. [CrossRef]
6. Khan, F.I.; Husain, T.; Hejazi, R. An overview and analysis of site remediation technologies. *J. Environ. Manag.* **2004**, *71*, 95–122. [CrossRef]
7. Pavel, L.V.; Gavrilescu, M. Overview of ex situ decontamination techniques for soil cleanup. *Environ. Eng. Manag. J.* **2008**, *7*, 815–834. [CrossRef]
8. Vocciante, M.; Reverberi, A.P.; Pietrelli, L.; Dovì, V.G. Improved remediation processes through cost-effective estimation of soil properties from surface measurements. *J. Clean. Prod.* **2017**, *167*, 680–686. [CrossRef]
9. da Silva, B.M.; Maranho, L.T. Petroleum-contaminated sites: Decision framework for selecting remediation technologies. *J. Hazard. Mater.* **2019**, *378*, 120722. [CrossRef]
10. Rosestolato, D.; Bagatin, R.; Ferro, S. Electrokinetic remediation of soils polluted by heavy metals (mercury in particular). *Chem. Eng. J.* **2015**, *264*, 16–23. [CrossRef]

11. Ferrucci, A.; Vocciante, M.; Bagatin, R.; Ferro, S. Electrokinetic remediation of soils contaminated by potentially toxic metals: Dedicated analytical tools for assessing the contamination baseline in a complex scenario. *J. Environ. Manag.* **2017**, *203*, 1163–1168. [CrossRef] [PubMed]
12. Vocciante, M.; Caretta, A.; Bua, L.; Bagatin, R.; Ferro, S. Enhancements in ElectroKinetic Remediation Technology: Environmental assessment in comparison with other configurations and consolidated solutions. *Chem. Eng. J.* **2016**, *289*, 123–134. [CrossRef]
13. Vocciante, M.; Dovì, V.G.; Ferro, S. Sustainability in ElectroKinetic Remediation Processes: A Critical Analysis. *Sustainability* **2021**, *13*, 770. [CrossRef]
14. Vocciante, M.; Caretta, A.; Bua, L.; Bagatin, R.; Franchi, E.; Petruzzelli, G.; Ferro, S. Enhancements in phytoremediation technology: Environmental assessment including different options of biomass disposal and comparison with a consolidated approach. *J. Environ. Manag.* **2019**, *237*, 560–568. [CrossRef]
15. U.S. Environmental Protection Agency. *Citizen's Guide to Thermal Desorption*; Office of Solid Waste and Emergency Response, Publication EPA, United States Environmental Protection Agency: Washington, DC, USA, 2012.
16. O'Connor, D.; Zheng, X.; Hou, D.; Shen, Z.; Li, G.; Miao, G.; O'Connell, S.; Guo, M. Phytoremediation: Climate change resilience and sustainability assessment at a coastal brownfield redevelopment. *Environ. Int.* **2019**, *130*, 104945. [CrossRef] [PubMed]
17. Song, Y.; Kirkwood, N.; Maksimović, Č.; Zheng, X.; O'Connor, D.; Jin, Y.; Hou, D. Nature based solutions for contaminated land remediation and brownfield redevelopment in cities: A review. *Sci. Total Environ.* **2019**, *663*, 568–579. [CrossRef]
18. Franchi, E.; Petruzzelli, G. Phytoremediation and the key role of PGPR. In *Advances in PGPR Research*; Singh, H.B., Sarma, B.K., Keswani, C., Eds.; CABI: Boston, MA, USA, 2017; pp. 306–329.
19. Franchi, E.; Rolli, E.; Marasco, R.; Agazzi, G.; Borin, S.; Cosmina, P.; Pedron, F.; Rosellini, I.; Barbafieri, M.; Petruzzelli, G. Phytoremediation of a multi contaminated soil: Mercury and arsenic phytoextraction assisted by mobilizing agent and plant growth promoting bacteria. *J. Soils Sediments* **2017**, *17*, 1224–1236. [CrossRef]
20. Franchi, E.; Cosmina, P.; Pedron, F.; Rosellini, I.; Barbafieri, M.; Petruzzelli, G.; Vocciante, M. Improved arsenic phytoextraction by combined use of mobilizing chemicals and autochthonous soil bacteria. *Sci. Total Environ.* **2019**, *655*, 328–336. [CrossRef]
21. Glick, B.R. Using soil bacteria to facilitate phytoremediation. *Biotechnol. Adv.* **2010**, *28*, 367–374. [CrossRef]
22. Barbafieri, M.; Pedron, F.; Petruzzelli, G.; Rosellini, I.; Franchi, E.; Bagatin, R.; Vocciante, M. Assisted phytoremediation of a multi-contaminated soil: Investigation on arsenic and lead combined mobilization and removal. *J. Environ. Manag.* **2017**, *203*, 316–329. [CrossRef] [PubMed]
23. Vocciante, M.; Reverberi, A.P.; Dovì, V.G. Approximate Solution of the Inverse Richards' Problem. *Appl. Math. Model.* **2016**, *40*, 5364–5376. [CrossRef]
24. Barbafieri, M.; Japenga, J.; Romkens, P.; Petruzzelli, G.; Pedron, F. Protocols for applying phytotechnologies in metal-contaminated soils. In *Plant based Remediation Processes*; Gupta, D.K., Ed.; Springer: Berlin/Heidelberg, Germany, 2013; pp. 19–37.
25. Loik, M.E.; Carter, S.A.; Alers, G.; Wade, C.E.; Shugar, D.; Corrado, C.; Jokerst, D.; Kitayama, C. Wavelength-Selective Solar Photovoltaic Systems: Powering Greenhouses for Plant Growth at the Food-Energy-Water Nexus. *Earth's Future* **2017**, *5*, 1044–1053. [CrossRef]
26. Thomas, G.W. Soil pH and soil acidity. In *Methods of Soil Analysis. Part 3. Chemical Methods*; Sparks, D.L., Ed.; Soil Science Society of America Inc.: Madison, WI, USA, 1996; pp. 475–490.
27. Sumner, M.E.; Miller, W.P. Cation exchange capacity and exchange coefficients. In *Methods of Soil Analysis. Part 3. Chemical Methods*; Sparks, D.L., Ed.; Soil Science Society of America Inc.: Madison, WI, USA, 1996; pp. 1201–1230.
28. Gee, G.W.; Bauder, J.W. Particle-size analysis. In *Methods of Soil Analysis. Part 1. Physical and Mineralogical Methods*; Klute, A., Ed.; American Society of Agronomy/Soil Science Society of America: Madison, WI, USA, 1986; pp. 383–411.
29. Bremner, J.M. Nitrogen total. In *Methods of Soil Analysis. Part 3. Chemical Methods*; Sparks, D.L., Ed.; Soil Science Society of America Inc.: Madison, WI, USA, 1996; pp. 1085–1121.
30. Kuo, S. Phosphorus. In *Methods of Soil Analysis. Part 3. Chemical Methods*; Sparks, D.L., Ed.; Soil Science Society of America Inc.: Madison, WI, USA, 1996; pp. 869–920.
31. Nelson, D.W.; Sommers, L.E. Total carbon, organic carbon and organic matter. In *Methods of Soil Analysis. Part 3. Chemical Methods*; Sparks, D.L., Ed.; Soil Science Society of America Inc.: Madison, WI, USA, 1996; pp. 961–1010.
32. Quartacci, M.F.; Argilla, A.; Baker, A.J.M.; Navari-Izzo, F. Phytoextraction of metals from a multiply contaminated soil by Indian mustard. *Chemosphere* **2006**, *63*, 918–925. [CrossRef]
33. Doumett, S.; Lamperi, L.; Checchini, L.; Azzarello, E.; Mugnai, S.; Mancuso, S.; Petruzzelli, G.; Del Bubba, M. Heavy metal distribution between contaminated soil and Paulownia tomentosa, in a pilot-scale assisted phytoremediation study: Influence of different complexing agents. *Chemosphere* **2008**, *72*, 1481–1490. [CrossRef] [PubMed]
34. Pereira, B.F.F.; de Abreu, C.A.; Herpin, M.F.U.; de Abreu, R.S. Berton. Phytoremediation of lead by jack beans on a Rhodic Hapludox amended with EDTA. *Sci. Agric.* **2010**, *67*, 308–318. [CrossRef]
35. Li, F.-L.; Qiu, Y.; Xu, X.; Yang, F.; Wang, Z.; Feng, J.; Wang, J. EDTA-enhanced phytoremediation of heavy metals from sludge soil by Italian ryegrass (*Lolium perenne* L.). *Ecotoxicol. Environ. Saf.* **2020**, *191*, 110185. [CrossRef]
36. Saleem, M.H.; Ali, S.; Rehman, M.; Rizwan, M.; Kamran, M.; Mohamed, I.A.A.; Khan, Z.; Bamagoos, A.A.; Alharby, H.F.; Hakeem, K.R.; et al. Individual and combined application of EDTA and citric acid assisted phytoextraction of copper using jute (*Corchorus capsularis* L.) seedlings. *Environ. Technol. Innov.* **2020**, *19*, 100895. [CrossRef]
37. Evert, R.F.; Eichhorn, S.E. *Raven Biology of Plants*, 8th ed.; W.H. Freeman/Palgrave Mcmillan: London, UK, 2013.

38. Sager, J.C.; Smith, W.O.; Edwards, J.L.; Cyr, K.L. Photosynthetic efficiency and phytochrome photoequilibria determination using spectral data. *Trans. ASAE* **1988**, *31*, 1882–1889. [CrossRef]
39. Zheng, J.; Hu, M.J.; Guo, Y.P. Regulation of photosynthesis by light quality and its mechanism in plants. *Ying Yong Sheng Tai Xue Bao* **2008**, *19*, 1619–1624.
40. Hogewoning, S.W.; Trouwborst, G.; Maljaars, H.; Poorter, H.; Van Leperen, W.; Harbinson, J. Blue light dose-responses of leaf photosynthesis, morphology, and chemical composition of Cucumis sativus grown under different combinations of red and blue light. *J. Exp. Bot.* **2010**, *61*, 3107–3117. [CrossRef] [PubMed]
41. Strid, A.; Chow, W.S.; Anderson, J.M. UV-B damage and protection at the molecular level in plants. *Photosynth. Res.* **1994**, *39*, 475–489. [CrossRef]
42. Bernardoni, P.; Vincenzi, D.; Mangherini, G.; Boschetti, M.; Andreoli, A.; Gjestila, M.; Samà, C.; Gila, L.; Palmery, S.; Tonezzer, M.; et al. Improved Healthy Growth of Basil Seedlings under LSC Filtered Illumination. In Proceedings of the 37th European Photovoltaic Solar Energy Conference and Exhibition, Lisbon, Portugal, 7–11 September 2020.
43. Strong, A.B. *Plastics: Materials and Processing*, 3rd ed.; Pearson Prentice Hall: Upper Saddle River, NJ, USA, 2006; p. 874.
44. Polymeric Composition Comprising a Fluorescent dye, Its Process of Preparation, Use and Object Comprising It, WO201821130. Available online: https://www.freepatentsonline.com/y2020/0172799.html (accessed on 18 January 2021).
45. U.S. Environmental Protection Agency (EPA). Method 3052: Microwave assisted acid digestion of siliceous and organically based matrices. In *Test Methods for Evaluating Solid Waste: Physical/Chemical Methods (SW-846)*; EPA: Washington, DC, USA, 1995; pp. 3052-1–3052-20.
46. Chidburee, A.; Bundittaya, W.; Suwanthad, C.; Ohtake, N.; Sueyoshi, K.; Ohyama, T.; Ruamrungsri, S. Effects of Red Light on Growth, Photosynthesis and Food Reserves in Curcuma alismatifolia Gagnep. *Thai J. Agric. Sci.* **2007**, *40*, 57–63.
47. Kim, S.J.; Hahn, E.J.; Heo, J.W.; Paek, K.Y. Effects of LEDs on net photosynthetic rate, growth and leaf stomata of Chrysanthemum plantlets in vitro. *Sci. Hortic.* **2004**, *101*, 143–151. [CrossRef]
48. Schoefs, B. Chlorophyll and carotenoid analysis in food products. Properties of the pigments and methods of analysis. *Trends Food Sci.Technol.* **2002**, *13*, 361–371. [CrossRef]
49. Jarrell, W.M.; Beverly, R.B. The dilution effect in plant nutrition studies. *Adv. Agron.* **1981**, *34*, 197–224.
50. Meyers, D.E.R.; Auchterlonie, G.J.; Webb, R.I.; Wood, B. Uptake and localisation of lead in the root system of Brassica juncea. *Environ. Pollut.* **2008**, *153*, 323–332. [CrossRef] [PubMed]
51. Mellem, J.; Baijanth, H.; Odhav, B. Translocation and accumulation of Cr, Hg, As, Pb, Cu and Ni by Amaranthus dubius (Amaranthaceae) from contaminated sites. *J. Environ. Sci. Health* **2009**, *44*, 568–575. [CrossRef]
52. Wenzel, W.W.; Unterbrunner, R.; Sommer, P.; Pasqualina, S. Chelate-assisted phytoextraction using canola (Brassica napus L.) in outdoors pot and lysimeter experiments. *Plant Soil* **2003**, *249*, 83–96. [CrossRef]
53. Chiu, K.K.; Ye, Z.H.; Wong, M.H. Enhanced uptake of As, Zn, and Cu by Vetiveria zizanioides and Zea mays using chelating agents. *Chemosphere* **2005**, *60*, 1365–1375. [CrossRef]
54. Tao, Y.; Zhang, S.; Jian, W.; Yuan, C.; Shan, X.Q. Effects of Oxalate and Phosphate on the Release of Arsenic from Contaminated Soils and Arsenic Accumulation in Wheat. *Chemosphere* **2006**, *65*, 1281–1287. [CrossRef]
55. Metsärinne, S.; Tuhkanen, T.; Aksela, R. Photodegradation of ethylenediaminetetraacetic acid (EDTA) and ethylenediamine disuccinic acid (EDDS) within natural UV radiation range. *Chemosphere* **2001**, *45*, 949–955. [CrossRef]

Article

Sedimentation of Fractal Aggregates in Shear-Thinning Fluids

Marco Trofa [†] and **Gaetano D'Avino** [*,†]

Dipartimento di Ingegneria Chimica, dei Materiali e della Produzione Industriale, Università di Napoli Federico II, Piazza Giorgio Ascarelli 80, 80125 Napoli, Italy; marco.trofa@unina.it
* Correspondence: gadavino@unina.it; Tel.: +39-081-7682241
† These authors contributed equally to this work.

Received: 10 April 2020; Accepted: 2 May 2020; Published: 8 May 2020

Abstract: Solid–liquid separation is a key unit operation in the wastewater treatment, generally consisting of coagulation and flocculation steps to promote aggregation and increase the particle size, followed by sedimentation, where the particles settle due to the effect of gravity. The sedimentation efficiency is related to the hydrodynamic behavior of the suspended particles that, in turn, depends on the aggregate morphology. In addition, the non-Newtonian rheology of sludges strongly affects the drag coefficient of the suspended particles, leading to deviations from the known settling behavior in Newtonian fluids. In this work, we use direct numerical simulations to study the hydrodynamic drag of fractal-shaped particles suspended in a shear-thinning fluid modeled by the power-law constitutive equation. The fluid dynamics governing equations are solved for an applied force with different orientations uniformly distributed over the unit sphere. The resulting particle velocities are interpolated to compute the aggregate dynamics and the drag correction coefficient. A remarkable effect of the detailed microstructure of the aggregate on the sedimentation process is observed. The orientational dynamics shows a rich behavior characterized by steady-state, bistable, and periodic regimes. In qualitative agreement with spherical particles, shear-thinning increases the drag correction coefficient. Elongated aggregates sediment more slowly than sphere-like particles, with a lower terminal velocity as the aspect ratio increases.

Keywords: sedimentation; drag; fractal aggregates; shear-thinning; non-Newtonian fluids; suspensions; numerical simulations

1. Introduction

Separation of solid particles suspended in liquids is a fundamental operation in the treatment of wastewater. This process generally consists of a sequence of steps, namely, coagulation and flocculation, followed by sedimentation [1]. The first operation aims at destabilizing the suspension through the addition of coagulants that neutralize the negative charges on fine solids. The small destabilized particles are able to come into contact and form larger particles called microflocs. In the flocculation step, the microflocs collide and stick together forming larger and larger particles (macroflocs). Once the suspended aggregates have reached a desired dimension, the suspension undergoes the sedimentation step, i.e., the particles are separated from the liquid through gravity or centrifugal force [2].

The knowledge of the hydrodynamic drag force, which counterbalances the sedimentation force, is crucial to the design and optimization of wastewater treatment plants. Such a force depends on the size and shape of the suspended particles, and on the rheological properties of the liquid. During the flocculation stage, the aggregates assume fractal shapes [3–6] described by the following equation [7],

$$N_{\mathrm{p}} = k_{\mathrm{f}} \left(\frac{R_{\mathrm{g}}}{a} \right)^{D_{\mathrm{f}}} \tag{1}$$

where N_p is the number of primary particles forming the aggregate, R_g is the radius of gyration, D_f is the fractal dimension, k_f is the fractal pre-factor, and a is the radius of the primary particles. Aggregates with a shape satisfying Equation (1) are referred to as fractal-like or quasi-fractal aggregates, as the scaling law relation is independent of whether the particle has a real scale-invariant (self-similar) morphology [8]. The fractal dimension provides a scaling law between the number of primary particles composing the aggregate and a characteristic cluster size (e.g., the gyration radius). It assumes values between 1 and 3 corresponding to rod-like and spherical-like particles, respectively. The fractal prefactor is a descriptor of the aggregate local structure and is related to the packing factor. Finally, the radius of gyration is a geometric measure of the spatial mass distribution about the aggregate center of mass.

The particle morphology has a relevant influence on the settling velocity, and thus it must be accounted for when dealing with the sedimentation process of flocculated particles [9,10]. Therefore, it is not surprising that the hydrodynamic drag of particles with complex shapes has been thoroughly studied in the literature. Several methodologies have been proposed to compute the hydrodynamic drag of a set of spherical particles in contact, based on expansions of analytical solutions for Stokes flows [11–13] or direct numerical simulations [14–16]. Due to the linearity of the creeping flow equations, the dynamics of a particle with arbitrary shape can be determined by the mobility tensor that univocally relates translational and angular velocities to the forces and torques acting on the particle [17]. The knowledge of the mobility tensor allows to completely predict the translational and orientational motion of the aggregate. In this regards, it is well known that the settling velocity depends on the aggregate orientation. The average velocity over all possible orientations can be linearly related to the applied force, with a proportionality constant given by the arithmetic mean of the three eigenvalues of the translational mobility tensor divided by the fluid viscosity [13,18,19]. A hydrodynamic radius can be defined as the radius of a sphere that gives the same drag force acting on the aggregate in a uniform flow [13]. It can be readily seen that the hydrodynamic radius is inversely proportional to the aforementioned mean of the eigenvalues of the translational mobility tensor. The knowledge of the hydrodynamic radius for an aggregate with arbitrary shape is, then, sufficient to characterize its average settling velocity. The ratio of the hydrodynamic radius and the gyration radius has been found to be a function of the parameters of the fractal Equation (1) [13,19,20]. Specifically, such a ratio is an increasing function of D_f, assuming a value of ~1 for $D_f = 2$, up to a limiting value of about 1.29 for $D_f = 3$ [13,15,20]. The number of primary particles strongly affects the ratio for low fractal dimensions (rod-like particles), whereas it has a weak influence for more spherical aggregates. Finally, increasing the fractal pre-factor moves the ratio to higher values without altering the dependence on D_f and N_p [13].

All the aforementioned studies consider a Newtonian suspending liquid. In addition, the developed methodologies used to compute the hydrodynamic drag are only applicable to Newtonian fluids. Active sludges, however, show a non-Newtonian rheology [21,22], which has a strong influence on the particle settling dynamics. For a spherical particle in an unbounded shear-thinning inelastic fluid, modeled with the power-law constitutive equation, several works are available showing that the drag force deviates from the Stokes' law [23–26]. A drag correction coefficient has been defined as the ratio between the applied force and the Stokes' drag law, where the viscosity is replaced by the power-law constitutive equation with a characteristic shear rate given by the terminal settling velocity divided by the particle diameter. The coefficient is 1 for a Newtonian fluid and increases as the flow index decreases (i.e., fluid shear-thinning increases). For non-spherical particles, Tripathi et al. [27] carried out finite element simulations to study the flow of a power-law fluid over prolate and oblate spheroidal particles aligned with the flow direction. In this particular orientation, the dependence of the total drag coefficient on the flow index was found to be qualitatively similar to that observed for spherical particles. As the aspect ratio of the prolate spheroid increases, the drag becomes relatively insensitive to the degree of shear-thinning. Oblate spheroids with high aspect ratio experience a lower drag as compared to spheres.

In summary, many works exist dealing with drag calculation of particles with non-spherical shape in Newtonian liquids. Concerning power-law fluids, the only results available are for spheres [25] and spheroids [27], the latter only considering particles oriented with a principal axis along the force direction. To the best of our knowledge, a study on the combined effect of non-Newtonian rheology and complex particle shape on the hydrodynamic drag is missing.

In this work, we investigate the hydrodynamic drag experienced by fractal aggregates suspended in a non-Newtonian fluid by numerical simulations. We assume that the aggregates are sufficiently large to neglect Brownian motion and that their concentration is low enough (less than 5% in volume) to avoid hydrodynamic interactions. This allows us to consider a single-particle problem. The suspending fluid is assumed to be inelastic and shear-thinning, and is modeled by the power-law constitutive equation. A map of particle velocities is precomputed by running finite element simulations for orientations of the applied force uniformly distributed over the unit sphere. Such velocities are then interpolated and used to reconstruct the aggregate dynamics by integrating the evolution equation of the particle position and orientation. The drag correction coefficient at long times is averaged over several initial orientations and particle shapes with the same fractal parameters. The effect of the fractal dimension, the number of primary particles forming the aggregate, and the flow index is investigated.

2. Mathematical Model and Numerical Method

2.1. Governing Equations

A rigid non-Brownian aggregate is suspended in a fluid and subjected to a constant force F. The fluid is at rest far from the aggregate. The computational domain, shown in Figure 1e, is a sphere with radius much larger than the maximum size of the particle. The aggregate is placed at the center of the sphere. A Cartesian reference frame is selected with x denoting the direction of the applied force F, i.e., $F = (F, 0, 0)$. The fluid velocity is set to zero on the external spherical surface whereas a rigid-body motion is imposed on the particle boundary. We denote by x_p and θ_p the position of the particle center of volume and the rotation angle, and by U_p and ω_p the translational and angular particle velocities, respectively. All the symbols used in this work are reported in Table 1.

We model the aggregate shape by a set of primary spherical particles with radius a arranged to satisfy the fractal Equation (1). The construction of aggregate shapes satisfying such equation can be done in several ways, for instance by iteratively adding spherical particles (particle–cluster methods) or by directly connecting clusters of particles (cluster–cluster methods) [28–32]. In this work, we adopt the particle–cluster aggregation method proposed in [33,34]. All the available algorithms are based on the generation of pseudorandom numbers. Therefore, infinite shapes for the same fractal parameters can be obtained by changing the seed of the random number generator. We report in Figure 1 two examples of structures generated with $N_p = 20$, $k_f = 1.3$, and $D_f = 1.5$ (Figure 1a) or $D_f = 2.5$ (Figure 1b).

Assuming negligible fluid and particle inertia, the fluid dynamics of the investigated system is governed by the following mass and momentum balance equations,

$$\nabla \cdot u = 0 \tag{2}$$

$$\nabla \cdot \sigma = 0 \tag{3}$$

$$\sigma = -pI + 2\eta(\dot{\gamma})D \tag{4}$$

where u, σ, p, I, η, and D are the velocity vector, the stress tensor, the pressure, the 3×3 unity tensor, the fluid viscosity, and the rate-of-deformation tensor $D = (\nabla u + (\nabla u)^{\mathrm{T}})/2$, respectively. We model the suspending fluid by the power-law constitutive equation:

$$\eta(\dot{\gamma}) = m\dot{\gamma}^{n-1} \tag{5}$$

where m is the consistency index, n is the flow index, and $\dot{\gamma} = \sqrt{2D : D}$ is the effective deformation rate. This model predicts shear-thinning for $n < 1$. For $n = 1$, a Newtonian fluid with (constant) viscosity m is recovered.

The fluid is at rest far from the aggregate and rigid-body motion is applied at the particle boundary, resulting in the following boundary conditions at the external spherical surface,

$$u = 0 \tag{6}$$

and at the surface of the aggregate,

$$u = U_\mathrm{p} + \omega_\mathrm{p} \times (x - x_\mathrm{p}) \tag{7}$$

with x a point of the particle boundary.

Table 1. List of symbols.

a	Radius of primary particles
C	Number of initial aggregate configurations
D	Rate-of-deformation tensor
D_f	Fractal dimension
f	Applied force in the body reference frame
f_0	Initial applied force in the body reference frame
F	Applied force in the fixed reference frame
I	3x3 unity tensor
k_f	Fractal pre-factor
m	Consistency index
n	Flow index
n	Unit vector normal to the particle surface
N_elem	Number of elements of the tetrahedral mesh
N_p	Number of primary particles composing the aggregate
N_seed	Number of seeds
p	Fluid pressure
R_eff	Radius of a sphere with the same volume of the aggregate
R_g	Radius of gyration
R_out	Radius of the external sphere
S	Particle surface
t	Time
T	Torque on the particle
u	Fluid velocity
U	Aggregate velocity along the direction of the applied force
U_p	Particle translational velocity
V	Volume of the aggregate
x	Point of particle boundary
x_p	Particle center of volume
$x_\mathrm{p,0}$	Initial particle position
X	Drag correction coefficient
$\langle X_\mathrm{R} \rangle$	Regime drag correction coefficient
$\langle X_\mathrm{R} \rangle_\mathrm{m}$	Ensemble-average drag correction coefficient
$\dot{\gamma}$	Effective deformation rate
Δx	Size of the elements on the aggregate
Δx_out	Size of the elements on the external surface
η	Fluid viscosity
θ	Polar spherical coordinate
θ_p	Particle rotation angle
$\theta_\mathrm{p,0}$	Initial particle rotation angle
σ	Fluid stress tensor
ϕ	Azimuthal spherical coordinate
ω_p	Particle angular velocity

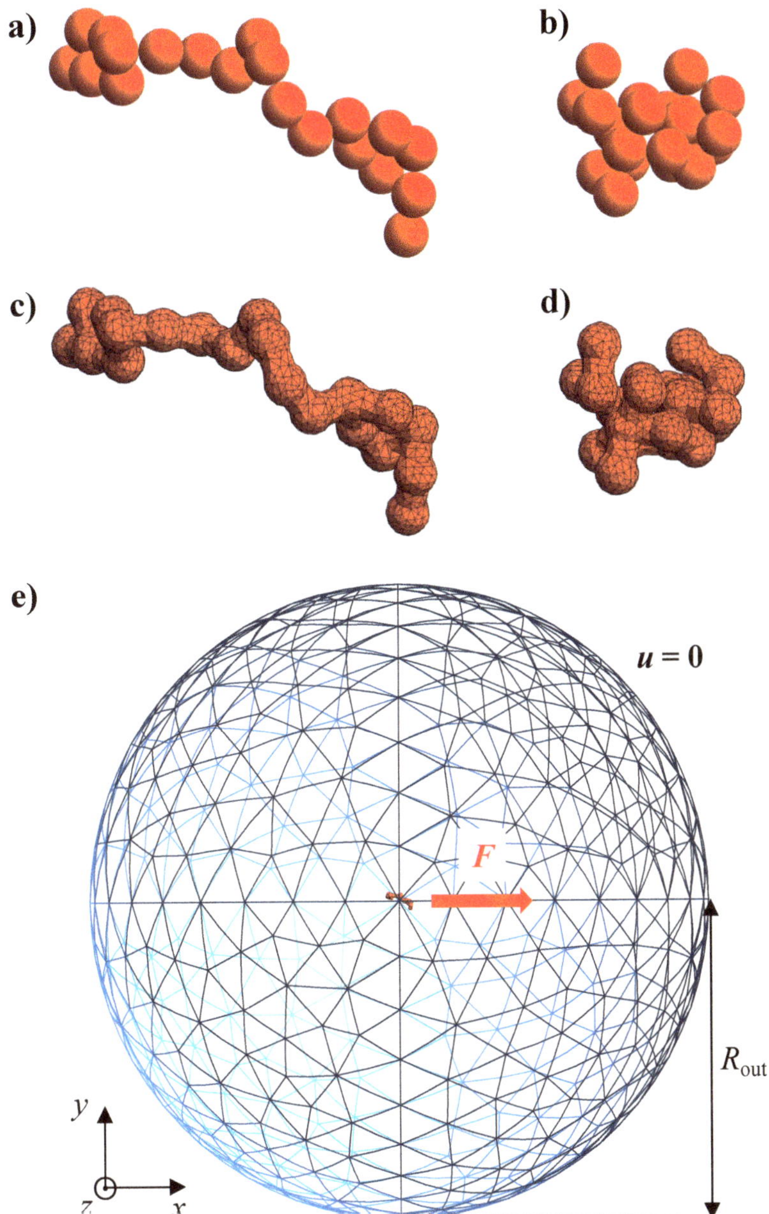

Figure 1. Examples of aggregate shapes obtained from the particle–cluster method for $N_{\text{p}} = 20$, $k_{\text{f}} = 1.3$, and $D_{\text{f}} = 1.5$ (**a**), and $D_{\text{f}} = 2.5$ (**b**). To avoid numerical issues due to the tangent point, the centers of the spheres in contact are connected with a set of cylinders with radius $0.732a$. In panels (**c**,**d**), the final geometry of the aggregates and the surface mesh are shown. The computational domain and the mesh on the external spherical surface is displayed in panel (**e**).

Finally, we need to specify the hydrodynamic total force and torque acting on the aggregate. Under the assumption of inertialess particle, the following equations hold,

$$F = \int_S \sigma \cdot n \, dS \tag{8}$$

$$T = \int_S (x - x_{\mathrm{p}}) \times (\sigma \cdot n) \, dS = 0 \tag{9}$$

where T is the total torque on the particle surface S and n is the unit vector normal to the particle surface pointing from the fluid to the boundary. Notice that the aggregate is torque-free, whereas the only external force is the applied force F.

The solution of the governing equations gives the fluid velocity and pressure fields, along with the particle translational and angular velocities. The translational and orientational dynamics can be computed by integrating the following equations,

$$\frac{dx_{\mathrm{p}}}{dt} = U_{\mathrm{p}} \tag{10}$$

$$\frac{d\theta_{\mathrm{p}}}{dt} = \omega_{\mathrm{p}} \tag{11}$$

with initial conditions $x_{\mathrm{p}}|_{t=0} = x_{\mathrm{p},0}$ and $\theta_{\mathrm{p}}|_{t=0} = \theta_{\mathrm{p},0}$. Notice that, for the problem under investigation (settling dynamics of a particle in an unbounded fluid), the evolution of the aggregate center of volume does not affect the orientational dynamics. Therefore, Equation (10) can be removed from the set of equations to be solved.

The governing equations can be made dimensionless by choosing appropriate characteristic quantities for length, time, and stress. As characteristic length, we choose the effective radius of the aggregate, defined as the radius of a sphere with the same volume V of the aggregate, $R_{\mathrm{eff}} = (\frac{3V}{4\pi})^{1/3}$. The characteristic time is chosen as the inverse of a characteristic shear rate $(U/R_{\mathrm{eff}})^{-1}$, where U is the velocity along the direction of the applied force. The characteristic stress is $m(U/R_{\mathrm{eff}})^n$. By using such characteristic quantities, we can recast the governing equations and boundary conditions in their dimensionless form. In these equations, only the flow index n appears as dimensionless parameter. Therefore, the investigated system is fully determined by specifying n and the geometry of the aggregate defined by the parameters in Equation (1). In this regard, the radius of the primary particles a is related to R_{eff} through N_{p}. Moreover, the radius of gyration R_{g} is determined once the other three parameters in Equation (1) are chosen. Therefore, the geometrical parameters that need to be specified are the number of particles N_{p}, the fractal dimension D_{f}, and the fractal pre-factor k_{f}.

To quantify the hydrodynamic resistance of the particle to the applied force, we introduce the drag correction coefficient [25]:

$$X = \frac{F}{6\pi \, m \left(\frac{U}{2R_{\mathrm{eff}}}\right)^{n-1} U \, R_{\mathrm{eff}}} \tag{12}$$

defined as the applied force divided by the modified Stokes drag coefficient where the Newtonian viscosity is replaced by the power-law model. Of course, $X = 1$ for a sphere in a Newtonian fluid. The value of X depends on the orientation of the aggregate and, as such, changes in time since the particle varies its orientation while sedimenting. As it will be discussed below, the aggregate orientation can achieve various regimes, leading to different regime drag correction coefficients X_{R}. As X_{R} is affected by the initial particle orientation, we average over C initial configurations:

$$\langle X_{\mathrm{R}} \rangle = \frac{1}{C} \sum_C X_{\mathrm{R}} \tag{13}$$

Finally, to make the results independent of the seed used in the algorithm to generate the aggregate, for each set of fractal parameters in Equation (1), the simulation is repeated with different seeds. The ensemble-average drag correction coefficient is computed as

$$\langle X_R \rangle_m = \frac{1}{N_{seed}} \sum_{N_{seed}} \langle X_R \rangle \tag{14}$$

with N_{seed} the number of seeds.

In this work, the fractal pre-factor is fixed to $k_f = 1.3$, which is a value commonly used in the literature to describe realistic aggregate shapes [8]. The sedimentation dynamics is studied by varying the flow index, the fractal dimension, and the number of primary spheres forming the particle.

2.2. Numerical Method

The calculation of the regime drag correction coefficient requires the knowledge of the orbit followed by the aggregate while pulled by the force. This would need, at each time step, the calculation of the particle angular velocity from the solution of the set of equations presented in the previous section (as discussed above, the translational dynamics is irrelevant). Instead of directly solving the governing equations to compute the orbit, we can greatly speed-up the calculations by precomputing a database of particle translational and angular velocities for different orientations of the aggregate [35,36]. The velocities needed at the right-hand side of Equations (10) and (11) are, then, obtained by interpolating the data of the database. In this way, the computational effort is only due to the construction of the database that can be used to compute the orbit for any initial orientation. Of course, the database must be re-computed for every particle morphology.

Due to the irregular particle shape, two orthogonal vectors fixed with the particle are needed to track the evolution of its orientation. However, it should be noted that any configuration obtained by rotating the aggregate around the applied force is equivalent in terms of particle dynamics and drag coefficient. Consequently, one orientation vector is sufficient to fully describe all the possible configurations of the aggregate with respect to the applied force. This can be readily seen if we consider the particle fixed in the laboratory frame and the applied force is rotated. All the possible configurations are, indeed, obtained by rotating the applied force around the unit sphere (so just considering one orientation vector). This is, in fact, the procedure we adopted to build the database, i.e., we fix the aggregate orientation as generated by the particle–cluster algorithm and solve the fluid dynamics problem for several orientations of the applied force uniformly distributed over the unit sphere. More specifically, we divide the unit sphere in a triangular mesh with icosahedral symmetry. The orientations of the force are taken as the directions connecting the center of the unit sphere and the vertices of the icosahedral mesh. We select an icosahedral subdivision with 162 vertices, verifying that this subdivision is sufficient to assure a good accuracy of the interpolation. Indeed, by computing the interpolating functions on an icosahedral grid with 42 vertices, the maximum relative error is ~3%.

Once the database of particle velocities (and the corresponding drag correction coefficients) has been computed, the rotational dynamics of the aggregate can be calculated by integrating Equation (11). It is, however, more convenient to compute the particle dynamics in the body reference frame, i.e., by fixing the orientation of the particle and rotating the applied force. The adopted procedure is summarized in Algorithm 1. We have denoted with f the (time-dependent) applied force vector acting on the aggregate, and with f_0 its initial value. In step 3, the search procedure described in [37] is used. A linear interpolation over the spherical triangle mesh is done in step 4 [37,38]. The update of the force f is carried out through quaternions [39]. Specifically, in step 5, a third-order Adams–Bashforth scheme is used to integrate in time the quaternions through the angular velocity computed at the previous step. A rotation matrix in terms of quaternions is constructed and used to update f. The procedure just described gives the time-evolution of the applied force f and of the drag correction coefficient $X(t)$ that can be used to calculate the regime drag correction coefficient X_R. Finally, the aggregate orbit can be reconstructed by transforming the force from the body to the lab reference frame, i.e., by rotating the aggregate according to the rotation matrix that, at every instant, transform f in $F = (F, 0, 0)$ (that is the applied force in the lab frame). The procedure summarized in Algorithm 1 has been adopted to simulate the rheology of a dilute suspension of fractal aggregates in shear-thinning fluids and validated by reproducing the dynamics of spheroids in shear flow [36].

Algorithm 1 Procedure used to update the applied force in the body reference frame

---

1:    $f_0 \leftarrow$ initial orientation of the force
2:    **for** $t \leftarrow 1, num\_time\_steps$ **do**
3:        Identify the spherical triangle containing f
4:        Compute ω_p and X through interpolation over the spherical triangle in f
5:        Update f using quaternions
6:    **end for**

The solution of the governing equations to build the database is done by the finite element method. The particle translational and angular velocities are treated as additional unknowns, and are included in the weak form of momentum equation. Lagrange multipliers in each node of the particle surface are employed to enforce the conditions in Equations (8) and (9) [40,41]. The fluid domain is discretized by tetrahedral elements. Mesh generation issues arise due to the contact points between the spheres generated by the particle–cluster algorithm. To overcome this problem, we perform a Boolean union operation of the spheres with a set of cylinders connecting the centers of the spheres in contact. The radius of the cylinders is $0.732a$. We checked that lower values of the cylinder radius do not significantly alter the results. The Boolean union, smoothing, and meshing of the aggregate surface is done by the library PyMesh [42]. Examples of the surface meshes for the aggregates in Figure 1a,b are shown in Figure 1c,d. The tetrahedral volume mesh is generated by Gmsh [43].

The mesh and geometrical parameters used in the simulations are reported in Table 2 for the three values of N_p considered in this work. The symbols Δx, Δx_{out}, and R_{out} denote the size of the elements on the aggregate surface and on the external domain (made dimensionless by the primary particle radius a), and the radius of the external sphere (see Figures 1e). The number of tetrahedral elements N_{elem} is reported in the last column of Table 2. Notice that, to neglect the effect of the boundary condition far from the particle, very large external domains are needed. Furthermore, as expected, bigger aggregates (i.e., higher values of N_p) require larger external domains. We verified mesh and domain size convergence by reducing both Δx and Δx_{out}, and by further increasing R_{out}. The (dimensionless) time-step size depends on the flow index n, ranging from about 0.01 for $n = 1$ to 0.005 for $n = 0.6$.

Table 2. Mesh and geometrical parameters.

N_p	Δx	R_{out}	Δx_{out}	N_{elem}
20	0.25	200	35	~25,000
50	0.25	400	60	~60,000
100	0.35	800	120	~60,000

The accuracy of the finite element solution is checked by comparing the results for a spherical particle in a power-law fluid with those reported in Dazhi and Tanner [25]. In Figure 2, the drag correction coefficient is reported as a function of the flow index n. The black circles are the simulation results by Dazhi and Tanner [25] and the triangles are obtained by our simulations for different mesh resolutions and size of the external domain (the radius of the spherical particle is 1, meshes M1 and M2 have approximately 50,000 and 70,000 elements). First of all, the superposition of the triangles denotes that the results are independent of the mesh and domain size used. A fair agreement between triangles and circles is observed for values of the flow index between 0.8 and 1. For lower *n*-values, deviations between the two sets of data are observed. We believe this is due to the coarser mesh used in Dazhi and Tanner that is particularly problematic for low values of the flow index due to large gradients of the velocity field around the particle. We have further examined this point by solving the same problem in a 2D axisymmetric geometry allowing for a much more refined mesh. The results show that the data for an extremely fine mesh overlap the triangles (deviations are lower than 1%). Furthermore, by progressively coarsening the mesh, the value of X moves towards the black circles. It should be noted, however, that the maximum deviation between the triangles and the circles (at $n = 0.4$) is ~4%, which is a relatively low value.

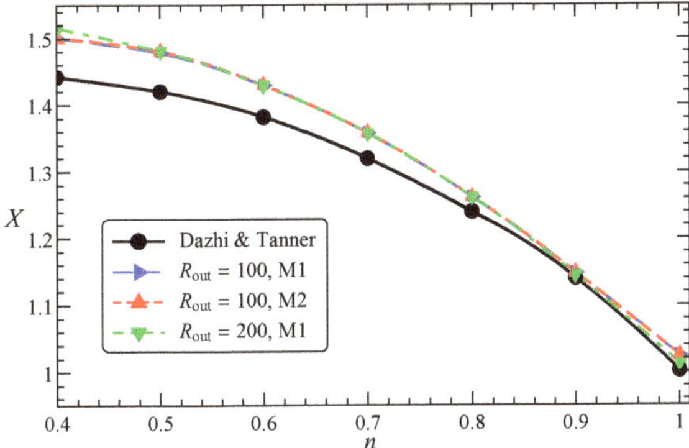

Figure 2. Drag correction coefficient X as a function of the flow index n for a spherical particle in an unbounded power-law fluid. The black circles are the simulation results by Dazhi and Tanner [25], and the triangles are obtained by our simulations for different mesh resolutions and size of the external domain.

3. Results

We investigate the aggregate dynamics and the resulting drag correction coefficient by varying the fractal dimension, the flow index, and the number of primary particles forming the aggregate. The values selected for the three parameters are $D_f = [1.5, 2.0, 2.5]$, $n = [1.0, 0.8, 0.6]$, and $N_p = [20, 50, 100]$. As discussed in Section 2.2, for each set of these parameters, we first run single-step simulations for different orientations of the applied force uniformly distributed over the unit sphere. Figure 3a–c reports the drag correction coefficient X as a function of the polar and azimuthal spherical coordinates $(0 \leq \theta \leq \pi, -\pi \leq \phi \leq \pi)$, identifying the orientation of the applied force for $n = 1$ (i.e., the Newtonian case), $n = 0.8$, and $n = 0.6$, respectively. The aggregate is the one shown in Figure 1a, i.e., with $N_p = 20$ and $D_f = 1.5$. It can be readily observed that (i) the drag correction coefficient depends on the orientation of the force; (ii) the distributions are symmetric since X is the same for a specific force orientation (θ, ϕ) and its opposite $(\pi - \theta, \phi \pm \pi)$; (iii) in agreement with the spherical particle case [25], the drag correction coefficient increases as the flow index decreases (see the bar legends on the right of the panels); and (iv) the distributions are not affected by the flow index (for instance, the maxima and minima are observed for the same orientations of the force). Previous results have evidenced a trend between the drag force experienced by a fractal aggregate and its area projected along the direction of the applied force [14], although this geometrical quantity is not sufficient to accurately predict the drag. The dimensionless area of the aggregate projected along the force direction is reported in Figure 3d. Specifically, we take the directions identified by the 162 vertices of the unity sphere and, for each of them, we compute the area of the aggregate projected on a plane orthogonal to that direction (identified by the spherical coordinates θ and ϕ). The comparison with panels (a)–(c) shows some similarities between the distributions, e.g., the position of the maxima and minima is approximately the same, although a strict correlation is not observed.

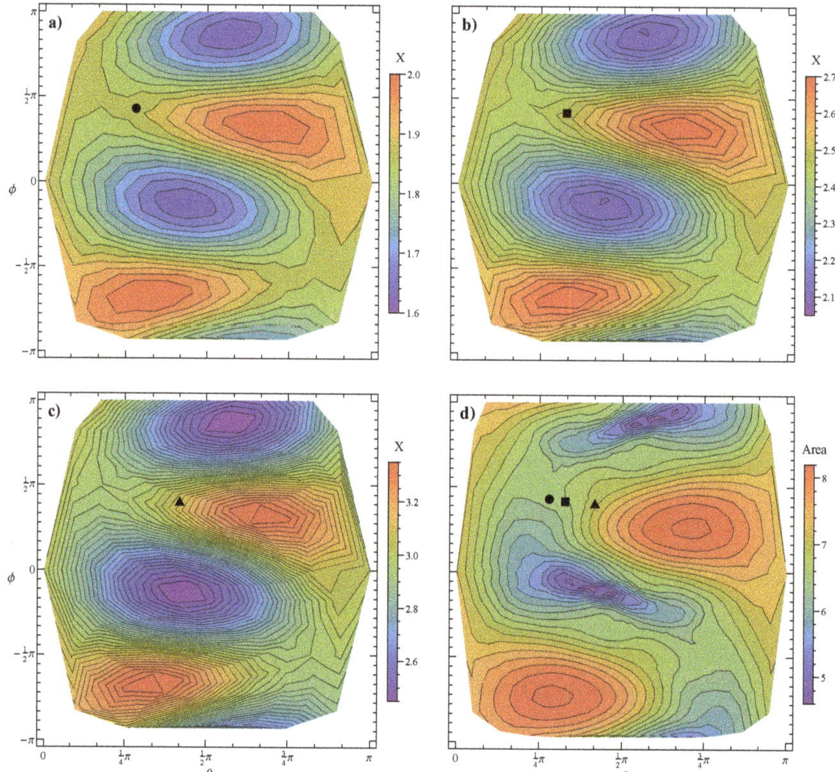

Figure 3. (**a–c**) Drag correction coefficient for the aggregate shown in Figure 1a ($N_\mathrm{p} = 20$, $D_\mathrm{f} = 1.5$) as a function of the force direction identified by the spherical coordinates (θ, ϕ) for the Newtonian fluid (a), the power-law fluid with $n = 0.8$ (b), and $n = 0.6$ (c). (**d**) Dimensionless area projected along the direction of the applied force for the same aggregate as in panels (a–c). The symbols denote the direction of the force attained at long times for the Newtonian fluid (circle), power-law fluid with $n = 0.8$ (square) and $n = 0.6$ (triangle).

The same quantities are reported in Figure 4 for the more sphere-like aggregate in Figure 1b ($N_\mathrm{p} = 20$, $D_\mathrm{f} = 2.5$). As for the previous case, similar distributions are observed as the flow index is varied, with higher values of X for more shear-thinning fluids. The projected area also shows a trend similar to the drag correction coefficient. Due to the higher value of the fractal dimension leading to an aggregate with a more spherical shape, the range of variation of both the projected area and the drag correction coefficient is narrower than in the case at $D_\mathrm{f} = 1.5$, i.e., the influence of the force orientation is weaker.

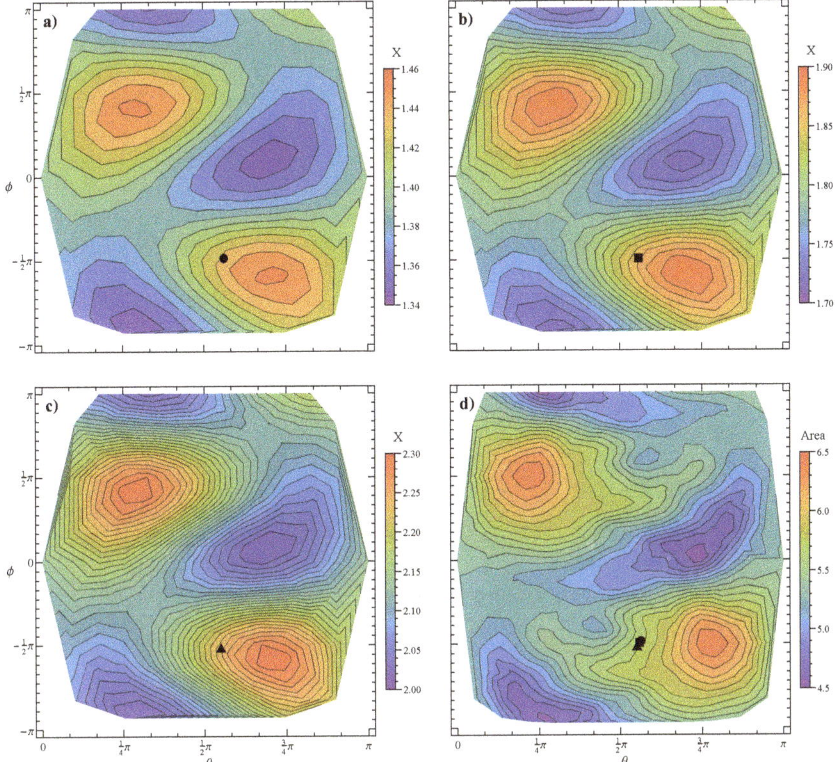

Figure 4. (**a–c**) Drag correction coefficient for the aggregate shown in Figure 1b ($N_\mathrm{p} = 20$, $D_\mathrm{f} = 2.5$) as a function of the force direction identified by the spherical coordinates (θ, ϕ) for the Newtonian fluid (a), the power-law fluid with $n = 0.8$, (b) and $n = 0.6$ (c). (**d**) Dimensionless area projected along the direction of the applied force for the same aggregate as in panels (a–c). The symbols denote the direction of the force attained at long times for the Newtonian fluid (circle), power-law fluid with $n = 0.8$ (square) and $n = 0.6$ (triangle).

The data presented so far refer to the instantaneous drag correction factor, i.e., the one obtained by solving the fluid dynamics equations for a fixed orientation of the force (or, equivalently, of the aggregate for a fixed force). The applied force, however, generates a rotation of the aggregate (and, of course, a translation) leading to a change of the orientation and, in turn, of the drag correction coefficient. The knowledge of the orientational dynamics of the aggregate is then crucial to determine the time evolution of the drag correction coefficient and the regime attained by the particle. By using the procedure described in the previous section, we compute the orientational dynamics of the applied force for different initial orientations. Figure 5 shows the orbits for the aggregates reported in Figure 1a (top row) and Figure 1b (bottom row), and for the Newtonian (left column) and power-law fluid with $n = 0.6$ (right column). Twelve initial orientations uniformly distributed over the unit sphere are considered (blue circles). For these sets of parameters, the orbits converge towards a unique equilibrium point (green circle) regardless of the initial orientation. In the fixed reference frame, this means that the aggregate achieves a stable orientation. Specifically, our simulations show that, once the regime is achieved, the particle still rotates around the applied force, although with a very small rotation rate (the resulting linear velocity, obtained as the angular velocity around the applied force times the effective radius, is 2–3 orders of magnitude smaller than the sedimentation velocity). It is important to point out, however, that this rotation does not influence the drag as any configuration around the force is equivalent (i.e., the force direction is a symmetry axis). The orientations of the force corresponding to the equilibrium points (green circles in Figure 5) are shown as symbols in the previous Figures 3 and 4. It can be readily observed that shear-thinning slightly affects the equilibrium orientation only for rod-like particles, whereas it has no influence for higher values of the fractal dimension (the symbols in Figure 4 overlap). Moreover, in both cases, the equilibrium orientation does not correspond to any special value of the projected area (for instance the minimum). Therefore, this quantity is not representative of the final orientation achieved by the aggregate and, as such, it is not helpful to estimate the drag correction factor at long times. On the contrary, the detailed microstructure of the aggregate needs to be considered to correctly predict the sedimentation dynamics.

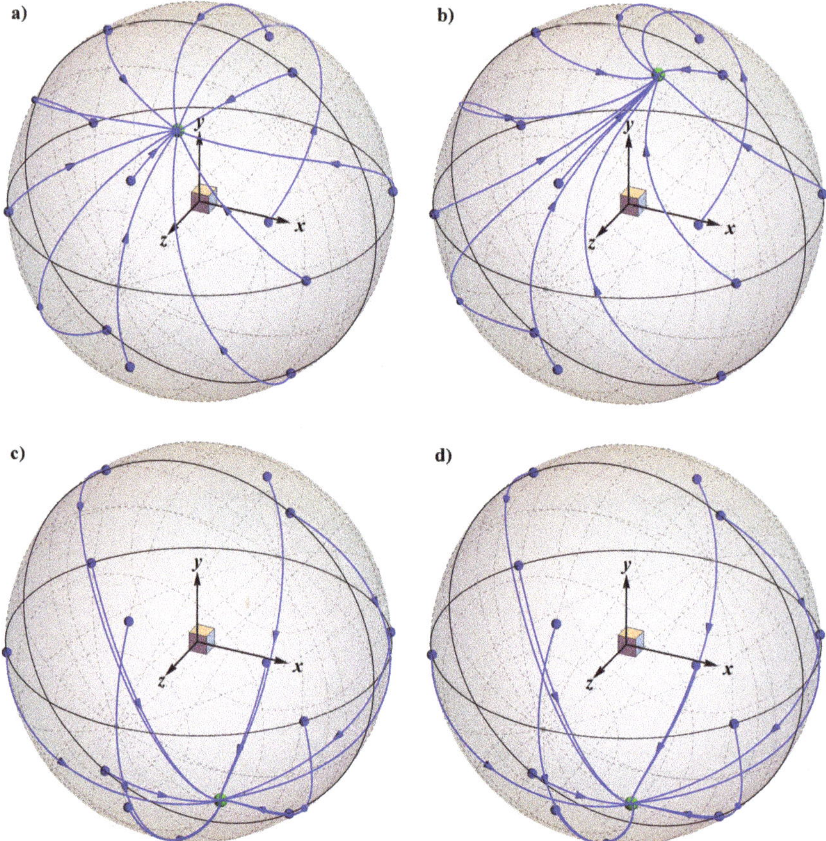

Figure 5. Orbits described by the orientation of the applied force for 12 initial orientations (blue circles) uniformly distributed over the unit sphere for: the aggregate shown in Figure 1a ($N_p = 20$, $D_f = 1.5$) in a Newtonian (**a**) or power-law fluid with $n = 0.6$ (**b**), the aggregate shown in Figure 1b ($N_p = 20$, $D_f = 2.5$) in a Newtonian (**c**) or power-law fluid with $n = 0.6$ (**d**). The equilibrium points are denoted by green circles.

To further investigate on the effect of aggregate morphology, we have repeated the calculations by varying the seed of the random number generator. We recall that, although the fractal parameters in Equation (1) are fixed, the morphologies obtained by varying the seed are different. In the leftmost panels of Figure 6, the regime drag correction coefficient is shown as a function of the seed for $N_p = 20$ and for different values of the fractal dimension and the flow index. If the force reaches an equilibrium point, regardless of the initial orientation like the orbits shown in Figure 5, X_R is taken as the steady-state value. These points are represented as solid circles in Figure 6. The data show that, for $N_p = 20$, the specific morphology (seed) has a relatively weak effect on X_R, with a maximum relative deviations of 7% from the average value. Furthermore, in all the investigated cases, a single equilibrium orientation is achieved, except in one case ($D_f = 2.5$, seed = 1, and $n = 0.6$) that will be discussed later.

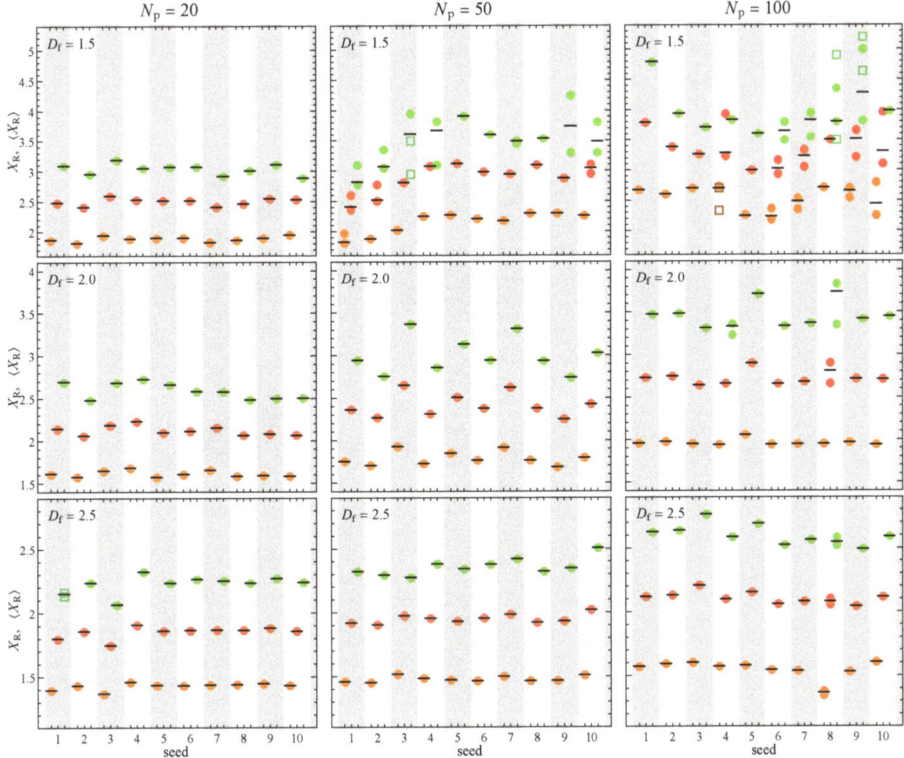

Figure 6. Regime drag correction coefficient X_R and its average over the initial orientations $\langle X_R \rangle$ for different number of primary particles (columns), fractal dimension (rows), random seed for the aggregate generation (bands), and flow index (orange $n = 1$, red $n = 0.8$, and green $n = 0.6$). Solid circles and open squares denote steady-state and periodic regimes. The black dashes represent $\langle X_R \rangle$.

A relevant quantity for the sedimentation process is the time t_R needed to achieve the final regime. For instance, with reference to Figure 5, this is the time needed to travel along the orbits from the initial orientation to the green circle. Of course, the time strongly depends on the initial orientation. Thus, we compute the orbits followed by the aggregate with orientation starting from the 162 vertices of the spherical triangle mesh discussed in the previous section and, for each orbit, we estimate the time needed to achieve the regime within a certain tolerance. In case a single steady-state regime exists, we calculate the time the force requires to align with the equilibrium orientation within an angle tolerance of 5°. The results for $N_p = 20$ and different values of the seed, fractal dimension, and flow index are shown as box plots in Figure 7. The lower and higher limits of each box plot represent the first and third quartile over the different initial orientations, whereas the black dash is the median. In general, the (dimensionless) time decreases as the flow index decreases, whereas it is rather unaffected by the fractal dimension, ranging between 10 and 100 for almost all the examined cases. There are, however, some exceptions leading to remarkably longer times.

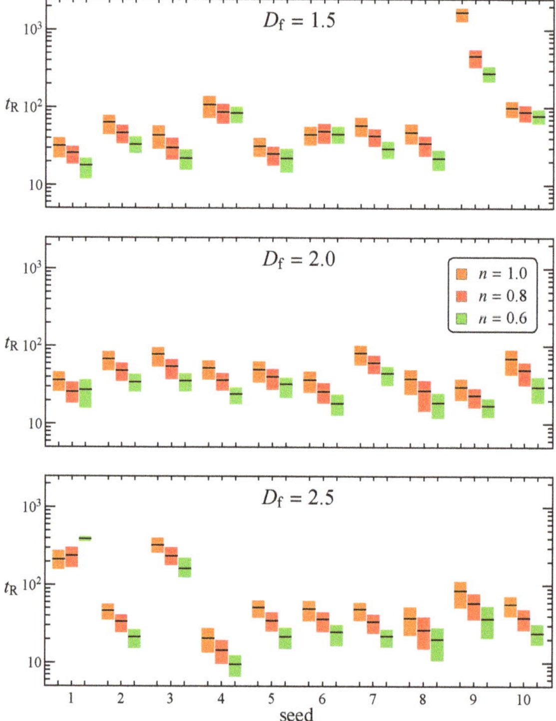

Figure 7. Box plot of the times needed for an aggregate to reach a stable regime as a function of particle random seed, flow index, and fractal dimension. The number of primary particles is $N_p = 20$. The black dash within each box represents the median of the distribution.

To investigate these particular cases more in detail , we show in Figure 8 the orbits described by the orientation of the force for (i) $n = 1$, $D_f = 1.5$, seed $= 9$ (Figure 8a corresponding to the highest box plot in the top panel of Figure 7); (ii) $n = 0.8$, $D_f = 2.5$, seed $= 1$ (Figure 8b corresponding to the leftmost red box plot in the bottom panel of Figure 7); and (iii) $n = 0.6$, $D_f = 2.5$, seed $= 1$ (Figure 8c corresponding to the leftmost green box plot in the bottom panel of Figure 7). In the first case, we still observe a dynamics similar to what reported in Figure 5 with all the orbits converging to a single equilibrium point. However, at variance with the previous cases where the orbits independently moved towards the equilibrium point, now each trajectory converges first towards a common orbit and then, very slowly, to the equilibrium orientation, resulting in a drastic increase of the time needed to reach the steady-state regime. A similar dynamics is also observed for the same seed and for $n = 0.8$ (red box plot) and $n = 0.6$ (green box plot), as well as for $D_f = 2.5$ and seed $= 3$. A different scenario is observed for the second case (Figure 8b), where the orientation of the force follows spiraling trajectories before reaching the equilibrium point, also resulting in a longer transient dynamics. In the third case reported in Figure 8c, the regime becomes periodic with the presence of a limit cycle. Therefore, while settling, the aggregate continuously changes its orientation around the applied force, coming back to the same configuration after a certain period. Notice that the cases in Figure 8b,c correspond to the same aggregate (the fractal parameters and the seed are the same) and differ for the flow index. Further, the same aggregate in a Newtonian fluid ($n = 1$) gives orbits like the ones shown in Figure 5. Thus, we conclude that, for this aggregate shape, a decrease of the flow index leads to the appearance of a bifurcation (specifically a Hopf bifurcation [44]) with a qualitative change in the regime attained by the aggregate. In the case of a periodic regime, the time reported in Figure 7 is evaluated as the time needed to reach the limit cycle within a tolerance of 5% on X. Notice that the appearance of the bifurcation inverts the trend of t_R with the flow index (the values of the box plots corresponding to seed = 1 in Figure 7c increase with decreasing n). As the number of primary particles of the aggregate increases, another possible scenario, depicted in Figure 8d for $N_p = 50$, $D_f = 1.5$, $n = 0.6$, appears. Two equilibrium regimes are observed, identified by the blue and red orbits. Therefore, depending on the initial configuration, the aggregate can orient along one of the two stable orientations.

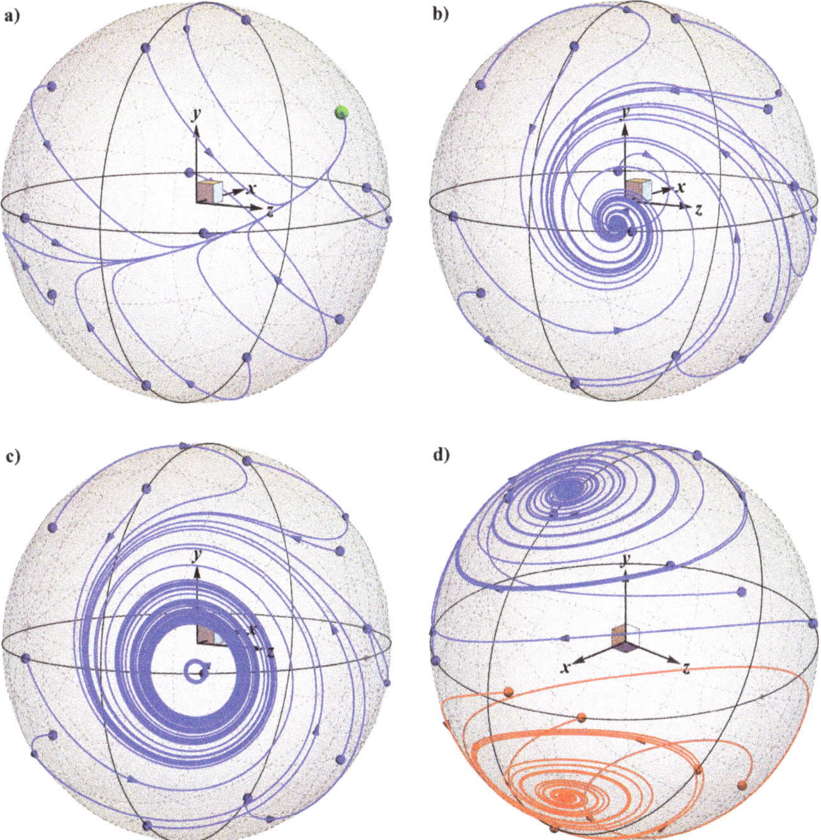

Figure 8. Orbits described by the orientation of the applied force for 12 initial orientations (blue circles) uniformly distributed over the unit sphere. The parameters are (**a**) $n = 1$, $N_p = 20$, $D_f = 1.5$, seed = 9; (**b**) $n = 0.8$, $N_p = 20$, $D_f = 2.5$, seed = 1; (**c**) $n = 0.6$, $N_p = 20$, $D_f = 2.5$, seed = 1; (**d**) $n = 0.6$, $N_p = 50$, $D_f = 1.5$, seed = 9.

By increasing the complexity of the shape, i.e., by increasing N_p and decreasing D_f, spiraling orbits, periodic, and bistable regimes are more frequent, leading to a substantial increase of the time needed to reach the final orientation, and, more importantly, to a significant effect of the detailed morphology (i.e., the seed used to generate the aggregate) on the settling dynamics. This is illustrated in the middle and right panels of Figure 6 where the regime drag correction coefficient is shown as a function of the seed for $N_\text{p} = 50$ and $N_\text{p} = 100$. The periodic regime is denoted by two open squares identifying the maximum and minimum values of the oscillation, with a corresponding X_R calculated by averaging X over a period. The bistability is indicated by two closed circles that represent the values of X_R for the two equilibrium points. In Figure 6, the average of the regime drag correction coefficient over all the initial orientations ($\langle X_\text{R} \rangle$ in Equation (13)) is also reported as a black dash. In case of a single equilibrium orientation, the unique solid circle coincides, in fact, with the dash. When multiple regimes coexist, the black dash is closer to the one that attracts more orbits. Notice that, in some cases (see, e.g., $N_\text{p} = 50$, $D_\text{f} = 1.5$, seed $= 9$, and $n = 0.6$), the values of X_R for the two equilibrium points are remarkably different, resulting in a relevant quantitative effect of the initial orientation on the terminal velocity. At variance with the case at $N_\text{p} = 20$, the effect of the microstructure (seed) is much more relevant, leading to deviations up to 25% from the value of the drag correction coefficient averaged over the seeds. In particular, maximum deviations are found for more elongated aggregates rather than sphere-like shapes. (Indeed, in the limiting case of a spherical aggregate, different seeds would produce the same shape.)

By averaging the data in Figure 6 over the seeds, we obtain the ensemble-average drag correction coefficient $\langle X_\text{R} \rangle_\text{m}$ reported in Figure 9. The data are shown as a function of the fractal dimension, where each panel refers to a fixed number of primary particles and the curves are parametric in the flow index. For the Newtonian case (orange symbols), $\langle X_\text{R} \rangle_\text{m}$ can be used to calculate the hydrodynamic radius, which is found to be quantitatively consistent with the one reported in [33]. In the investigated range of D_f, the drag correction factor is a linear decreasing function of the fractal dimension, i.e., an aggregate with a more spherical shape sediments faster than one with a rod-like morphology. In agreement with previous results for spheres [25], shear-thinning increases the drag correction coefficient. As already noted in Figure 6, higher values of $\langle X_\text{R} \rangle_\text{m}$ are observed as N_p increases. This effect is more evident for low fractal dimensions where a variation of the number of primary particles affects the aspect ratio of the aggregate, in turn altering the drag experienced by the particle. On the contrary, as previously remarked, for aggregates with a sphere-like shape (high D_f), the number of primary particles mainly affects the "resolution" of the microstructure, without substantially changing the main geometrical features. For the same reason, the error bars are larger for low D_f and high N_p. As a final note, we recall that, especially for low fractal dimension, a variation of the number of particles and flow index may lead to different dynamics followed by the aggregate while sedimenting. In some cases, our simulations evidenced a qualitative change of the regime attained by the particle (e.g., a bifurcation) as one of these parameters is varied, with obvious consequences on the drag correction factor and on the time needed to achieve such regime. These observations prevent us to derive a simple scaling of $\langle X_\text{R} \rangle_\text{m}$ with N_p and n. In this regard, also the averaging of $\langle X_\text{R} \rangle$ over different seeds is, in some sense, misleading as it combines drag correction coefficients of aggregates that can experience very different dynamics. Therefore, we point out once again the importance of considering the detailed microstructure of the aggregate to correctly predict its sedimentation dynamics.

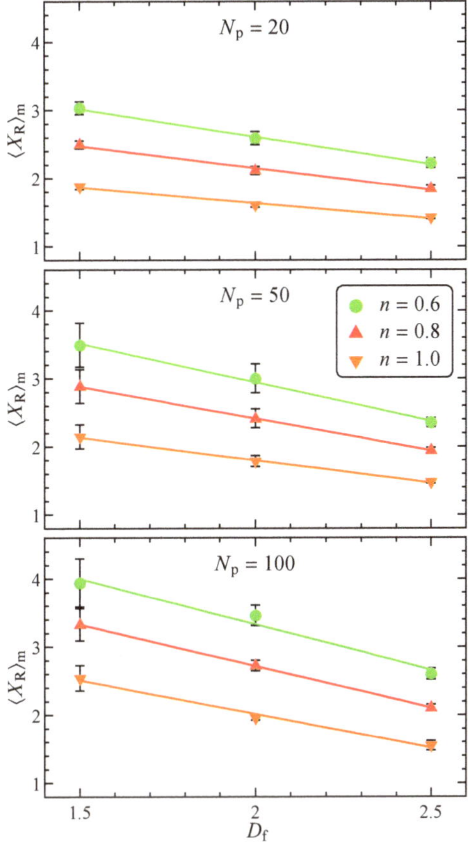

Figure 9. Ensemble-average drag correction coefficient as a function of fractal dimension, parametric in the number of aggregate primary particles and for different flow indexes. The data standard deviation and trend line are also reported.

4. Conclusions

In this work, the hydrodynamic drag experienced by a fractal aggregate suspended in a non-Newtonian fluid is studied by numerical simulations. The aggregate shape is generated through a particle–cluster method combining equally-sized spherical particles. The power-law constitutive equation is used to model the suspending fluid. Finite element simulations are employed to solve the fluid dynamics governing equations, for orientations of the applied force uniformly distributed over the unit sphere. These velocities are interpolated and used to reconstruct the translational and orientational aggregate dynamics. The drag correction coefficient at long times is averaged over several initial orientations and particle shapes with the same fractal parameters.

The results show a relevant effect of the aggregate morphology and shear-thinning on the sedimentation dynamics. Depending on the fractal dimension, the number of primary particles forming the aggregate, and the flow index, the aggregate can undergo a variety of rotational dynamics while settling. These can lead to a stable orientation, periodic oscillations around the force direction, or coexistence of multiple equilibrium orientations, with relevant implications on the terminal velocity and the time needed to achieve the long-time regime.

The ensemble-average drag correction coefficient linearly decreases by increasing the fractal dimension in the investigated range, i.e., rod-like aggregates sediment more slowly than particles with an isotropic shape. Shear-thinning further reduces the settling velocity. At low fractal dimensions, the number of primary spheres has a relevant influence on the drag correction coefficient as it affects the aggregate aspect ratio. On the contrary, a weak effect is observed for aggregates with a sphere-like shape as an increase of the number of spheres does not produce a relevant change of the overall morphology.

The results reported in the present work highlight that the detailed particle shape needs to be considered to properly predict the sedimentation dynamics. As a matter of fact, a variation of the morphology, even with the same fractal characteristics, may lead to different transients and final regimes. Therefore, to properly understand the settling phenomenon, the connection between the shape of the aggregate and the resulting translational and rotational dynamics needs to be investigated. This will be the subject of future works.

Finally, we would like to point out that the present results, although discussed in the context of the sedimentation process, apply to any system in which a particle of fractal shape moves in a shear-thinning liquid in a uniform flow field. Indeed, regardless of the nature of the applied force, the particle experiences an hydrodynamic resistance that can be predicted from the results reported in this work. In this regard, neglecting the details of the specific morphology, our calculation can be exploited to derive a drag correlation model to be included in a computational fluid dynamic solver for simulating particle laden flows [45].

Author Contributions: Conceptualization, M.T. and G.D.; methodology, M.T. and G.D.; software, M.T. and G.D.; validation, M.T. and G.D.; formal analysis, M.T. and G.D.; writing—original draft preparation, M.T. and G.D.; writing—review and editing, M.T. and G.D.; visualization, M.T. and G.D.; funding acquisition, G.D. All authors have read and agreed to the published version of the manuscript.

Funding: This work was carried out in the context of the VIMMP project (www.vimmp.eu). The VIMMP project has received funding from the European Union's Horizon 2020 research and innovation programme under grant agreement No. 760907.

Conflicts of Interest: The authors declare no conflicts of interest.

References

1. Bratby, J. *Coagulation and Flocculation in Water and Wastewater Treatment*; IWA Publishing: London, UK, 2016.
2. Stuetz, R.; Stephenson, T. *Principles of Water and Wastewater Treatment Processes*; IWA Publishing: London, UK, 2009.
3. Spicer, P.T.; Pratsinis, S.E. Shear-induced flocculation: The evolution of floc structure and the shape of the size distribution at steady state. *Water Res.* **1996**, *30*, 1049–1056, doi:10.1016/0043-1354(95)00253-7.
4. Gregory, J. Monitoring particle aggregation processes. *Adv. Colloid Interface Sci.* **2009**, *147–148*, 109–123, doi:10.1016/j.cis.2008.09.003.
5. He, W.; Nan, J.; Li, H.; Li, S. Characteristic analysis on temporal evolution of floc size and structure in low-shear flow. *Water Res.* **2012**, *46*, 509–520, doi:10.1016/j.watres.2011.11.040.
6. Moruzzi, R.B.; de Oliveira, A.L.; da Conceição, F.T.; Gregory, J.; Campos, L.C. Fractal dimension of large aggregates under different flocculation conditions. *Sci. Total Environ.* **2017**, *609*, 807–814, doi:10.1016/j.scitotenv.2017.07.194.
7. Oh, C.; Sorensen, C. The Effect of Overlap between Monomers on the Determination of Fractal Cluster Morphology. *J. Colloid Interface Sci.* **1997**, *193*, 17–25, doi:10.1006/jcis.1997.5046.
8. Melas, A.D.; Isella, L.; Konstandopoulos, A.G.; Drossinos, Y. Morphology and mobility of synthetic colloidal aggregates. *J. Colloid Interface Sci.* **2014**, *417*, 27–36, doi:10.1016/j.jcis.2013.11.024.
9. Johnson, C.P.; Li, X.; Logan, B.E. Settling Velocities of Fractal Aggregates. *Environ. Sci. Technol.* **1996**, *30*, 1911–1918, doi:10.1021/es950604g.
10. Vahedi, A.; Gorczyca, B. Predicting the settling velocity of flocs formed in water treatment using multiple fractal dimensions. *Water Res.* **2012**, *46*, 4188–4194, doi:10.1016/j.watres.2012.04.031.
11. Durlofsky, L.; Brady, J.F.; Bossis, G. Dynamic simulation of hydrodynamically interacting particles. *J. Fluid Mech.* **1987**, *180*, 21, doi:10.1017/s002211208700171x.

12. Hassonjee, Q.; Ganatos, P.; Pfeffer, R. A strong-interaction theory for the motion of arbitrary three-dimensional clusters of spherical particles at low Reynolds number. *J. Fluid Mech.* **1988**, *197*, 1–37, doi:10.1017/s0022112088003155.

13. Filippov, A. Drag and Torque on Clusters of N Arbitrary Spheres at Low Reynolds Number. *J. Colloid Interface Sci.* **2000**, *229*, 184–195, doi:10.1006/jcis.2000.6981.

14. Binder, C.; Feichtinger, C.; Schmid, H.J.; Thürey, N.; Peukert, W.; Rüde, U. Simulation of the hydrodynamic drag of aggregated particles. *J. Colloid Interface Sci.* **2006**, *301*, 155–167, doi:10.1016/j.jcis.2006.04.045.

15. Chopard, B.; Nguyen, H.; Stoll, S. A lattice Boltzmann study of the hydrodynamic properties of 3D fractal aggregates. *Math. Comput. Simul* **2006**, *72*, 103–107, doi:10.1016/j.matcom.2006.05.024.

16. Schlauch, E.; Ernst, M.; Seto, R.; Briesen, H.; Sommerfeld, M.; Behr, M. Comparison of three simulation methods for colloidal aggregates in Stokes flow: Finite elements, lattice Boltzmann and Stokesian dynamics. *Comput. Fluids* **2013**, *86*, 199–209, doi:10.1016/j.compfluid.2013.07.005.

17. Happel, J.; Brenner, H. *Low Reynolds Number Hydrodynamics: With Special Applications to Particulate Media*; Springer: Dordrecht, The Netherlands, 1981.

18. Kim, A.S.; Stolzenbach, K.D. The Permeability of Synthetic Fractal Aggregates with Realistic Three-Dimensional Structure. *J. Colloid Interface Sci.* **2002**, *253*, 315–328, doi:10.1006/jcis.2002.8525.

19. Fellay, L.S.; Twist, C.; Vanni, M. Motion of rigid aggregates under different flow conditions. *Acta Mech.* **2013**, *224*, 2225–2248, doi:10.1007/s00707-013-0928-9.

20. Harshe, Y.M.; Ehrl, L.; Lattuada, M. Hydrodynamic properties of rigid fractal aggregates of arbitrary morphology. *J. Colloid Interface Sci.* **2010**, *352*, 87–98, doi:10.1016/j.jcis.2010.08.040.

21. Seyssiecq, I.; Ferrasse, J.H.; Roche, N. State-of-the-art: Rheological characterisation of wastewater treatment sludge. *Biochem. Eng. J.* **2003**, *16*, 41–56, doi:10.1016/s1369-703x(03)00021-4.

22. Ratkovich, N.; Horn, W.; Helmus, F.; Rosenberger, S.; Naessens, W.; Nopens, I.; Bentzen, T. Activated sludge rheology: A critical review on data collection and modelling. *Water Res.* **2013**, *47*, 463–482, doi:10.1016/j.watres.2012.11.021.

23. Wasserman, M.L.; Slattery, J.C. Upper and lower bounds on the drag coefficient of a sphere in a power-model fluid. *AIChE J.* **1964**, *10*, 383–388, doi:10.1002/aic.690100320.

24. Chhabra, R.P.; Uhlherr, P.H.T. Creeping motion of spheres through shear-thinning elastic fluids described by the Carreau viscosity equation. *Rheol. Acta* **1980**, *19*, 187–195, doi:10.1007/bf01521930.

25. Dazhi, G.; Tanner, R. The drag on a sphere in a power-law fluid. *J. Non-Newton. Fluid Mech.* **1985**, *17*, 1–12, doi:10.1016/0377-0257(85)80001-X.

26. Chhabra, R.P. *Bubbles, Drops, and Particles in Non-Newtonian Fluids*; Taylor & Francis Inc: Boca Raton, FL, USA, 2006.

27. Tripathi, A.; Chhabra, R.P.; Sundararajan, T. Power law fluid flow over spheroidal particles. *Ind. Eng. Chem. Res.* **1994**, *33*, 403–410, doi:10.1021/ie00026a035.

28. Thouy, R.; Jullien, R. A cluster–cluster aggregation model with tunable fractal dimension. *J. Phys. A Math. Gen.* **1994**, *27*, 2953–2963, doi:10.1088/0305-4470/27/9/012.

29. Meakin, P. A Historical Introduction to Computer Models for Fractal Aggregates. *J. Sol-Gel Sci. Technol.* **1999**, *15*, 97–117, doi:10.1023/a:1008731904082.

30. Vormoor, O. Large scale fractal aggregates using the tunable dimension cluster–cluster aggregation. *Comput. Phys. Commun.* **2002**, *144*, 121–129, doi:10.1016/s0010-4655(02)00142-x.

31. Brown, M.; Errington, R.; Rees, P.; Williams, P.; Wilks, S. A highly efficient algorithm for the generation of random fractal aggregates. *Phys. D* **2010**, *239*, 1061–1066, doi:10.1016/j.physd.2010.02.018.

32. Mroczka, J.; Woźniak, M.; Onofri, F.R. Algorithms and methods for analysis of the optical structure factor of fractal aggregates. *Metrol. Meas. Syst.* **2012**, *19*, 459–470, doi:10.2478/v10178-012-0039-2.

33. Filippov, A.; Zurita, M.; Rosner, D. Fractal-like Aggregates: Relation between Morphology and Physical Properties. *J. Colloid Interface Sci.* **2000**, *229*, 261–273, doi:10.1006/jcis.2000.7027.

34. Skorupski, K.; Mroczka, J.; Wriedt, T.; Riefler, N. A fast and accurate implementation of tunable algorithms used for generation of fractal-like aggregate models. *Phys. A* **2014**, *404*, 106–117, doi:10.1016/j.physa.2014.02.072.

35. Trofa, M.; D'Avino, G.; Maffettone, P.L. Numerical simulations of a stick-slip spherical particle in Poiseuille flow. *Phys. Fluids* **2019**, *31*, 083603, doi:10.1063/1.5109305.

36. Trofa, M.; D'Avino, G. Rheology of a Dilute Suspension of Aggregates in Shear-Thinning Fluids. *Micromachines* **2020**, *11*, 443, doi:10.3390/mi11040443.

37. Lawson, C.L. C^1 surface interpolation for scattered data on a sphere. *Rocky Mt. J. Math.* **1984**, *14*, 177–202, doi:10.1216/rmj-1984-14-1-177.

38. Carfora, M.F. Interpolation on spherical geodesic grids: A comparative study. *J. Comput. Appl. Math.* **2007**, *210*, 99–105, doi:10.1016/j.cam.2006.10.068.

39. Rapaport, D.C. *The Art of Molecular Dynamics Simulation*; Cambridge University Press: Cambridge, UK, 2004.

40. Hu, H.H.; Patankar, N.; Zhu, M. Direct Numerical Simulations of Fluid–Solid Systems Using the Arbitrary Lagrangian–Eulerian Technique. *J. Comput. Phys.* **2001**, *169*, 427–462, doi:10.1006/jcph.2000.6592.

41. D'Avino, G.; Maffettone, P.; Greco, F.; Hulsen, M. Viscoelasticity-induced migration of a rigid sphere in confined shear flow. *J. Non-Newton. Fluid Mech.* **2010**, *165*, 466–474, doi:10.1016/j.jnnfm.2010.01.024.

42. Zhou, Q. PyMesh - Geometry Processing Library for Python. 2018. Available online: https://github.com/qnzhou/PyMesh (accessed on 10th April 2020).

43. Geuzaine, C.; Remacle, J.F. Gmsh: A 3-D finite element mesh generator with built-in pre- and post-processing facilities. *Int. J. Numer. Methods Eng.* **2009**, *79*, 1309–1331, doi:10.1002/nme.2579.

44. Strogatz, S. *Nonlinear Dynamics and Chaos: With Applications to Physics, Biology, Chemistry, and Engineering*; Westview Press: Boulder, CO, USA, 2015.

45. Zhou, Z.; Kuang, S.; Chu, K.; Yu, A. Discrete particle simulation of particle–fluid flow: Model formulations and their applicability. *J. Fluid Mech.* **2010**, *661*, 482–510, doi:10.1017/s002211201000306x.

 © 2020 by the authors. Licensee MDPI, Basel, Switzerland. This article is an open access article distributed under the terms and conditions of the Creative Commons Attribution (CC BY) license (http://creativecommons.org/licenses/by/4.0/).

applied sciences

Article

Chromium(III) Removal from Wastewater by Chitosan Flakes

Loris Pietrelli [1,2,*], **Iolanda Francolini** [2] , **Antonella Piozzi** [2], **Maria Sighicelli** [1], **Ilaria Silvestro** [2] and **Marco Vocciante** [3]

1   ENEA, CR Casaccia, Via Anguillarese 301, 00060 Roma, Italy; maria.sighicelli@enea.it
2   Dipartimento di Chimica, Università di Roma Sapienza, 00185 Roma, Italy; iolanda.francolini@uniroma1.it (I.F.); antonella.piozzi@uniroma1.it (A.P.); ilaria.silvestro@uniroma1.it (I.S.)
3   Dipartimento di Chimica e Chimica Industriale, Università degli Studi di Genova, Via Dodecaneso 31, 16146 Genova, Italy; marco.vocciante@unige.it
*   Correspondence: loris.pietrelli@uniroma1.it

Received: 30 January 2020; Accepted: 7 March 2020; Published: 11 March 2020

Featured Application: The ability of chitosan as a low-cost and environmentally friendly Cr(III) adsorbent was studied to evaluate its potential application in the field of tannery wastewater treatment, in terms of the removal of chromium ions avoiding their conversion into Cr(VI), the compounds of which exert highly toxic and carcinogenic effects on biological systems.

Abstract: Chitosan is very effective in removing metal ions through their adsorption. A preliminary investigation of the adsorption of chromium(III) by chitosan was carried out by means of batch tests as a function of contact time, pH, ion competition, and initial chromium(III) concentration. The rate of adsorption was rather rapid ($t_{1/2} < 18$ min) and influenced by the presence of other metal ions. The obtained data were tested using the Langmuir and Freundlich isotherm models and, based on R^2 values, the former appeared better applicable than the latter. Chitosan was found to have an excellent loading capacity for chromium(III), namely 138.0 mg Cr per g of chitosan at pH = 3.8, but metal ions adsorption was strongly influenced by the pH. About 76% of the recovered chromium was then removed simply by washing the used chitosan with 0.1 M EDTA (Ethylenediaminetetraacetic acid) solution. This study demonstrates that chitosan has the potential to become an effective and low-cost agent for wastewater treatment (e.g., tannery waste) and in situ environmental remediation.

Keywords: chitosan; chromium; heavy metals; adsorption; kinetics; low-cost adsorbent; tannery; ion exchange

---

1. Introduction

The removal of heavy metal ions from aqueous solutions, both for pollution control and for the recovery of raw materials, has assumed increasing importance in recent years. Among the many metals potentially harmful to the environment and human health, chromium pollution is of considerable concern, as the metal is widely used in many industrial activities such as electroplating, leather tanning, nuclear power plants, and textile industries [1,2].

To address this problem, numerous separation techniques are currently available (e.g., adsorption, ion exchange, selective precipitation, nanofiltration, etc.), the selection of which, however, is far from trivial and deserves extensive attention to avoid a suboptimal choice or the failure of the reclamation activity [3]. In general, adsorption-based technologies have proven to be among the most viable alternatives proposed for the treatment of industrial wastewater contaminated by a wide variety of pollutants, both organic [4] and inorganic [5,6], due to the low processing and instrumentation costs, the simplicity of operation, and the availability of different types of low-cost and environmentally friendly adsorbents. A wide range of materials, including activated carbon [7], metal oxides, carbon nanotubes, polymers, agricultural residues [8], and natural and modified clays [9], have been used successfully to adsorb heavy metals from aqueous solutions. This is even more interesting when it is possible to exploit low-cost adsorbent materials from industrial waste [10], with a double advantage for the environment, in line with operational guidelines such as the Circular Economy and the "near-zero discharge" of hazardous waste [11] established by the most recent European laws.

In this context, a very promising and inexpensive material is chitosan (poly-β-(1→4)-2-amino-2-deoxy-D-glucose), a nitrogenous polysaccharide prepared from chitin by partially deacetylating its acetoamine groups using strong alkaline solutions at about 70 °C. Chitosan has a high potential for the adsorption of metal ions, since it has both amino and hydroxyl groups that can act as chelation sites for metal ions. One of the most interesting advantages of chitosan is its versatility, since the material can be easily physically modified to obtain different forms of polymers such as beads [12], membranes [13], or sponges [14] for different applications. Chitosan can also be easily chemically modified to increase its applications [15]. Recently, several critical reviews have been published on the many applications of chitosan as an environmentally friendly biomaterial [16], ranging from the medical field [17] to food technology [18] and environmental protection [19].

Chromium can be found in the environment in the forms Cr(III) and Cr(VI), as its other oxidation states are not stable in aerated aqueous media [20]. The trivalent state is the most stable form in reducing conditions and is present as a cationic species ($Cr(OH)^{2+}$, $Cr(OH)_2^+$), with the first or second hydrolysis products dominating at pH values from 4 to 8. The low solubility of $Cr(OH)_3$ (log k = −16.19) considerably limits the concentration of Cr(III) for pH values above about 5.

Given its high danger to biological systems, many studies have focused on the removal of Cr(VI), while very few articles deal with the adsorption of Cr(III) by chitosan. Maruca et al. [21] reported the uptake of Cr(III) ions by chitosan flakes and the effect of PO_4^{3-} on the adsorption mechanism. Chui and collaborators [22] studied the removal of Cr(III) using a packed column filled with crab chitosan. More recently, Singh & Nagendran [23] reported comparative studies on the sorption of Cr(III) on chitin and chitosan in terms of a comparison between Langmuir and Freundlich isotherms. Overall, the papers often concern the application of chitosan membranes or beads, and not of chitosan flakes.

Considering that equilibrium analysis is the most important fundamental study required to evaluate the affinity of a sorbent, the ability of chitosan to remove chromium(III) by adsorption was studied in the present work to evaluate its potential application in the field of tannery wastewater treatment [1]. Numerous adsorption tests of chromium(III) on chitosan flakes were conducted to investigate the effects of contact time, pH, initial Cr(III) concentration, and, using real wastewater, ion competition. The thermodynamic behavior was assessed using the well-known Langmuir and Freundlich isotherm models.

2. Materials and Methods

2.1. Material and Reagents

Chitosan (molecular weight: 400 k, 66.9% <40 mesh, degree of deacetylation: 84–86%) was provided by Merck KGaA (Darmstadt, Germany) and used without further purification. Its surface area, estimated with the nitrogen adsorption method (BET, Quantachrome Nova 2200, Quantachrome Instruments, Boynton Beach, FL, U.S.), was equal to 1.578 m^2 g^{-1}. The water content of this commercial chitosan, determined by thermogravimetric analysis (TGA1, Mettler Toledo, OH, U.S.), was 12.7%, while its decomposition temperature was 292.12 °C (Figure 1). In order to study the effect of particle size on the adsorption of metal ions, two granulometric fractions (<0.42 mm and >0.42 mm) were obtained from the starting material by using a sieve shaker. Analytical grade chemicals were supplied by Merck Co. (Kenilworth, NJ, U.S.); aqueous solutions were prepared dissolving $Cr(NO_3)_3$ at different concentrations in deionized water (Millipore Milli-Q, Merck KGaA, Darmstadt, Germany), and the initial pH was adjusted by adding a few drops of HNO_3 and NaOH solutions.

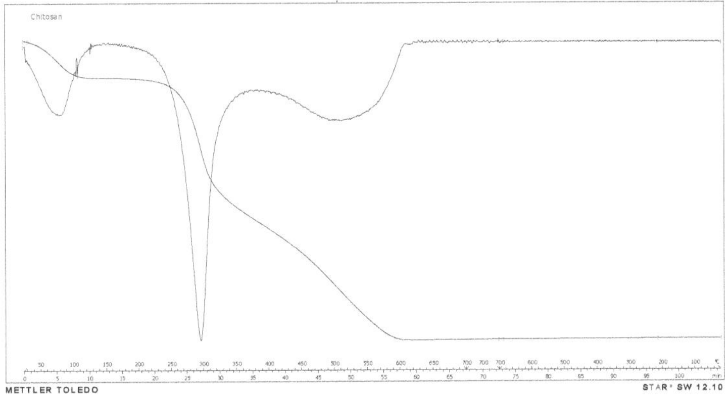

Figure 1. Thermogravimetric analysis (TGA) of the chitosan sample.

2.2. Adsorption Tests

Batch experiments (in triplicate) were carried out using 100 and 500 mg of adsorbent each time; chitosan was added to 50 mL of Cr solution in a conical flask. The stirring rate was set at 120 rpm for all adsorption/desorption tests using a temperature-controlled magnetic stirrer. Analysis of metal ions was carried out using an ICP-optical emission spectrometer (Inductively Coupled Plasma, Optima 2000 DV, Perkin Elmer, Waltham, MA, U.S.).

For experiments on the pH effect, the solutions were initially adjusted with aqueous solutions of acid or base (0.01 M HNO_3 and/or 0.01 M NaOH) to reach pH values between 0.5 and 5, and thus avoid the precipitation of $Cr(OH)_3$. Isotherms were recorded during the execution of adsorption experiments with various initial metal concentrations (C_0 = 50–2000 mg L^{-1}) at 20 °C. Kinetic tests were performed using 100 mg of chitosan flakes, 100 and 500 mg L^{-1} as the initial metal concentration at 20 °C, and pH = 3.8 for fixed time intervals during adsorption (t = 0–24 h). The effect of the granulometry of the flakes on the adsorption capacity was investigated using the two granulometric fractions of chitosan (<0.42 mm and >0.42 mm) under the same conditions. For all adsorption batch tests, a contact time of 120 min was set.

The equilibrium amount of metal in the solid phase, expressed as Q_e (mg g^{-1}), was determined with reference to the mass balance equation: $Q_e = (C_0 - C_e) \times (V/m)$, where C_0 and C_e (mg L^{-1}) are the initial and equilibrium metal concentrations, respectively, V (L) is the volume of the aqueous solutions, and m (g) is the mass of the adsorbent.

Dynamic tests were performed using a glass column with an internal diameter of 0.9 cm and a bed high of 40 cm, filled with 2 g of chitosan. Tests were performed using both a real wastewater solution and a 500 mg L^{-1} chromium(III) solution at pH = 3.5, imposing a flow rate of 23.6 mL h^{-1} (1 BV h^{-1}) by using a peristaltic pump.

The real wastewater, a tannery washing solution, had the following composition: pH = 3.2, Chemical Oxygen Demand, COD = 9.1 g L^{-1}, Total Suspended Solids, TSS < 1 g L^{-1}, Cr^{3+} = 635 mg L^{-1}, Na^+ = 1050 mg L^{-1}, Mg^{2+} = 760 mg L^{-1}, Ca^{2+} = 300 mg L^{-1}, Zn^{2+} = 115 mg L^{-1}, Cd^{2+} = 87 mg L^{-1}, SO_4^{2-} = 1.820 mg L^{-1}, and Cl^- = 818 mg L^{-1}.

Desorption experiments were performed in batch mode (T = 20 °C, t = 24 h). In particular, after the end of the adsorption phase, the adsorbent material was separated from the supernatant using filtration membranes (0.22 μm). Then, desorption tests were performed using 50 mL solutions of H_2SO_4 and EDTA as desorption reagents at a concentration of 0.1 M and 0.05 M, respectively.

The quantitative evaluation of desorption was carried out using desorption percentages calculated from the difference between the amount of metal loaded on the adsorbent after adsorption and the amount of metal in solution after desorption. To investigate the reuse capacity of the adsorbents, the above procedure was repeated 5 times under the same conditions (first adsorption and then desorption).

3. Results

3.1. Adsorption Dynamics

The uptake of metal ions from the solution involves several steps, necessary for the transfer of the solute from the liquid phase to the specific sites within the chitosan particles (e.g., external diffusion and intraparticle diffusion).

In the case of chitosan, its chains have a large number of the $-NH_2$ and $-OH$ groups distributed throughout the structure, making the kinetic or mass transfer representation likely to be global. The $-NH_2$ groups are the most important binding sites for metal ions [24], yet the hydroxyl groups can also contribute as coordinator groups, especially those in C-3 position [15,22]. To examine the adsorption mechanism of the metal ion of interest, two kinetic models were tested:

i.   the pseudo-first-order equation described by Lagergren [25], which can be rearranged to obtain a linear form as shown by Equation (1):

$$Log(q_e - q_t) = Log(q_e) - (k_1/2.303)t \tag{1}$$

ii.   a pseudo-second-order equation based on the equilibrium adsorption capacity, which can be expressed as in Equation (2):

$$t/q_t = \left(1/k_2 q_e^2\right) + (1/q_e)t \tag{2}$$

In the above equations, q_e (mg g^{-1}) represents the quantity of Cr(III) adsorbed when the system is at equilibrium, q_t (mg g^{-1}) is the quantity of Cr(III) adsorbed at time t, and k_1 (min^{-1}) and k_2 (g mg^{-1} min^{-1}) are the rate constants of the pseudo-first and pseudo-second order kinetic models, respectively.

Given that Equations (1) and (2) are not able to provide information on the adsorption mechanism, the simplified intraparticle diffusion model [26] was also tested, being k_i (g mg^{-1} min^{-1}) the rate constant of the model:

$$q_t = k_i\, t^{1/2} \tag{3}$$

The validity of these models was assessed by analyzing the slopes and intercepts of $Log(q_e - q_t)$ vs. t, t/q_t vs. t, and q_t vs. $t^{1/2}$ for each of the linearized equations.

The results obtained with different concentrations of chromium(III) are shown in Table 1 in terms of correlation coefficients (R^2) as well as calculated and experimental adsorption capacity values.

Table 1. Values of the adsorption kinetic constants at T = 20 °C, pH = 3.8, C_0 = 0.5 g L^{-1}.

1st Order Kinetic Model			2nd Order Kinetic Model			Intraparticle Diffusion Model	
k_1 (min^{-1})	$q_{e,\,cal}$ (mg g^{-1})	R^2	k_2 (g mg^{-1} min^{-1})	$q_{e,\,cal}$ (mg g^{-1})	R^2	k_i (g mg^{-1} min^{-1})	R^2
1.07×10^{-2}	9.32	0.7838	5.07×10^{-3}	23.89	0.9987	1.711	0.9199

The correlation coefficient R^2 for the pseudo-second-order adsorption model was the highest and, in fact, its estimate of the equilibrium adsorption capacity $q_{e,\,cal}$ was quite close to the experimental q_t values (23–28.5 mg g^{-1} as shown in Figure 2 for the two grain-sizes). These results suggest that a second-order mechanism is predominant and that the overall Cr(III) adsorption rate is controlled by a chemisorption process.

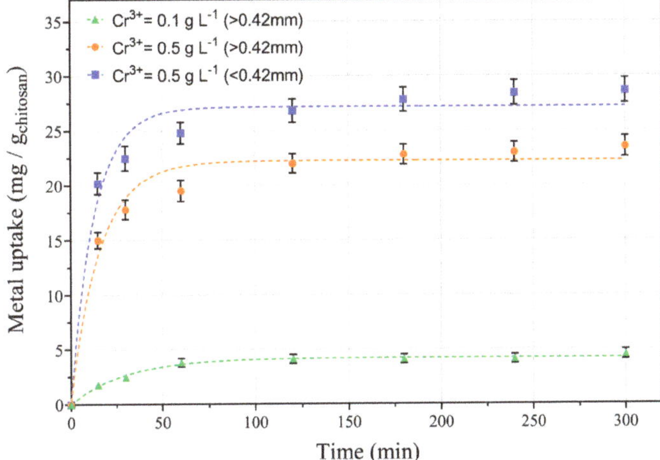

Figure 2. Effect of the grain size of the flakes on the adsorption of Cr(III) on chitosan (100 mg), at pH = 3.8, T = 20 °C, and C_0 = 100 and 500 mg L^{-1} of Cr(III).

3.2. Grain Size Effect

Figure 2 shows the effect of the grain size of the flakes on the adsorption capacity at pH = 3.8 and T = 20 °C. It can be observed that the metal uptake was higher on particles with a small size (<0.42 mm). This is likely due to the higher surface area exposed by these particles, which favors the removal of Cr(III) from the solution in the initial stages of the adsorption process. This phenomenon, previously reported for the adsorption on chitin [21], chitosan [21,27,28], and Neem sawdust [29], was further improved by the ability of metal ions to penetrate into the internal structure of chitosan.

Figure 2 also confirms that the adsorption process was rather rapid, with $t_{1/2}$ < 18 min and the maximum adsorption obtained in about 120 min. The reference time for the subsequent equilibrium tests was thus set at 120 min, an adequate compromise between accuracy and speed in the execution of experimental tests.

3.3. Effect of pH

Figure 3 shows the effect of pH on Cr(III) adsorption on chitosan. Notably, the pH of the solution strongly affected the adsorption of metal ions, with the latter increasing with the pH of the solution. Under acidic conditions, the amino groups (R–NH$_3^+$) and the hydroxyl groups (R–OH$_2^+$) are protonated and the molecule is a sort of polycation, with a reduced number of binding sites available for the adsorption of Cr(III); according to [30], the pKa of the amine groups is 6.3. In addition, the positive surface charge may hinder the adsorption of metal ions. On the contrary, a high pH will favor their adsorption since the nitrogen free electron doublet is responsible for cations coordination. Considering that K_{ps} = [Cr^{3+}] × [OH]^3 = 6.7 × 10^{-31}, chromium(III) hydroxide begins to precipitate at pH ≈ 6.5; for pH values higher than 3.8, there is a significant reduction of the Cr(III) fraction, with formation of Cr(OH)$^{2+}$ and Cr(OH)$_2^+$ hydrolyzed complex species [20]. The result is an increase in chromium adsorption due mainly to hydrolyzed forms. Therefore, the adsorption of metal ions is mainly due to the electrostatic interactions between counter ions.

As previously reported [21], the final pH values of the equilibrated solutions were higher as the Cr(III) concentration became smaller (Table 2). This is probably due to the fact that Cr ions are Lewis acids; therefore, the lower the concentration, the higher the pH (Figure 3). Moreover, the chromium adsorption capacity increases by increasing the metal ions concentration (see Section 3.4), which causes a greater competition with H$^+$ protons.

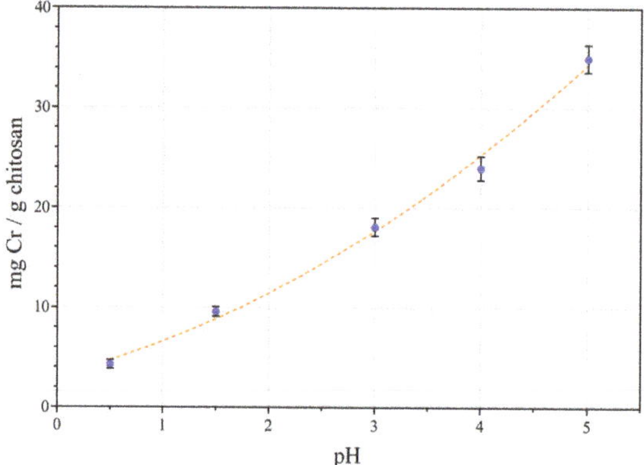

Figure 3. Effect of pH (initial value) on chromium(III) adsorption on chitosan (C_0 = 500 mg L^{-1}).

Table 2. pH variation vs. Cr(III) concentration (pH$_{in}$ = 3.8).

Cr(III) (ppm)	pH
50	5.16
100	5.02
200	4.98
400	4.67
500	4.47

As already stated, it is accepted that chitosan amino groups are the main reactive sites for metal ions and that hydroxyl groups (in particular in C-3 position) may contribute to sorption. Metal sorption may involve different mechanisms (chelating, electrostatic attraction) depending on the pH, the solution, and the metal (concentration, speciation, etc.). Protonation of the amino groups at acidic pH increases the adsorption of anionic species, while cations interactions increase with the pH due to deprotonation of amino/hydroxyl groups. Moreover, increasing the ions competition, due to increase in metal concentration, the pH decreases. The fraction of free (accessible) amine groups is the key parameter.

3.4. Adsorption Isotherms

To determine the maximum adsorption capacity of Cr(III) on chitosan, a study was carried out on the adsorption isotherm by comparing the most common models; in particular, data were analyzed using the Langmuir and Freundlich equations:

$$q_e = Q^\circ k_L C_{eq}/\left(1 + k_L C_{eq}\right) \quad \text{or, linearized} \quad 1/q_e = 1/\left(Q^\circ k_L\right)\left(1/C_{eq}\right) + 1/Q^\circ \tag{4}$$

$$q_e = k_F C_{eq}^{1/n} \quad \text{or, linearized} \quad Log\left(q_e\right) = Log\left(k_F\right) + 1/n \, Log\left(C_{eq}\right) \tag{5}$$

where q_e (mg g^{-1}) is the amount of Cr(III) on the solid phase at equilibrium, and C_{eq} (mg L^{-1}) is the equilibrium concentration of Cr(III) in the aqueous phase. According to Langmuir's equation, Q° (mg g^{-1}) is the amount of Cr(III) required for a complete coverage of available adsorption sites, while k_L is an empirical coefficient related to the affinity of adsorption sites for the adsorbed species. With reference to Freundlich's equation, k_F and n are empirical constants representing the adsorption capacity and adsorption intensity, respectively; all parameters can be estimated through the intercepts and slopes of the linearized forms of isotherm equations.

The essential characteristics of the Langmuir equation can be expressed in terms of a dimensionless separation factor R_L, which has been defined in [27] as:

$$R_L = 1/[1 + (k_L C_o)] \qquad (6)$$

where C_o is the highest initial Cr(III) ion concentration (mg L^{-1}). R_L is related to the shape of the isotherm: the adsorption is unfavorable if $R_L > 1$, favorable if $0 < R_L < 1$, irreversible if $R_L = 0$, and linear if $R_L = 1$. All the estimated isotherm parameters are reported in Table 3. The R_L value confirms the affinity between chitosan and chromium ions, and the adsorption equilibrium data correlate well with the Langmuir isotherm equation, with a maximum adsorption capacity estimated at 138.04 mg g^{-1}. This implies a monolayer interaction of chromium on the adsorbent [31]. Ngah and colleagues [23] found that $Q°$ was 30.03 mg g^{-1} using Cr(III) in the range 4–14 mg L^{-1} and identified the Langmuir isotherm as the best model for the adsorption on cross-linked chitosan. However, the chromium concentrations considered were much lower than those investigated in the present work, and the number of amino groups available for ions coordination in cross-linked chitosan is limited, which obviously results in a lower adsorption capacity. Eiden and colleagues [32] found that $Q°$ was 62 mg g^{-1} for chitosan flakes at pH = 4, but no information regarding chitosan characteristics was provided (especially regarding its degree of deacetylation).

Table 3. Langmuir's and Freundlich's isotherm parameters at 20 °C, C_0 = 0.5 g L^{-1}, and pH = 3.8.

Langmuir				Freundlich		
$Q°$ (mg g^{-1})	k_L	R^2	R_L	k_F (mg g^{-1})	$1/n$	R^2
138.04	3.7×10^{-4}	0.9925	0.575	0.035	1.061	0.9694

Indeed, the basis of the high adsorption capacity found in our study lies precisely in the fact that the investigated chromium concentrations are much higher than those reported in other works (see Figure 2, which shows that adsorption on chitosan increases with the concentration of the target species). Although the different experimental conditions make it difficult to compare the results obtained, it can be observed that at lower concentrations, the adsorption is in line with the values found in the literature (Figures 2 and 4).

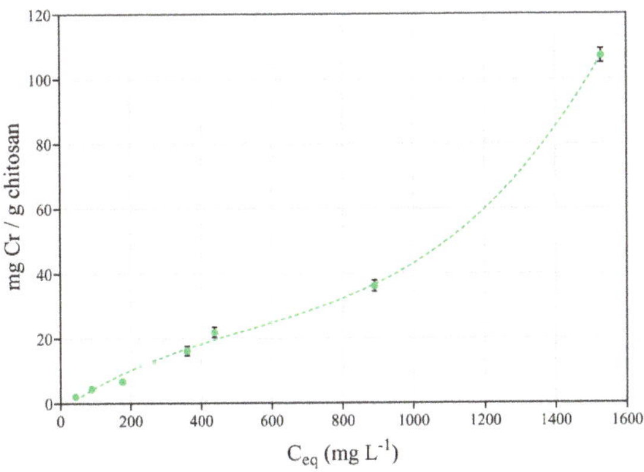

Figure 4. Adsorption isotherm at 20 °C, C_0 = 50–2000 mg L^{-1}, and pH = 3.8.

3.5. Desorption

After Cr(III) adsorption, the chitosan flakes were washed thoroughly with deionized water and treated with the desorption agents. Desorption tests were performed using H_2SO_4 and EDTA as desorption reagents; the chromium desorption efficiency of chitosan flakes for different washing solutions is reported in Table 4.

Table 4. Desorption efficiency (%) for each adsorption cycle.

Reagent	1st Cycle	2nd Cycle	3rd Cycle	4th Cycle	5th Cycle
EDTA 0.05 M	53.3	-	-	-	-
EDTA 0.1 M	76.5	73.4	70.6	71.7	72.1
H_2SO_4 0.05 M	37.8	-	-	-	-
H_2SO_4 0.1 M	45.3	46.5	40.7	43.1	39.7

As reported in [23], EDTA is an efficient desorption agent: being an hexadentate chelating agent, it is capable to form a strong complex with Cr(III) ions. Considering the pH effect on the chromium adsorption, sulfuric acid (such as other acidic media) was also considered as a reagent potentially able to remove Cr(III) ions from the chitosan polymer.

Despite being efficient in terms of chitosan regeneration, EDTA persists in municipal wastewater treatments, making its use in technical applications potentially unwelcome. Although future research may investigate other chelating agents that are biodegradable and more suitable for wastewater treatment plants, the current approach has considered the possibility of treating the spent solution to recover EDTA by precipitation in acidic media as a precautionary measure.

3.6. Dynamic Tests

The effect of competing ions can be observed from the breakthrough curves shown in Figure 5. The inhibition effect of anions such as chlorides and sulfates has been reported for Cr(VI) [33]; in addition, it has been reported that chitosan forms complexes with transition metal ions, but not with complexes with alkali and alkali earth metal ions due to the absence of d and f unsaturated orbitals [34,35]. Therefore, the presence of Na, Mg, and Ca does not reduce the chromium adsorption. In contrast, the removal of chromium is influenced by the presence of zinc and cadmium due to their affinity with chitosan [36]. When real wastewater was considered, chitosan was able to absorb 104.7 mg g^{-1} of chromium, while 127.5 mg g^{-1} were adsorbed using a chromium solution.

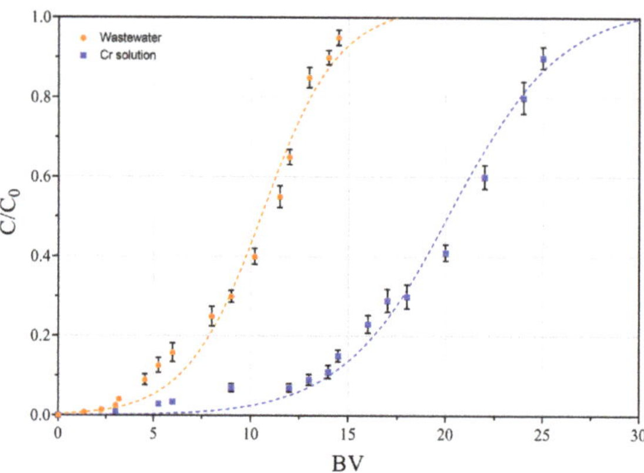

Figure 5. Breakthrough tests performed with real wastewater solution (pH = 3.2) and 500 mg L^{-1} chromium(III) solution (at pH = 3.5), using a flow rate of 1 BV h^{-1} (BV = 23.6 mL).

The breakthrough curve should be symmetrical and sigmoid in shape; however, a deviation from a symmetrical S-shape and not very steep curves can be observed in Figure 5. The packed column was probably not very homogeneous (chitosan flakes seem to be inappropriate to be used in a packed column) and the flow-rate was too fast, causing physical non-equilibrium processes and high mass transfer zone [37].

4. Conclusions

The present study indicated the suitability of chitosan in applications aimed at removing chromium(III) ions from aqueous solutions. Chitosan is a low-cost reagent and its utilization is novel, non-toxic, and environmentally compatible. Although its effectiveness has been proven in the adsorption of several metal ions, such as Cu, Ag, Cd, Zn, Mo, V, Pb, and Cr(VI), Cr(III) has been poorly investigated. Moreover, among the usable forms of chitosan, flakes are the least studied. In this work, the adsorption of Cr(III) on chitosan flakes was investigated considering concentrations (up to 2000 mg L^{-1}) that are much higher than those reported in the few studies present in literature. This made it possible to highlight a much higher adsorption capacity of chitosan (up to 138.45 mg g^{-1}), well beyond the values observed so far.

The pH significantly influences the adsorption capacity of the biopolymer, the latter increasing with the pH of the solution. The adsorption process is rather rapid: it was found that 2 h of contact time are sufficient to reach about 95% of the adsorption equilibrium; the Langmuir equation provided the best fit over the entire concentration range, thus suggesting a monolayer interaction of chromium on the adsorbent.

The adsorption capacity for Cr(III) in dynamic tests was found to be penalized by the competition between ions, and particularly influenced by the presence of zinc and cadmium due to their strong affinity with chitosan. The recovery of Cr(III) ions (and the consequent regeneration of chitosan) can be obtained by desorption with a 0.1 M EDTA solution, with the latter being recoverable by precipitation in an acidic medium.

Since chitosan is basically a low-cost and environmentally friendly material, its use as it is produced (i.e., flakes) is desirable, especially considering the greater theoretical adsorbent capacity due to a major availability of both amino and hydroxyl groups, but involves some critical issues. In particular, since the shape and size characteristics of commercial chitosan flakes introduce hydrodynamic limitations such as column clogging, a batch reactor is probably more appropriate as a experimental setup. Further investigations are therefore underway to quantify the effectiveness of alternative operating solutions.

Author Contributions: Conceptualization, L.P.; methodology, L.P. and A.P.; validation, L.P. and I.F.; investigation, I.S. and M.S.; data curation, M.V.; writing—original draft preparation, L.P. and M.V.; writing—review and editing, L.P. and M.V.; project administration, L.P.; funding acquisition, L.P. All authors have read and agree to the published version of the manuscript.

Funding: This research received no external funding.

Conflicts of Interest: The authors declare no conflict of interest.

References

1. Pietrelli, L.; Ippolito, N.M.; Reverberi, A.P.; Vocciante, M. Heavy Metals Removal and Recovery from Hazardous Leather Sludge. *Chem. Eng. Trans.* **2019**, *76*, 1327–1332.
2. Rengaraj, S.; Yeon, K.H.; Moon, S.H. Removal of chromium from water and wastewater by ion exchange resins. *J. Hazard. Mater.* **2001**, *87*, 273–287. [CrossRef]
3. Vocciante, M.; Reverberi, A.P.; Pietrelli, L.; Dovì, V.G. Improved remediation processes through cost-effective estimation of soil properties from surface measurements. *J. Clean. Prod.* **2017**, *167*, 680–686. [CrossRef]
4. Vocciante, M.; Finocchi, A.; De Folly, D.; Auris, A.; Conte, A.; Tonziello, J.; Pola, A.; Reverberi, A.P. Enhanced Oil Spill Remediation by Adsorption with Interlinked Multilayered Graphene. *Materials* **2019**, *12*, 2231. [CrossRef] [PubMed]
5. Fu, F.; Wang, Q. Removal of heavy metal ions from wastewaters: A review. *J. Environ. Manag.* **2011**, *92*, 407–418. [CrossRef] [PubMed]

6. Vocciante, M.; De Folly, D.; Auris, A.; Finocchi, A.; Tagliabue, M.; Bellettato, M.; Ferrucci, A.; Reverberi, A.P.; Ferro, S. Adsorption of ammonium on clinoptilolite in presence of competing cations: Investigation on groundwater remediation. *J. Clean. Prod.* **2018**, *198*, 480–487. [CrossRef]

7. El-Shafey, E.I.; Cox, M.; Pichugin, A.A.; Appleton, Q. Application of a carbon sorbent for the removal of cadmium and other heavy metal ions from aqueous solution. *J. Chem. Technol. Biotechnol. Intern. Res. Process Environ. Clean Technol.* **2002**, *77*, 429–436. [CrossRef]

8. Garcia-Reyes, R.B.; Rangel-Mendez, J.R. Adsorption kinetics of chromium (III) ions on agro-waste materials. *Bioresour. Technol.* **2010**, *101*, 8099–8108. [CrossRef] [PubMed]

9. Chen, L.F.; Liang, H.W.; Lu, Y.; Cui, C.H.; Yu, S.H. Synthesis of an attapulgite clay@ carbon nanocomposite adsorbent by a hydrothermal carbonization process and their application in the removal of toxic metal ions from water. *Langmuir* **2011**, *27*, 8998–9004. [CrossRef]

10. Pietrelli, L.; Ippolito, N.M.; Ferro, S.; Dovì, V.G.; Vocciante, M. Removal of Mn and As from drinking water by red mud and pyrolusite. *J. Environ. Manag.* **2019**, *237*, 526–533. [CrossRef]

11. Pietrelli, L.; Ferro, S.; Vocciante, M. Raw materials recovery from spent hydrochloric acid-based galvanizing wastewater. *Chem. Eng. J.* **2018**, *341*, 539–546. [CrossRef]

12. Chiou, M.S.; Li, H.Y. Adsorption behavior of reactive dye in aqueous solution on chemical cross-linked chitosan beads. *Chemosphere* **2003**, *50*, 1095–1105. [CrossRef]

13. Pietrelli, L.; Xingrong, L. Chitosan membrane: Tool for chromium (III) recovery from aqueous solutions. *Ann. Chim. J. Anal. Environ. Cult. Herit. Chem.* **2004**, *94*, 389–398. [CrossRef] [PubMed]

14. Ko, J.A.; Kim, B.K.; Park, H.J. Preparation of acetylated chitosan sponges (chitin sponges). *J. Appl. Polym. Sci.* **2010**, *117*, 1618–1623. [CrossRef]

15. Guibal, E. Interactions of metal ions with chitosan-based sorbents: A review. *Sep. Purif. Technol.* **2004**, *38*, 43–74. [CrossRef]

16. Bakshia, P.S.; Selvakumara, D.; Kadirvelub, K.; Kumara, N.S. Chitosan as an Environment Friendly Biomaterial—A Review on Recent Modifications and Applications. *Int. J. Biol. Macromol.* **2019**, in press. [CrossRef] [PubMed]

17. Wei, S.; Ching, Y.C.; Chuah, C.H. Synthesis of chitosan aerogels as promising carriers for drug delivery: A review. *Carbohydr. Polym.* **2019**, *231*, 115744. [CrossRef]

18. Mujtaba, M.; Morsi, R.E.; Kerch, G.; Elsabee, M.Z.; Kaya, M.; Labidi, J.; Khawar, K.M. Current advancements in chitosan-based film production for food technology: A review. *Int. J. Biol. Macromol.* **2019**, *121*, 889–904. [CrossRef]

19. Vakili, M.; Deng, S.; Cagnetta, G.; Wang, W.; Meng, P.; Liu, D.; Yu, G. Regeneration of chitosan-based adsorbents used in heavy metal adsorption: A review. *Sep. Purif. Technol.* **2019**, *224*, 373–387. [CrossRef]

20. Fendorf, S.E. Surface reactions of chromium in soils and waters. *Geoderma* **1995**, *67*, 55–71. [CrossRef]

21. Maruca, R.; Suder, B.J.; Wightman, J.P. Interaction of heavy metals with chitin and chitosan. III. Chromium. *J. Appl. Polym. Sci.* **1982**, *27*, 4827–4837. [CrossRef]

22. Chui, V.W.D.; Mok, K.W.; Ng, C.Y.; Luong, B.P.; Ma, K.K. Removal and recovery of copper (II), chromium (III), and nickel (II) from solutions using crude shrimp chitin packed in small columns. *Environ. Int.* **1996**, *22*, 463–468. [CrossRef]

23. Ngah, W.W.; Kamari, A.; Fatinathan, S.; Ng, P.W. Adsorption of chromium from aqueous solution using chitosan beads. *Adsorption* **2006**, *12*, 249–257. [CrossRef]

24. Lerivrey, J.; Dubois, B.; Decock, P.; Micera, G.; Urbanska, J.; Kozłowski, H. Formation of D-glucosamine complexes with Cu (II), Ni (II) and Co (II) ions. *Inorg. Chim. Acta* **1986**, *125*, 187–190. [CrossRef]

25. Lagergren, S. Zur theorie der sogenannten adsorption geloster stoffe. *Kungliga Svenska Vetenskapsakademiens Handlingar* **1898**, *24*, 1–39.

26. Weber, W.J.; Morris, J.L. Kinetics of adsorption on carbon from solutions. *J. Sanit. Eng. Div. Proc. Am. Soc. Civ. Eng.* **1963**, *89*, 31–60.

27. McKay, G. Adsorption of dyestuffs from aqueous solutions with activated carbon I: Equilibrium and batch contact-time studies. *J. Chem. Technol. Biotechnol.* **1982**, *32*, 759–772. [CrossRef]

28. Annadurai, G.; Ling, L.Y.; Lee, J.F. Adsorption of reactive dye from an aqueous solution by chitosan: Isotherm, kinetic and thermodynamic analysis. *J. Hazard. Mater.* **2008**, *152*, 337–346. [CrossRef]

29. Annadurai, G.; Chellapandian, M.; Krishnan, M.R.V. Adsorption of reactive dye on chitin. *Environ. Monit. Assess.* **1999**, *59*, 111–119. [CrossRef]

30. Udaybhaskar, P.; Iyengar, L.; Rao, A.P. Hexavalent chromium interaction with chitosan. *J. Appl. Polym. Sci.* **1990**, *39*, 739–747. [CrossRef]

31. Singh, P.; Nagendran, R. A comparative study of sorption of chromium (III) onto chitin and chitosan. *Appl. Water Sci.* **2016**, *6*, 199–204. [CrossRef]

32. Eiden, C.A.; Jewell, C.A.; Wightman, J.P. Interaction of lead and chromium with chitin and chitosan. *J. Appl. Polymer Sci.* **1980**, *25*, 1587–1599. [CrossRef]

33. Boddu, V.M.; Abburi, K.; Talbott, J.L.; Smith, E.D. Removal of hexavalent chromium from wastewater using a new composite chitosan biosorbent. *Environ. Sci. Technol.* **2003**, *37*, 4449–4456. [CrossRef] [PubMed]

34. Hsien, T.Y.; Rorrer, G.L. Effects of acylation and crosslinking on the material properties and cadmium ion adsorption capacity of porous chitosan beads. *Sep. Sci. Technol.* **1995**, *30*, 2455–2475. [CrossRef]

35. Kubota, N.; Kikuchi, Y. Macromolecular complexes of chitosan. In *Polysccharides*; Dumitriu, S., Ed.; Dekker: New York, NY, USA, 1998; pp. 595–628.

36. Bassi, R.; Prasher, O.; Simpson, B.K. Removal of selected metal ions from aqueous solutions using chitosan flakes. *Sep. Sci. Technol.* **2000**, *35*, 547–560. [CrossRef]

37. Patel, H. Fixed-bed column adsorption study: A comprehensive review. *Appl. Water Sci.* **2019**, *9*, 45. [CrossRef]

© 2020 by the authors. Licensee MDPI, Basel, Switzerland. This article is an open access article distributed under the terms and conditions of the Creative Commons Attribution (CC BY) license (http://creativecommons.org/licenses/by/4.0/).

Article

Remediation of Copper Contaminated Soils Using Water Containing Hydrogen Nanobubbles

Dongchan Kim [1] and Junggeun Han [2,*]

[1] Candidate Ph.D Department of Civil Engineering Chung-Ang University, Seoul 06974, Korea; hero081123@nate.com

[2] School of Civil and Environmental Engineering, Urban Design and Study, Chung-Ang University, Seoul 06974, Korea

* Correspondence: jghan@cau.ac.kr

Received: 5 February 2020; Accepted: 18 March 2020; Published: 23 March 2020

Featured Application: Nanobubbles were manufactured by using pressurized hydrogen gas and were applied to soils contaminated with heavy metals, to confirm their function as enhancers. The manufactured nanobubbles remained for an extended period of time. A batch-test experiment revealed that the nanobubbles improved desorption of heavy metals. In addition, when the nanobubbles were used as enhancers in electrokinetic experiments, the nanobubbles had better remediation effects than distilled water (DW). The remediation of heavy metals is expected to have a significant impact when using the nanobubbles.

Abstract: This basic research study was undertaken to use ecofriendly nanobubbles that can improve the electrokinetic remediation of copper-contaminated soil, as well as to determine that remediation efficiency. The nanobubbles were generated by using pressurized hydrogen gas, and the quantity of hydrogen gas bubble that remained over 14 days was measured. The generated nanobubbles were used as an enhancer to remove a heavy metal on contaminated soil, and their applicability was confirmed. A batch test was used to compare the remediation effects of nanobubbles and distilled water on copper-contaminated soil. The results proved that the nanobubbles are a proper desorption agent for copper-contaminated sand and clay specimens. The solid–liquid ratio and the contact time for desorption of the sand and clay were then respectively determined. A large amount of effluent was obtained from electrokinetic remediation of the sand sample after applying the nanobubbles as an enhancer. The remediation efficiency demonstrated with sand proved to be higher than that for clay. This greater efficiency was attributed to a wider specific surface area, demonstrating the potential use of the nanobubbles as an enhancer for soil contaminated by copper with a large amount of effluent outflow. It was also assumed to be affected by the moving capability of the nanobubbles in the soil layer. Thus, the nanobubbled water can be used to improve the removal of heavy metals from contaminated soils. An ecofriendly enhancer for electrokinetic remediation with a relatively large void ratio and fast flowrate was confirmed by the nanobubbles.

Keywords: nanobubbles; contaminated soil; electrokinetic; in situ; remediation

---

1. Introduction

The use of chemical substances is increasing, and the resulting outflows from factories, gas stations, and industrial complexes are also increasing. Soil contamination, an obstacle to development, needs to be addressed, and contaminated land needs to be reclaimed. While the buffering capacity of land is comparatively large compared to the volume of contaminants, much depends on the soil characteristics and environmental factors, and so the buffering capacity of the land varies [1]. In particular, soil contamination sourced from chemical facility effluents has characteristics that can be difficult to effectively treat as compared to water or air pollution.

Soil remediation methods can be categorized into in situ or ex situ, depending on where treatment takes place. Offsite processing, the ex situ method of restoration, is more effective and faster, but it is also more expensive. By contrast, onsite processing, the in situ method, has a lower cost and uses technologies pertinent to the characteristics of the site [2]. While some in situ remediation methods employ ecofriendly enhancers, chemical solvents are also employed, due to the inherent difficulty in desorbing heavy metals from fine-grained soil. Toxic heavy metals exist in water media, and effective removal represents a continuing challenge for the scientific community as a whole.

To remedy this, numerous studies have been published discussing the use of adsorbent materials for aqueous media decontamination from heavy metal ions [3–6]. However, the improper extraction of enhancers employed for remediation may result in enhancers remaining in the soil at the conclusion of treatment, causing secondary contamination [7].

The diameter of the nanobubbles ranges from 1 to 999 nm [8]. Bubbles with a diameter greater than 50 μm are classified as microbubbles, while bubbles with a diameter size between 10 and 50 μm are classified as microbubbles, and bubbles with a diameter under 200 nm are nanobubbles [9]. The microbubbles have a fast rate of rise due to their large buoyant force. It does not last long in the water and rapidly rises to the surface, and it bursts when it reaches the surface. On the other hand, the microbubbles become smaller over time as they stay in water due to their slow rate of increase. If the diameter of the microbubble is 10 μm, the internal pressure of the microbubble is about 0.3 atm. As the diameter of the microbubble decreases to 1 μm and 100 nm, the internal pressure of the microbubbles rises up to 3 and 30 atm, respectively.

During this process, the internal gas spreads according to Henry's Law, so all gases in the microbubble spread into the surrounding water until the microbubble disappears [10]. On the other hand, the nanobubbles exist in water for a long time once they are created. This is known to be due to the strong hydrogen bonds found in ice and gas that hydrate the surface of the nanobubbles. These bonds prevent the spread of internal gases and maintain an adequate mechanical balance against high internal pressures, as observed via Attenuated Total Reflectance Infrared (ATR-IR) [11]. These various characteristics of the nanobubbles have been observed by many researchers in different areas, including water purification, drag reduction, purification of surface contamination, removal of tumors using shock waves caused by the collapse of bubbles, growth promotion of animals and plants, and contrast medium.

The characteristics of the nanobubbles are that they have a wide specific area, they generate radicals on their boundary surfaces, and they permit high internal pressure in the liquids. Thus, studies looking into how to exploit these beneficial characteristics of the nanobubbles are being actively pursued in various fields, including surface cleaning, remediation of poor water quality, and as a supersonic contrast agent [9]. Kyzas [12] investigated the effect of the nanobubbles (NBs) on dissolved heavy metal adsorption with activated carbon and confirmed that the inherent property of the NBs to accept charged particles onto their interface assists in the diffusion and penetration of lead ions into the activated carbon pores. Choi [13] sought to determine the removal efficiency and degradation rate for total petroleum hydrocarbons (TPH). Thus, the objective of this study was to perform batch tests to evaluate the efficacy of applying nanobubbles as enhancers. The solid–liquid ratio and contact time required for desorption of the sand and clay were then calculated. Furthermore, electrokinetic in situ remediation using nanobubbles as an enhancer for remediation of heavy metals was performed for both copper-contaminated sand soils and copper-contaminated clay soils, in an attempt to determine the optimal remediation method.

2. Experimental Methods and Condition

2.1. Soils

To simulate soils contaminated with heavy metals, standard Jumunjin sand and clay specimens were used. These clay specimens are comprised of a mixture of seaside sand and cohesive shore sediment soils. Their physical characteristics are summarized in Table 1. Copper was selected as a contaminant, and the soil specimens were contaminated, using copper nitrate ($Cu(NO_3)_2$). The soil properties were measured, using the Korea Standard, and the pH measurements followed, using KSI ISO 10390.

Table 1. Material properties of soils.

Soil Content	Sand	Silty Clay (<0.075 mm)
Soil Classification	SW	CL
Liquid Limit (%)	19.81	44.20
Plastic Limit (%)	NP	23.82
Specific Gravity (G_s)	2.59	2.46
pH	7.6	9.4
Cation Exchange Capacity (Cmol/kg)	-	18.23
BET Surface Area (m^2/g)	-	16.0
Organic Matter (%)	0.22	0.85
Water content (%)	17.3	40

※ SW: well-graded sand, CL: low plastic clay, NP: non plastic.

2.2. Nanobubbles

The hydrogen nanobubbles were created by using the equipment shown in Figure 1. Nanobubbles were created with hydrogen gas that was pressurized and filtered through a ceramic filter, so as to create tiny bubbles. These tiny bubbles were transferred to the surface of the water and crushed, thereby creating other nanobubble nuclei [14]. The nanobubbles were created for a period of 24 h. Upon completion, their zeta-potential was measured so as to appraise their stability. The parameters for the zeta-potential measurements and the measured value are presented in Table 2. The prepared nanobubbles were measured by using zeta-potential analysis equipment (ZetaPALS, Brookhaven Corp., USA), to measure the zeta potential of the nanobubbles (Figure 2c). The size and quantity of the generated nanobubbles were measured, using a laser within the Nanoparticles Tracking Analysis (NATA) equipment, as shown in Figure 1. Nanobubbles with an average diameter of 171 ± 6.11 nm, a mode of 130 ± 9.85 nm, and a concentration of $1.5 \pm 0.03 \times 10^8$ particles/mL were thus obtained. The pH measured after the manufacture of the nanobubbles was confirmed to be neutral 7.

Table 2. Parameters for ζ-potential measurement and value of zeta-potential of nanobubble.

	Measurement	Values
Parameters	Refractive Index	1.331
	Dielectric Constant	78.54
	Electric Field (V/cm)	27.79
	Viscosity (cP)	0.890
Zeta-potential (mV)		−20.13

The stability of the nanobubbles was confirmed over two months. For the experiment, stable 14-day-old nanobubbles were used [15]. Figure 3 shows how long nanobubbles are maintained. Figure 3a confirms the persistence by making nanobubbles in gasoline, and the results confirm that the nanobubbles are maintained for 120 h. Figure 3b shows a comparison of the changes in the concentration of the nanobubbles over 14 days, and it is the same as the nanobubbles used in this experiment. After day 14, approximately 15% of the nanobubbles were lost. Thus, the long-term existence of the nanobubbles was confirmed.

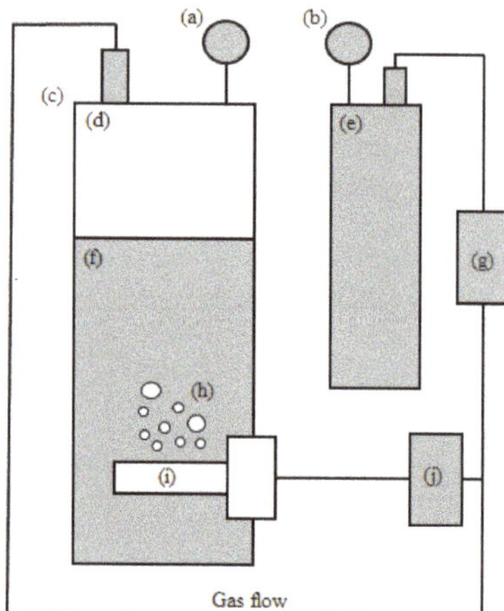

Figure 1. Schematic view of a nanobubble Generator [16]. (**a**) Pressure gauge (air pressure in water tank), (**b**) pressure gauge (air pressure in air tank), (**c**) water tank, (**d**) air, (**e**) gas tank, (**f**) water, (**g**) pressure gauge (outflow), (**h**) bubble, (**i**) filter, and (**j**) pressure gauge (inflow).

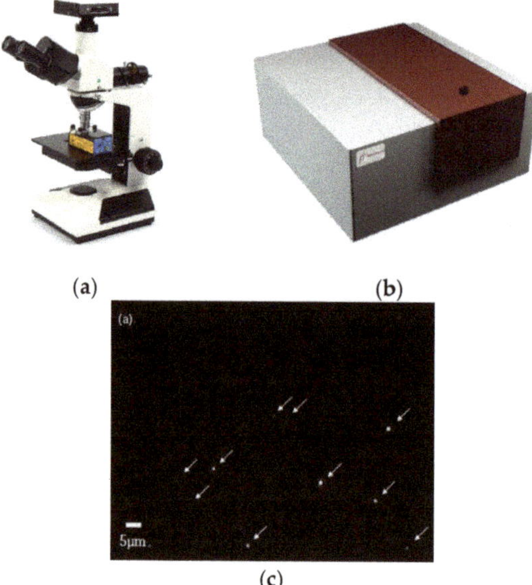

Figure 2. Nanobubble (**a**) NTA Instrument (Malvern Nanosight LM10, Malvern Panalytical Ltd, Unites Kingdom), (**b**) image of nanobubbles [17], and (**c**) Zeta Potential Analyzer (ZetaPALS).

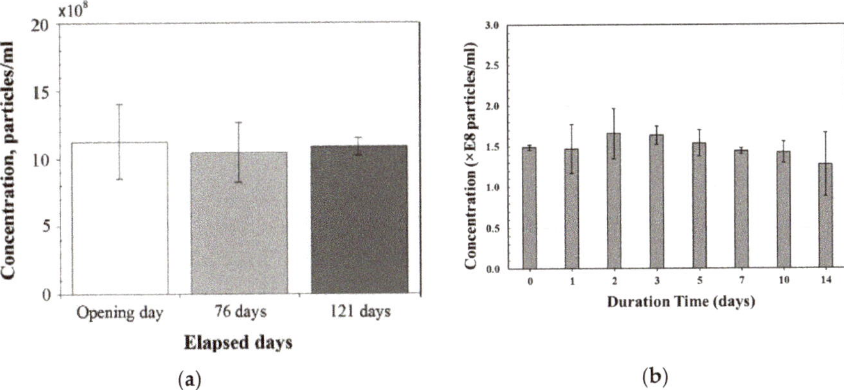

Figure 3. Average particle concentration of nanobubbles: (**a**) concentration of nanobubbles in gasoline [15]; (**b**) concentration of nanobubbles [16].

2.3. Batch Test

Batch testing was carried out in order to identify desorption characteristics of the nanobubbles as an enhancer. The experimental conditions set for the batch test are summarized in Table 3. The solid–liquid ratio, contact time, and desorption of heavy metals by the nanobubbles used as an enhancer were tested. The copper-contaminated solution was prepared, using copper nitrate powder ($Cu(NO_3)_2$) in distilled water. The contamination concentration of the copper was referred to as the Korea Standard. An agitator was used for the batch test, while a centrifuge was used to collect the supernatant solution for analysis by means of inductively coupled plasma (ICP), following the Korea Standard Test Method. Different soil–liquid ratios and contact times for contaminant desorption were determined respectively to attain the optimal conditions of desorption.

Table 3. Batch-test conditions.

Parameters	Units	Conditions
Soil	-	Sand, Clay
Contaminant	-	Copper
Contaminant Concentration	ppm	500
Agents	-	Distilled Water (DW) H_2-Nanobubble Water (NBW)
Solid–Liquid Ratio	g:mL	1:1, 1:2, 1:3, 1:5, 1:10, 1:20, 1:50
Contact Time	Hr	0.5, 1, 2, 3, 6, 12, 24
Contact Speed	rpm	150

2.4. Electrokinetic Test

A laboratory test was carried out to apply an electrokinetic remediation process to soils contaminated with copper. The same soil specimens from the batch test, sand and clay, were used to construct contaminated and uncontaminated soil. The copper-contaminated solution was prepared, using copper nitrate powder ($Cu(NO_3)_2$) in DW. Then, 500 ppm of copper was contaminated to soil, at a water content of 40%. Clay was tested immediately after contamination. Sand was used as a simulation of the ground condition, and it was contaminated with 500 ppm of copper solution and dried in the shade. Dry sand contaminated with copper was installed with the maximum consolidation and tested. The contamination concentration of copper was referred to as the Korea Standard. To identify the remediation characteristics, different conditions of soils with Jumunjin standard sand and clay were used. In addition, nanobubbles identified as an appropriate enhancer from the batch test were placed into a Mariotte bottle and injected constantly. The experimental device designed for electrokinetic remediation is shown in Figure 4. The cell and Mariotte bottle are made of acrylic, and the electrode plate used graphite. The nanobubbles were injected into the cell, using a hydraulic gradient. An electrode water tank was installed at each end of the cell, to provide unrestricted outflow of electro-osmosis during the experiments. The testers were designed to conduct five simultaneous tests. The experimental conditions are summarized in Table 4.

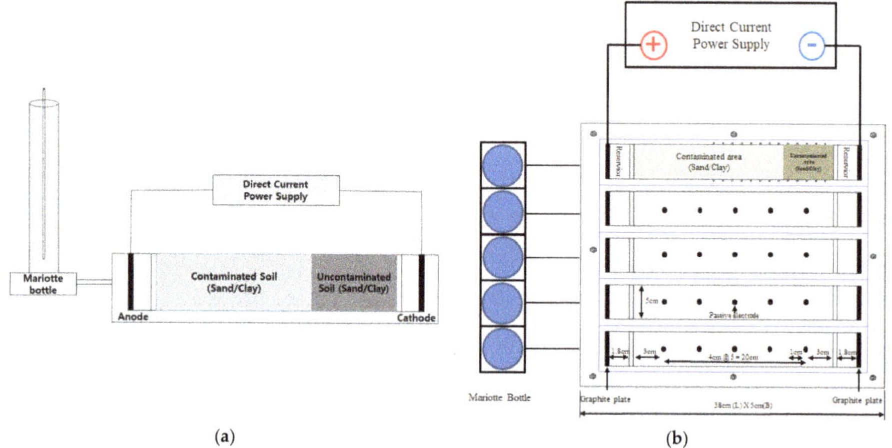

Figure 4. (**a**) Side view. (**b**) Floor plan. Diagram of the Electrokinetic system.

Table 4. Electrokinetic test conditions.

	Parameters	Units	Conditions	
Fixed Factor	Concentration (C_0)	ppm	500 (Copper)	
	Electrokinetic Gradient	V/cm	1	
	Contaminated Soil Bed	x/L	0.75	
	Soil	-	Sand, Clay	
Variable Factor	Enhancer	-	Distilled Water (DW) H_2-Nanobubble Water (NBW)	
	Duration Time	Day	10 (Clay)	Case 1 (DW) Case 2 (NBW)
			5 (Sand)	Case 3 (DW) Case 4 (NBW)

3. Results

3.1. Batch Test

Batch testing was carried out by varying the solid-liquid ratio and the contact time for different conditions of copper-contaminated soils, so as to allow samples to be purified, using nanobubbles as an enhancer. The results obtained from the batch test are as shown below.

Figure 5 shows the incremental desorption percentage of copper according to changes in the solid-liquid ratios of soils contaminated with copper. After agitation for 24 h, the level of desorption of the heavy metal by the nanobubbles was found to be greater in sand than in clay. The increase in solid-liquid ratios in both sand and clay also increased the efficiency with which heavy metals were removed than compared to the use of distilled water alone. However, contrary to the proportional increase of the removal efficiency of the copper to solid-liquid ratio in sand, a solid-liquid ratio greater than 1:20 in clay rendered no significant difference in the efficiency of the copper removal by DW and NBW. Taking both efficiency and economy into account, the optimal solid–liquid ratio, while using nanobubbles as an enhancer in sand, was determined to be 1:20. This finding was attributed to the comparatively larger specific surface area and zeta-potential of the nanobubbles, which is known to be advantageous for desorption and transfer of copper from the surface of particles [15,18].

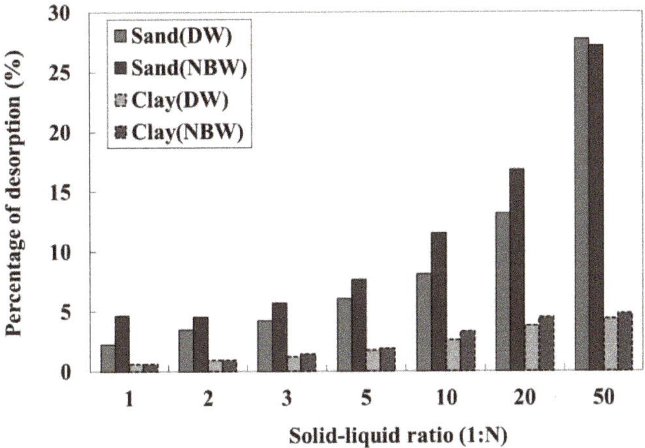

Figure 5. Solid-liquid ration test.

The results of the batch test, according to variations in the reaction time with a constant solid-liquid ratio of 1:20, are presented in Figure 6. The nanobubbles demonstrated a greater efficiency for removing copper than distilled water. The efficiency of the copper removal varied according to the reaction time. Both distilled water and the nanobubbles initially increased desorption of copper. In the case of sand with 6 h of contact time, the highest removal efficiency of copper, by both distilled water and the nanobubbles, was 15.61% and 18.48% respectively. Thereafter, the removal efficiency of both distilled water and the nanobubbles tended to converge. In the case of clay, the highest removal efficiency (desorption of copper) of both distilled water and the nanobubbles was observed within the initial first hour. The appropriate interval for contact time by the nanobubbles was found to be 6 h for sand and 1 h for clay. The explanation for this variation could be that the characteristics of re-adsorption of copper from the surface of clay are greatly influenced by the clay's surface charge.

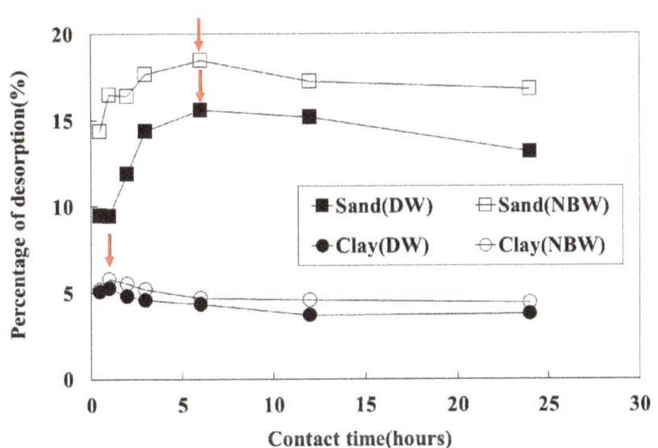

Figure 6. Contact time test (S:L = 1:20).

3.2. Electrokinetic Test

An electrokinetic remediation test was carried out, with the nanobubbles acting as an enhancer for copper-contaminated soil. The current density and effluent changes, according to the soil type, were then determined. The copper and pH in soil after completion of the experiment were also determined.

Figure 7 illustrates changes in the current density according to different types of soil. Clay (DW) corresponds to a peak current density of 9.21 (mA/cm^2) at 5 h from the start of the experiment. Clay (NBW) corresponds to a peak current density of 9.81 (mA/cm^2) at 8 h from the start of the experiment. Sand (DW) corresponds to a peak current density of 0.88 (mA/cm^2) at 24 h from the start of the experiment, and Sand (NBW) corresponds to a peak current density of 0.74 (mA/cm^2) at 5 h from the start of the experiment. Both sand and clay exhibited an initial increase, followed by a decrease in the current density. The result of all the tests tended to converge as time went by. The migration of heavy metal ions increased initially and then decreased. Han et al. [19] postulate that this might be attributable to the migration and sedimentation of the heavy metal ions.

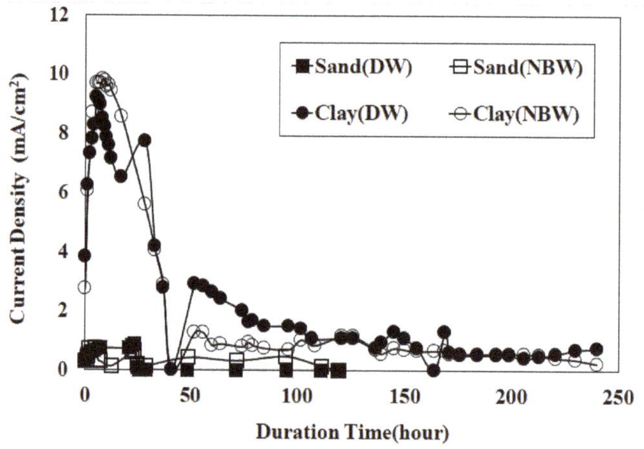

Figure 7. Current density.

Figure 8 illustrates the amount of effluent according to the remediation time for sand and clay as part of the remediation experiment, where nanobubbles were used as an enhancer. Effluents of Clay (DW), Clay (NBW), Sand (DW), and Sand (NBW) were 729, 662, 5855, and 1198 mL, respectively. The amount of effluent from the one treated with distilled water was found to be greater than the outflow observed with the nanobubbles treatment. This was attributed to bubbles of hydrogen gas being generated at the cathode by electrokinetic remediation with nanobubbles that disturbed the discharge of the effluent. The amount of effluent from sand appeared greater than that outflowing from clay. The removal of heavy metal from sand with greater effluent discharge was found to be highly efficient. This could be attributed to the comparatively larger specific surface area and zeta-potential of the nanobubbles that resulted in remediation being more efficient.

Figure 9 illustrates the concentrations of residual heavy metals and pH level measured at each position in the cell upon completion of the experiment. In the case of sand and clay, all exhibited an increased pH due to the collision of acid and base at the point of 0.75 x/L by electrokinetic remediation.

Remediation of copper contamination occurred, as compared to the initial state of the specimen. The migration and accumulation of heavy metal by electrokinetic remediation were identified at a point of 0.75 (x/L) [20]. The efficiency of the removal of heavy metals from the specimens contaminated by heavy metal increased by the contact of the nanobubbles and by the high ratio of the void in sand, which increased the flow of effluent through the copper-contaminated soil under basic conditions [21,22].

The nanobubbles were made of hydrogen gas in distilled water and have an initial neutral pH 7. In the case of sand, the initial pH was influenced by the initial pH due to the inflow of nanobubbles in water, and it can be seen that the pH change at the point of 0.1 to 0.7 (x/L) is small. Sand has a smaller void than clay, and heavy metals adhere to the surface, so the sand is considered to be desorbed when passing through the void of the nanobubbles.

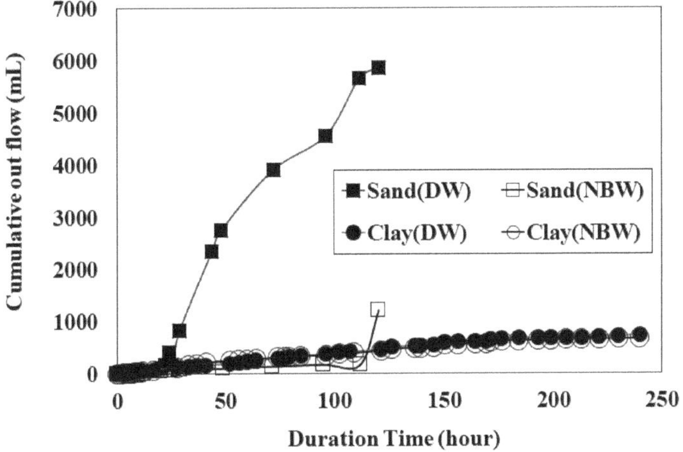

Figure 8. Cumulative flow.

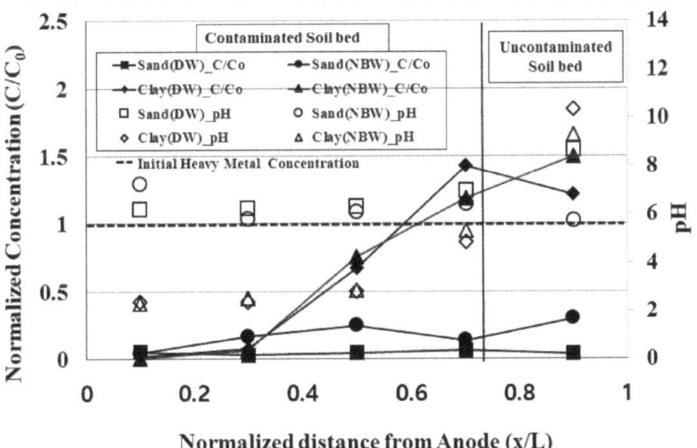

Figure 9. Final Copper Profile and pH at the end of Electrokinetic Test.

4. Conclusions

The present study carried out tests to determine the effect of applying nanobubbles to copper-contaminated sand and clay specimens for heavy metal remediation. An enhancer can be employed to solve the problems associated with sedimentation of heavy metals due to the basicity of soils in electrokinetic remediation. The enhancer is mainly comprised of surfactants and solutions with acidity or basicity. However, any enhancer that remains in the soil can produce secondary contamination. Thus, the objective of the present study was to use ecofriendly hydrogen gas, which would not lead to secondary contamination, as an enhancer for in situ remediation. The results of the experiment carried out in the present study are as follows.

Batch testing was performed to determine the optimal solid–liquid ratio and contact time to remediate copper-contaminated sand and clay, using nanobubbles. The optimal solid-liquid ratio was found to be 1:20 for both sand and clay, while the optimal contact time was 6 h for sand and 1 h for clay. The shorter contact time for clay was attributed to the re-adsorption of desorbed copper according to the varied surface charge of clay. The nanobubbles exhibited a higher copper removal efficiency than distilled water, regardless of the soil conditions.

The electrokinetic remediation tests for sand and clay specimens were carried out, using nanobubbles. The current density was found to increase initially, after which it converged at a certain point. The amount of effluent from sand was found to exceed that of clay due to its higher permeability. The remediation efficiency for heavy metals in sand was also found to be higher than that for clay, due to more frequent contact with the nanobubbles. Additionally, the pH in soils at the cathode was found to be low, suggesting less influence by the sedimentation of the heavy metals.

Based on the results of the present study, the use of nanobubbles is expected to be applicable for electrokinetic remediation of the heavy-metal-contaminated sand and clay. By considering the engineering characteristics of soils and contaminants corresponding to diverse soil conditions, nanobubbles are a promising approach applicable for the in situ remediation of land contaminated by heavy metals. Based on this study, a remediation experiment using nanobubbles is needed for application to field soil.

Author Contributions: All authors contributed to this study; J.H., conceptualization, funding acquisition, project administration, supervision, writing of review, and editing; D.K., investigation, formal analysis, and writing of the original draft. All authors have read and agreed to the published version of the manuscript.

Funding: This research was funded by the Korea Agency for Infrastructure Technology Advancement under the Ministry of Land, Infrastructure and Transport of the Korean government. (Project Number: 19SCIP-B108153-05) and by a grant from the National Research Foundation (NRF) of Korea, funded by the Korea government (MSIP) (NRF-2019R1A2C2088962).This research was funded by the Human Resources Development (No. 20204030200090) of the Korea Institute of Energy Technology Evaluation and Planning (KETEP) grant funded by the Korea government Ministry of Trade, Industry and Energy by X-mind Corps program of National Research Foundation of Korea (NRF) funded by the Ministry of Science, ICT & Future Planning (2017H1D8A1030599). And The APC was funded by a grant from the National Research Foundation (NRF) of Korea, funded by the Korea government (MSIP) (NRF-2019R1A2C2088962).

Acknowledgments: This research was supported by the Korea Agency for Infrastructure Technology Advancement under the Ministry of Land, Infrastructure and Transport of the Korean government. (Project Number: 19SCIP-B108153-05) and by a grant from the National Research Foundation (NRF) of Korea, funded by the Korea government (MSIP) (NRF-2019R1A2C2088962).

This research was supported by the Human Resources Development (No. 20204030200090) of the Korea Institute of Energy Technology Evaluation and Planning (KETEP) grant funded by the Korea government Ministry of Trade, Industry and Energy by X-mind Corps program of National Research Foundation of Korea (NRF) funded by the Ministry of Science, ICT & Future Planning (2017H1D8A1030599).

We are grateful to Sohee Jeong and Heewon Chae for help in publishing our paper.

Conflicts of Interest: The authors declare no conflict of interest.

References

1. Park, K.W. A Study on Soil Washing Efficiency of Copper(Cu) Contaminated Soil in Military. Master's Thesis, Kyungpook National University Graduate School of Industry, Daegu, Korea, 2013.
2. Lima, A.T.; Hofmann, A.; Reynolds, D.; Ptacek, C.; Van Cappellen, P.; Ottosen, L.M.; Pamukcu, S.; Alshawabekh, A.; O'Carroll, D.; Riis, C.; et al. Environmental Electrokinetics for a sustainable subsurface. *Chemosphere* **2017**, *181*, 122–133. [CrossRef] [PubMed]
3. Naushad, M.; Ahamad, T.; Al-Maswari, B.; Alqadami, A.A.; AlShehri, S.M. Nickel ferrite bearing nitrogen-doped mesoporous carbon as efficient adsorbent for the removal of highly toxic metal ion from aqueous medium. *Chem. Eng. J.* **2017**, *330*, 1351–1360. [CrossRef]

4. Yuan, G.; Tian, Y.; Liu, J.; Tu, H.; Liao, J.; Yang, J.; Yang, Y.; Wang, D.; Liu, N. Insight into the influences of pH value on Pb(II) removal by the biopolymer extracted from activated sludge. *Chem. Eng. J.* **2017**, *308*, 1098–1104.

5. Kim, J.; Kwak, S.-Y. Efficient and selective removal of heavy metals using microporous layered silicate AMH-3 as sorbent. *Chem. Eng. J.* **2017**, *313*, 975–982. [CrossRef]

6. Lingamdinne, L.P.; Chang, Y.-Y.; Yang, J.-K.; Singh, J.; Choi, E.-H.; Shiratani, M.; Koduru, J.R.; Attri, P. Biogenic reductive preparation of magnetic inverse spinel iron oxide nanoparticles for the adsorption removal of heavy metals. *Chem. Eng. J.* **2017**, *307*, 74–84. [CrossRef]

7. Temesgen, T.; Bui, T.T.; Han, M.; Kim, T.-I.; Park, H. Micro and nanobubble technologies as a new horizon for water-treatment techniques: A review. *Adv. Colloid Interface Sci.* **2017**, *246*, 40–51. [CrossRef] [PubMed]

8. Alheshibri, M.; Qian, J.; Jehannin, M.; Craig, V.S.J. A History of Nanobubbles. *Langmuir* **2016**, *32*, 11086–11100. [CrossRef] [PubMed]

9. Agarwal, A.; Ng, W.J.; Liu, Y. Principle and applications of microbubble and nanobubble technology for water treatment. *Chemosphere* **2011**, *84*, 1175–1180. [CrossRef] [PubMed]

10. Takahashi, M.; Chiba, K.; Li, P. Free-Radical Generation from Collapsing Microbubbles in the Absence of a Dynamic Stimulus. *J. Phys. Chem. B* **2007**, *111*, 1343–1347. [CrossRef] [PubMed]

11. Ohgaki, K.; Khanh, N.Q.; Joden, Y.; Tsuji, A.; Nakagawa, T. Physicochemical approach to nanobubble solutions. *Chem. Eng. Sci.* **2010**, *65*, 1296–1300. [CrossRef]

12. Kyzas, G.Z.; Bomis, G.; Kosheleva, R.I.; Efthimiadou, E.K.; Favvas, E.P.; Kostoglou, M.; Mitropoulos, A.C. Nanobubbles effect on heavy metal ions adsorption by activated carbon. *Chem. Eng. J.* **2019**, *356*, 91–97. [CrossRef]

13. Choi, Y.I.; Choi, H.E.; Jung, J.H. and Jung, B.G. Changes in the Removal Efficiency and Degradation Rate of Total Petroleum Hydrocarbons in Accordance with Changes in the Air Flowrate into Micro-nano Bubbles. *J. Korea Soc. Environ. Technol.* **2016**, *17*, 458–463.

14. Jang, J.K.; Kim, M.Y.; Sung, J.H.; Chang, I.S.; Kim, T.; Kim, H.W.; Kang, Y.K.; Kim, Y.H. Effect of the Application of Microbubbles and/or Catalyst on the Sludge Reduction and Organic matter of Livestock Wastewater. *J. Korean Soc. Environ. Eng.* **2015**, *37*, 558–562. [CrossRef]

15. Oh, S.H.; Han, J.G.; Kim, J.-M. Long-term stability of hydrogen nanobubble fuel. *Fuel* **2015**, *158*, 399–404. [CrossRef]

16. Jeong, S.-H.; Kim, D.-C.; Han, J. The Fundamental Study on th e Soil Remediation for Copper Contaminated Soil using Nanobubble Water. *J. Korean Geosynth. Soc.* **2017**, *16*, 31–39. [CrossRef]

17. Chae, H.W.; Jeong, S.H.; Kim, D.C.; Han, J.G. A Study on Nanobubble's Remediation Effect on Contaminated Soft Soil with Different Amount of Fine Content. In Proceedings of the 2017 Spring Geosynthetics Conference, Seoul, Korea, 20 April 2017; pp. 89–90.

18. Li, H.; Hu, L.; Xia, Z. Impact of Groundwater Salinity on Bioremediation Enhanced by Micro-Nano Bubbles. *Materials* **2013**, *6*, 3676–3687. [CrossRef] [PubMed]

19. Han, J.-G.; Hong, K.-K.; Kim, Y.-W.; Lee, J.-Y. Enhanced electrokinetic (E/K) remediation on copper contaminated soil by CFW (carbonized foods waste). *J. Hazard. Mater.* **2010**, *177*, 530–538. [CrossRef] [PubMed]

20. Narasimhan, B.; Ranjan, R.S. Electrokinetic barrier to prevent subsurface contaminant migration: Theoretical model development and validation. *J. Contam. Hydrol.* **2000**, *42*, 1–17. [CrossRef]

21. Han, J.G.; Yoon, W.I.; Jung, D.H.; Kim, Y.S.; Lee, J.Y. Column Tests for the Design of PRB System using CFW. *J. Korean Geosynth. Soc.* **2011**, *10*, 35–43.

22. Kim, Y.W.; Lee, J.Y.; Lee, Y.K.; Han, J.G. A Study on Heavy Metals sedimentation by pH. In Proceedings of the 2009 Korean Society of Civil Engineers Conference, Hoengseong-gun, Korea, 21–23 October 2009; pp. 2037–2040.

© 2020 by the authors. Licensee MDPI, Basel, Switzerland. This article is an open access article distributed under the terms and conditions of the Creative Commons Attribution (CC BY) license (http://creativecommons.org/licenses/by/4.0/).

 *applied sciences*

 MDPI

Article

Isolation and Characterization of a Novel Bacterium from the Marine Environment for Trichloroacetic Acid Bioremediation

Mahshid Heidarrezaei [1,2,*], **Hoofar Shokravi** [3], **Fahrul Huyop** [4,5], **Seyed Saeid Rahimian Koloor** [6] and **Michal Petrů** [6]

1   Department of Bioprocess Engineering, School of Chemical and Energy Engineering, Faculty of Engineering, Universiti Teknologi Malaysia, Johor Bahru 81310, Malaysia
2   Institute of Bioproduct Development (IBD), Universiti Teknologi Malaysia, Johor Bahru 81310, Malaysia
3   School of Civil Engineering, Faculty of Engineering, Universiti Teknologi Malaysia, Johor Bahru 81310, Malaysia; shoofar2@graduate.utm.my
4   Department of Biosciences, Faculty of Science, Universiti Teknologi Malaysia, Johor Bahru 81310, Malaysia; fahrul@utm.my
5   Enzyme Technology and Green Synthesis Research Group, Department of Chemistry, Faculty of Science, Universiti Teknologi Malaysia, Johor Bahru 81310, Malaysia
6   Institute for Nanomaterials, Advanced Technologies and Innovation (CXI), Technical University of Liberec (TUL), Studentska 2, 461 17 Liberec, Czech Republic; s.s.r.koloor@gmail.com (S.S.R.K.); michal.petru@tul.cz (M.P.)
*   Correspondence: mahshid@ibd.utm.my

Received: 23 May 2020; Accepted: 30 June 2020; Published: 2 July 2020

Abstract: Halogenated compounds are an important class of environmental pollutants that are widely used in industrial chemicals such as solvents, herbicides, and pesticides. Many studies have been carried out to explore the biodegradation of these chemicals. Trichloroacetic acid (TCA) is one of the main halogenated compounds that are carcinogenic to humans and animals. The bacterium was isolated from the northern coastline of Johor Strait. In this study, the ability of strain MH2 to biodegrade TCA was evaluated by a growth experiment and dehalogenase enzyme assay. The growth profile of the isolated strain was examined. The doubling time for *L. boronitolerans* MH2 was found to be 32 h. The release of chloride ion in the degradation process was measured at $0.33 \times 10^{-3} \pm 0.03 \text{ mol·L}^{-1}$ after 96 h when the growth curve had reached its maximum within the late bacterial exponential phase. The results showed that the strain had a promising ability to degrade TCA by producing dehalogenase enzyme when cell-free extracts were prepared from growth on TCA as the sole carbon source with enzyme-specific activity, $1.1 \pm 0.05 \text{ } \mu \text{molCl}^- \text{min}^{-1} \cdot \text{mg}^{-1}$ protein. Furthermore, the morphological, and biochemical aspects of the isolated bacterium were studied to identify and characterize the strain. The morphological observation of the isolated bacterium was seen to be a rod-shaped, Gram-positive, motile, heterotrophic, and spore-forming bacterium. The amplification of the 16S rRNA and gene analysis results indicated that the isolated bacterium had 98% similarity to *Lysinibacillus boronitolerans*. The morphological and biochemical tests supported the 16S rRNA gene amplification. To the best of the authors' knowledge, this is the first reported case of this genus of bacteria to degrade this type of halogenated compound.

Keywords: biodegradation; dehalogenase-producing bacteria; haloalkanoic acids; trichloroacetic acid; 16S rRNA; *Lysinibacillus*

1. Introduction

Xenobiotic compounds are synthetic molecules which include halogenated hydrocarbon, polyaromatic hydrocarbons, polycyclic biphenyls, and lignin [1,2]. They are biologically active compounds that are widely used in several drug and pesticide industries [3,4]. Hence, xenobiotic compounds often appear in industrial wastewaters and aquatic ecosystems [5,6]. Xenobiotic compounds can be considered stable, in the thermodynamic sense. Moreover, they are fairly resistant to biodegradation by the native microorganisms, and persistent in the environment [7]. Damage caused by xenobiotic compounds poses significant health and ecological risks in developing countries [8]. The parent xenobiotic compounds are not directly toxic but they can be transformed into harmful oxy-radicals or carbon-centered radicals that attack the double bonds of cellular macromolecules generating oxidative damage [9]. Xenobiotic compounds have the potential to cause toxic effects on humans and animals with consequent acute carcinogenic, teratogenicity, and mutagenic effects [7]. Nonetheless, a diverse group of aerobic and anaerobic groups of bacteria often found in various habitats are capable of degrading xenobiotic compounds [10]. Several researches exist about biodegradation of xenobiotic compounds using dehalogenase-producing bacteria such as organofluorine [11,12], organochlorine [13,14], and organobromine [15,16].

Halogenated hydrocarbons are organic compounds that have many significant industrial applications such as multipurpose solvents, plasticizers, pesticides, fuel additives, flame retardants, and anesthetics [17–19]. Several halogenated hydrocarbon like polychlorinated biphenyl (PCB) residues and chlorinated hydrocarbon pesticide residues were detected in human adipose tissue, milk, and blood serum. Transfer and accumulation of these chemicals in the human body will consequently cause serious health problems [20]. One of the most well-known applications of halogenated aliphatics in the industry is related to the mixed substituted chlorofluorocarbons, CFC [21]. Millions of tons of halogenated aliphatics are produced annually and mainly have been used in a variety of manufacturing processes of solvents and cleaning agents [22]. These chemical compounds are common pollutants of aquatic habitats, as they are relatively water-soluble and can migrate to underground water-supplies and therefore threaten groundwater quality [23].

Haloacetic acids (HAAs) are recognized as oxidation products of airborne C_2-halocarbons which have been demonstrated to be fast in volatilization [24]. They are carcinogenic in humans even at low concentrations [25]. Several studies show that haloacetic acids are biodegradable in anaerobic environments including soil and, wastewater, [26]. HAAs come either from brominated and/or chlorinated organic halogen compounds. These organic halogenated compounds are important by-products of the reaction between organic compounds present in water and chlorine in water-treatment plants [27]. There are five contaminating by-products of halogenated organic compounds, which exceed the maximum levels established by the US Environmental Protection Agency (60 $\mu g \cdot L^{-1}$). These by-products include monochloroacetic acid (MCA), dichloroacetic acid (DCA), trichloroacetic acid (TCA), monobromoacetic acid (MBA), and dibromoacetic acid (DBA). As a result, the literature reporting on the analysis of HAAs mostly is focused on these five species [28]. TCA is a member of a very small group of "moderately strong" acids with ionization constants in the range 0.1–10 that is the most commonly used agent for chemical peeling [29,30]. TCA is one of the main contaminants of the environment that are carcinogenic to humans and animals [31]. The DCA and TCA are the most abundant HAAs detected in water resources containing nearly 80% of existing HAAs [32]. Hence, the decontamination of water resources from these compound is one of the major challenges for the preservation of the aquatic ecosystem and wastewater treatment [33].

Biodegradation is one of the main natural processes for the removal of xenobiotics such as chloroaliphatic compounds from the environment using microorganisms [34,35]. The major chemical processes for the metabolism of xenobiotic compounds are oxygenation, oxidation, reduction, hydrolysis, and conjugation, which are catalyzed by enzymes [36]. Dehalogenases are known as enzymes that catalyze the degradation reaction of halogen atoms from halogenated organic compounds. Dehalogenase-producing bacteria can detoxify harmful halogenated compounds from their substrates by cleaving carbon–halogen bonds that link halogens in aliphatic compounds [37,38]. Based on cleavage nature, dehalogenase enzymes can be classified as haloalkane, dichloromethane, halohydrin, and L- and D-haloalkanoic acid dehalogenases [39]. Each specific group of these enzymes has its enantioselectivity and product inhibition characteristics [40]. Dehalogenase enzyme is present in bacterial systems and plays an important role in the biodegradation of environmental pollution caused by halogenated hydrocarbons [41].

Several research projects have been conducted to study the degradation mechanism of dehalogenase-producing bacteria for bioremediation of halogenated hydrocarbons and related hazardous compounds in soil and water ecosystems. Muslem et al. [42] characterized *Bacillus cereus* WH2 strain for its ability to metabolize β-haloalkanoic acids as carbon and energy sources. Alrawas and Huyop [35] isolated *Ralstonia* sp. TCAA-2 native to Danga Bay coast, Malaysia, to biodegrade TCA. Hamid et al. [43] further characterized dehalogenase enzyme functions of *Rhizobium* sp. RC1 bacteria. It was reported that RC1 can grow in α-haloalkanoic acid as the sole carbon source. α-haloalkanoic acids are halogenated compounds that are widely liberated into the ecosystem through the use of weed herbicides in the agricultural sector. Adamu et al. [44] studied the catalytic properties and mechanistic analysis of the microbial dehalogenases specific in *Rhizobium* sp. RC1. The isolated strain was reported to have outstanding performance in detoxifying L-2-haloacid. Selvamani et al. [45] characterized deploying *Trichoderma asperellum* SD1 to degrade 3-chloropropionic acid in the terrestrial ecosystem. Bagherbaigi et al. [46] used *Arthrobacter* sp. S1 for degradation of α and β-haloalkanoic acids.

As such, most of the research to date has focused on bioremediation of aquatic and terrestrial habitats. Nonetheless, there is a very little study to decontaminate TCA using dehalogenase- producing bacteria and their focus is on terrestrial bacteria or the identified bacteria are non-native to the coastal region of Johor strait. To the best of the authors' knowledge, this is the first reported case that shows *Lysinibacillus* bacteria could degrade halogenated compounds such as TCA. Because of the issues described above, this study intends to characterize a native bacterium with the capability of degrading TCA. To this end, several bacterial strains capable of aerobically degrading TCA were isolated from the northern coastline of Johor Strait in Malaysia and evaluated for their morphological and biochemical characteristics to degradation capabilities of TCA. The phylogenetic tree of the isolated strain was constructed using 16S-rRNA sequencing and it was recognized that the characterized bacterium was a new subspecies of the genus *Lysinibacillus*. The new bacterium was given the scientific name of *Lysinibacillus boronitolerans* MH2.

2. Materials and Methods

2.1. Sampling, Enrichments, and Growth

The seawater sample was collected from the northern coastline of Johor Strait in Malaysia. The water sample (1 mL) was added into 250 mL flasks containing 100 mL of minimal medium and TCA as a carbon source. Minimal medium contained basal salts solution, trace metals solution; and distilled water. The composition and proportion of the constituent of the minimal medium are shown in Table 1. The medium was adjusted to pH 7 before autoclaving. Then, the bacterial culture was incubated aerobically at 30 °C conditions in an orbital shaker set at 200 rpm. In order to prepare a solidified medium, two flasks (labeled Flask 1 and Flask 2) were prepared. Flask 1 contained 1.6 g oxide agar added into 48 mL distilled water (DW). Flask 2 was filled with 10 mL of trace metal salts and 10 mL of basal salts added into 30 mL DW. Flask 1 and flask 2 were autoclaved separately at 121 °C, for 15 min at 101.3 kPa. After autoclaving, solution in both flasks 1 and 2 were mixed and 1 mL of 1 M concentration TCA was added to the mixture, and the obtained medium was poured on agar plates. After the solidification of the agar, bacterial dilution was streaked onto the plates. Among the isolated strains, three different bacterial colonies were detected that the colony with the highest growth rate was retained for further purification. No further study was carried out for the identification of the two other isolated bacteria strains. The selected colony was picked and sub-cultured by repeated streaking on a fresh agar plate to purify for further analysis. The selected colony was diluted in a drop of distilled water and transferred into minimal medium containing the ingredients shown in Table 1.

Table 1. Composition and proportion of the constituents in trichloroacetic acid (TCA)-enriched minimal medium.

Composition	Proportion	Composition	Name	Formula	Reference
Basal salts	10 mL	42.5 g·L^{-1}	Dipotassium hydrogen phosphate 3-hydrate	$K_2HPO_4 \cdot 3H_2O$	Hareland et al. [47]
		10 g·L^{-1}	Sodium dihydrogen phosphate 2-hydrate	$NaH_2PO_4 \cdot 2(H_2O)$	
		25 g·L^{-1}	Ammonium sulfate	$(NH_4)_2 SO_4$	
		1.0 g·L^{-1}	Nitrilotriacetic acid	$N(CH_2COOH)_3$	
		2.0 g·L^{-1}	Magnesium sulphate	$MgSO_4$	
Trace metals solution	10 mL	0.12 g·L^{-1}	Iron sulphateheptahydrate	$FeSO_4 \cdot 7H_2O$	Hareland et al. [47]
		0.03 g·L^{-1}	Magan sulfate heptahydrate	$MnSO_4 \cdot 7H_2O$	
		0.03 g·L^{-1}	Zinc sulphate monohydrate	$ZnSO_4 \cdot H_2O$	
		0.01 g·L^{-1}	Cobalt chlorine hexahydrate	$CoCl_2 \cdot 6H_2O$	
TCA	1 mL	10 mm	Trichloroacetic acid	$C_2HCl_3O_2$	Fisher Scientific
Distilled water	79 mL	-	Distilled water	H_2O	-
Total volume	100 mL	-	-	-	-

The bacteria were grown in minimal medium containing TCA as the sole carbon source. Chlorine atoms are eliminated by hydrolytic dehalogenation from the compound [48]. Chloride ion liberation had been monitored by the halide ion assay to prove the degradation of TCA [49]. The concentration of chloride ions was determined by converting absorbance value to concentration in mmol·L^{-1} based on the standard curve constructed using sodium chloride as the standard measurement of soluble chloride [50,51]. The suspension of the selected colony was used to evaluate the growth profile and TCA degradation. Several direct and indirect methods can be used for the evaluation of the bacterial growth profile. A colony-forming unit (CFU) is a direct method to evaluate the actual colony forming units per volume of culture in given bacterial incubation. The enumeration of CFU is one of the most accurate methods to find out the number of viable bacteria strains growing in the presence of harsh chemicals [52]. On the other hand, the spectrophotometric method is a widely used laboratory technique to indirectly estimate the bacterial concentration and quantitative analysis of the released chloride ions [44,53]. Spectrophotometry methods are based on the absorption and scattering of light beams passing through the culture medium. These methods are sensitive and selective; however, enumeration of CFU by plate count might be a disadvantage due to the fact that cells grown in clusters into a single colony and assumption that each colony arises from one cell is not accurate. In addition, commercial TCA may contain impurities and colony count may due to the growth of impurities rather than on TCA. In this study, absorbance A_{680nm} and A_{460nm} are used for analysis of the bacterial growth and released chloride ions, respectively. Chloride was liberated during the growth of TCA. The absorbance of the released chloride ions was recorded every 6 h using Jenaway spectrophotometer at A_{460nm}. Growth experiment and chloride ion released analysis were carried out in triplicate.

2.2. Preparation of Cell-Free Extracts and Dehalogenase Enzyme Assay

Apart from measuring cell growth (A_{680nm}) and chloride ions released (A_{460nm}) in the growth medium, the presence of dehalogenase was assessed as the liberation of halide in cell-free extracts in vitro. Cells from a 100 mL culture grown in 10 mm TCA and nutrient broth were harvested by centrifugation at 20,000× *g* for 10 min at 4 °C during the late-growth phase, between 72–96 h (Figure 1). The pellets were resuspended in 10 mL of 0.1 M tris-acetate, 1 mm ethylenediaminetetraacetic acid (EDTA), 10% (mass·vol^{-1}.) glycerol, pH 7.5, and washed three times with 0.1 M tris-acetate buffer by centrifugation at 20,000× *g* for 10 min at 4 °C. Finally, the cells were then resuspended in 3 mL of the same buffer and maintained at 0 °C for sonication in an MSE Soniprep 150 W ultrasonic disintegrator at a peak amplitude (λ = 10 microns) for 30 s. Any unbroken cells and cell wall materials were removed by centrifugation at 25,000× g for 20 min at 4 °C. The cell-free extracts activity was assayed immediately after the preparation as previously described by Mesri et al. [54].

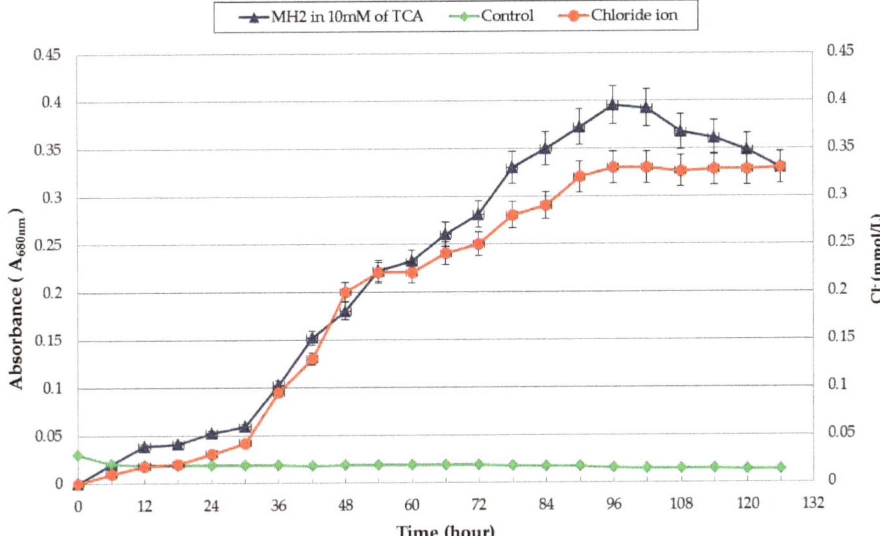

Figure 1. The growth curve of the MH2 bacterium measured at A_{680nm} and chloride ions released measured at A_{460nm}. The control culture without strain MH2 was incubated under the same growth conditions. All readings were based on triplicate analysis.

2.3. Biochemical and Morphological Characterization

Morphological tests, namely Gram staining, spore staining, and motility tests as well as biochemical tests including catalase, oxidase, urease, gelatin liquefaction, citrate, MacConkey agar, casein hydrolysis, starch, indole, and triple sugar iron (TSI) tests, were carried out for basic identification of bacteria generics and inferring their properties following the standard outline in Bergey's manual [55]. Table 2 shows the morphological and biochemical characteristics of the selected bacteria.

Table 2. Morphological and biochemical characterization of *L. boronitolerans* MH2.

Objective	Test	Composition and Materials	Incubation Condition	Aim	Description	Reference
Characterization of bacteria	Gram staining	• Crystal violet • Iodine • Alcohol • Safranin	-	To determine the type of bacteria (Gram-positive or gram-negative)	Smear color	[55]
	Spore staining	• Safranin • Malachite	-	To determine the spore production of bacterium	Retaining cell color	
	Motility	• 5.0 g peptone, • 2.0 g agar • 2.5 g NaCl • 1.5 g beef extract • 200 mL distilled water (DW)	Incubated at 30–37 °C for 24–48 h.	To determine the motility and to explore the capability of bacteria to migrate away from a line of inoculation	Turbidity diffusion from the stab line	
Biochemical tests	Catalase	• Hydrogen peroxide	-	To differentiate aerobic and obligate anaerobic bacteria	Bubble formation	[56,57]
	Oxidase	• N′N′N′N′-tetramethyl-p -phenylenediamine36dihydrochloride (TMPD)	-	To determine the taxonomic status and identity of pathogenic bacteria	Color change	
	Urease	• 5.5 g of Christen's urea agar • 50 mL of 40% urea solution • 250 mL DW	Autoclaved at 121 °C for 20 min. Incubated at 37 °C 24–48 h.	To determine the ability of bacteria to decompose urea which produces the enzyme urease	Color change	
	Gelatin Liquefaction	• 0.75 g of beef extract • 1.25 g of peptone • 30 g of gelatin in • 250 mL DW	Incubated at 33 °C for four days	To explore the ability of an organism to produce extracellular proteolytic enzyme	Solidification and hydrolysis of gelatin	

Table 2. *Cont.*

Objective	Test	Composition and Materials	Incubation Condition	Aim	Description	Reference
Biochemical tests	Citrate	• 3.5 g of Simmon's citrate in • 150 mL DW	Autoclaved at 121 °C for 20 min. Incubated at 37 °C for 24–48 h.	To look at the ability of enteric organisms based on the ability to ferment citrate as a carbon source	Color change	[56,57]
	MacConkey Agar	• 17 g peptone • 3 g proteose peptone • 10 g lactose monohydrate • 1.5 g bile salts • 5 g sodium chloride • 0.03 g neutral red • 0.001 g crystal violet • 13.5 g agar • DW (to adjust 1 liter)	Autoclaved at 121 °C for 20 min. Incubated at 37 °C for 24–48 h.	To isolate Gram-negative enteric bacteria and the differentiation of lactose fermenting from lactose non- fermenting Gram-negative bacteria	Colony observation	
	Casein hydrolysis	• 3.6 g nutrient agar • 0.5 g casein • 200 mL DW	Incubated at 33 °C for three days	To identify bacteria capable of hydrolyzing casein with casease enzyme.	Observation of clear zone	
	Starch	• 1.5 g soluble starch • 2.5 g bacteriological agar • 150 mL DW	Autoclaved at 121 °C for 20 min. Incubated at 37 °C. for 24–48 h	To determine the amylase enzyme with the ability to degrade starch	Color formation	
	Indole	• 1 g sodium chloride • 2 g tryptophan or peptone • 200 mL DW	Autoclaved at 121 °C for 20 min at pH 7.5	Tryptophanase enzyme	Color formation	
	Triple Sugar Iron (TSI)	• 1.5 g soluble starch • 2.5 g bacteriological agar • 150 mL DW	Autoclaved at 121 °C for 20 min; incubate for 24–72 h	To test a microorganism's ability to ferment sugars and to produce acid	Color formation	

2.4. Molecular Identification

Genomic DNA from bacterial cells of strain MH2 was isolated using the Wizard® Genomic Purification Kit (Promega). Universal polymerase chain reaction (PCR) forward and reverse primers Fd1 (5′-aga gtt tga tcc tgg ctc ag-3′) and Rp1 (5′-acg gtc ata cct tgt tac gac tt-3′), respectively, were synthesized by 1st Base® Laboratory Malaysia Sdn. Bhd., 43300 Seri Kembangan, Selangor, Malaysia [58]. The amplified 16S rRNA gene has to be purified before sequencing. QIAquick polymerase chain reaction (PCR) purification kit (Qiagen) was used to purify the DNA. The sequencing was carried out by ABI PRISM® 377 DNA sequencer (1st Base® Laboratory Malaysia Sdn. Bhd., 43300 Seri Kembangan, Selangor, Malaysia). The thermocycling conditions for PCR amplification are presented in Table 3.

Table 3. Thermocycling conditions for polymerase chain reaction (PCR) amplification.

Step	Temperature	Time	Cycles
Initial Denaturation	94 °C	5 min	1
Denaturation	94 °C	1 min	
Annealing	55 °C	1 min	30
Elongation	72 °C	4 min	
Final Elongation	72 °C	10 min	1

The phylogram of unidentified bacteria was rebuilt using Mega5 Molecular software and compared with the sequences from the 16S rRNA gene stored in the Gene Bank by the National Center for Biotechnology (NCBI) using the Basic Local Alignment Search Tool (BLASTn) for nucleotides [59]. Sequences were aligned using Bioedit Sequence Alignment Editor X and a neighbor-joining tree was constructed [60]. Bootstrap consensus tree was inferred from 500 replicates.

3. Results

A seawater sample was taken from the marine environment and three strains of bacteria were isolated using minimal media. Once the bacterial strains were purified, only one isolate (MH2) was selected from three bacterial samples due to its distinctively superior growth and survival abilities in solid minimal media compared to the other two bacteria for further analysis. The doubling time for MH2 strain was found 32 h and the maximum released chloride ion due to dehalogenase enzyme activity was approximately $0.33 \times 10^{-3} \pm 0.03$ mol·L^{-1}. 16S rRNA phylogenetic analysis was conducted to determine the phylogenetic relatedness and the species affiliation of the MH2 strain.

3.1. Growth of L. boronitolerans MH2

The growth profile of the isolated *L. boronitolerans* MH2 strain bacteria was examined in fixed time intervals. The growth profile exhibiting the lag, exponential, stationary, and decline phase as it was expected. The growth of the bacteria by measuring through light absorbance of the solutions measured at 680 nm (A_{680nm}) by using a JENWAY, 6300 ultraviolet (UV)-visible spectrophotometer at every 6 h. The measured values for the growth of MH2 were recorded based on 10 mm of the TCA-enriched medium. In the lag phase, the bacterium growth was very little because the bacterium does not immediately adapt to the growth condition and the carbon source which was toxic to the bacteria. In the exponential phase, the growth rate of the MH2 bacterium increased. At the end of the exponential phase, the growth curve reached its maximum rate at 96 h. After a long exponential phase, the growth curve reached a stationary phase and then the bacterium enters the decline phase. The growth profile shows that MH2 was able to grow in minimal medium containing 10 mm of TCA. According to Slater et al. [51], the bacterium absorbs TCA from the compound as the sole carbon source and releases chloride ions. The liberated chloride ion during the metabolism of MH2 strain was monitored by a halide ion assay order as an indicator of TCA degradation in the medium [51].

The results of chloride ion released and the growth curve of the bacterium was plotted as shown in Figure 1.

Figure 1 shows that both the halide ion assay and growth curves of the bacterium in minimal medium exhibited identical trends indicating that the concentration of the released chloride ion and the growth are directly related to each other. The curves for halide ion assay and growth of MH2 were in contrast with the control experiment. The concentration of the chloride ion in the cultivation solution was determined by converting absorbance corresponding to the concentration using the standard curve. The standard curve was constructed using NaCl as a typical standard for measuring soluble chloride concentration [61]. Doubling time of the bacteria in 10 mm TCA-enriched minimal medium was estimated to be 32 h.

3.2. Enzyme Activity in Cell-Free Extracts from Cell Growth in Trichloroacetic Acid (TCA) Medium and Nutrient Broth

Dehalogenase activity in cell-free extracts was assessed by the release of free halide in an enzyme-substrate reaction in vitro. Cell-free extracts from strain MH2 grown on 10 mm TCA medium as the sole carbon source was prepared. Halide liberation was measured from extracts of cells grown in TCA using TCA as a substrate. However, no halide was liberated by extracts of the same bacteria grown in nutrient broth. This suggests that in nutrient broth no expression of the bacterial dehalogenase gene takes place. The enzyme specific activity of the dehalogenase for TCA was $1.1 \pm 0.05\ \mu molCl^-min^{-1}\cdot mg^{-1}$ protein.

3.3. Morphology and Biochemical Characteristics

The result of the bacteria characterization based on morphological and biochemical analysis shows that the isolated bacterium grew well at aerobic conditions and can readily metabolize TCA and produce chloride ions by dechlorination treatment. The morphological observations for Gram-staining and spore staining using light microscopy 1000× magnification are shown in Figures 2 and 3, respectively. The result of Gram-staining analysis indicated that the MH2 strain is a Gram-positive bacterium and produced creamy to pink colonies on solid minimal medium. *L. boronitolerans* MH2 was further examined for biochemical characteristics.

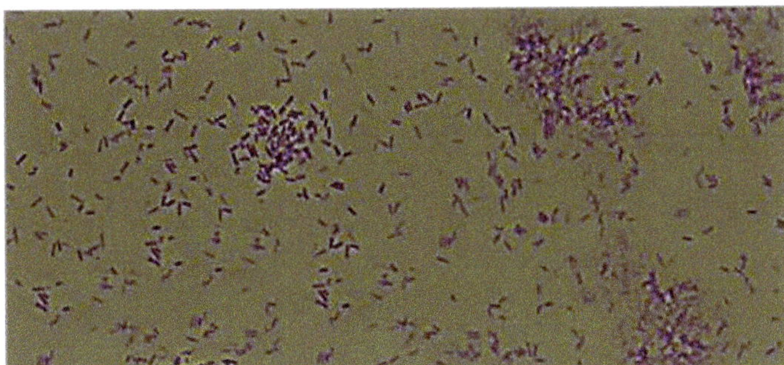

Figure 2. The Gram-staining of strain MH2 observation under light microscope (×1000 magnification). Purple colour bacteria appeared as the result of Gram staining which indicated MH2 was a Gram-positive bacterium.

A spore staining test was performed to determine if MH2 is an endospore-forming bacterium. Spores were observed during the microscopic examination that proves that bacterium MH2 was capable of producing spores. Microscopic observation of the spore staining is shown in Figure 3.

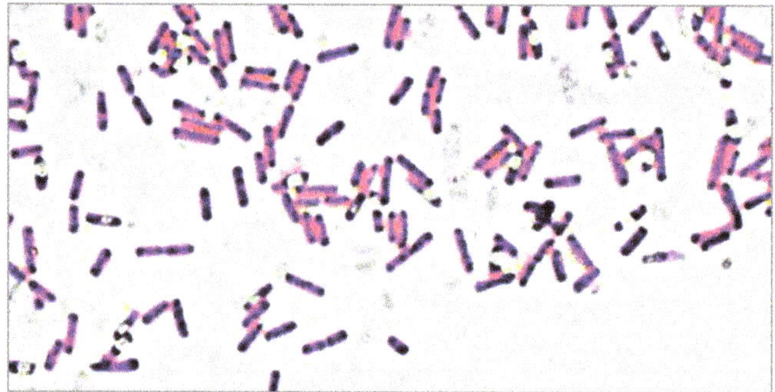

Figure 3. Spore staining for strain MH2 observed under a light microscope (×1000 magnification). Spores were seen as black spots in each cell.

Tables 4 and 5 showed the summary of morphological characteristics of MH2 and the results of biochemical tests, respectively. The oxidase, urease, casein, and starch hydrolysis tests showed positive results and catalase, gelatin, citrate, MacConkey agar, indole, and TSI demonstrated negative results. Morphological and biochemical tests were conducted based on Bergey's manual systematic bacteriology [56].

Table 4. Morphological character of MH2.

Characteristic	Parameter
Identity	*L. boronitolerans* MH2
Cell shape	Rod shape
Size	0.5~1 mm
Colony	Cream
O_2 requirement	Aerobic
Colony morphology	Smooth, mucoid, a little elevated

Table 5. Morphological and biochemical characteristics of MH2.

Test	Spore Forming	Gram Staining	Motility	Catalase	Oxidase	Urease	Gelatin	Citrate	MacConkey Agar	Casein Hydrolysis	Starch	Indole	TSI
Result	+	+	+	−	+	+	−	−	−	+	+	−	−

Positive (+); Negative (−).

3.4. Polymerase Chain Reaction (PCR) Amplification of 16s rRNA Gene Analysis

Colony PCR (cPCR) is performed to verify the construct of the DNA-sequence. 16S rRNA genes were amplified using universal primers Fd1 and Rp1, respectively. The amplified genes were monitored with agarose gel electrophoresis. In Figure 4, the unique amplified band (1500 bp) was observed and compared with the 1 Kb DNA ladder.

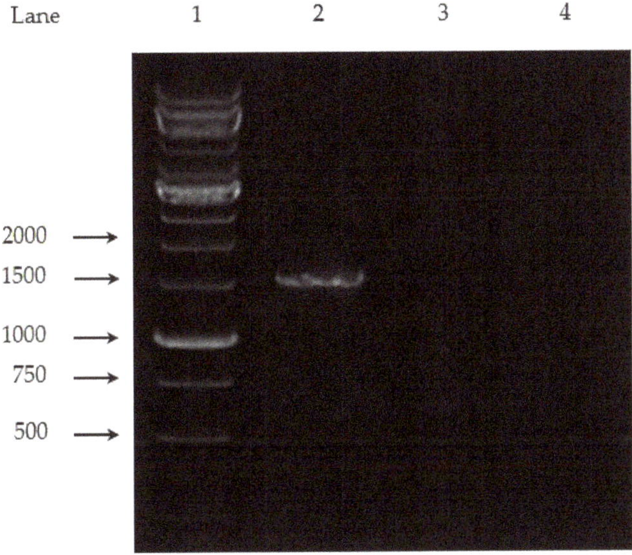

Figure 4. Amplification of the 16S rRNA gene of strain MH2 showing 1500 bp DNA fragment on an agarose gel (1%) (Lane 2). Lane 1: 1kb DNA ladder (Promega); Lanes 3 and 4, negative controls, polymerase chain reaction (PCR) mixture without forward (Fd1) and reverse (Rp1) primer respectively, showing no amplification.

The continuous stretch (1274 bp) of 16S rRNA gene was investigated to determine the closest phylogenetic neighbors. The BLAST program was employed and it was revealed that the MH2 has 98% identity matches in sequence with *L. boronitolerans*. Amplification and direct sequence analysis of partial length of 16S rRNA indicate that bacterium MH2 is phylogenetically related to *L. boronitolerans*. The results of the 16S rRNA analysis supported the biochemical characteristics of the bacterium belong to the same genus and species.

3.5. Phylogeny Tree Analysis

The phylogenetic tree was constructed using the neighbor-joining method and the maximum composite likelihood method. The bootstrap consensus tree offered from 500 replicates is taken to represent the evolutionary history of the taxa analyzed. Branches corresponding to partitions reproduced in less than 50% bootstrap replicates are excluded. The percentage of replicate trees is shown next to the branches. Evolutionary analyses were conducted in MEGA5 [59]. Figure 5 presented the phylogeny tree of the MH2 strain.

The maximum composite likelihood method was used to estimate the evolutionary distances. The units of evolutionary distances were based on the number of base substitutions per site; 11 nucleotide sequences were involved in the analysis. All gaps and missing data were excluded for tree building before analysis. Only 1238 positions were considered in the final dataset.

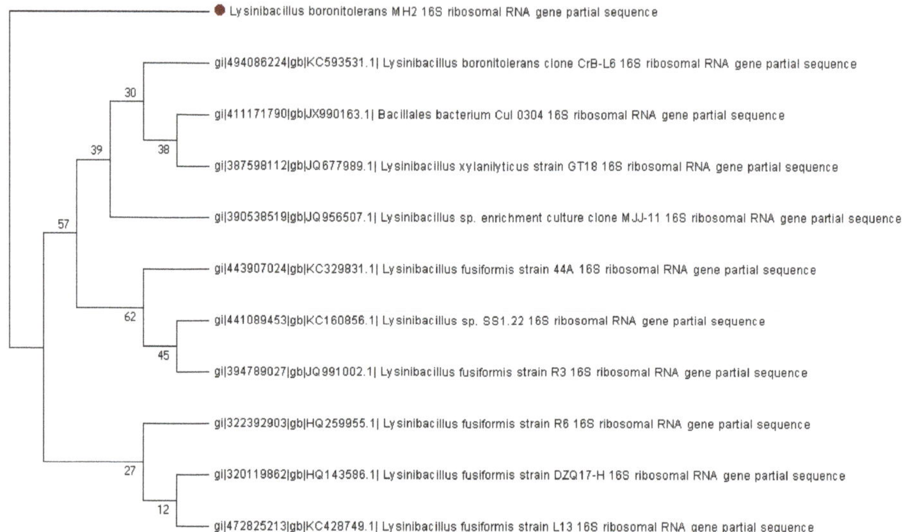

Figure 5. Neighbor-joining tree of strains M2 based on 16S rRNA gene sequences. Scale Bar 0.001 indicates sequence divergence.

4. Discussion

Biodegradation is an effective biological process to clean up polluted environments through the isolation of efficacious halo-organic compound biodegraders. Notably, the BLASTn search on the 16S rRNA gene sequence revealed that MH2 has the 98% identity match with *Lysinibacillus boronitolerans* (accession number: KC59351.1). Hence, to identify the capability of MH2 to degrade the sample pollutants, this study grew the MH2 bacterial isolate in TCA-enriched minimal media as the sole carbon source (pollutants).

The initial composition of the culture medium and the concentration of the substrate (TCA) are two important parameters that can affect the degrading ability of the bacteria. Using artificial seawater or sterilized seawater as a medium for cultivation of sea-isolated bacterium was refrained from due to the potential interference of natural chloride ions in seawater with those liberated during the growth experiment. Minimal medium contained basal salts solution, trace metals solution, and distilled water was used in this study to provide the required growth culture for the bacteria. The existing nitrilotriacetic acid ($N(CH_2COOH)_3$) in the trace metals solution is low enough not to interfere significantly with the results of the experiment and to be consistent with the hypothesis of utilizing TCA as the sole carbon and energy source in the bacteria growth process. The optimal concentration of the TCA pollutant sample was selected 10 mm in the study of the growth profile and the ion liberation experiments. Using TCA in concentrations higher than 20 mm can stifle the cell growth due to toxic effects while the lower concentration of pollutants was not enough to observe the induced catalytic reaction of the dehalogenase-producing bacteria. Growth was strictly monitored by measuring the cells' turbidity and the amount of chloride ions released at appropriate time intervals. An uninoculated flask treated in the same way was used as a control. This is important to make sure the chloride measured in the growth medium was due to the cells using the TCA rather than the auto-degradation of the substrate in the growth medium.

MH2 strain was capable of degrading hydrocarbon compounds. The nature and type of carbon sources are among the most important factor to determine bacterial growth. Haloacetates (i.e., MCA, DCA, and TCA) are common classes of water chlorination by-products. Several bacteria are available that can grow on MCA. *Burkholderia* sp. DehCL1 [62], *Bacillus* sp. TW1 [63], *Xantobacter autotrophicus*

GJ10 [64,65], and *Pseudomonas* sp. R1 [66] are examples of bacteria that can degrade MCA haloacetates. Additionally, Meusel and Rehm [67] described *Xanthobacter autotrophicus* GJ10 that can degrade DCA. To the best of the authors' knowledge, this is the first report on *Lysinibacillus* sp. isolated from seawater adapted in metabolizing TCA.

The results showed that the highest growth in *L. boronitolerans* MH2 bacterium was achieved at 96 h. Low absorbance was observed for the lag phase due to the low intensity and slow growth of the bacterial biomass produced. In broth, the doubling time of the *L. boronitolerans* MH2 was 32 h. Chloride ion concentration during the biodegradation process was monitored by a chloride ion assay [61,68]. Based on the standard curve of the NaCl, the released chloride ion by *L. boronitolerans* MH2 strain after 96 h was 0.33×10^{-3} mol·L^{-1}. The study of the growth curve shows that the *L. boronitolerans* MH2 was capable of growing in a minimal medium having TCA as the sole source of carbon and energy. Growth on TCA was further analyzed by preparing the cell-free extracts from growth on TCA as described in Figure 1 and Section 3.2. The presence of dehalogenase was detected by measuring the enzyme-specific activity. The *L. boronitolerans* MH2 dehalogenase described herein appears to be inducible because MH2 cells grown in nutrient broth lacking TCA exhibited no dehalogenase activity.

Hydrolytic dehalogenation of TCA produces oxalate as a final product where oxalate serves as the carbon source and it can be immediately converted to CO_2 [69]. The number of bacteria capable of using oxalate as the sole source of carbon is very limited [70]. Biochemical and morphological experiments were carried out according to Bergey's manual [56] aimed at verifying the obtained result of 16S rRNA gene analysis. The results of biochemical characteristics supported the findings of the bacteria suggested by the 16S rRNA analysis. The growth profile of the isolated strain was examined and the results showed that *L. boronitolerans* MH2 bacterium has a promising ability as a dehalogenase-producing bacterium.

5. Conclusions

Using dehalogenase enzymes to detoxify chlorinated organic compounds was envisaged as a promising biological control method. The applicability of *L. boronitolerans* MH2 is an important outcome discovered in current analysis suggesting dehalogenase-producing bacteria is important for the future to the exploitation of the bacterium for in situ efforts to detoxify halogen-contaminated environments. More importantly, these findings further add to the limited body of knowledge with regards to the degrading of halogenated compounds by the bacteria.

Author Contributions: Resources, M.H., H.S., F.H., S.S.R.K., and M.P.; investigation, M.H.; writing—original draft preparation, M.H., H.S.; writing—review and editing, M.H., H.S., F.H.; visualization, M.H., H.S., F.H., M.P. and S.S.R.K.; supervision, F.H.; project administration, M.H., H.S., F.H., M.P., and S.S.R.K.; funding acquisition, F.H., M.P. and S.S.R.K.; All authors have read and agreed to the published version of the manuscript.

Funding: The APC is founded by Ministry of Education, Youth, and Sports of the Czech Republic and the European Union (European Structural and Investment Funds Operational Program Research, Development, and Education) in the framework of the project "Modular platform for autonomous chassis of specialized electric vehicles for freight and equipment transportation", Reg. No. CZ.02.1.01/0.0/0.0/16_025/0007293.

Acknowledgments: The authors would like to thank the Department of Biosciences, Faculty of Science, Universiti Teknologi Malaysia, for facilities and services given and acknowledged Ministry of Education, Youth, and Sports of the Czech Republic and the European Union (European Structural and Investment Funds Operational Program Research, Development, and Education) for founding APC in the framework of the project "Modular platform for autonomous chassis of specialized electric vehicles for freight and equipment transportation", Reg. No. CZ.02.1.01/0.0/0.0/16_025/0007293.

Conflicts of Interest: The authors declare no conflict of interest.

References

1. Ekambaram, P.; Narayanan, M. Upregulation of HSP70 extends Cytoprotection to fish brain under xenobiotic stress. *J. FisheriesSciences.com* **2017**, *11*, 11. [CrossRef]
2. Mishra, V.K.; Singh, G.; Shukla, R. Impact of xenobiotics under a changing climate scenario. *Clim. Chang. Agric. Ecosyst.* **2019**, 133–151. [CrossRef]
3. George, S.G. Enzymology and molecular biology of phase II xenobiotic-conjugating enzymes in fish. In *Aquatic Toxicology: Molecular, Biochemical and Cellular Perspectives*; Donald, C., Ed.; CRC Press: Boca Raton, FL, USA, 1994; pp. 37–85.
4. Klont, F.; Jahn, S.; Grivet, C.; König, S.; Bonner, R.; Hopfgartner, G. SWATH data independent acquisition mass spectrometry for screening of xenobiotics in biological fluids: Opportunities and challenges for data processing. *Talanta* **2020**, *211*, 120747. [CrossRef] [PubMed]
5. Khani, M.; Amin, N.A.S.; Hosseini, S.N.; Heidarrezaei, M. Kinetics study of the photocatalytic inactivation of Escherichia coli. *Int. J. Nano Biomater.* **2016**, *6*, 139–150. [CrossRef]
6. Pehlivanoglu-Mantas, E.; Insel, G.; Karahan, O.; Cokgor, E.U.; Orhon, D. Case studies from Turkey: Xenobiotic-containing industries, wastewater treatment and modeling. *Water Air Soil Pollut. Focus* **2008**, *8*, 519–528. [CrossRef]
7. Agrawal, N.; Shahi, S.K. An environmental cleanup strategy-microbial transformation of xenobiotic compounds. *Int. J. Curr. Microbiol. Appl. Sci.* **2015**, *4*, 429–461.
8. Wu, R.S.S. Eutrophication, water borne pathogens and xenobiotic compounds: Environmental risks and challenges. *Mar. Pollut. Bull.* **1999**, *39*, 11–22. [CrossRef]
9. George, S.; Riley, C.; McEvoy, J.; Wright, J. Development of a fish in vitro cell culture model to investigate oxidative stress and its modulation by dietary vitamin E. *Mar. Environ. Res.* **2000**, *50*, 541–544. [CrossRef]
10. Johnson, G.R.; Spain, J.C. Evolution of catabolic pathways for synthetic compounds: Bacterial pathways for degradation of 2, 4-dinitrotoluene and nitrobenzene. *Appl. Microbiol. Biotechnol.* **2003**, *62*, 110–123. [CrossRef]
11. Murphy, C.D. Biodegradation and biotransformation of organofluorine compounds. *Biotechnol. Lett.* **2010**, *32*, 351–359. [CrossRef]
12. Kiel, M.; Engesser, K.-H. The biodegradation vs. biotransformation of fluorosubstituted aromatics. *Appl. Microbiol. Biotechnol.* **2015**, *99*, 7433–7464. [CrossRef] [PubMed]
13. Barragan-Huerta, B.E.; Costa-Pérez, C.; Peralta-Cruz, J.; Barrera-Cortés, J.; Esparza-García, F.; Rodríguez-Vázquez, R. Biodegradation of organochlorine pesticides by bacteria grown in microniches of the porous structure of green bean coffee. *Int. Biodeterior. Biodegrad.* **2007**, *59*, 239–244. [CrossRef]
14. Kong, L.; Zhu, S.; Zhu, L.; Xie, H.; Su, K.; Yan, T.; Wang, J.; Wang, J.; Wang, F.; Sun, F. Biodegradation of organochlorine pesticide endosulfan by bacterial strain Alcaligenes faecalis JBW4. *J. Environ. Sci.* **2013**, *25*, 2257–2264. [CrossRef]
15. Waaijers, S.L.; Parsons, J.R. Biodegradation of brominated and organophosphorus flame retardants. *Curr. Opin. Biotechnol.* **2016**, *38*, 14–23. [CrossRef] [PubMed]
16. Holman, H.Y.N.; Nieman, K.; Sorensen, D.L.; Miller, C.D.; Martin, M.C.; Borch, T.; McKinney, W.R.; Sims, R.C. Catalysis of PAH biodegradation by humic acid shown in synchrotron infrared studies. *Environ. Sci. Technol.* **2002**, *36*, 1276–1280. [CrossRef] [PubMed]
17. Heidarrezaei, M. *Degradation of Trichloroacetic Acid (TCA) by Bacteria Isolated from Marine Environment*; Universiti Teknologi Malaysia: Johor, Malaysia, 2013.
18. Horvath, A.L.; Getzen, F.W.; Maczynska, Z. IUPAC-NIST solubility data series 67. Halogenated ethanes and ethenes with water. *J. Phys. Chem. Ref. Data* **1999**, *28*, 395–627. [CrossRef]
19. Raucy, J.L.; Kraner, J.C.; Lasker, J.M. Bioactivation of halogenated hydrocarbons by cytochrome P4502E1. *Crit. Rev. Toxicol.* **1993**, *23*, 1–20. [CrossRef]
20. Wong, W.-Y.; Huyop, F. Molecular identification and characterization of Dalapon-2, 2-dichloropropionate (2, 2DCP)-degrading bacteria from a Rubber Estate Agricultural area. *Afr. J. Microbiol. Res.* **2012**, *6*, 1520–1526.
21. Fetzner, S. Bacterial dehalogenation. *Appl. Microbiol. Biotechnol.* **1998**, *50*, 633–657. [CrossRef]
22. van den Wijngaard, A.J.; Wind, R.D.; Janssen, D.B. Kinetics of bacterial growth on chlorinated aliphatic compounds. *Appl. Environ. Microbiol.* **1993**, *59*, 2041–2048. [CrossRef]

23. Janssen, D.B.; Poelarends, G.J.; Rink, R. Engineering hydrolases for the conversion of halogenated aliphatic hydrocarbons and epoxides. In *Novel Approaches for Bioremediation of Organic Pollution*; Springer: Boston, MA, USA, 1999; pp. 105–116.

24. Frank, H.; Scholl, H.; Renschen, D.; Rether, B.; Laouedj, A.; Norokorpi, Y. Haloacetic acids, phytotoxic secondary air pollutants. *Environ. Sci. Pollut. Res.* **1994**, *1*, 4–14. [CrossRef] [PubMed]

25. Aziz, K.H.H. Application of different advanced oxidation processes for the removal of chloroacetic acids using a planar falling film reactor. *Chemosphere* **2019**, *228*, 377–383. [CrossRef] [PubMed]

26. Zhang, P.; Hozalski, R.M.; Leach, L.H.; Camper, A.K.; Goslan, E.H.; Parsons, S.A.; Xie, Y.F.; LaPara, T.M. Isolation and characterization of haloacetic acid-degrading Afipia spp. from drinking water. *FEMS Microbiol. Lett.* **2009**, *297*, 203–208. [CrossRef]

27. Mourão, A.O.; Silva, D.F.; Rodriguez, M.; Torres, T.S.; Franco, E.S.; Pádua, V.L.; da Silva Faria, M.C.; Maia, L.F.O.; Rodrigues, J.L. Degradation of haloacetic acids with the Fenton-like and analysis by GC-MS: Use of bioassays for monitoring of genotoxic, mutagenic and cytotoxic effects. *Environ. Monit. Assess.* **2019**, *191*, 513. [CrossRef] [PubMed]

28. Ma, Y.-C.; Chiang, C.-Y. Evaluation of the effects of various gas chromatographic parameters on haloacetic acids disinfection by-products analysis. *J. Chromatogr. A* **2005**, *1076*, 216–219. [CrossRef]

29. Bonner, O.D.; Prichard, P.R. The ionization of trichloroacetic acid and evidence for an unusual type of ion pairing. *J. Solut. Chem.* **1979**, *8*, 113. [CrossRef]

30. Terzi, E.; Guvenc, U.; Türsen, B.; Kaya, T.İ.; Erdem, T.; Türsen, Ü. The effectiveness of matrix cauterization with trichloroacetic acid in the treatment of ingrown toenails. *Indian Dermatol. Online J.* **2015**, *6*, 4. [PubMed]

31. Pereira, M.A.; Kramer, P.M.; Conran, P.B.; Tao, L. Effect of chloroform on dichloroacetic acid and trichloroacetic acid-induced hypomethylation and expression of the c-myc gene and on their promotion of liver and kidney tumors in mice. *Carcinogenesis* **2001**, *22*, 1511–1519. [CrossRef] [PubMed]

32. Kadmi, Y.; Favier, L.; Ionut, S.A.; Matei, E.; Wolbert, D. Improved Determination of Dichloroacetic and Trichloroacetic Acids in Water by Solid Phase Extraction Followed by Ultra-high Performance Liquid Chromatography–Tandem Mass Spectrometry. *Anal. Lett.* **2016**, *49*, 433–443. [CrossRef]

33. Pinna, M.V.; Baronti, S.; Miglietta, F.; Pusino, A. Photooxidation of foramsulfuron: Effects of char substances. *J. Photochem. Photobiol. A Chem.* **2016**, *326*, 16–20. [CrossRef]

34. Oyewusi, H.A.; Wahab, R.A.; Kaya, Y.; Edbeib, M.F.; Huyop, F. Alternative bioremediation agents against haloacids, haloacetates and chlorpyrifos using novel halogen-degrading bacterial isolates from the Hypersaline Lake Tuz. *Catalysts* **2020**, *10*, 651. [CrossRef]

35. Alrawas, A.A.; Huyop, F. Biodegradation of Trichloroacetic acid by Ralstonia sp.(TCAA-2) strain isolated from Danga Bay coastal area in Malaysia. *J. Biotechnol. Sci. Res.* **2015**, *2*, 103–108.

36. Guengerich, F.P. Effects of nutritive factors on metabolic processes involving bioactivation and detoxication of chemicals. *Annu. Rev. Nutr.* **1984**, *4*, 207–231. [CrossRef] [PubMed]

37. Ismail, S.N.F.; Wahab, R.A.; Huyop, F. Microbial isolation and degradation of selected haloalkanoic aliphatic acids by locally isolated bacteria: A review. *Malays. J. Microbiol.* **2017**, *13*, 261–272.

38. Slater, J.H. Microbial dehalogenation of haloaliphatic compounds. In *Biochemistry of Microbial Degradation*; Springer: Dordrecht, the Netherlands, 1994; pp. 379–421.

39. Satpathy, R.; Konkimalla, V.B.; Ratha, J. In silico phylogenetic analysis and molecular modelling study of 2-haloalkanoic acid dehalogenase enzymes from bacterial and fungal origin. *Adv. Bioinform.* **2016**, *2016*. [CrossRef]

40. Fabritz, S.; Maaß, F.; Avrutina, O.; Heiseler, T.; Steinmann, B.; Kolmar, H. A sensitive method for rapid detection of alkyl halides and dehalogenase activity using a multistep enzyme assay. *AMB Express* **2012**, *2*, 51. [CrossRef]

41. Harvey, S.P. *Enzymatic Degradation of HD*; U.S. Army Edgewood Chemical Biological Center (ECBC): Gunpower, MD, USA, 2001.

42. Muslem, W.H.; Edbeib, M.F.; Aksoy, H.M.; Kaya, Y.; Hamid, A.A.A.; Hood, M.H.M.; Wahab, R.A.; Huyop, F. Biodegradation of 3-chloropropionic acid (3-CP) by Bacillus cereus WH2 and its in silico enzyme-substrate docking analysis. *J. Biomol. Struct. Dyn.* **2019**, 1–10. [CrossRef]

43. Hamid, A.A.A.; Tengku Abdul Hamid, T.H.; Wahab, R.A.; Huyop, F. Identification of functional residues essential for dehalogenation by the non-stereospecific α-haloalkanoic acid dehalogenase from Rhizobium sp. RC1. *J. Basic Microbiol.* **2015**, *55*, 324–330. [CrossRef]

44. Adamu, A.; Wahab, R.A.; Aliyu, F.; Aminu, A.H.; Hamza, M.M.; Huyop, F. Haloacid dehalogenases of Rhizobium sp. and related enzymes: Catalytic properties and mechanistic analysis. *Process Biochem.* **2020**, *92*, 437–446. [CrossRef]

45. Selvamani, S.; Ab Wahab, R.; Huyop, F. A novel putative non-ligninolytic dehalogenase activity for 3-chloropropionic acid (3CP) utilization by Trichoderma asperellum strain SD1. *Malays. J. Microbiol.* **2015**, *11*, 265–272. [CrossRef]

46. Bagherbaigi, S.; Gicana, R.G.; Lamis, R.J.; Nemati, M.; Huyop, F. Characterisation of Arthrobacter sp. S1 that can degrade α and β-haloalkanoic acids isolated from contaminated soil. *Ann. Microbiol.* **2013**, *63*, 1363–1369. [CrossRef]

47. Hareland, W.A.; Crawford, R.L.; Chapman, P.J.; Dagley, S. Metabolic function and properties of 4-hydroxyphenylacetic acid 1-hydroxylase from Pseudomonas acidovorans. *J. Bacteriol.* **1975**, *121*, 272–285. [CrossRef] [PubMed]

48. Leisinger, T.; Bader, R. Microbial dehalogenation of synthetic organohalogen compounds: Hydrolytic dehalogenases. *Chim. Int. J. Chem.* **1993**, *47*, 116–121.

49. Bhatt, P.; Kumar, M.S.; Mudliar, S.; Chakrabarti, T. Biodegradation of chlorinated compounds—A review. *Crit. Rev. Environ. Sci. Technol.* **2007**, *37*, 165–198. [CrossRef]

50. Slater, J.H.; Bull, A.T.; Hardman, D.J. Microbial dehalogenation. *Biodegradation* **1995**, *6*, 181–189. [CrossRef]

51. Slater, J.H.; Bull, A.T.; Hardman, D.J. Microbial dehalogenation of halogenated alkanoic acids, alcohols and alkanes. *Adv. Microb. Physiol.* **1996**, *38*, 133–176.

52. Nichele, L.; Persichetti, V.; Lucidi, M.; Cincotti, G. Quantitative evaluation of ImageJ thresholding algorithms for microbial cell counting. *OSA Contin.* **2020**, *3*, 1417–1427. [CrossRef]

53. Salim, A.A.; Bidin, N.; Lafi, A.S.; Huyop, F.Z. Antibacterial activity of PLAL synthesized nanocinnamon. *Mater. Des.* **2017**, *132*, 486–495. [CrossRef]

54. Mesri, S.; Wahab, R.A.; Huyop, F. Degradation of 3-chloropropionic acid (3CP) by Pseudomonas sp. B6P isolated from a rice paddy field. *Ann. Microbiol.* **2009**, *59*, 447–451. [CrossRef]

55. Cappuccino, J.G.; Sherman, N. *Microbiology: A Laboratory Manual*; Pearson/Benjamin Cummings: San Francisco, CA, USA, 2005.

56. Holt, J.G.; Krieg, N.R.; Sneath, P.H.A. *Bergey's Manual of Determinative Bacterology*; Springer: New York, NY, USA, 1994.

57. Reddy, N.G.; Ramakrishna, D.P.N.; Raja Gopal, S.V. A morphological, physiological and biochemical studies of marine Streptomyces rochei (MTCC 10109) showing antagonistic activity against selective human pathogenic microorganisms. *Asian J. Biol. Sci.* **2011**, *4*, 1–14. [CrossRef]

58. Fulton, C.K.; Cooper, R.A. Catabolism of sulfamate by Mycobacterium sp. CF1. *Environ. Microbiol.* **2005**, *7*, 378–381. [CrossRef] [PubMed]

59. Tamura, K.; Peterson, D.; Peterson, N.; Stecher, G.; Nei, M.; Kumar, S. MEGA5: Molecular evolutionary genetics analysis using maximum likelihood, evolutionary distance, and maximum parsimony methods. *Mol. Biol. Evol.* **2011**, *28*, 2731–2739. [CrossRef] [PubMed]

60. Saitou, N.; Nei, M. The neighbor-joining method: A new method for reconstructing phylogenetic trees. *Mol. Biol. Evol.* **1987**, *4*, 406–425. [PubMed]

61. Bergmann, J.G.; Sanik, J. Determination of trace amounts of chlorine in naphtha. *Anal. Chem.* **1957**, *29*, 241–243. [CrossRef]

62. Horisaki, T.; Yoshida, E.; Sumiya, K.; Takemura, T.; Yamane, H.; Nojiri, H. Isolation and characterization of monochloroacetic acid-degrading bacteria. *J. Gen. Appl. Microbiol.* **2011**, *57*, 277–284. [CrossRef]

63. Zulkifly, A.H.; Roslan, D.D.; Hamid, A.A.A.; Hamdan, S.; Huyop, F. Biodegradation of low concentration of monochloroacetic acid-degrading bacillus sp. TW 1 Isolated from Terengganu Water Treatment and Distribution Plant. *J. Appl. Sci.* **2010**, *10*, 2940–2944.

64. Vasileva, E.; Petrov, K.; Beschkov, V. Fed batch strategy for biodegradation of monochloroacetic acid by immobilized Xantobacter autotrophicus GJ10 in polyacrylamide gel. *C. R. L'Academie Bulg. Des. Sci.* **2009**, *62*, 1241–1246.

65. Vasileva, E.; Petrov, K.; Beschkov, V. Biodegradation of monochloroacetic acid by immobilization of xanthobacter autotrophicus gj10 in polyacrylamide gel. *Biotechnol. Biotechnol. Equip.* **2009**, *23*, 788–790. [CrossRef]

66. Ismail, S.N.; Taha, A.M.; Jing, N.H.; Wahab, R.A.; Hamid, A.A.; Pakingking, R.V., Jr.; Huyop, F. Biodegradation of monochloroacetic acid by a presumptive Pseudomonas sp. strain R1 bacterium isolated from Malaysian paddy (Rice) field. *Biotechnology* **2008**, *7*, 481–486. [CrossRef]
67. Meusel, M.; Rehm, H.-J. Biodegradation of dichloroacetic acid by freely suspended and adsorptive immobilized Xanthobacter autotrophicus GJ10 in soil. *Appl. Microbiol. Biotechnol.* **1993**, *40*, 165–171. [CrossRef]
68. Pang, B.C.M.; Tsang, J.S.H. Mutagenic analysis of the conserved residues in dehalogenase IVa of Burkholderia cepacia MBA4. *FEMS Microbiol. Lett.* **2001**, *204*, 135–140. [CrossRef] [PubMed]
69. Weightman, A.L.; Weightman, A.J.; Slater, J.H. Microbial dehalogenation of trichloroacetic acid. *World J. Microbiol. Biotechnol.* **1992**, *8*, 512–518. [CrossRef] [PubMed]
70. Jayasuriya, G.C.N. The isolation and characteristics of an oxalate-decomposing organism. *Microbiology* **1955**, *12*, 419–428. [CrossRef] [PubMed]

© 2020 by the authors. Licensee MDPI, Basel, Switzerland. This article is an open access article distributed under the terms and conditions of the Creative Commons Attribution (CC BY) license (http://creativecommons.org/licenses/by/4.0/).

Article

The Use of MgO Obtained from Serpentinite in the Synthesis of a Magnesium Potassium Phosphate Matrix for Radioactive Waste Immobilization

Svetlana A. Kulikova [1,*], Sergey E. Vinokurov [1], Ruslan K. Khamizov [1], Natal'ya S. Vlasovskikh [2], Kseniya Y. Belova [1], Rustam K. Dzhenloda [1], Magomet A. Konov [2] and Boris F. Myasoedov [1]

[1] Vernadsky Institute of Geochemistry and Analytical Chemistry, Russian Academy of Sciences, 19 Kosygin st., 119991 Moscow, Russia; vinokurov.geokhi@gmail.com (S.E.V.); khamiz@mail.ru (R.K.K.); ksysha_3350@mail.ru (K.Y.B.); dzhenloda@gmail.com (R.K.D.); bfmyas@mail.ru (B.F.M.)

[2] JSC Scientific Production Enterprise "Radiy", 28 Chasovaya st., 125315 Moscow, Russia; newchemtechnology@mail.ru (N.S.V.); info@npp-radiy.ru (M.A.K.)

* Correspondence: kulikova.sveta92@mail.ru; Tel.: +7-495-939-7007

Featured Application: MgO is used for the synthesis of a magnesium potassium phosphate matrix as a material for immobilization of radioactive waste in order to ensure radiation safety during storage or disposal of waste.

Abstract: Magnesium oxide is a necessary binding agent for the synthesis of a magnesium potassium phosphate (MPP) matrix based on $MgKPO_4 \times 6H_2O$, which is promising for the solidification of radioactive waste (RW) on an industrial scale. The performed research is devoted to finding a cost-effective approach to the synthesis of MPP matrix by using MgO with an optimal ratio of the quality of the binding agent and the cost of its production. A method for obtaining MgO from the widely available natural mineral serpentinite was proposed. The phase composition, particle morphology, and granulometric composition of MgO were studied. It was found that the obtained MgO sample, in addition to the target periclase phase, also contains impurities of brucite and hydromagnesite; however, after calcining at 1300 °C for 3 h, MgO transforms into a monophase state with a periclase structure with an average crystallite size of 62 nm. The aggregate size of the calcined MgO powder in an aqueous medium was about 55 μm (about 30 μm after ultrasonic dispersion), and the specific surface area was 5.4 m^2/g. This powder was used to prepare samples of the MPP matrix, the compressive strength of which was about 6 MPa. The high hydrolytic stability of the MPP matrix was shown: the differential leaching rate of magnesium, potassium, and phosphorus from the sample on the 91st day of its contact with water does not exceed 1.6×10^{-5}, 4.7×10^{-4} 8.9×10^{-5} g/(cm^2·day), respectively. Thus, it was confirmed that the obtained MPP matrix possesses the necessary quality indicators for RW immobilization.

Keywords: serpentinite; magnesium oxide; calcination; particle size distribution; specific surface area; magnesium potassium phosphate matrix; radioactive waste; immobilization; hydrolytic stability; strength

Citation: Kulikova, S.A.; Vinokurov, S.E.; Khamizov, R.K.; Vlasovskikh, N.S.; Belova, K.Y.; Dzhenloda, R.K.; Konov, M.A.; Myasoedov, B.F. The Use of MgO Obtained from Serpentinite in the Synthesis of a Magnesium Potassium Phosphate Matrix for Radioactive Waste Immobilization. *Appl. Sci.* **2021**, *11*, 220. https://doi.org/10.3390/app11010220

Received: 2 December 2020
Accepted: 23 December 2020
Published: 28 December 2020

Publisher's Note: MDPI stays neutral with regard to jurisdictional claims in published maps and institutional affiliations.

Copyright: © 2020 by the authors. Licensee MDPI, Basel, Switzerland. This article is an open access article distributed under the terms and conditions of the Creative Commons Attribution (CC BY) license (https://creativecommons.org/licenses/by/4.0/).

1. Introduction

Industrial activities associated with the production and use of materials containing radioactive substances inevitably lead to the generation of radioactive waste (RW) of various activity levels. The largest amount of RW is generated during the operation of nuclear fuel cycle enterprises and exploitation of nuclear power reactors of various purposes. In some countries, including the USA, Sweden, and Finland, spent nuclear fuel (SNF) of nuclear power reactors is classified as RW and is stored without reprocessing. In other countries, including Russia, France, and Japan, SNF is subject to reprocessing for the purpose of extraction of uranium and plutonium for reuse, as well as a number of valuable components from fission products and actinides, and the residue part is considered to be RW. In case of uncontrolled spread of radioactive substances, they have a negative impact on humans and environmental objects. Therefore, solving the problem of RW management in order to ensure their reliable isolation from the human environment is key for the further development of nuclear energy and industry. For controlled storage and/or final disposal of RW, it should be converted to a stable solidified form using stable matrices.

Earlier, we showed in [1–5] that magnesium potassium phosphate (MPP) matrix $MgKPO_4 \times 6H_2O$ is an effective and multipurpose mineral-like material for immobilization of different RW, and it possesses the number of benefits over cement and glass-like compounds. Therefore, MPP matrix was investigated for solidification of liquid intermediate level waste (ILW) [1], high level waste (HLW) [2,3], and also RW containing radiocarbon ^{14}C [4] and spent electrolyte [5], obtained as a result of pyrochemical reprocessing of mixed nitride uranium-plutonium SNF. This matrix is obtained by the acid-base reaction (1) of magnesium oxide (MgO) with potassium dihydrogen phosphate (KH_2PO_4) in an aqueous solution at room temperature, and it is an analog of the natural mineral K-struvite [6].

$$MgO + KH_2PO_4 + 5H_2O \rightarrow MgKPO_4 \times 6H_2O \tag{1}$$

Magnesium oxide is a necessary binding agent for reaction (1) for the synthesis of MPP matrix; it is usually produced by calcination of carbonate minerals—magnesite ($MgCO_3$) [7] and dolomite ($CaMg(CO_3)_2$) [8]. It is widely used to obtain the refractories in the production of steel and cement (70–80% of world consumption) in electrical engineering, agriculture, for wastewater treatment, and gas absorption. Magnesium oxide on the market with a purity of at least 99 wt% has a high cost—up to $5000 per ton. At the same time, it is obvious that for the competitiveness of the technology of RW solidification using the MPP matrix, cheaper raw materials should be used; for example, so that the cost of MgO is at the level of the cost of Portland cement ($300–400 per ton). For this reason, the aim of the study was to find a cost-effective approach to synthesizing the MPP matrix through the use of MgO with low production cost. In this regard, a natural mineral, serpentinite ($Mg_3Si_2O_5(OH)_4$), which contains 32–38% of MgO [9], is of special interest. There are various technologies for reprocessing of serpentinite, primarily methods using mineral acids: sulfuric acid [10], nitric acid [9], and hydrochloric acid [11].

Earlier in [12], while testing several commercial MgO samples obtained from magnesite, we recommended the use of MgO powder with an average particle size of about 50 μm, which has a high degree of crystallinity (the average crystallite size is not less than 40 nm), without impure readily soluble magnesium phases (first of all, magnesium hydroxide) to obtain a homogeneous compound based on MPP matrix with a high compressive strength (up to 15 MPa). The specific surface of the conditioned MgO powder was no higher than $7 \text{ m}^2/\text{g}$ [1]. It was also noted that the impurities in MgO of metal compounds, primarily silicon, calcium, and iron, do not significantly influence the synthesis conditions and the mechanical strength of the compound.

This article presents the results of studying the characteristics of MgO powder obtained from serpentinite, as well as determination of composition, mechanical strength, and hydrolytic stability of the prepared MPP matrix samples.

2. Materials and Methods

2.1. Obtaining MgO from Serpentinite

The flow chart for obtaining MgO from serpentinite is presented in Figure 1. In this work, an average sample of serpentinite from the "Bedenskoye" deposit, having a composition presented in Table 1, was used as a source of magnesium oxide.

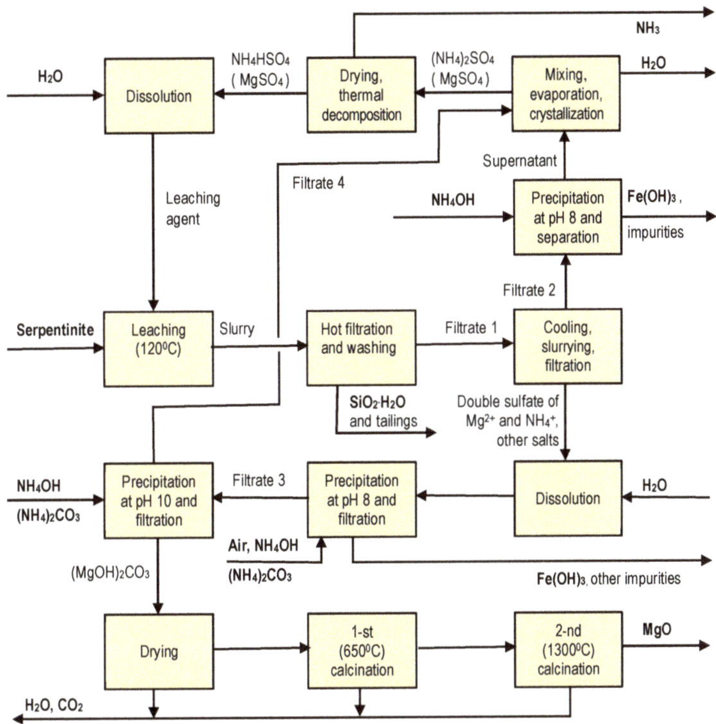

Figure 1. Flow chart for obtaining MgO from serpentinite.

Table 1. Chemical composition of serpentinite.

Content	MgO	SiO$_2$	Fe$_3$O$_4$	CaO	NiO	Al$_2$O$_3$	Cr$_2$O$_3$	MnO	SO$_3$	K$_2$O	Na$_2$O	LOI *
(wt%)	38.05	39.96	8.55	0.63	0.36	0.33	0.32	0.11	0.06	0.02	0.01	11.30

* Loss on Ignition.

Several repeated experiments were carried out to decompose serpentinite and remove magnesium hydroxide. In each of them, 100 g of finely ground serpentinite powder (with a grain boundary size d < 45 μm), obtained in a ball mill with ceramic balls, were placed in a laboratory autoclave with a mechanical stirrer, and 650 g (520 mL) of circulating solution containing 35% NaHSO$_4$, up to 7% MgSO$_4$ and up to 10% (NH$_4$)$_2$SO$_4$ were added. The mixture was stirred for 4 h at 120 °C. The obtained hot suspension was filtered under reduced pressure created by a water-jet pump on a Buchner filter with a thermostatic shirt at 95 °C to obtain filtrate No. 1, and then at the same temperature, the filter precipitate was washed by 250 mL of hot deionized water. Filtrate No. 1 and wash water were combined, and 912–925 g (730–740 mL) of mixed solution, containing no less than 3.4% of Mg(II) and no more than 0.08% of Fe(II, III), were obtained in different experiments that corresponded to the degree of magnesium removal—no less than 80%, and iron—up to 11.5%. The solution was placed in a cold-storage, where it was cooled to +4 °C and kept there for 2 h. As a result, a suspension was obtained, which was filtered on a thermostatted Buechner filter at 5 °C. Filtrate No. 2 (no more than 490 g with up to 0.5% magnesium and 0.1% iron) and a glassy precipitate with a pale yellowish-pink color were obtained. Chemical analysis of such precipitate (no less than 370 g), dried at 105 °C for 1 h until the loss of more than 10% moisture, showed a magnesium content of 6.5% and a molar ratio of Mg^{2+}:NH$_4$$^+$:SO$_4$$^{2-}$ = 1:2:2 that corresponded to the double salt—magnesium ammonium sulfate (Boussingaultite). The iron content in the dry sediment did not exceed 0.06%.

The precipitate was dissolved in 650 mL of hot deionized water (95 °C), and a 25% ammonia solution (purity qualification "analytical grade") was added to the obtained solution up to pH of 8, and 2 g of ammonium carbonate, $(NH_4)_2CO_3$ ("reagent grade"), was also added. A hot suspension of light green color was obtained, which was placed in an autoclave, in which the temperature was maintained at 95 °C. Air was passed through the suspension, adjusting the outlet of gases from the autoclave so that an excess pressure is created in it. The mixture was kept for 4 h with stirring. A light brown suspension was discharged from the autoclave, which was filtered through hot thermostatted Buchner filter with a water-jet pump using a filter with pore size 1–2.5 nm. The resulting filtrate No. 3 was again placed in an autoclave, previously washed with deionized water, and then filtrate No. 3 was cooled to room temperature. 80 mL of the solution of 25% ammonia were poured into the contents of the autoclave, and 45 g of ammonium carbonate were added. The autoclave was closed, the temperature was raised up to 65 °C, and the mixture was stirred under excess pressure for 4 h. A suspension of white color with a bluish sheen was obtained, which was filtered through a thermostatted Buchner filter using a paper filter with pore size 1–2.5 nm. Filtrate No. 4 and precipitate of magnesium hydroxide-carbonate were obtained. The precipitate was dried at 120 °C and then subjected to preliminary calcination at a temperature of 650 °C for 3 h.

Filtrate No. 2 was heated up to 50 °C, a 25% ammonia solution was added to it to pH 8, and the mixture was stirred for 5 h on a magnetic stirrer with heating. A finely dispersed light brown suspension was formed, which was separated in a laboratory centrifuge to obtain a supernatant.

The supernatant and filtrate No. 4 were mixed, and the resulting solution was subjected to vacuum evaporation on a Buchi R-124 rotary evaporator (Bunker Lake Blvd., Ramsey, MN, USA) at 85 °C and a pressure of 0.15 bar until a wet crystalline mass was obtained. This mass was placed in a muffle furnace and first dried at 120 °C for 1 h; then the temperature was raised to 310 °C and decomposition was carried out for another 7 h. In repeated experiments, at least 335 g of precipitate were obtained, to which 320 mL of deionized water and 5 g of ammonium sulfate were added after cooling. A recycled solution was obtained with a content of at least 35% $NaHSO_4$, up to 7% $MgSO_4$, and up to 10% $(NH_4)_2SO_4$, which was used to decompose the next portion of serpentinite.

In each of the consecutive carried out experiments, at least 30 g of the preproduct was obtained. The samples of magnesium oxide obtained in the course of six successive experiments were combined, and an average composition of precalcined magnesium oxide was obtained, in which the content of iron and manganese oxides did not exceed 0.01 wt%.

2.2. Preparation of the MPP Matrix

The synthesis of the MPP matrix was carried out according to reaction (1) at the MgO: H_2O:KH_2PO_4 weight ratio of 1:2:3. Previously in studies [1,13–16], it was shown that to reduce the rate of reaction (1) and, accordingly, to ensure a technologically acceptable setting time of the mixture for the purpose of high-quality mixing and tight packing of the resulting mixture into containers for subsequent storage, MgO powder should be used after preliminary heat treatment at 1300–1500 °C. Thus, MgO obtained from serpentinite in accordance with the method in Section 2.1 and precalcined at temperatures of 1300 °C for 3 h (hereinafter referred to as calcined MgO) in a muffle furnace (SNOL 30/1300, AB UMEGA GROUP, Utena, Lithuania) and KH_2PO_4 ("Rushim" LLC, Moscow, Russia) crushed to a particle size of 0.15–0.25 mm were used for synthesis of MPP matrix. The excess of MgO in relation to the stoichiometry of reaction (1) was 10 wt% [1]. To decrease the rate of reaction (1), boric acid was added to the initial mixture in an amount corresponding to its 1.5 wt% content in the sample. The obtained mixture was placed in PTFE molds.

Cubic samples of the MPP matrix with dimensions of $2 \times 2 \times 2$ cm^3 were prepared and kept for at least 15 days to cure at ambient atmospheric conditions.

2.3. Investigation of MgO and MPP Matrix Samples

The phase composition of MgO and MPP matrix samples was determined by X-ray diffraction (XRD) method using an Ultima-IV X-ray diffractometer (Rigaku, Tokyo, Japan). The XRD data were interpreted using the Jade 6.5 software package (MDI, Livermore, CA, USA) with PDF-2 powder database. The average crystallite size of the studied MgO samples was calculated by the Scherrer [12]. The composition of MgO was determined using the Rietveld method [17], with a PROFEX GUI software package for BGMN [18].

The microstructure of MgO and MPP matrix samples was investigated by the scanning electron microscopy (SEM) using a JSM-6700F (Jeol, Tokyo, Japan) and Vega 3 (Tescan, Brno, Czech Republic) microscopes; the electron probe microanalysis of the samples was performed by energy-dispersive X-ray spectroscopy (EDS) using an X-ACT analyzer (Oxford Inst., High Wycombe, UK).

The elemental composition of MgO powder was studied using an Axious Advanced PW 4400/04 X-ray fluorescence (XRF) spectrometer (PANalytical B.V., Almelo, Netherlands).

Adsorption measurements of MgO powder samples were carried out in an ASAP 2000 automatic sorbtometer (Micromeritics, Norcross, GA, USA); the specific surface area was calculated using the Micromeritics software package.

The particle size distribution of MgO samples was determined using a SALD-7500 nano laser diffraction granulometer (Shimadzu, Kyoto, Japan), including the use of 60 W ultrasound for 5 min.

The compressive strength of MPP matrix samples was determined using a Cybertronic 500/50 kN test machine (Testing Bluhm & Feuerherdt GmbH, Germany). At least three compound samples were used in experiment.

The hydrolytic stability of MPP matrix samples was determined in accordance with the semi-dynamic standard test GOST R 52126-2003 [19]. Before leaching, monolithic cubic samples were immersed in ethanol for 5–7 s, then the samples were dried in air for 30 min. Next, the samples were placed in a PTFE container, and double-distilled water was poured in as a leaching agent (pH 6.6 $\pm$ 0.1, volume 100 mL), which was replaced at regular time intervals. The samples were removed from the container at the set time, washed with double-distilled water (volume 100 mL), and combined with the leachate, and the content of the matrix components in the solutions after leaching was determined by ICP–AES (iCAP-6500 Duo, Thermo Scientific, Waltham, MA, USA). The calculations of the differential leaching rate LR_{dif} (g/(cm^2·day)) of the matrix components from samples were made according to Equation (2).

$$LR = \frac{c \cdot V}{S \cdot f \cdot t},\qquad(2)$$

where c—element concentration in solution after leaching, g/L; V—the volume of leaching agent, L; S—the open geometric surface area of the monolithic samples, cm^2; f—element content in matrix, g/g; and t—leaching time, days (for calculating the differential leaching rate t—duration of the n-th leaching period between shifts of contact solution).

Leaching mechanism of MPP matrix components was assessed according to a de Groot and van der Sloot model [20], which can be presented as Equation (3), where values of the coefficient A (slope of the line) correspond to the following mechanisms: <0.35—surface wash-off (or a depletion if it is found in the middle or at the end of the test); 0.35–0.65—diffusion transport; and >0.65—surface dissolution [21]. The calculation of B_i was carried out according to Equation (4).

$$\log(B_i) = A \log(t_n) + const \qquad(3)$$

$$B_i = A_n \cdot \frac{V}{S} \cdot \frac{\sqrt{t_n}}{\left(\sqrt{t_n} - \sqrt{t_{n-1}}\right)} \qquad(4)$$

3. Results and Discussion

3.1. Features of the Obtaining Process of MgO from Serpentinite

The advantage of the bisulfate method of serpentinite reprocessing provides the possibility of creating a closed-loop technological process, where all reagents are recovered. This process was first proposed in [22]. In contrast to the complex chemical formulas used in this work, we introduced simplifications that make it possible to better understand the discussed cyclic process. If we present the formula of serpentinite in the form of a ratio of the main macrocomponents, oxides of magnesium, silicon, and iron, in accordance with the composition presented in Table 1, then the process of decomposition of serpentinite with ammonium bisulfate can be presented as reaction (5).

$$3\,MgO \times 2\,SiO_2 \times 0.04\,Fe_3O_4 + 6.4\,NH_4HSO_4 \rightarrow 2\,SiO_2 \times H_2O{\downarrow} + 3\,Mg(NH_4)_2(SO_4)_2 + $$
$$0.08\,FeNH_4(SO_4)_2 + 0.08\,Fe(NH_4)_2(SO_4)_2 + H_2O + 0.08\,(NH_4)_2SO_4 \tag{5}$$

In reaction (5), in contrast to [22], it was taken into account that in concentrated solutions containing sulfate and ammonium, Mg (II) sulfate and Fe (II) and Fe (III) sulfates form double salts with ammonium sulfate [23,24]. From the mixture obtained in accordance with reaction (5), it is possible to separate the silicic acid phase (as well as the undecomposed part of serpentinite, which takes place under the usually used real conditions), but this separation is possible only at elevated temperature, since the solubility of double magnesium sulfate and ammonium decreases sharply with temperature decreasing. Therefore, the separation in the work presented by us is carried out at temperatures above 90 °C. In this case, a simple cooling operation allows most of the magnesium to be removed from the system in the form of Boussingaultite, reaction (6).

$$3\,Mg(NH_4)_2(SO_4)_2 + 18\,H_2O \rightarrow 3\,Mg(NH_4)_2(SO_4)_2 \times 6H_2O{\downarrow} \tag{6}$$

This process, similar to recrystallization, significantly reduces the problems of further purification of the magnesium product. However, it should be taken into account that all Tutton salts (double sulfates formed by two and singly charged cations) with a monoclinic crystal structure are capable of cocrystallization; therefore, Mohr's salt Fe ($Fe(NH_4)_2(SO_4)_2 \times 6H_2O$), like a similar manganese salt, can pollute the Boussingaultite. That is why at the stage of purification of a magnesium compound not an ammonia solution is used, but a mixture of an ammonia solution and ammonium carbonate, since it is impossible to achieve quantitative precipitation of manganese hydroxide at pH 8. Within the framework of the processes presented for macrocomponents, the purification of double magnesium and ammonium sulfate from iron occurs in the following simple way, reaction (7).

$$0.08\,Fe(NH_4)_2(SO_4)_2 + 0.16\,NH_4OH + 0.04\,H_2O + 0.02\,O_2 \rightarrow 0.08\,Fe(OH)_3{\downarrow} + 0.16\,(NH_4)_2SO_4 \tag{7}$$

The Fe (III) double salt (ferric ammonium alum) predominantly remains in the filtrate after the Boussingaultite is separated. The filtrate is purified from iron by the reaction (8).

$$0.08\,FeNH_4(SO_4)_2 + 0.24\,NH_4OH \rightarrow 0.08\,Fe(OH)_3{\downarrow} + 0.16\,(NH_4)_2SO_4 \tag{8}$$

A very important problem is the problem of magnesium precipitation from double ammonium magnesium sulfate. Magnesium hydroxide is practically impossible to quantitatively precipitate with just ammonia from solutions containing excessive amounts of ammonium sulfate. We used the process of precipitation of a hydroxide-carbonate complex, similar to the analogous process in [22], according to reaction (9).

$$3\,Mg(NH_4)_2(SO_4)_2 + 3\,NH_4OH + 1.5\,(NH_4)_2CO_3 \rightarrow 1.5\,(MgOH)_2CO_3{\downarrow} + 6\,(NH_4)_2SO_4 \tag{9}$$

Reactions (5), (7)–(9) give a total of 6.4 mol of ammonium sulfate. Removal in solid form by evaporation and subsequent heat treatment of ammonium sulfate leads to the production of all reagents necessary for the implementation of the cyclic process, reaction (10).

$$6.4\ (NH_4)_2SO_4 \rightarrow 6.4\ NH_4HSO_4 + 6.4\ NH_3 \tag{10}$$

In particular, when this amount of ammonia is dissolved, 6.4 mol of its aqueous solution can be obtained, of which 3.4 mol isfor carrying out reactions (7)–(9), as well as 3 mol for obtaining 1.5 mol of $(NH_4)_2CO_3$ for reaction (9). A real technological process can also be closed in terms of carbon dioxide, which is released during the calcination of magnesium hydroxide-carbonate obtained in accordance with reaction (9).

While carrying out sequentially repeated laboratory experiments, as can be seen from the description in Section 2.1, a closed process for ammonium bisulfate was carried out. To standardize the conditions for laboratory experiments, the collection of gas components and recuperation of ammonia solution on a small scale were not carried out.

According to our estimates, the cost of the enlarged proposed process for the production of MgO from serpentinite should not exceed $400 per ton, which corresponds to the cost of Portland cement.

3.2. Effect of Calcination of MgO Powder

The obtained X-ray diffraction patterns of MgO powder samples are shown in Figure 2. It was established that the dominant phase in the studied samples is the target phase with the periclase structure, which is identified by the reflexes 2.43, 2.11, and 1.49 Å (Figure 2). The average crystallite size of MgO and calcined MgO was 40 and 62 nm, respectively, which corresponds to the requirements [12]. It should be noted that MgO samples prepared by high-temperature processing do not contain an impurity (Figure 2b) of $Mg(OH)_2$ (brucite) and $Mg_5(CO_3)_4(OH)_2 \times 4H_2O$ (hydromagnesite), which present in amounts about 21 and about 7 wt%, respectively, in the initial MgO sample (Figure 2a). It was previously noted that the presence of such phases in MgO during the synthesis of the MPP matrix is extremely undesirable, because it leads to an unacceptable increase in the rate of reaction (1) and produces an inhomogeneous compound with low strength.

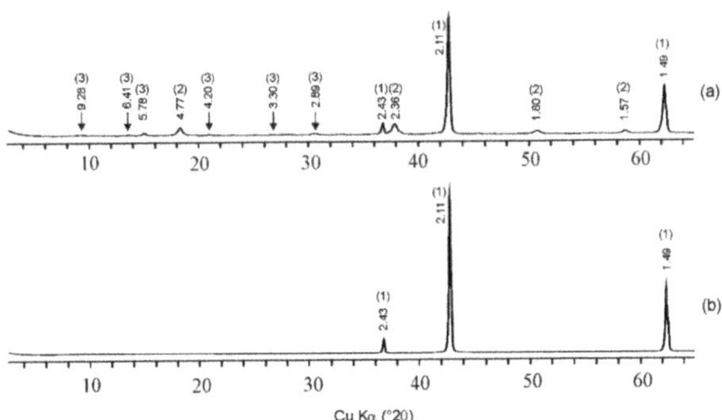

Figure 2. X-ray diffraction patterns of MgO (**a**) and calcined MgO (**b**). 1—MgO (periclase); 2—$Mg(OH)_2$ (brucite), 3—$Mg_5(CO_3)_4(OH)_2 \times 4H_2O$ (hydromagnesite).

According to XRF analysis, the calcined sample of magnesium oxide contains 0.22 wt% impurities (Table 2), that corresponds to the chemical purity of reagent "analytical grade" in accordance with Russian standard [25].

Table 2. Chemical composition of calcined MgO.

Component Content (wt%)	
MgO	**Impurities**
99.78	SiO_2—0.10; CaO—0.08; Fe_2O_3—0.01; MnO—0.02; P_2O_5—0.01

When studying the morphology of MgO powder particles, it was found that the initial sample consists of particles of irregular shape with a size of several to tens of μm (Figure 3a), and the surface structure of this powder presents staggered flake layer, which is typical of $Mg(OH)_2$ (Figure 3b). The morphology of MgO changed from a flake appearance into cubic crystals (Figure 3c) after its calcination; the discovered effect was previously also observed in [16].

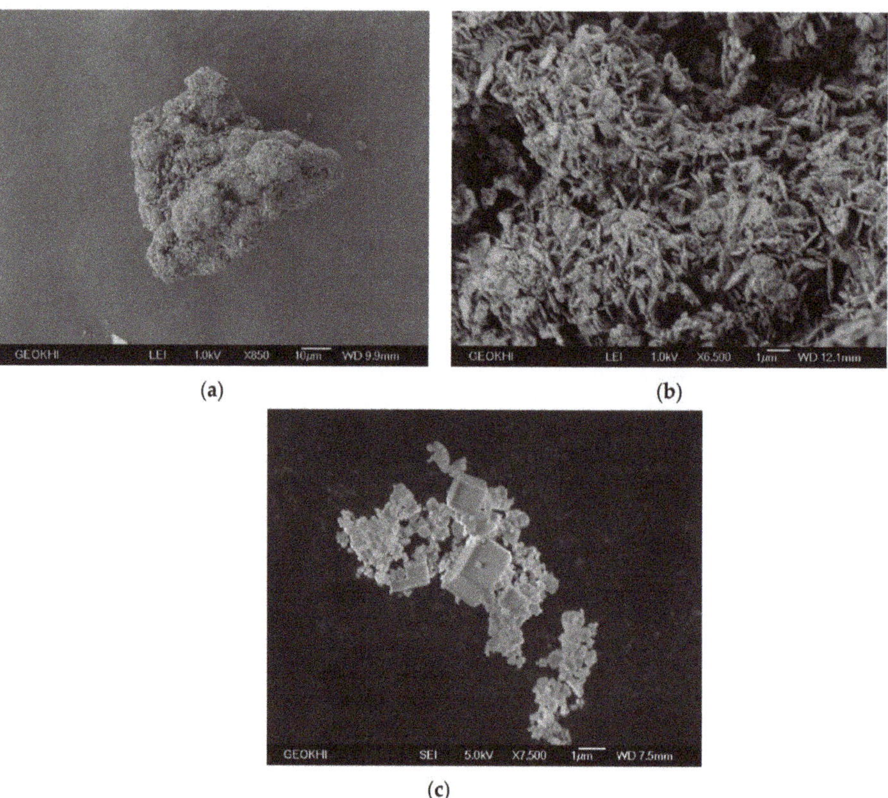

Figure 3. Scanning electron microscope (SEM) images of MgO (**a**,**b**) and calcined MgO (**c**).

The obtained data on the granulometric composition of MgO powders are presented in Figure 4 and in Table 3. The distribution of particle agglomerates by size of the initial and calcined powder can be characterized as monomodal with a value of about 55 μm and a complication in the region of small sizes with values less than 1 μm (Figure 4a) and less than 3 μm (Figure 4c), respectively. The effect of ultrasound on MgO samples leads to the destruction of large agglomerates. For example, as a result of the influence of ultrasound on the initial powder, about 56% of the particles acquire a size of less than 8 μm (Figure 4a,b), and in the case of a calcined powder, about 79%—less than 31 μm (Figure 4c,d). It is noted that, in general, calcination leads to partial agglomeration of MgO particles, for example, to an increase of the number of aggregates with a size of about 100 μm (Figure 4d).

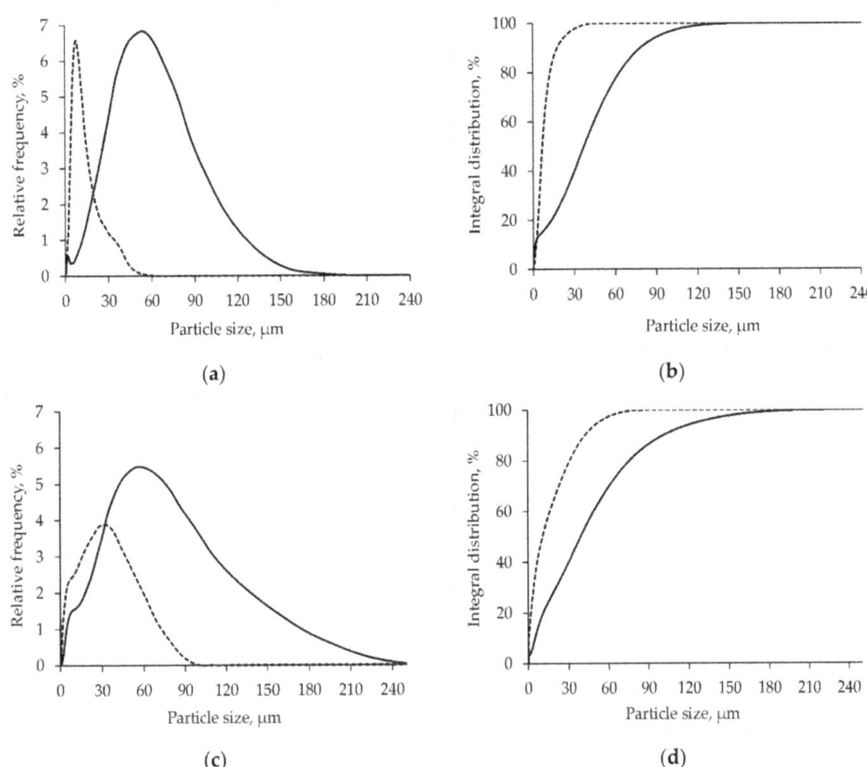

Figure 4. Size distribution of MgO (**a,b**) and calcined MgO (**c,d**) (dashed curve—size distribution after the effects of ultrasound on powders).

Table 3. The results of granulometric analysis of MgO.

Sample	Mode (μm)	Median (μm)	>90% (μm)	<0.1 μm	<1 μm	<10 μm	<100 m
MgO	54.92	37.09	87.51	0.99%	8.04%	18.20%	98.12%
Calcined MgO	54.92	38.92	110.47	0.71%	3.51%	19.77%	92.35%
MgO *	7.58	6.81	19.26	no	3.70%	74.13%	100%
Calcined MgO *	30.68	10.58	43.51	4.73%	14.64%	50.32%	100%

* After exposure by ultrasound on powders.

It was found that the initial MgO powder has a large specific surface area (64.5 m^2/g), apparently due to the flake layer structure. In this case, as a result of calcining MgO powder at 1300 °C for 3 h, the specific surface area of magnesium oxide decreases to 5.4 m^2/g, which corresponds to the previously obtained data [1].

3.3. Study of the Obtained Samples of the MPP Matrix

For the synthesis of the MPP matrix, we used a precalcined MgO powder, the characteristics of which are given in the previous Section. When studying the phase composition of the synthesized samples of the MPP matrix, it was found that their main crystalline phase is the target phase $MgKPO_4 \times 6H_2O$, which is an analog of K-struvite [6] with main reflections at 5.86; 5.56; 5.40; 4.25; 4.13 Å etc.) (Figure 5). The samples also contain phase of MgO (periclase), which is associated with the excess of the used MgO in relation to the stoichiometry of reaction (1).

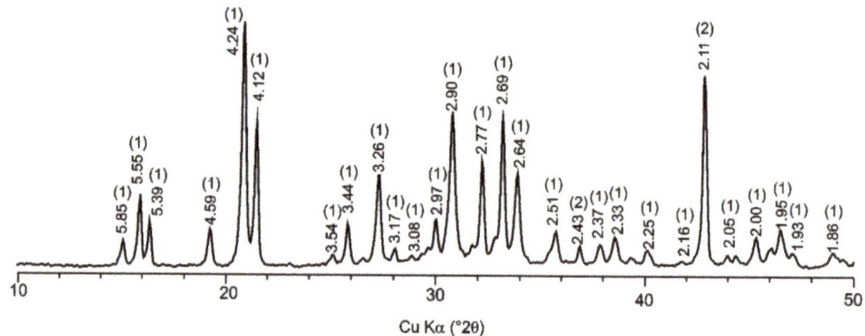

Figure 5. X-ray diffraction pattern of the magnesium potassium phosphate (MPP) matrix. 1—$MgKPO_4 \times 6H_2O$ (K-struvite); 2—MgO (periclase).

The SEM micrograph of the surface of the MPP compound is shown in Figure 5. The elemental composition of the predominant phases in the compound sample includes matrix components of the basic composition $MgKPO_4 \times 6H_2O$ with insignificant variations in the Mg/K ratio, as we noted earlier in [1]. Open pores with a linear size of about 100 μm are also observed (Figure 6).

Figure 6. SEM image of the MPP matrix.

The compressive strength of the MPP matrix obtained using MgO after calcining at 1300 °C for 3 h was 6.19 ± 0.45 MPa, which meets the regulatory requirements for a cement compound: no less than 4.9 MPa [26].

The determination results of hydrolytic stability of MPP compound to the leaching of matrix-forming components are shown in Figure 7a,b. Data in Figure 7a shows that the differential leaching rate of magnesium, potassium, and phosphorus from the compound on the 91st day of contact of the sample with water is 1.6×10^{-5}, 4.7×10^{-4} 8.9×10^{-5} g/(cm² day), respectively. It was found that the leaching of the matrix-forming elements for 91 days of contact of the sample with water is controlled by various mechanisms. Leaching of potassium and phosphorus from the MPP matrix in the first seven days occurs due to its wash-off of from the surface of the compound, followed by depletion of the surface layer (Figure 7b, coefficients A in Equation (3) are for potassium −0.43 and 0.17, and for phosphorus −0.50, 0.04, and −0.66). Leaching of magnesium in the first 10 days occurs due to its wash-off of from the surface of the compound, and in the next 81 days due to diffusion from its inner layers (Figure 7b, −0.81 and 0.48). The obtained results on the hydrolytic stability (rate and mechanism of leaching) of the MPP matrix obtained using MgO obtained from serpentinite are consistent with the previously obtained data for the MPP matrix obtained from commercial MgO [1].

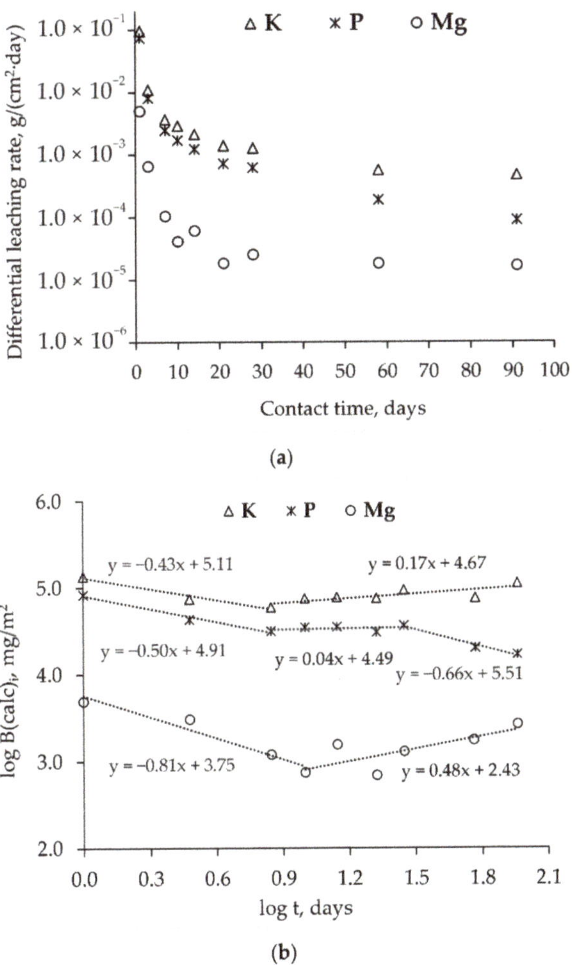

Figure 7. Kinetic curve of the leaching rate (**a**) and logarithmic dependence of the release (**b**) of the matrix components from the MPP matrix.

4. Conclusions

The applicability of MgO obtained during the reprocessing of widely available mineral raw materials—serpentinite by almost waste-free and economically profitable way for the synthesis of the MPP matrix was established. The characteristics of the obtained matrix correspond to the requirements for the material for RW immobilization and for preventing of release of highly toxic radionuclides into the environment.

Author Contributions: Conceptualization: S.A.K., S.E.V., and R.K.K.; methodology: S.A.K., S.E.V., N.S.V., and R.K.K.; validation: S.A.K., S.E.V., and N.S.V.; formal analysis: S.A.K., K.Y.B., and S.E.V.; investigation: S.A.K., N.S.V., K.Y.B., and R.K.D.; resources: R.K.D.; writing—original draft preparation: S.A.K., S.E.V., and R.K.K.; writing—review and editing: M.A.K. and B.F.M.; supervision: S.E.V., R.K.K., M.A.K., and B.F.M.; project administration: S.E.V.; funding acquisition: S.E.V. All authors have read and agreed to the published version of the manuscript.

Funding: This research was funded by the GEOKHI RAS state assignment (0137-2019-0022).

Institutional Review Board Statement: Not applicable.

Informed Consent Statement: Not applicable.

Data Availability Statement: Data sharing not applicable.

Acknowledgments: The authors thank V.V. Krupskaya and I.A. Morozov (Lomonosov Moscow State University; Institute of Geology of Ore Deposits, Petrography, Mineralogy, and Geochemistry of Russian Academy of Sciences) for the opportunity provided to use Ultima-IV X-ray diffractometer (Rigaku) and I.N. Gromyak (Laboratory of Methods for Investigation and Analysis of Substances and Materials, GEOKHI RAS) for performing the ICP-AES analysis.

Conflicts of Interest: The authors declare no conflict of interest. The funders had no role in the design of the study; in the collection, analyses, or interpretation of data; in the writing of the manuscript; or in the decision to publish the results.

References

1. Vinokurov, S.E.; Kulikova, S.A.; Krupskaya, V.V.; Myasoedov, B.F. Magnesium potassium phosphate compound for radioactive waste immobilization: Phase composition, structure, and physicochemical and hydrolytic durability. *Radiochemistry* **2018**, *60*, 70–78. [CrossRef]
2. Vinokurov, S.E.; Kulikova, S.A.; Myasoedov, B.F. Solidification of high level waste using magnesium potassium phosphate compound. *Nucl. Eng. Technol.* **2019**, *51*, 755–760. [CrossRef]
3. Kulikova, S.A.; Vinokurov, S.E. The influence of zeolite (Sokyrnytsya deposit) on the physical and chemical resistance of a magnesium potassium phosphate compound for the immobilization of high-level waste. *Molecules* **2019**, *24*, 3421. [CrossRef] [PubMed]
4. Dmitrieva, A.V.; Kalenova, M.Y.; Kulikova, S.A.; Kuznetsov, I.V.; Koshcheev, A.M.; Vinokurov, S.E. Magnesium-potassium phosphate matrix for immobilization of ^{14}C. *Russ. J. Appl. Chem.* **2018**, *91*, 641–646. [CrossRef]
5. Kulikova, S.A.; Belova, K.Y.; Tyupina, E.A.; Vinokurov, S.E. Conditioning of spent electrolyte surrogate LiCl-KCl-CsCl using magnesium potassium phosphate compound. *Energies* **2020**, *13*, 1963. [CrossRef]
6. Graeser, S.; Postl, W.; Bojar, H.-P.; Berlepsch, P.; Armbruster, T.; Raber, T.; Ettinger, K.; Walter, F. Struvite-(K), $KMgPO_4 \cdot 6H_2O$, the potassium equivalent of struvite—A new mineral. *Eur. J. Miner.* **2008**, *20*, 629–633. [CrossRef]
7. Sasaki, K.; Moriyama, S. Effect of calcination temperature for magnesite on interaction of MgO-rich phases with boric acid. *Ceram. Int.* **2014**, *40*, 1651–1660. [CrossRef]
8. Yu, J.; Qian, J.; Wang, F.; Li, Z.; Jia, X. Preparation and properties of a magnesium phosphate cement with dolomite. *Cem. Concr. Res.* **2020**, *138*, 106235. [CrossRef]
9. Sirota, V.; Selemenev, V.; Kovaleva, M.; Pavlenko, I.; Mamunin, K.; Dokalov, V.; Yapryntsev, M. Preparation of crystalline $Mg(OH)_2$ nanopowder from serpentinite mineral. *Int. J. Min. Sci. Technol.* **2018**, *28*, 499–503. [CrossRef]
10. Zhao, Q.; Liu, C.-J.; Jiang, M.-F.; Saxén, H.; Zevenhoven, R. Preparation of magnesium hydroxide from serpentinite by sulfuric acid leaching for CO_2 mineral carbonation. *Miner. Eng.* **2015**, *79*, 116–124. [CrossRef]
11. Teir, S.; Kuusik, R.; Fogelholm, C.-J.; Zevenhoven, R. Production of magnesium carbonates from serpentinite for long-term storage of CO_2. *Int. J. Miner. Process.* **2007**, *85*, 1–15. [CrossRef]
12. Vinokurov, S.E.; Kulikova, S.A.; Krupskaya, V.V.; Tyupina, E.A. Effect of characteristics of magnesium oxide powder on composition and strength of magnesium potassium phosphate compound for solidifying radioactive waste. *Russ. J. Appl. Chem.* **2019**, *92*, 490–497. [CrossRef]
13. Wagh, A.S. *Chemically Bonded Phosphate Ceramics: Twenty-First Century Materials with Diverse Application*, 2nd ed.; Elsevier: Amsterdam, The Netherlands, 2016; pp. 1–422.
14. Tan, Y.; Yu, H.; Li, Y.; Wu, C.; Dong, J.; Wen, J. Magnesium potassium phosphate cement prepared by the byproduct of magnesium oxide after producing Li_2CO_3 from salt lakes. *Ceram. Int.* **2014**, *40*, 13543–13551. [CrossRef]

15. Viani, A.; Sotiriadis, K.; Šašek, P.; Appavou, M.-S. Evolution of microstructure and performance in magnesium potassium phosphate ceramics: Role of sintering temperature of MgO powder. *Ceram. Int.* **2016**, *42*, 16310–16316. [CrossRef]
16. Dong, J.; Yu, H.; Xiao, X.; Li, Y.; Wu, C.; Wen, J.; Tan, Y.; Chang, C.; Zheng, W. Effects of calcination temperature of boron-containing magnesium oxide raw materials on properties of magnesium phosphate cement as a biomaterial. *J. Wuhan Univ. Technol. Sci. Ed.* **2016**, *31*, 671–676. [CrossRef]
17. Post, J.E.; Bish, D.L. Rietveld refinement of crystal structures using powder X-ray diffraction data. In *Modern Powder Diffraction, Reviews in Mineralogy*; MSA: Washington, DC, USA, 1989; pp. 277–308.
18. Döbelin, N.; Kleeberg, R. Profex: A graphical user interface for the Rietveld refinement program BGMN. *J. Appl. Crystallogr.* **2015**, *48*, 1573–1580. [CrossRef] [PubMed]
19. Russian Federation. *Radioactive Waste. Long Time Leach Testing of Solidified Radioactive Waste Forms*; GOST R 52126-2003; Standardinform: Moscow, Russia, 2003; pp. 1–8.
20. De Groot, G.; van der Sloot, H. Determination of leaching characteristics of waste materials leading to environmental product certification. In *Stabilization and Solidification of Hazardous, Radioactive, and Mixed Wastes*; Gilliam, T., Wiles, C., Eds.; ASTM International: West Conshohocken, PA, USA, 1992; Volume 2, pp. 149–170.
21. Torras, J.; Buj, I.; Rovira, M.; de Pablo, J. Semi-dynamic leaching tests of nickel containing wastes stabilized/solidified with magnesium potassium phosphate cements. *J. Hazard. Mater.* **2011**, *186*, 1954–1960. [CrossRef] [PubMed]
22. Pundsack, F.L. Recovery of Silica, Iron Oxide and Magnesium Carbonate from the Treatment of Serpentine with Ammonium Bisulfate. U.S. Patent 3,338,667, 29 August 1967.
23. Khamizov, R.K.; Zaitsev, V.A.; Gruzdeva, A.N.; Krachak, A.N.; Rarova, I.G.; Vlasovskikh, N.S.; Moroshkina, L.P. Feasibility of acid–salt processing of alumina-containing raw materials in a closed-loop process. *Russ. J. Appl. Chem.* **2020**, *93*, 1059–1067. [CrossRef]
24. Nduagu, E.; Highfield, J.; Chen, J.; Zevenhoven, R. Mechanisms of serpentine–ammonium sulfate reactions: Towards higher efficiencies in flux recovery and Mg extraction for CO_2 mineral sequestration. *RSC Adv.* **2014**, *4*, 64494–64505. [CrossRef]
25. Russian Federation. *Reagents. Magnesium oxide. Specifications*; GOST 4526-75; Standardinform: Moscow, Russia, 1975; pp. 1–11.
26. Russian Federation. Collection, processing, storage and conditioning of liquid radioactive waste. Safety requirements. In *Federal Norms and Rules in the Field of Atomic Energy Use*; NP-019-15; Rostekhnadzor: Moscow, Russia, 2015; pp. 1–22.

Article

Electrochemical Determination of Lead Using A Composite Sensor Obtained from Low-Cost Green Materials:Graphite/Cork

Iasmin B. Silva [1], Danyelle Medeiros de Araújo [2], Marco Vocciante [3], Sergio Ferro [4,*],
Carlos A. Martínez-Huitle [1,*] and Elisama V. Dos Santos [1,*]

[1] Laboratório de Eletroquímica Ambiental e Aplicada, Instituto de Química, Universidade Federal do Rio Grande do Norte, Lagoa Nova, Natal 59072-900, Brazil; iasminbds@gmail.com
[2] Laboratório de Eletroquímica e Química Analítica, Departamento de Química, Universidade do Estado do Rio Grande do Norte, Mossoró 59610-210, Brazil; danyellearaujo@uern.br
[3] Department of Chemistry and Industrial Chemistry, University of Genova, 16124 Genova, Italy; marco.vocciante@unige.it
[4] Ecas4 Australia Pty Ltd., Mile End South 5031, Australia
*   Correspondence: sergio@ecas4.com.au (S.F.); carlosmh@quimica.ufrn.br (C.A.M.-H.); elisamavieira@ect.ufrn.br (E.V.D.S.)

Abstract: The purpose of this study was to develop an inexpensive, simple, and highly selective cork-modified carbon paste electrode for the determination of Pb(II) by differential pulse anodic stripping voltammetry (DPASV) and square-wave anodic stripping voltammetry (SWASV). Among the cork–graphite electrodes investigated, the one containing 70% w/w carbon showed the highest sensitivity for the determination of Pb(II) in aqueous solutions. Under SWASV conditions, its linear range and relative standard deviation are equal to 1–25 µM and 1.4%, respectively; the limit of detection complies with the value recommended by the World Health Organization. To optimize the operating conditions, the selectivity and accuracy of the analysis were further investigated by SWASV in acidic media. Finally, the electrode was successfully applied for the determination of Pb(II) in natural water samples, proving to be a sensitive electrochemical sensor that meets the stringent environmental control requirements.

Keywords: cork–graphite electrode; electrochemistry; lead; environmental application

check for **updates**

Citation: Silva, I.B.; de Araújo, D.M.; Vocciante, M.; Ferro, S.; Martínez-Huitle, C.A.; Dos Santos, E.V. Electrochemical Determination of Lead Using A Composite Sensor Obtained from Low-Cost Green Materials:Graphite/Cork. *Appl. Sci.* **2021**, *11*, 2355. https://doi.org/10.3390/app11052355

Academic Editor: Fethi Bedioui

Received: 13 January 2021
Accepted: 1 March 2021
Published: 6 March 2021

Publisher's Note: MDPI stays neutral with regard to jurisdictional claims in published maps and institutional affiliations.

Copyright: © 2021 by the authors. Licensee MDPI, Basel, Switzerland. This article is an open access article distributed under the terms and conditions of the Creative Commons Attribution (CC BY) license (https://creativecommons.org/licenses/by/4.0/).

1. Introduction

Lead is a highly toxic heavy metal that causes serious environmental problems due to its non-biodegradability. It is commonly released into the environment because of mining activities, natural processes, and the development of new technological devices [1,2], being frequently used by the automotive, plastics, paints, and ceramics industries for its corrosion resistance [3].

Since the nitrate and chloride salts of lead show excellent solubility in water [4,5], lead is normally present in soil and aquatic ecosystems in ionic form, as Pb(II). According to the World Health Organization (WHO), a Pb(II) concentration as low as 0.24 µmol L^{-1} can cause decreased intelligence in children, behavioral difficulties, and learning problems. For this reason, the concentration of lead in water and soils should always be below the WHO limit and, consequently, must be monitored.

Nowadays, several analytical methods are employed for lead detection, such as spectroscopy [6], optical colorimetry [7], inductively coupled plasma mass spectrometry (ICP-MS) [8], atomic absorption spectrometry (AAS) [9], and fluorescence spectrometry [10]. However, these analytical methods are expensive (they require trained operators, complex equipment, solvents or gases, and so on) and, in some cases, sample preparation procedures are required. In this context, electrochemical techniques have been investigated because of their significant advantages such as simplicity of operation, high sensitivity, low cost, and easy handling [11–15]. In general, electrochemical sensors are rapid, portable, inexpensive, and highly sensitive and offer a low limit of detection, good reproducibility, good signal-to-noise ratio, and selective detection [12]. Consequently, electrochemical sensors have been applied for the determination of heavy metals in the environment, industrial products, food matrices, electronic waste, and clinical materials [16–18].

Among the electrochemical sensors used, graphite-modified electrodes have been extensively developed due to their higher selectivity, sensitivity, high specific area, unique electrical conductivity, self-assembly behavior, mechanical flexibility, extreme resistance to oxidation, natural origin, and low cost [19–22]. However, these properties can be improved by including other modifiers in their composition.

Recently, cork has emerged as a promising low-cost and efficient green material for various environmental applications (e.g., compound detection [13,23,24], soil and water remediation [25,26]). Cork is a natural organic polymeric material, which has modest electrical, magnetic, and optical properties and exhibits self-cleaning behavior and antibacterial activity. For raw cork (RAC), electrical conductivity (σ) values of approximately 1.2×10^{-10} and 1.67×10^{-13} S m^{-1} were registered at 25 and 50 °C, respectively [27]. Based on the existing literature, two types of cork are often used: raw cork (RAC) and regranulated cork (RGC). Their differences are mainly due to their composition, which depends on the thermal pretreatment applied to RAC to produce RGC.

In the present communication, cork–graphite composite electrodes to be used as electrochemical sensors for the detection of lead ions are discussed. The effects of the cork composition, the cork–graphite ratio, and the supporting electrolyte for detecting Pb(II) were investigated. The performance in Pb(II) detection of two voltammetric techniques (differential pulse adsorptive stripping voltammetry (DPASV) and square-wave adsorptive stripping voltammetry (SWASV)) was also evaluated. Finally, the applicability of the cork–graphite voltammetric device was successfully demonstrated by detecting Pb(II) in real water matrixes (groundwater, tap water, and "produced water") as well as verifying the selectivity, repeatability, reproducibility, and stability of the sensors.

2. Materials and Methods

The highest quality commercially available chemicals were used. Graphite powder and Pb(NO$_3$)$_2$ were sourced from Sigma (Brazil); the former was used without further purification. Acetate buffer, NaNO$_3$, CdCl$_2$, H$_2$SO$_4$, NaCl, FeCl$_2$, KCl, CaCl$_2$, MgSO$_4$, ZnCl$_2$, AlCl$_3$, and MnSO$_4$ were sourced from Merck (Brazil). The raw cork (RAC) used in the experimental studies was provided by Corticeira Amorim S.G.P.S., S.A. (Portugal); the granules were washed twice with distilled water in cycles of 2 h at 60 °C to remove impurities and other water-extractable components that could interfere with the electrochemical analysis. Before use, the RAC was dried at 60 °C in an oven for 24 h [23]. Aqueous solutions were prepared using ultrapure water obtained from a Millipore Milli-Q direct-0.3 purification system.

2.1. Preparation of Cork-Modified Electrodes

The RAC granules were reduced in size using a ball mill and sieved to obtain the finest fractions. The fraction below 150 μm (designated as RAC powder) was selected for use in this work. The cork–graphite composite sensor (working electrode) was prepared by mechanical homogenization of RAC and graphite (Gr) in different proportions (10:90, 70:30, and 90:10 %w/w), using 0.3 mL of paraffin oil as a binder and mixing everything in an agate mortar for about 30 minutes, as previously reported [23]. The paste was packed in a polypropylene nozzle (model K31-200Y) used as a support, and the sensor surface was smoothed over a tissue paper. Before use, the sensor was electroactivated by cyclic voltammetry between -1.1 and 0 V (scan rate: 100 mV s^{-1}) in 0.5 M H$_2$SO$_4$. The different sensors are referred to as GrRAC-X, where X is the amount of cork (RAC) expressed as %w/w. The unmodified graphite sensor (Gr) was prepared as described for the GrRAC-X sensors, but in the absence of RAC powder. Electrode stability was also determined by repetitive determinations of Pb (25 µM) in 0.5 M H_2SO_4.

2.2. Electrochemical Measurements

The electrochemical tests were performed using an Autolab PGSTAT302N (Metrohm) controlled with NOVA 1.8 software, and a three-electrode cell including an Ag/AgCl (3.0 M KCl) reference electrode, a Pt wire auxiliary electrode, and one of the cork–graphite sensors (GrRAC) as the working electrode (geometrical area of approximately 0.45 mm^2). Differential pulse anodic stripping voltammetry (DPASV) measurements were performed with different concentrations of Pb(II) ions in acetate buffer solutions (pH 4.5), 0.5 M $NaNO_3$, and 0.5 M H_2SO_4. The accumulation of Pb(II) ions on the surface of the composite sensor was achieved by applying a potential of -1.2 V (vs. Ag/AgCl) for different preconcentration times (40, 70, 100, 130, and 160 s), during which the stirring conditions were kept constant for 30, 60, 90, 120, and 150 s; the remaining 10 s were considered as an equilibration time, without stirring. Subsequently, the anodic stripping scan was performed at 50 mV s^{-1}, with a modulation amplitude of +0.05 V, and a modulation time of 0.04 s. Square-wave anodic stripping voltammetry (SWASV) measurements were performed in 0.5 M H_2SO_4. In this case, a preconcentration potential of -1.2 V was applied to the working electrode for 120 s under continuous magnetic stirring, with a scanning frequency of 80 Hz, an amplitude of 50 mV, and a step potential of 5 mV. All electrochemical studies were conducted without deaerating and performed at room temperature (25 $\pm$ 2 °C). Each measurement was performed in triplicate, and obtained data were subjected to statistical analysis and reported as mean $\pm$ standard deviation (SD). For the determination of Pb(II) in different water matrices (tap water, groundwater, and produced water), the water samples were spiked with a known quantity of a standard solution of Pb(II) and the determination of Pb(II) was performed using the standard addition method.

3. Results and Discussion

3.1. Effect of the Supporting Electrolyte

In order to evaluate the voltammetric response of the proposed modified sensor, the quantification of Pb(II) was carried out in different supporting electrolytes. Figure 1a–c show the DPASV response for the determination of Pb(II) using a GrRAC-70% sensor (this composition was selected for preliminary analysis based on the results reported in the literature) from solutions of 0.1 M acetate buffer (pH 4.5), 0.5 M $NaNO_3$, and 0.5 M H_2SO_4, respectively, using a preconcentration time of 30 s. The sulfuric acid solution proved to be the most suitable electrolytic solution because it provided a well-defined voltammetric signal and the response increased linearly without significant deviations (Figure 1). The limit of detection (LOD), for each of the supporting electrolytes used, was estimated by the equation $LOD = 3 \times S_{y/x}/b$, where $S_{y/x}$ is the residual standard deviation and b is the slope of the calibration plot, in accordance with IUPAC recommendations of the mean value for samples analyzed in triplicate. This approach allows the control of both false positive and negative errors ($\alpha = \beta = 0.05$), as recommended by IUPAC [28,29], and has been confirmed and recommended by experts in the field [30,31]. For the 0.1 M acetate buffer solution, no significant current response was obtained, resulting in an LOD of 1.8 µM; a similar outcome was obtained in 0.01 M acetate buffer solution, where an LOD of approximately 3.2 µM confirmed the poor performance of the sensor in acetate buffer solutions. Conversely, the analytical approach significantly improved when 0.5 M $NaNO_3$ and 0.5 M H_2SO_4 solutions were used, which resulted in LOD values of 2.8 (Figure 1b) and 1.6 µM (Figure 1c). Comparing the values obtained with an unmodified graphite electrode and with the modified cork–graphite electrodes in H_2SO_4, the LOD is 3 times higher on graphite ($\approx$4.8 µM) than that obtained with the GrRAC-70% electrode. The best result obtained using the composite material shows that the cork–graphite mixture is able to influence the intensity of the current signals.

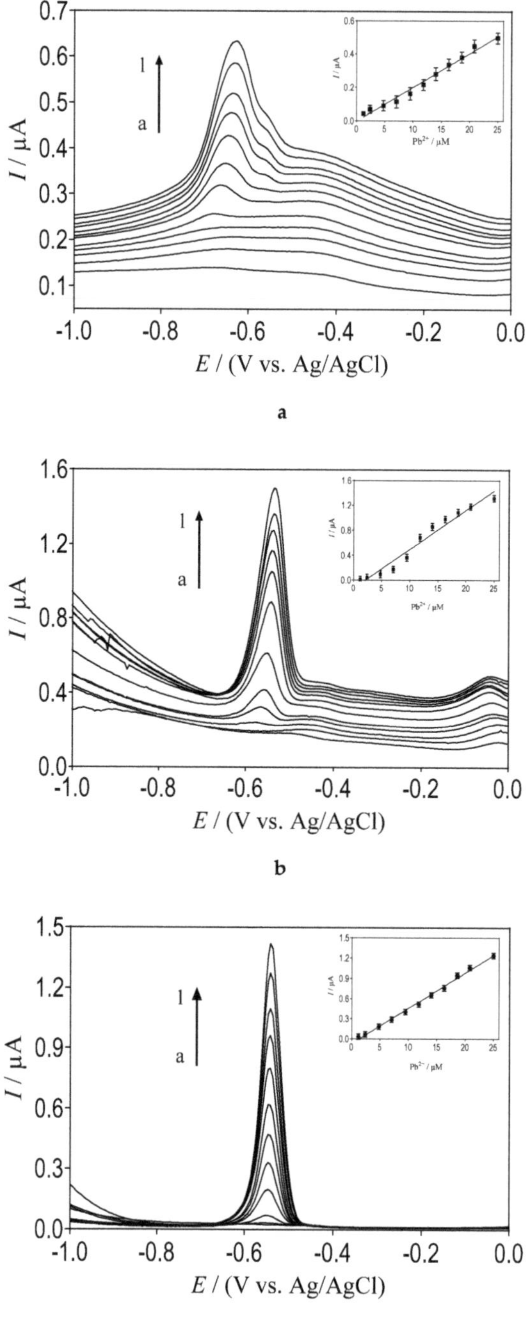

Figure 1. DPASV curves recorded for different concentrations of Pb(II) in (**a**) 0.1 M acetate buffer (pH 4.5), (**b**) 0.5 M NaNO$_3$, and (**c**) 0.5 M H$_2$SO$_4$. Lead concentrations: (a) 0, (b) 1.2, (c) 2.4, (d) 4.8, (e) 7.1, (f) 9.5, (g) 11.8, (h) 14.0, (i) 16.3, (j) 18.6, (k) 20.8, (l) 25 μM. Inserts: plots of the electrochemical response, in terms of current, as a function of the lead concentration.

According to the literature [32], cork has a great adsorption capacity, which is assumed to occur through a so-called biosorption mechanism consisting of an initial physical adsorption (rapid metal uptake) and then a slower chemisorption. The biosorption mechanism is the result of several kinds of interactions, such as complexation, coordination, chelation, ion exchange, inorganic microprecipitation, and hydrolysis products of metal ions; in the case of metal ion biosorption, ion exchange is usually the main mechanism. Hence, the type of cork, the pH conditions, and the contact time determine the interactions that can occur between the target compound and the cork surface. Depending on the cork used, specific active sites may predominate in its surface composition (phenolic, carboxylic, sulfonic, phosphate, and amino groups as well as coordination sites), in addition to the cork surface charge depending on pH conditions [33]. It is also important to consider that carbonaceous materials have micropores and mesopores, the accessibility of which will be increased following the inclusion of cork as a surface modifier. Thus, an improvement in voltammetric current signals can be achieved. Another important feature to consider is that the surface morphology of GrRAC-70% is more homogeneous, as evidenced by the SEM micrographs, which can positively influence its current response [34].

In the case of the acetate buffer as the supporting electrolyte, the formation of $Pb(CH_3COO)_2$ can decrease the availability of Pb(II) in solution; in addition, the acetate ions can compete with Pb(II) ions for the active sites available on the graphite–cork surface. As a result, a poor current response is achieved, with limitations on the selectivity and sensitivity of the modified electrode. Indeed, lead concentrations below 10 μM (Figure 1a), which affect the LOD, cannot be efficiently detected. In the case of $NaNO_3$, the lack of complexing activity by nitrate anions and the possibility of preferential interactions between the composite material (cork–graphite) and Pb(II) ions in solution allow significant improvements in the current response, with consequent benefits in terms of the linearity of the response, although superficial adsorption phenomena may be highlighted for lead concentrations below 10 μM (Figure 1b). Finally, well-defined voltammetric signals were observed at the GrRAC-70% electrode when H_2SO_4 was used (Figure 1c). Due to the acidic conditions, lead is present in solution in its cationic form, Pb^{2+}, and the cork surface is also completely protonated and positively charged. When the working electrode is negatively polarized, the lead ions compete with protons for surface sites; however, the surface accumulation of Pb ions is favored due to ion exchange mechanisms with active sites. In fact, the current response increased linearly without any significant deviations. Therefore, H_2SO_4 was selected as the supporting electrolyte for the subsequent experiments.

3.2. Influence of the Preconcentration Time

The effect of the preconcentration time (40, 70, 100, 130, and 160 s) on the voltammetric response for Pb(II) detection was studied in 15 mL of 0.5 M H_2SO_4 by using the GrRAC-70% sensor. The results indicate that the peak current increased with the preconcentration time, from 40 to 160 s, as illustrated in Figure 2. For all tests, the initial potential was held constant at -1.2 V under stirring conditions for different times (30, 60, 90, 120, and 150 s), with an additional resting time of 10 s without stirring; subsequently, the stripping voltammetry was carried out at 50 mV s^{-1}. As seen in Figure 2, a decrease in current was achieved when the preconcentration time was extended to 160 s; thus, 130 s was chosen as the most suitable preconcentration time for further analysis. Ten replicates were considered in order to study the effect of preconcentration. According to Student's *t*-test at a confidence level of 95% (parameter denominated as *p*), there were no significant differences between the experimental value (23.5 μM of Pb) and the theoretical value (25 μM of Pb). The observed trend can be motivated by the high adsorption rate due to the porous structure of the cork [33]; for preconcentration times greater than 130 s (120 s under stirring conditions and 10 s in rest conditions), the GrRAC-70% sensor plausibly reached the maximum adsorption capacity on its surface.

3.4. SWASV Analysis

The GrRAC-70% composite sensor produced the best DPASV results for Pb(II) detection and was therefore chosen as the working electrode for evaluating Pb(II) traces by SWASV. Figure 4 shows the SWASV voltammetric responses of Pb(II) under pre-selected experimental conditions: 0.5 M H_2SO_4 as the supporting electrolyte, -1.2 V as the preconcentration potential, 120 s of preconcentration time plus 10 s of resting time. The stripping voltammetric peaks of Pb(II) ions appeared at -0.44 V for the GrRAC-70% sensor. The peak current (I_p) increased linearly with the concentration of Pb(II) in the range from 1 to 25 μM; the linear regression equation (I_p vs. C) was obtained as

$$I_p \ (\mu A) = (0.4 \pm 0.1) \times C - (0.8 \pm 0.2); \ R^2 = 0.98$$

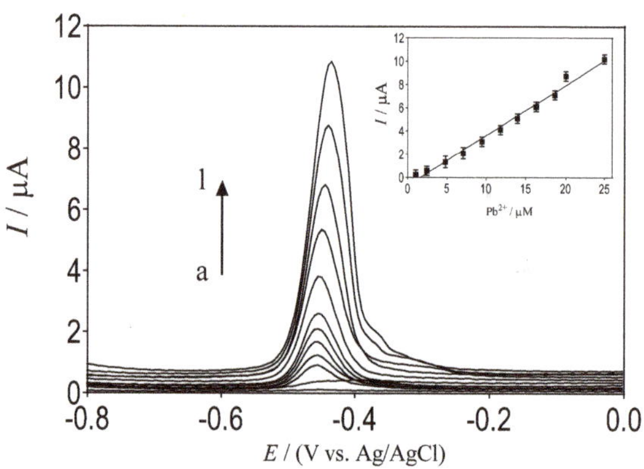

Figure 4. SWASV curves of GrRAC-70% recorded for different concentrations of Pb(II) in 0.5 M H_2SO_4: (a) 0, (b) 1.2, (c) 2.4, (d) 4.8, (e) 7.1, (f) 9.5, (g) 11.8, (h) 14.0, (i) 16.3, (j) 18.6, (k) 20.8, (l) 25 μM. Inserts: plots of the electrochemical response, in terms of current, as a function of the lead concentration.

The LOD was found to be 0.3 μM. Compared with DPASV, SWASV produced much better results: the regression residuals are randomly distributed around zero and the linearity is practically perfect. Another important point is that no noticeable alterations were noted in the calibration curves recorded on different days, confirming the stability of the GrRAC-70% composite sensor. The GrRAC-70% sensor used in this work remained stable for at least two months of intensive use.

Table 1 collects the results available in the literature and relating to the analysis of Pb(II) with different electrodes and allows a comparison with the results obtained in this study. The ease of sensor preparation and the analytical protocol suggested here offer advantages over the other methods reported.

Table 1. Comparison of the analytical parameters of the sensors reported in the literature for the determination of Pb(II).

Electrodes	Method	Electrolyte	Linear Range μM	LOD/μM	Ref.
MTZ-PMO-S-S [1]	SWASV	0.2 M HCl	0.01–10	0.024	[35]
5-Br-PADAP/MWCNT [2]	DPA	0.1 M acetate buffer	0.9–114	0.5	[36]
PPy/CNFs/CPE [3]	SWASV	0.1 M acetate buffer	0.2–130	0.05	[37]
SRE [4]	DPASV	0.01 M HNO_3 + 0.01 M KCl	0.01–0.1	0.02	[38]
Glassy carbon	ADSV	0.5 M $NaNO_3$		0.5	[12]
ErGO–MWNTs–L-cys [5]	DPASV	0.1 M acetate buffer	0.2–40	0.05	[39]
CPE modified with IIP-MWCNTs [6]	DPSV	0.1 M acetate buffer	3–55	0.5	[40]
IJP-MW-CNT [7]	SWASV	0.1 M acetate buffer	5–20	0.05	[41]
GrRAC	**SWASV**	**0.5 M H_2SO_4**	**1–25**	**0.3**	**This work**

[1] Mercaptothiazoline-disulfide-bridged periodic mesoporous organosilica. [2] 2-(5-bromo-2-pyridylazo)-5-diethylaminophenol modified multiwalled carbon nanotube electrode. [3] Nanocomposite of polypyrrole (PPy) and carbon nanofibers (CNFs)-modified carbon paste electrode (CPE). [4] Silver ring electrode. [5] Electrochemically reduced graphene oxide–multiwalled carbon nanotubes–L-cysteine. [6] Carbon paste electrode (CPE) modified with ion-imprinted polymer nanoparticles and multiwalled carbon nanotubes. [7] Inkjet-printed multiwalled carbon nanotubes.

3.5. Study of the Interferences

In order to evaluate the specificity of the suggested approach, the sensor response in the presence of several potentially interfering species was investigated. In particular, the response to Pb(II) was evaluated in solutions containing 10 μmol/L of the following cations: Fe^{2+}, Na^+, K^+, Ca^{2+}, Mg^{2+}, Zn^{2+}, Al^{3+}, Mn^{2+}, Cu^{2+}, and Cd^{2+}. No additional signals were recorded when Fe^{2+}, Na^+, K^+, Ca^{2+}, Mg^{2+}, Zn^{2+}, Al^{3+}, Mn^{2+}, or Cu^{2+} ions were present in solution during the determination of Pb at different concentrations (Figure 5). Conversely, a well-defined peak signal for Cd^{2+} was observed at -0.8 V. However, no changes in the Pb current peak were observed for the GrRAC-70% composite sensor in the presence of Cd^{2+} ions in solution (Figure 5). Therefore, the GrRAC-70% sensor can be used to detect Pb(II) even in the presence of other metals.

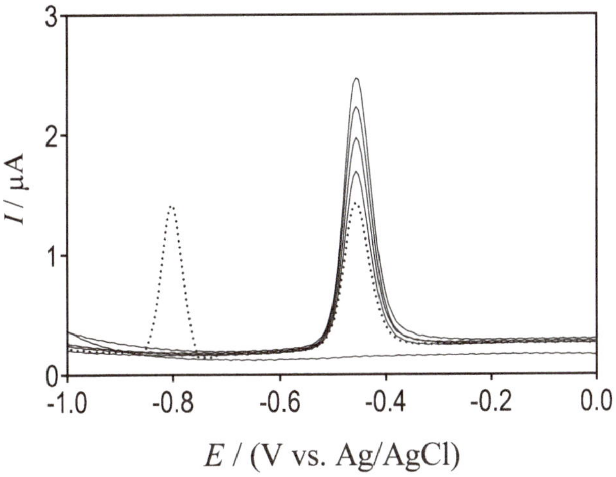

Figure 5. SWASV curves of GrRAC-70% recorded for different concentrations of Pb(II) in 0.5 M H_2SO_4 in the presence of a fixed concentration of interfering metal ions (10 μmol/L).

3.6. Stability

The stability of GrRAC was previously examined by determining caffeine, obtaining results of good consistency with a relative standard deviation (RSD) of 1.41% ($n = 3$); this outcome suggested that the cork–graphite composite sensor can be reused. A similar stability assessment was carried out in the case of Pb(II) by recording a series of voltammetric analyses over 10 days. No significant changes in the Pb peak current were observed after five measurements over that time period (Pb(II) = 15 μM; RSD = 1.52%, $n = 5$). The cork–graphite composite sensor was washed and stored at 25 °C after each experiment.

3.7. Analytical Applications

The effectiveness of the proposed method, for the detection of Pb(II) in real samples, was also tested by analyzing tap water, groundwater, and "produced water" (a brackish water that is extracted as a by-product from underground during the process of oil and natural gas extraction). The electrochemical determination of Pb(II) was based on SWASV in acidic medium. For this, a quantity of Pb(II) was added to the water samples to obtain a well-known concentration between 10 and 50 μM for each sample. Then, the prepared samples were analyzed using the GrRAC-70% composite sensor under the optimized experimental conditions reported in Section 3.4. The validity of the proposed method for the determination of Pb(II) was evaluated using the standard addition method and recovery studies were conducted on the samples. As shown in Table 2, the recoveries ranged from 89 to 115 ($n = 3$), indicating that the proposed method can be efficiently applied for the detection of Pb(II) in real water samples. In all cases, the relative standard deviation (RSD) values ranged from 0.3 to 2.1%, which confirms that the developed detection approach is potentially applicable.

Table 2. Pb(II) content in real water samples, measured with SWASV using the GrRAC-70% composite sensor.

Sample	Present Method [1]	Pb^{2+} Added (µM)	Pb^{2+} Found (µM) [1]	Recovery (%)
Groundwater	Not detected	10	11.5 ± 0.3	115
		50	52.1 ± 1.8	104
Tap water	Not detected	10	10.8 ± 0.5	108
		50	50.4 ± 1.5	100
Produced water	12.0 ± 0.4 µM	10	19.6 ± 0.5	89
		50	60.3 ± 2.1	97

[1] Mean of three determinations ± standard deviation.

4. Conclusions

Cork–graphite-based sensors offer a fast, reliable, cost-effective, and simple way to determine Pb(II) in real samples. The composite sensor exhibits higher sensitivity and reproducibility than conventional unmodified graphite sensors, and the low LOD allows for reduced matrix effects in dilute solutions. As for the materials tested, the affinity of the cork with the analyte allowed a substantial improvement in sensitivity. According to the results reported in this work, the sensor obtained by mixing 70% w/w of cork with 30% w/w of graphite allowed obtaining higher voltammetric responses and a rapid detection of Pb(II). The proposed approach is precise, with a limit of quantification of 0.3 µM, reproducible, and less expensive, both in terms of time and materials, compared to other analytical methods. The composite electrode can be applied effectively for the determination of Pb(II) in acidic media. As for the physical and chemical properties, which favor the interactions with the analytes to be detected or/and quantified, more experiments are needed to better understand the chemical and electrochemical processes that occur on the cork–graphite surface when the current is applied or when cork participates as a mediator.

Finally, even if the LOD reported in this work (0.3 µmol L^{-1}) is slightly above the limit established by the WHO (Pb: 0.24 µmol L^{-1}), there is room for improvement; for example, the size of the electrochemical sensor could be reduced, in order to approximate a micro-electrode, while the use of other carbon-based modifiers could allow improving the sensitivity.

Author Contributions: Conceptualization, E.V.D.S. and C.A.M.-H.; methodology, M.V.; formal analysis, I.B.S., S.F., and C.A.M.-H.; investigation, I.B.S. and C.A.M.-H.; resources, C.A.M.-H.; data curation, I.B.S., E.V.D.S., and C.A.M.-H.; writing—original draft preparation, I.B.S. and D.M.d.A.; writing—review and editing, M.V., S.F., and C.A.M.-H.; funding acquisition, D.M.d.A. and C.A.M.-H. All authors have read and agreed to the published version of the manuscript.

Funding: This research was funded by Conselho Nacional de Desenvolvimento Científico e Tecnológico (Brazil), grant number 306323/2018-4, and Fundação de Amparo à Pesquisa do Estado de São Paulo (Brazil), grant numbers 2014/50945-4 and 2019/13113-4.

Acknowledgments: I.B. Silva acknowledges the Coordenação de Aperfeiçoamento de Pessoal de Nível Superior for her fellowship. Financial supports from the Conselho Nacional de Desenvolvimento Científico e Tecnológico (CNPq–306323/2018-4) and from Fundação de Amparo à Pesquisa do Estado de São Paulo (Brazil) (FAPESP 2014/50945-4 and 2019/13113-4) are gratefully acknowledged. The authors also thank V.J.P. Vilar from the University of Porto (Portugal) for providing the cork samples used in this study.

Conflicts of Interest: The authors declare no conflict of interest.

References

1. Gupta, R. *Reproductive and Developmental Toxicology*, 1st ed.; Ramesh Gupta, Ed.; Academic Press: Cambridge, MA, USA, 2017.
2. Khoshbin, Z.; Housaindokht, M.R.; Verdian, A.; Bozorgmehr, M.R. Simultaneous detection and determination of mercury(II) and lead(II) ions through the achievement of novel functional nucleic acid-based biosensors. *Biosens. Bioelectron.* **2018**, *116*, 130–147. [CrossRef]
3. Pinho, S.; Ladeiro, B. Phytotoxicity by lead as heavy metal focus on oxidative stress. *J. Bot.* **2012**, *2012*, 1–10. [CrossRef]
4. Tangahu, B.V.; Sheikh Abdullah, S.R.; Basri, H.; Idris, M.; Anuar, N.; Mukhlisin, M. A review on heavy metals (As, Pb, and Hg) uptake by plants through phytoremediation. *Int. J. Chem. Eng.* **2011**, *2011*. [CrossRef]
5. Fan, R.D.; Liu, S.Y.; Du, Y.J.; Reddy, K.R.; Yang, Y.L. Impacts of presence of lead contamination on settling behavior and microstructure of clayey soil—calcium bentonite blends. *Appl. Clay Sci.* **2017**, *142*, 109–119. [CrossRef]
6. Chisholm-brause, C.J.; Hayes, K.I.M.F.; Lawrence, A. Spectroscopic investigation of Pb(II) complexes at the $\gamma A1_2O_3$/water interface. *Geochim. Cosmochim. Acta* **1990**, *54*, 1897–1909. [CrossRef]
7. Roto, R.; Mellisani, B.; Kuncaka, A.; Mudasir, M.; Suratman, A. Colorimetric sensing of Pb^{2+} ion by using ag nanoparticles in the presence of dithizone. *Chemosensors* **2019**, *7*, 28. [CrossRef]
8. Krachler, M.; Irgolic, K.J. The potential of inductively coupled plasma mass spectrometry (ICP-MS) for the simultaneous determination of trace elements in whole blood, plasma and serum. *J. Trace Elem. Med. Biol.* **1999**, *13*, 157–169. [CrossRef]
9. L'vov, B.V. Fifty years of atomic absorption spectrometry. *J. Anal. Chem.* **2005**, *60*, 382–392. [CrossRef]
10. Beltrán, B.; Leal, L.O.; Ferrer, L.; Cerdà, V. Determination of lead by atomic fluorescence spectrometry using an automated extraction/pre-concentration flow system. *J. Anal. At. Spectrom.* **2015**, *30*, 1072–1079. [CrossRef]
11. Dai, X.; Wu, S.; Li, S. Progress on electrochemical sensors for the determination of heavy metal ions from contaminated water. *J. Chin. Adv. Mater. Soc.* **2018**, *6*, 91–111. [CrossRef]
12. Silva, K.N.O.; Paiva, S.S.M.; Souza, F.L.; Silva, D.R.; Martínez-Huitle, C.A.; Dos Santos, E.V. Applicability of electrochemical technologies for removing and monitoring Pb^{2+} from soil and water. *J. Electroanal. Chem.* **2018**, *816*, 171–178. [CrossRef]
13. Monteiro, M.K.S.; Santos, E.C.M.M.; Silva, D.R.; Martínez-Huitle, C.A.; Dos Santos, E.V. Simultaneous determination of paracetamol and caffeine in pharmaceutical formulations and synthetic urine using cork-modified graphite electrodes. *J. Solid State Electrochem.* **2020**, 18–20. [CrossRef]
14. Araújo, E.G.; Jhones dos Santos, A.; da Silva, D.R.; Salazar, R.; Martínez-Huitle, C.A. Cysteic acid-modified glassy carbon electrode for monitoring oxalic acid (OA) concentration during its electrochemical oxidation at Ti/Pt anode. *Electroanalysis* **2014**, *26*, 748–755. [CrossRef]
15. Araújo, E.G.; Oliveira, G.R.; Santos, E.V.; Martínez-Huitle, C.A.; Panizza, M.; Fernandes, N.S. Applicability of electroanalysis for monitoring oxalic acid (OA) concentration during its electrochemical oxidation. *J. Electroanal. Chem.* **2013**, *701*. [CrossRef]
16. March, G.; Nguyen, T.D.; Piro, B. Modified electrodes used for electrochemical detection of metal ions in environmental analysis. *Biosensors* **2015**, *5*, 241–275. [CrossRef] [PubMed]
17. Palisoc, S.; Vitto, R.I.M.; Natividad, M. Determination of heavy metals in herbal food supplements using bismuth/multi-walled carbon nanotubes/Nafion modified graphite electrodes sourced from waste batteries. *Sci. Rep.* **2019**, *9*, 1–13. [CrossRef] [PubMed]
18. Yantasee, W.; Lin, Y.; Hongsirikarn, K.; Fryxell, G.E.; Addleman, R.; Timchalk, C. Electrochemical sensors for the detection of lead and other toxic heavy metals: The next generation of personal exposure biomonitors. *Environ. Health Perspect.* **2007**, *115*, 1683–1690. [CrossRef]
19. Koirala, K.; Santos, J.H.; Tan, A.L.; Ali, M.A.; Mirza, A.H. Chemically modified carbon paste electrode for the detection of lead, cadmium and zinc ions. *Sens. Rev.* **2016**, *36*, 339–346. [CrossRef]
20. Nery, E.W.; Kundys-Siedlecka, M.; Furuya, Y.; Jönsson-Niedziółka, M. Pencil lead as a material for microfluidic 3D-electrode assemblies. *Sensors* **2018**, *18*, 4037. [CrossRef] [PubMed]
21. Palisoc, S.; Gonzales, A.J.; Pardilla, A.; Racines, L.; Natividad, M. Electrochemical detection of lead and cadmium in UHT-processed milk using bismuth nanoparticles/Nafion®-modified pencil graphite electrode. *Sens. Bio-Sens. Res.* **2019**, *23*, 100268. [CrossRef]
22. Batista, L.C.D.; Santos, T.I.S.; Santos, J.E.L.; da Silva, D.R.; Martínez-Huitle, C.A. Metal Organic Framework-235 (MOF-235) modified carbon paste electrode for catechol determination in water. *Electroanalysis* **2021**, *33*, 57–65. [CrossRef]
23. Monteiro, M.K.S.; Paiva, S.S.M.; da Silva, D.R.; Vilar, V.J.P.; Martínez-Huitle, C.A.; Dos Santos, E.V. Novel cork-graphite electrochemical sensor for voltammetric determination of caffeine. *J. Electroanal. Chem.* **2019**, *839*, 283–289. [CrossRef]
24. Henrique, J.M.M.; Monteiro, M.K.S.; Cardozo, J.C.; Martínez-Huitle, C.A.; da Silva, D.R.; Dos Santos, E.V. Integrated-electrochemical approaches powered by photovoltaic energy for detecting and treating paracetamol in water. *J. Electroanal. Chem.* **2020**, *876*, 114734. [CrossRef]
25. Pintor, A.M.A.; Ferreira, C.I.A.; Pereira, J.C.; Correia, P.; Silva, S.P.; Vilar, V.J.P.; Botelho, C.M.S.; Boaventura, R.A.R. Use of cork powder and granules for the adsorption of pollutants: A review. *Water Res.* **2012**, *46*, 3152–3166. [CrossRef]
26. Gil, L. New cork-based materials and applications. *Materials* **2015**, *8*, 625–637. [CrossRef] [PubMed]
27. Marat-Mendes, J.N.; Neagu, E.R. The study of electrical conductivity of cork. *Ferroelectrics* **2003**, *294*, 123–131. [CrossRef]
28. Danzer, K.; Currie, L.A. Guideline for calibration in analytical chemistry—Part 1. Fundamentals and single component calibration. *Pure Appl. Chem.* **1998**, *70*, 993–1014. [CrossRef]

29. Currie, L.A. International Union of Pure and Applied Chemistry Nomenclature in Evaluation of Analytical Methods Including Detection and Quantification Capabilities. *Pure Appl. Chem.* **1995**, *67*, 1699–1723. [CrossRef]
30. Desimoni, E.; Brunetti, B. About estimating the limit of detection of heteroscedastic analytical systems. *Anal. Chim. Acta* **2009**, *655*, 30–37. [CrossRef] [PubMed]
31. Brunetti, B.; Desimoni, E. Voltammetric determination of vitamin B6 in food samples and dietary supplements. *J. Food Compos. Anal.* **2014**, *33*, 155–160. [CrossRef]
32. Pintor, A.M.A.; Vieira, B.R.C.; Brandaõ, C.C.; Boaventura, R.A.R.; Botelho, C.M.S. Complexation mechanisms in arsenic and phosphorus adsorption onto iron-coated cork granulates. *J. Environ. Chem. Eng.* **2020**, *8*, 104184. [CrossRef]
33. Pintor, A.M.A.; Silvestre-Albero, A.M.; Ferreira, C.I.A.; Pereira, J.P.C.; Vilar, V.J.P.; Botelho, C.M.S.; Rodríguez-Reinoso, F.; Boaventura, R.A.R. Textural and surface characterization of cork-based sorbents for the removal of oil from water. *Ind. Eng. Chem. Res.* **2013**, *52*, 16427–16435. [CrossRef]
34. Monteiro, M.K.S.; Da Silva, D.R.; Quiroz, M.A.; Vilar, V.J.P.; Martínez-Huitle, C.A.; Dos Santos, E.V. Applicability of cork as novel modifiers to develop electrochemical sensor for caffeine determination. *Materials* **2021**, *14*, 37. [CrossRef]
35. Morante-Zarcero, S.; Pérez-Quintanilla, D.; Sierra, I. A disposable electrochemical sensor based on bifunctional periodic mesoporous organosilica for the determination of lead in drinking waters. *J. Solid State Electrochem.* **2015**, *19*, 2117–2127. [CrossRef]
36. Salmanipour, A.; Taher, M.A. An electrochemical sensor for stripping analysis of Pb(II) based on multiwalled carbon nanotube functionalized with 5-Br-PADAP. *J. Solid State Electrochem.* **2011**, *15*, 2695–2702. [CrossRef]
37. Oularbi, L.; Turmine, M.; El Rhazi, M. Electrochemical determination of traces lead ions using a new nanocomposite of polypyrrole/carbon nanofibers. *J. Solid State Electrochem.* **2017**, *21*, 3289–3300. [CrossRef]
38. Baś, B.; Jakubowska, M. The renovated silver ring electrode in determination of lead traces by differential pulse anodic stripping voltammetry. *Anal. Chim. Acta* **2008**, *615*, 39–46. [CrossRef] [PubMed]
39. Al-Gahouari, T.; Bodkhe, G.; Sayyad, P.; Ingle, N.; Mahadik, M.; Shirsat, S.M.; Deshmukh, M.; Musahwar, N.; Shirsat, M. Electrochemical sensor: L-cysteine induced selectivity enhancement of electrochemically reduced graphene oxide–multiwalled carbon nanotubes hybrid for detection of lead (Pb^{2+}) ions. *Front. Mater.* **2020**, *7*. [CrossRef]
40. Ghanei-Motlagh, M.; Taher, M.A. An electrochemical sensor based on novel ion imprinted polymeric nanoparticles for selective detection of lead ions. *Anal. Bioanal. Chem. Res.* **2017**, *4*, 295–306. [CrossRef]
41. Rahm, C.E.; Torres-Canas, F.; Gupta, P.; Poulin, P.; Alvarez, N.T. Inkjet printed multi-walled carbon nanotube sensor for the detection of lead in drinking water. *Electroanalysis* **2020**, *32*, 1533–1545. [CrossRef]

Article

Economic Valuation of Reducing Submerged Marine Debris in South Korea

Se-Jun Jin [1], Young-Ju Kwon [1] and Seung-Hoon Yoo [2,*] 

[1] Ocean Science and Technology Policy Research Section, Korea Institute of Ocean Science and Technology, 385 Haeyang-Ro, Yeongdo-Gu, Busan 49111, Korea; sjjin@kiost.ac.kr (S.-J.J.); yjkwon@kiost.ac.kr (Y.-J.K.)

[2] Department of Energy Policy, Graduate School of Energy & Environment, Seoul National University of Science & Technology, 232 Gongreung-Ro, Nowon-Gu, Seoul 01811, Korea

* Correspondence: shyoo@seoultech.ac.kr; Tel.: +82-2-970-6802

Received: 27 July 2020; Accepted: 31 August 2020; Published: 2 September 2020

Abstract: Submerged marine debris (SMD) scattered between sea level and the bottom of the sea damages the habitats of marine life and threatens its growth in South Korea. The collection of SMD is more difficult and expensive than that of coastal and floating debris. The government is trying to achieve a 33% reduction in SMD by 2023 by expanding its collection, which requires huge additional investments and additional information about the economic value or benefits of the reduction. This article seeks to conduct an economic valuation of the reduction by employing contingent valuation (CV), which asks people to indicate their willingness to pay (WTP) for the reduction. A dichotomous choice CV survey was undertaken with 1000 households by a professional survey firm through person-to-person interviews during July 2019. Overall, people understood the CV questions well and reported the WTP responses for a hypothetical market successfully created with CV. Although 37.9% of interviewees stated zero WTP, the average of the yearly household WTP was estimated as 5523 Korean won (KRW) (USD 4.92). This value ensures statistical significance. The population's WTP for the reduction would be KRW 110.30 billion (USD 99.75 million) per year over the next five years. It was found that the reduction is socially beneficial since the value was greater than the costs involved in the reduction.

Keywords: submerged marine debris; economic valuation; contingent valuation; economic benefit; willingness to pay

---

1. Introduction

The "plastic-free" movement is taking place around the world with plastic waste emerging as a serious environmental problem. In particular, the ocean often becomes the final destination for plastic waste, resulting in problems such as the creation of an island made up of plastic waste [1]. Waste containing plastics that flows into the ocean is called marine debris. The marine debris containing plastic is being pointed out as a global pollution problem [2], and various discussions and international agreements to strengthen regulations to reduce marine debris in the future are under way [3–7].

In line with this international situation, South Korea has tried to reduce marine debris. Marine debris in the past was mostly composed of decomposable materials, but postindustrial marine debris consisting of synthetic materials such as plastics cannot be decomposed. Due to its high buoyancy, plastic goods move thousands of miles into the ocean current and threaten marine ecosystems and wildlife [2]. Marine debris also causes damage to the economy and the marine environment [8]. Thus, in order for the marine ecosystem to provide sustainable services, it is urgent that the marine debris be proactively managed [9,10].

Three sides of the Korean Peninsula are surrounded by the sea, with the result that the country actively engages in fishing and trading activities. There are numerous ports and fishing ports on the East Sea, the South Sea, and the West Sea (the Yellow Sea). Moreover, rivers are connected to the sea. Inevitably, the country's geographical situation provides a route for land-based waste to flow into the sea. The inflow of waste through the rivers flows into the stream due to increased flow rates caused by a rainy season or typhoon. Currently, the country is actively carrying out policies and projects to mitigate floating marine debris on the coast [11], which is one type of marine debris.

It is difficult to identify the distribution and inflow path of the other type of marine debris, known as submerged marine debris (SMD), because it is located at the bottom of the sea. It is relatively easy and cheap to collect floating marine debris, but collection of SMD is expensive because it requires divers and special equipment. In addition, the disposal cost of marine waste in South Korea, approximately KRW 2.24 million (USD 1995) per ton in the case of sunken fishing net, is about eight times higher than that of land waste, which is approximately 270,000 Korean won (KRW) (USD 240) per ton. The SMD from the ocean as well as SMD flowing from land into the ocean has a negative impact on the marine environment. For instance, destruction of habitats of marine life, deterioration of the quality of marine products, threat to maritime safety, and damage to marine resources can arise [12,13]. In summary, although SMD is not classified as special waste, the collection of SMD is more difficult and expensive than that of coastal and floating debris because it is submerged at the bottom of the sea and requires special equipment and diving personnel to collect.

The South Korean government has established a legal basis for marine debris management at the national level and has pursued various polices to reduce marine debris, such as prevention, collection, and publicity. In particular, the government is trying to achieve a 33% reduction in SMD by 2023 by expanding its collection. The government's intent is to develop a collection system that considers the effects of SMD on the marine ecosystem and to pursue the collection of SMD in a way that reflects the use of space and ecological characteristics. Since the reduction requires a considerable amount of investment, the government is interested in the value people place on reducing SMD [1,14]. From an economics' point of view, people's willingness to pay (WTP) for the reduction is interpreted as the economic value or benefits ensuing from the reduction [15,16]. Whether or not the reduction is socially beneficial can be determined through comparing the economic benefits with the costs involved in the reduction.

This paper attempts to determine the economic value or benefits of the reduction of SMD by collecting and exploring people's WTP for the reduction. For this purpose, a survey-based economic technique called a contingent valuation (CV) method was adopted, and the results from a CV survey of 1000 interviewees are reported.

In South Korea, SMD occurs mainly through three channels. First, waste from the land or riverside flows into the river when heavy rain falls, and then into the sea through the estuary. If the nature of these wastes is investigated, more than 90% consists of trees and grass, although some of it is household waste such as waste appliances, waste plastics, and waste vinyl. It is impossible to identify the polluters in these cases. Therefore, it is difficult to find the polluters and make them pay for the cost of collecting SMD. In fact, the polluters may be ordinary people.

The second SMD channel consists of fishing-related waste that is intentionally or accidentally thrown into the ocean. Fishermen's fishing gear, fishing nets, waste nets, Styrofoam for buoys, and feed bins for cultivating aquatic products are flowing into the ocean. In the case of fishing-related waste, fishermen could be charged with SMD collection costs since they are clearly the polluters. However, not all fishermen discharge waste into the ocean. It is not easy to accurately identify the polluters. In addition, in South Korea, fishermen are exempt from various taxes and are given subsidies because their income levels are lower than those of other occupations. Therefore, imposing a financial burden on fishermen is not a very feasible alternative.

Third, waste intentionally or accidentally dumped by coastal inhabitants or islanders also flows into the ocean. Likewise, it is not easy to figure out who among these residents has leaked waste, and their income levels are low, making it difficult for them to bear new burdens even if they are polluters.

In summary, identifying the causal provider of SMD is not easy and, even if it is possible, imposing collection costs is not socially acceptable. In the end, the central government has no choice but to collect and utilize SMD through the funds raised from taxes paid by the general public. Thus, the framework of this study, which randomly selected 1000 households nationwide and asked WTP questions, was reasonable. This is because the polluters are not economic players such as certain companies, but an unspecified majority of the general public.

The effects of SMD belong to the category of negative externalities, which have many other economic effects. The economic effects of SMD are summarized in three ways. First, there is a decrease in tourism income owing to marine environment pollution by SMD. When SMD flows onto the beach, the number of visitors decreases, which, in turn, reduces the income of accommodations and shops near the beach. Second, there is economic damage that occurs to fishermen. SMD caught in a net causes damage to fish catches or fishing nets. In addition, SMD caught in a ship screw can lead to a ship accident, resulting in huge economic losses. The third effect of SMD is a national economic loss due to transboundary pollution. When SMD crosses the border along the current, transboundary pollution occurs, which can cause conflicts between countries.

These various economic effects of SMD can affect the public's WTP for reducing SMD. Therefore, in order to control these effects, it was assumed in the study that everything remained in its current state except for the change in the goods to be assessed. The assessed goods presented to respondents in this study reflected a 33% reduction of SMD by 2023 as compared to the business-as-usual (BAU) state, assuming that there are no economic effects on SMD.

There are three sections in the subsequent content of the paper. Section 2 reports materials and methods. Section 3 shows the main results of the analysis. Conclusions are presented in Section 4.

2. Materials and Methods

2.1. Survey Implementation and Data Collection

In order to collect CV data, the method of deriving the WTP from respondents; the payment vehicle, unit, and period; the method of survey; and the sample size had to be determined [17]. First, out of four methods of open-ended questions, bidding game questions, payment card questions, and dichotomous choice (DC) questions, which have been used in the literature as methods of eliciting WTP, this study adopted the DC question method. This was because the DC question method has been most frequently employed in the literature and possesses various merits, such as incentive compatibleness and mitigation of the respondents' cognitive burden [18]. In addition, Korea Development Institute [19] and Arrow et al. [20] present methodological guidance to be followed in applied CV research. For example, the survey correctly explained to respondents that there exist substitutes for the good, that the respondents' income is limited, and that consumption of other goods should be reduced to pay the WTP they have reported. In addition, this survey evaluated one of the many projects that the government should undertake. As will be explained below, this study tried to follow most of these guidelines.

Six important points had to be determined in order to conduct an actual field survey with a well-made CV questionnaire. First, the method of survey should be determined. This study adopted a person-to-person interview, which can facilitate the delivery of information rather than utilizing a relatively low-cost telephone, mail, or Internet method. In addition, the survey was conducted by experienced interviewers belonging to a professional opinion research institute.

Second, the size of the sample had to be determined. In this study, the population was all households in South Korea, and the population size is 19,971,359. The appropriate sample size had to be determined from this, with a 95% confidence level usually considered. In addition, a sample error of 5% is widely applied in South Korea, but a sample error of 3.1% was adopted in this study for more rigidity. The appropriate size of the sample was thus derived as approximately 1000, and 1000 observations were collected for the final analysis. Although a larger sample is better, it is important to size the sample at an appropriate level because the cost of the survey increases accordingly. In this regard, the size of the sample was set at 1000 following the suggestion of Korea Development Institute [19] and Arrow et al. [20]. In particular, since the costliest method of a person-to-person individual interview was conducted in this study, the sample size of 1000 was considered large enough and appropriate.

Third, the unit of the survey had to be determined. In this study, households were chosen out of individuals and households. This was because Korea Development Institute [19] proposed the use of households as a unit of the CV survey. In addition, conducting surveys of individuals may cause an issue as to what to do with the population when expanding the sample value to the population value. In other words, whether to include people under 20 or over 65 years of age who may lack economic ability can have a significant impact on the analysis results. On the other hand, conducting a survey of households is free of this issue. To improve the reliability of the data, the participants from households were limited to the household owner and his/her spouse, who have the actual burden of tax payment.

Fourth, the payment period must be determined. Naturally, the longer the payment period, the larger the total WTP, and the shorter the payment period, the smaller the total WTP. Therefore, it is important to reasonably determine the payment period. In this survey, payment was due for the next 10 years. This was because this period has been used in most applied CV works conducted in South Korea.

Fifth, the payment vehicle had to be fixed. The payment vehicle was decided as the yearly household income tax. Income tax is the most widely applied payment vehicle in empirical CV research for South Korea as it has the advantage of being relatively familiar to interviewees and not tied to everyday spending. In addition, Korea Development Institute [19] suggests as a guideline for applied CV studies that yearly household income tax should be used as a means of payment.

Sixth, the method of eliciting WTP responses had to be determined. Instead of the direct open-ended question method, the close-ended question method most widely applied in the literature was adopted. Of several close-ended questions, a DC question asking if an interviewee is willing to pay a specific amount was employed. The main part of the survey questionnaire in this study is given in Appendix A.

Various versions of the DC question are actually found in the literature. This study tried to apply a one-and-one-half-bounded (1.5B) model. Cooper et al. [21] proposed the model, which has several advantages [22–26]. The procedure of applying the model can be explained in the following manner. First, two bid amounts, D^L and D^H ($D^L < D^H$), should be determined through preliminary investigation. Half of all respondents were asked to agree on payment after presenting the smaller (D^L) of the two amounts first. If "yes" was responded, the higher bid (D^H) was presented and an additional question about its payment was asked. If "no" was stated to D^L, an additional question was not needed. The remaining respondents were given a higher amount (D^H) first. If "yes" was reported, an additional question was not asked. However, if "no" was answered, the lower amount (D^L) was presented to the respondent. Thus, six responses were possible: "no," "yes-no," "yes-yes," "no-no," "no-yes," and "yes."

This study sought to take a closer look at the cases of "no" and "no-no" responses among these. Responses that indicated no intention of paying a lower amount (D^L) were further classified into zero WTP and WTP greater than zero and less than a lower amount (D^L). Therefore, a further question was asked to identify to which of the two classifications the "no" and "no-no" responses belonged. To this end, a further question was presented about whether the interviewee had no intention of paying a dime, i.e., to check for a zero WTP. The final number of cases was, therefore, eight:

$$
\begin{cases}
I_q^{YY} = K\!\left(D_q^H < E_q\right) = K(q\text{th answer is "yes-yes"}) \\
I_q^{YN} = K\!\left(D_q^L < E_q \le D_q^H\right) = K(q\text{th answer is "yes-no"}) \\
I_q^{LY} = K\!\left(E_q \le D_q^L\right) = K(q\text{th answer is "no-yes"}) \\
I_q^{NN} = K\!\left(E_q = 0\right) = K(q\text{th answer is "no-no"}) \\
J_q^{Y} = K\!\left(D_q^H < E_q\right) = K(q\text{th answer is "yes"}) \\
J_q^{NY} = K\!\left(D_q^L < E_q \le D_q^H\right) = K(q\text{th answer is "no-yes"}) \\
J_q^{NNY} = K\!\left(E_q \le D_q^L\right) = K(q\text{th answer is "no-no-yes"}) \\
J_q^{NNN} = K\!\left(E_q = 0\right) = K(q\text{th answer is "no-no-no"})
\end{cases}
\tag{1}
$$

where I and J are binary variables with zero or one, q indicates qth interviewee, and $K(\cdot)$ is an indicator function. If the proposition in parenthesis is true, the function has a value of one. Otherwise, the function has a value of zero.

The WTP data obtained from a CV survey conducted on 1000 households during July 2019 is summarized in Table 1.

Table 1. Willingness-to-pay data obtained and used in this study.

Bid Amount [a]		Lower Bid is Suggested First (%) [b]				Higher Bid is Suggested First (%) [b]				Number of Observations
		"yes-yes"	"yes-no"	"no-yes"	"no-no"	"yes"	"no-yes"	"no-no-yes"	"no-no-no"	
1000	3000	20 (14.0)	19 (13.3)	7 (4.9)	26 (18.2)	31 (21.7)	12 (8.4)	3 (2.1)	25 (17.5)	143 (100.0)
2000	4000	26 (18.2)	10 (7.0)	8 (5.6)	27 (18.9)	29 (20.3)	11 (7.7)	5 (3.5)	27 (18.9)	143 (100.0)
3000	6000	18 (12.6)	13 (9.1)	10 (7.0)	30 (21.0)	30 (21.0)	13 (9.1)	9 (6.3)	20 (14.0)	143 (100.0)
4000	8000	20 (14.0)	12 (8.4)	14 (9.8)	26 (18.2)	15 (10.5)	12 (8.4)	18 (12.6)	26 (18.2)	143 (100.0)
6000	10,000	14 (9.9)	13 (9.2)	18 (12.7)	26 (18.3)	20 (14.1)	5 (3.5)	16 (11.3)	30 (21.1)	142 (100.0)
8000	12,000	16 (11.3)	15 (10.6)	13 (9.2)	27 (19.0)	16 (11.3)	10 (7.0)	17 (12.0)	28 (19.7)	142 (100.0)
10,000	15,000	12 (8.3)	12 (8.3)	20 (13.9)	28 (19.4)	11 (7.6)	8 (5.6)	20 (13.9)	33 (22.9)	144 (100.0)
Totals		126 (12.6)	94 (9.4)	90 (9.0)	190 (19.0)	152 (15.2)	71 (7.1)	88 (8.8)	189 (18.9)	1000 (100.0)

Notes: [a] Unit is Korean won (USD 1.0 = KRW 1122.8 at the time of the survey). [b] Numbers reported in parentheses mean the percentage of the number of observations.

2.2. Method: CV

The CV method has various advantages and disadvantages. Three advantages are as follows. First, unlike revealed preference approaches, such as the hedonic price technique and the travel cost technique, CV is a stated preference technique and can be used to estimate the economic value that explicitly includes non-use value. Second, from an economic point of view, the CV technique can theoretically provide an accurate estimate of the economic value or benefits from the supply of a certain good, while the revealed preference techniques have room for underestimation or overestimation. Third, since the validity and reliability of the CV approach is proven to some extent in the literature [22–26], the CV approach has been a widely applied one.

The CV method also has three disadvantages. First, the application of CV is more costly than that of other economic techniques because a survey of many respondents is essentially needed. For researchers facing budget constraints, the application of CV may be restrictive. Second, a valuation through CV based on the data collected using the questionnaire can be subject to various biases, as it can be influenced by the content of the questionnaire, the attitude of the interviewer, and the operation of the survey method. Third, since CV techniques are based on stated responses gathered from people instead of human behavior, people are less likely to believe in the value obtained by using CV.

The DC question method has several merits and demerits. There are two typical merits to the DC method. First, it is quite familiar to respondents. Even if a person has not experienced a referendum, the type of question is similar to deciding whether to buy a good on the market. Therefore, people can answer DC questions without much difficulty. Second, it is incentive compatible for people to respond.

People buy a certain good if their utility from the purchase and consumption of it is greater than or equal to the price of the good; they do not otherwise buy the good. If a person's WTP is greater than the presented bid, she/he will answer "yes" and otherwise "no." There is no reason to take a strategic behavior when a person is faced with the DC question.

The DC method has two demerits. First, the use of the DC question results in discrete interval data rather than continuous point data. This makes an econometric analysis of the CV data less statistically efficient than other value elicitation methods such as open-ended questions. Thus, the use of the DC method requires the collection of a large number of observations and demands a large survey cost. Second, a pretest survey is required to determine a list of bid amounts to be presented to the respondents, resulting in costs associated with the pretest survey and longer application period than other methods.

Reducing SMD is a typical nonmarket good. A nonmarket good means that it cannot be traded in the usual market. A good traded in the market is easily valued, but a nonmarket good does not have a market and its value is not well observed [27]. Therefore, for the purpose of assessing the economic benefits ensuing from the reduction, it is necessary to create a hypothetical market for the reduction and hypothetically trade the reduction in the market. A CV method is a typical economic method that can be done in this way, and it has been widely utilized in the literature [28–34]. The CV technique uses a questionnaire to explain the good to be assessed to the randomly chosen potential consumers and then to make a hypothetical transaction, leading them to reveal their WTP for consuming the good [35–39]. Next, the researcher estimates the WTP model and calculates the average WTP by applying an econometric model to the WTP data collected from a survey.

Therefore, the first thing a researcher should do for the application of CV is to carefully make the questionnaire. CV questionnaires usually have three components. The first component addresses questions about potential consumer perceptions and experiences toward the good being assessed. The second component presents an explanation of the good to be evaluated and a question about the WTP. Questions about the individual characteristics of the consumers are shown in the third component.

The most important part of the CV questionnaire is the explanation of the good under investigation. The good should be identical for all respondents and should be accurately described in the questionnaire. The BAU state, which can be a reference for valuation, and the target state to be assessed should be clearly described. In other words, the object of valuation in CV is the amount of the WTP to obtain a change from the BAU state to the target state. Moreover, the policy measures associated with how to get the change should be fully explained. The policy measures presented in the CV survey were the establishment of a scientific forecasting system for early prediction of inflows and travel paths of marine debris, and research and development on how to collect and treat the SMD.

2.3. Modeling the CV Data

As mentioned earlier, this study attempted to explicitly deal with zero WTP in analyzing the DC CV data. To this end, the spike model given in Kriström [40], Habb and McConell [41] and Yoo and Kwak [42] was applied. Hanemann's [43] approach to modeling DC CV data was also used. Therefore, the CV data model used in this study was the 1.5B DC spike model. The 1.5B DC spike model can explicitly reflect zero WTP as well as positive WTP responses. The mean WTP estimate obtained from analyzing the 1.5B DC spike model using a total of 1000 observations was considered reasonable in its use for information about the benefits ensuing from the reduction. For the application of this model, the cumulative distribution function (cdf) of WTP had to first be defined. This study adopted a logistic function that is almost always applied in the spike model. Thus, the cdf, $M_E(\cdot)$, can be specified as:

$$M_E(E; \tau_0, \tau_1) = \begin{cases} [1 + \exp(\tau_0 - \tau_1 E)]^{-1} & \text{if } E \geq 0 \\ 0 & \text{if } E < 0 \end{cases} \tag{2}$$

where τ_0 and τ_1 are parameters of $M_E(\cdot)$. If $E = 0$, the equation in the first line on the right side becomes the spike. Thus, the spike means $\Pr(E = 0)$. E is a bid presented to respondents and $\Pr(\cdot)$ means a probability.

Concerning the model, the log-likelihood function can be specified as:

$$
\begin{aligned}
\ln L = \ \sum_{q=1}^{1000} & \left\{ \left(I_q^{YY} + J_q^{Y}\right) \ln\left[1 - M_E\left(E_q^{H}; \tau_0, \tau_1\right)\right] \right. \\
& + \left(I_q^{YN} + J_q^{NY}\right) \ln\left[M_E\left(E_q^{H}; \tau_0, \tau_1\right) - M_E\left(E_q^{L}; \tau_0, \tau_1\right)\right] \\
& + \left(I_q^{NY} + J_q^{NNY}\right) \ln\left[M_E\left(E_q^{L}; \tau_0, \tau_1\right) - M_E\left(0; \tau_0, \tau_1\right)\right] \\
& \left. + \left(I_q^{NN} + J_q^{NNN}\right) \ln M_E\left(0; \tau_0, \tau_1\right) \right.
\end{aligned}
\tag{3}
$$

Maximum likelihood (ML) estimation method relates to obtaining parameter estimates that maximize the log-likelihood function. This study employed the ML estimation method. Thus, the estimates for τ_0 and were obtained by maximizing Equation (3). In addition, the mean WTP was derived from Equation (2) as [44,45]:

$$
(1/\tau_1) \ln[1 + \exp(\tau_0)]
\tag{4}
$$

3. Results and Discussion

3.1. Estimation Results

The results obtained through an application of maximum likelihood estimation to the model, Equation (3), are given in Table 2. The dependent variable is the probability of responding "yes" to a suggested bid. As the value of the bid becomes greater, the probability should be reduced. The coefficient estimate for bid amount is negative and statistically significant at the 1% level. This is quite reasonable. Given that the sample proportion of "no-no" and "no-no-no" responses, that is, zero WTP was 37.9%, the estimated spike of 0.3859 implies that the data were well represented by the spike model. Moreover, the spike secures statistical significance at the 1% level.

Table 2. Estimation results of the model and the mean willingness to pay (WTP).

Variables	Coefficient Estimates (*t*-Values)
Constant	0.4645 (7.31) [#]
Bid amount [a]	−0.1724 (−21.16) [#]
Spike	0.3859 (25.62) [#]
Yearly household mean WTP	KRW 5,523 (USD 4.92)
t-value	13.70 [#]
95% CI [b]	KRW 5055 to 6076 (USD 4.90 to 5.41)
99% CI [b]	KRW 4896 to 6284 (USD 4.36 to 5.60)
Sample size	1000
Log-likelihood	−1330.22
Wald statistic (*p*-value) [c]	656.28 (0.000)

Notes: [a] The unit is 1000 Korean won (USD 1.0 = 1122.8 at the time of the survey). [b] CI indicates confidence interval. [c] The null hypothesis is that all the parameter estimates are jointly zero. The [#] means statistical significance at the 1% level.

The Wald test can be employed for the specification test of the model. The null hypothesis is that the estimated coefficients for bid amount as well as constant terms are not distinguishable from zero. In other words, the hypothesis implies the meaninglessness of the model. The statistic was 656.28. Since this value is sufficiently large, the hypothesis could be rejected without deficiency. In addition, the *p*-value for the statistic was 0.000. Thus, the model possessed statistical significance.

The yearly household average WTP for the reduction was obtained as KRW 5523 (USD 4.92). Uncertainties can be involved in the estimation of the average. In such cases, it may be a good idea to report confidence intervals together rather than just point estimate. To this end, this study adopted a parametric estimation technique developed by Krinsky and Robb [46] to present a confidence interval for the average. This method assumed that the estimates of the constant term and coefficient for the bid amount, given in Table 2, followed a bivariate normal distribution and produced an empirical distribution of the mean WTP by extracting the coefficients from this distribution and calculating the mean WTP 5000 times. Cutting the appropriate proportion from the left and right sides of this empirical distribution can find 95% and 99% confidence intervals. In other words, to find 95% and 99% confidence intervals, 2.5% and 0.5% are cut from the left and right sides of the distribution, respectively. Table 2 also reports them.

The results presented in Table 2 do not contain other covariates related to the interviewee's characteristics. However, other factors could influence the likelihood of reporting "yes" to an offered bid. For instance, some variables concerning the interviewee, such as gender, household income, and education level, can be introduced. A model including some covariates can be considered for investigating the possible effects of such variables. For this purpose, four variables were reflected in the model with covariates. Basic information about the covariates is described in Table 3.

Table 3. Information about some variables considered in this study.

Variables	Definitions	Mean	Standard Deviation
Education	Education level of the respondent in years	14.10	2.23
Income	Monthly income of the respondent's household (unit: million Korean won = USD 891)	4.91	2.14
Head	Dummy for the respondent's being head of household (0 = no; 1 = yes)	0.53	0.50
Age	Age of the respondent (unit: years)	47.78	9.20

Table 4 shows the estimation results of the model containing the variables described in Table 3. The Wald statistic for the specification test of the model was 607.79. This indicates that the null hypothesis of the model's being meaningless was rejected, considering that its *p*-value became 0.000. One of the purposes of estimating the model with covariates is to check for internal consistency or theoretical validity. Except for the estimated coefficient for Head variable, all the coefficient estimates were statistically significant. Thus, it seems that the model employed in this paper secured internal consistency. As explained above, the sign of the coefficient means the direction of the effect of the variable on the likelihood of responding "yes" to a provided bid. The coefficient estimates for Education, Age, and Income variables had statistical significance. Respondents with higher levels of education had a higher possibility than respondents with lower levels of education. The age of the respondents was negatively correlated with the possibility. An interviewee with a higher income was more likely to answer "yes" to a proposed bid than an interviewee with a lower income. Contrary to our prior expectations, the estimates of the coefficient for the Head variable was not statistically significant at a level of 10%. Whether or not the respondent was the head of the household did not affect the respondent's determination of WTP for the reduction. This was quite an interesting finding because it is often thought in the country that this variable will affect the respondent's decision on WTP.

The important purpose of estimating a model containing covariates is to verify the theoretical validity or internal consistency of the model. The model used in this study appeared to be meeting these since the estimation results were overall significant. Table 4 also presents the yearly household average WTP estimate and its confidence intervals. The mean WTP was KRW 5412 (USD 4.82), which is not much different from the results given in Table 2 (KRW 5523 or USD 4.92). In addition, the confidence intervals were much the same as those given in Table 2.

Table 4. Estimation results of the model containing the variables described in Table 3.

Variables	Estimates	*t*-Values
Constant	−1.1772	−1.80 *
Bid amount [a]	−0.1803	−21.29 **
Education	0.1252	3.90 **
Income	0.1221	4.32 **
Head	−0.1881	−1.54
Age	−0.0122	−1.66 *
Spike	0.3769	24.65 **
Yearly household mean willingness to pay	KRW 5412 (USD 4.82)	
t-value	21.01 **	
95% CI [b]	KRW 4946 to 5965 (USD 4.41 to 5.31)	
95% CI [b]	KRW 4778 to 6123 (USD 4.26 to 5.45)	
Sample size	1000	
Log-likelihood	−1297.06	
Wald statistic (*p*-value) [c]	607.79 (0.000)	

Notes: [a] The unit is 1000 Korean won (USD 1.0 = 1122.8 at the time of the survey). [b] CI indicates confidence interval. [c] The null hypothesis is that all the parameter estimates are jointly zero. The * and ** indicate statistical significance at the 10% and 5% levels, respectively.

3.2. Discussion of the Results

This study investigated people's WTP for reducing SMD in South Korea by 33% by 2023 by means of expanding SMD collection. It is possible to compare three sample characteristics with three population characteristics given in Statistics Korea [47]. First, the sample proportion of female persons can be compared with the population proportion of female persons. The first (50.0%) is not different from the second (49.9%). Second, three areas with a large number of households can be investigated. The sample proportions of Gyeonggi, Seoul, and Busan respondents were 23.9%, 20.1%, and 7.2%, while the population proportions of Gyeonggi, Seoul, and Busan respondents were 23.7%, 19.4%, and 7.0%, making no significant difference. Third, the average monthly income of households can be examined. The sample value was KRW 4.86 million (USD 4136) and the population value was KRW 4.92 million (USD 4187), almost the same. The authors also think that the population could be reasonably represented by the sample because sampling was entrusted to a specialized survey institute. Therefore, extending the results for the previously presented sample to the population would not be a problem.

One of the most important purposes of the applied CV study was to expand the location value of WTP obtained from the sample to the population. In this regard, we should have determined whether to use the location values for the sample. Usually, mean, median, and mode are used for location value. The median WTP obtained in this study was zero and the mean WTP was estimated to be positive. The mode WTP could not be computed because we used DC WTP question. It was necessary to determine which, of mean or median, to use. Median is known to be all the more robust than mean because mean is vulnerable to outliers but median is not. Median can be useful for identifying the central tendency of the sample, but it is not used for expanding a sample value to a population value because it can cause underestimation to overestimation in the expansion. Therefore, the mean WTP has been almost always employed in the literature to estimate population value using sample value.

As explained earlier, we conducted stratified random sampling using 16 strata. The sample size allocated to each stratum was decided based on the Census implemented by Statistics Korea in 2015. The sample size of each stratum was, thus, consistent with the population. In this study, the total value was calculated by multiplying the mean WTP by the number of households in the population instead of using a mean formula applied for stratified sampling.

The average of the household's yearly WTP for the reduction was computed as KRW 5523 (USD 4.92). This sample value can be extended to the population. The population's total WTP was derived as the multiplication of the relevant number of households by the average WTP. Since the

CV survey was conducted throughout the country, the relevant population became the entire country. There were 19,971,359 households when the survey was implemented [47]. The yearly population value would be KRW 110.30 billion (USD 99.75 million). Comparison of this value with the costs involved in the reduction is an interesting task.

An economic feasibility analysis of the reduction was tried as a final exercise. To this end, some prerequisites needed to be examined and determined. First, the period for the analysis had to be set. It was determined as five years, beginning from 2019 when the survey was implemented and when the reduction begins in 2023. Second, a social discount rate should be set. Concerning this, the government-run Korea Development Institute announced the suggested use of 4.5% as a social discount rate. This study adopted the value. Third, the time of "present" as a baseline for calculating the present value had to be set. In this study, this was set to be 2019, the time when the survey was conducted.

The next important information that was needed was benefits and costs arising from the reduction. As presented earlier, the economic benefits ensuing from the reduction would occur annually for 10 years, from 2019 to 2023. The costs largely relate to collection and treatment of SMD and were taken from "The Third Marine Debris Master Plan (2019–2023)" contained in Korea Ministry of Oceans and Fisheries [48]. The costs amounted to about KRW 6.69 billion (USD 0.60 million), KRW 10.20 billion (USD 0.91 million), KRW 11.22 billion (USD 1.00 million), KRW 12.35 billion (USD 1.10 million), and KRW 12.66 billion (USD 1.13 million) over the period 2019-2023, respectively. All benefits and costs are expressed in a 2019 constant price. The ratio of benefit over cost (B/C), which is one of the indicators for cost-benefit analysis (CBA), can be calculated from the constant values of benefits and costs. It is a simplification of the real CBA.

The present value of the benefits arising from the reduction was computed as KRW 1103.02 billion (USD 98.24 million) and that of the costs arising from the reduction became KRW 48.16 billion (USD 4.29 million). Thus, the net present value of the reduction became KRW 863.89 billion (USD 76.94 million), which is larger than zero and thus implies that the reduction passed the economic feasibility analysis. Furthermore, the ratio of benefit over cost was computed to be 18.93, which is larger than one, confirming the finding that the reduction secures economic feasibility. In conclusion, the reduction of SMD in South Korea is socially beneficial. Therefore, a continuous and stable reduction must be conducted.

4. Conclusions

South Korea is trying to reduce SMD by 33% by 2023 through expanding the collection of it. The implication from this study is all the more interesting since no research that has evaluated people's WTP for reducing SMD is found in the literature. Therefore, this study can be a useful contribution to the literature. In particular, the information about the value people place on the reduction is widely demanded to determine whether the reduction has sufficient public support. For the purpose of providing this information to policymakers, this article empirically looked into people's WTP for the reduction, employing data collected through a survey of 1000 households through person-to-person interviews during July 2019. In addition, the spike model was estimated not only using the entire sample data, including observations with zero WTP, but also only using observations with positive WTP. In this regard, the mean WTP estimate obtained for the sample can be extended over the population.

Judging from the comments of the supervisor and interviewers, the survey was implemented without difficulty and successfully enough to collect opinions representative of the population. Overall, people understood the CV questions well and reported the WTP responses in a hypothetical market successfully created with CV. Respondents stated a significant amount of WTP for the reduction. This study can provide three important policy implications. First, people's WTP for reducing SMD was quantitatively assessed. Although 37.9% of interviewees stated zero WTP, the average of the yearly household WTP was calculated to be KRW 5523 (USD 4.92), which is not big compared to the average household's annual income (KRW 58.9 million or USD 52.5 thousand). However, it has statistical significance and can be utilized as a logical basis for the government to continue to push for reduction of SMD. In fact, the various CV empirical studies conducted in South Korea have often encountered too many zero WTP responses [22–24,26]. For example, Lim and Yoo [22], Kim et al. [23], Kim et al. [24], and Kim and Yoo [26] reported that the proportion of zero WTP responses was 56.5%, 46.5%, 61.7%, and 63.6%, respectively. The zero WTP response rate in this study is fairly small compared with the preceding studies. Therefore, it can be judged that the public is giving considerable value to the reduction.

Second, according to microeconomics, WTP means the economic benefits that arise from the reduction. The study evaluated the economic benefits that ensue from a 33% reduction in SMD by 2023 through an expansion of its collection and found that the population's total WTP was KRW 110.30 billion (USD 99.75 million).

Third, the economic benefits can be compared with the costs caused by the reduction. In this regard, a cost-benefit analysis of the reduction indicated that the reduction is socially beneficial and, therefore, the investment on the reduction can be economically justified. This is because the net present value was larger than zero. Moreover, the benefit/cost ratio was greater than one. Thus, the reduction of SMD should be stably and continuously performed.

As addressed above, the CV approach has some limitations due to using a survey of respondents. For example, in a hypothetical market created with CV, payments of a certain amount are not actually made but are made hypothetically. Therefore, it would be useful to conduct a test for the respondents' sincerity that adopts some statistical methods. One method is to include questions which gather ordinal data on a Likert-type scale in the CV questionnaire and then to compute a Cronbach's alpha to test for internal consistency. Unfortunately, this study did not contain the questions in the CV questionnaire. This point needs to be appropriately handled in future CV studies.

To the best of the authors' knowledge, there have not been many cases studies that applied CV to reducing marine debris, specifically SMD. Thus, one purpose of this study was to add a case study of South Korea to the literature. In particular, the implications of this study are all the more useful because there have been no related studies for the country. Nevertheless, this study needs to be improved in several respects to ensure that it is distinct from previous studies. First, if more observations are obtained through additional budgeting, the respondents can be segmented and the analysis could be made according to various criteria, such as geolocation of the respondent, whether the respondent had knowledge of campaigns around the issue prior to the survey, and relevance of the local issue, so as to obtain differentiated implications for each segmented group. Second, because the presented results are preliminary or partial, they can be supplemented by including enterprises in the survey. Since households consume products produced by businesses, there is a view that businesses are the ultimate polluting sources of the seas and oceans. Therefore, a follow-up study can be carried out by including enterprises as interviewees and using taxes paid by entrepreneurs and companies as the payment vehicle. In addition, future research should be conducted on companies that can respond to the polluters pay principle. Only then will this be helpful in establishing a more detailed policy on SMD.

Author Contributions: This paper was written through the collaboration of the three authors. S.-J.J. collected data, analyzed the collected data statistically, and compiled the results; Y.-J.K. compiled the data and wrote a significant portion of the paper; and S.-H.Y. proposed research ideas, secured the necessary budget for the survey, and supervised the entire process of the study. All authors have read and agreed to the published version of the manuscript.

Funding: This research was a part of the project titled A Study on the Integrated Management of Marine Space (PE99843), funded by the Korea Institute of Ocean Science and Technology (KIOST).

Acknowledgments: The authors are grateful to the three anonymous reviewers for their valuable comments and suggestions for improving this manuscript. This research was supported by the project titled A Study on the Integrated Management of Marine Space (PE99843), awarded by the Korea Institute of Ocean Science and Technology (KIOST).

Conflicts of Interest: The authors declare no conflict of interest.

Appendix A. Main Part of the Survey Questionnaire

Part 1. Questions about Socio-Economic Characteristics

The interviewees were asked to respond to their socio-economic characteristics, such as the gender of the individual, the number of family members, the level of education, and the monthly income per household (before tax deduction). Questions about the number of family and income were open-ended questions, while the question about the level of education was as follows:

Table A1. Please check with √ your education level in years.

Education Level	Uneducated	Elementary School	Middle School	High School	University	Graduate School
Education level in years	0	1, 2, 3, 4, 5, 6	7, 8, 9	10, 11, 12	13, 14, 15, 16	17, 18, 19, 20

Part 2. Questions about Willingness to Pay for Reducing the Submerged Marine Debris (SMD) in South Korea

Type A. Q1. Is your household willing to pay additional income tax of 1000 Korean won (lower bid amount) annually for the next 10 years for reducing SMD in South Korea, supposing that the protection is certain to succeed?

a.   Yes—go to Type A. Q2.
b.   No—go to Q3.

Type A. Q2. Is your household willing to pay additional income tax of about 3000 Korean won (upper bid amount) annually for the next 10 years for reducing SMD in South Korea, supposing that the protection is certain to succeed?

a.   Yes—finish this survey.
b.   No—finish this survey.

Type B. Q1. Is your household willing to pay additional income tax of about 3000 Korean won (upper bid amount) annually for the next 10 years for reducing SMD in South Korea, supposing that the protection is certain to succeed?

a.   Yes—finish this survey.
b.   No—go to Type B. Q2.

Type B. Q2. Is your household willing to pay additional income tax of about 1000 Korean won (lower bid amount) annually for the next 10 years for reducing SMD in South Korea, supposing that the protection is certain to succeed?

a.   Yes—finish this survey.
b.   No—go to Q3.

Q3. Then, is your household not willing to pay anything for reducing SMD in South Korea?

a.   Yes, our household is willing to pay something less than 1000 Korean won.
b.   No, our household is not willing to pay anything. In other words, our household's willingness to pay is zero.

References

1.   Zambrano-Monserrate, M.A.; Ruano, M.A. Estimating the damage cost of plastic waste in Galapagos Islands: A contingent valuation approach. *Mar. Policy* **2020**, *117*, 103933. [CrossRef]
2.   Sheavly, S.B.; Register, K.M. Marine Debris & Plastics: Environmental Concerns, Sources, Impacts and Solutions. *J. Polym. Environ.* **2007**, *15*, 301–305.

3. Chen, C.L. Regulation and management of marine litter. In *A Marine Anthropogenic Litter*; Bergmann, M., Gutow, L., Klages, M., Eds.; Springer: Dordrecht, The Netherlands, 2015; pp. 395–428.

4. United Nations Environment Programme. *Marine Plastic Debris and Microplastics Global Lessons and Research to Inspire Action and Guide Policy Change*; United Nations Environment Programme: Nairobi, Kenya, 2016. (In English)

5. Ellen MacArthur Foundation. The New Plastics Economy: Rethinking the Future of Plastics & Catalysing Action. 2017. Available online: http://www.ellenmacarthurfoundation.org/publications (accessed on 4 June 2020).

6. National Oceanic and Atmospheric Administration. What is Marine Debris? 2018. Available online: https://oceanservice.noaa.gov/facts/marinedebris.html (accessed on 4 June 2020).

7. Olivelli, A.; Hardesty, D.; Wilcox, C. Coastal margins and backshores represent a major sink for marine debris: Insights from a continental-scale analysis. *Environ. Res. Lett.* **2020**, in press. Available online: https://doi.org/10.1088/1748-9326/ab7836 (accessed on 4 June 2020). [CrossRef]

8. McIlgorm, A.; Campbell, H.F.; Rule, M.J. The economic cost and control of marine debris damage in the Asia-Pacific Region. *Ocean Coast. Manag.* **2011**, *54*, 643–651. [CrossRef]

9. United Nations. The Sustainable Development Goals Report. 2017. Available online: https://unstats.un.org/sdgs/report/2017/ (accessed on 4 June 2020).

10. Chiba, S.; Saito, H.; Fletcher, R.; Yogi, T.; Kayo, M.; Miyagi, S.; Ogido, M.; Fujikura, K. Human footprint in the abyss: 30 year records of deep-sea plastic debris. *Mar. Policy* **2018**, *96*, 204–212. [CrossRef]

11. Jang, Y.C.; Lee, J.M.; Hong, S.W.; Mok, J.Y.; Kim, K.S.; Lee, Y.J.; Choi, H.W.; Kang, H.M.; Lee, S.H. Estimation of the annual flow and stock of marine debris in South Korea for management purposes. *Mar. Pollut. Bull.* **2014**, *86*, 505–511. [CrossRef]

12. Cho, D.O. Challenges to marine debris management in Korea. *Coast. Manag.* **2005**, *33*, 389–409. [CrossRef]

13. Hong, S.W.; Lee, J.M.; Jang, Y.C.; Kim, Y.J.; Kim, H.J.; Han, D.; Hong, S.H.; Kang, D.; Shim, W.J. Impacts of marine debris on wild animals in the coastal area of Korea. *Mar. Pollut. Bull.* **2013**, *66*, 117–124. [CrossRef]

14. Choi, E.C.; Lee, J.S. The willingness to pay for removing the microplastics in the ocean—The case of Seoul metropolitan area, South Korea. *Mar. Policy* **2018**, *93*, 93–100. [CrossRef]

15. Dribek, A.; Voltaire, L. Contingent valuation analysis of willingness to pay for beach erosion control through the stabiplage technique: A study in Djerba (Tunisia). *Mar. Policy* **2017**, *86*, 17–23. [CrossRef]

16. Nieminen, E.; Ahtiainen, H.; Lagerkvist, C.J.; Oinonen, S. The economic benefits of achieving Good Environmental Status in the Finnish marine waters of the Baltic Sea. *Mar. Policy* **2019**, *99*, 181–189. [CrossRef]

17. Champ, P.A. Collecting nonmarket valuation data. In *A Primer on Nonmarket Valuation*, 2nd ed.; Champ, P.A., Boyle, K.J., Brown, T.C., Eds.; Springer: Dordrecht, The Netherlands, 2017.

18. Mitchell, R.C.; Carson, R.T. *Using Surveys to Value Public Goods: The Contingent Valuation Method*; Resources for the Future: Washington, DC, USA, 1989.

19. Korea Development Institute. *Guidelines for Preliminary Feasibility Study Using Contingent Valuation Method*; Korea Development Institute: Sejong, Korea, 2012. (In Korean)

20. Arrow, K.; Solow, R.; Portney, P.R.; Leamer, E.E.; Radner, R.; Schuman, H. Report of the NOAA panel on contingent valuation. *Fed. Regist.* **1993**, *58*, 4601–4614.

21. Cooper, J.C.; Hanemann, M.; Signorello, G. One-and-one-half bound dichotomous choice contingent valuation. *Rev. Econ. Stat.* **2002**, *84*, 742–750. [CrossRef]

22. Lim, S.Y.; Yoo, S.H. Will South Korean residential consumers accept the renewable heat incentive scheme? A stated preference approach. *Energies* **2019**, *12*, 1910. [CrossRef]

23. Kim, J.H.; Kim, H.J.; Yoo, S.H. Public value of enforcing the $PM_{2.5}$ concentration reduction policy in South Korean urban areas. *Sustainability* **2018**, *10*, 1144. [CrossRef]

24. Kim, H.J.; Lee, H.J.; Yoo, S.H. Public willingness to pay for endocrine disrupting chemicals-free labelling policy in Korea. *Appl. Econ.* **2019**, *51*, 131–140. [CrossRef]

25. Mahmoodi, A.; Ghashti, M.J.; Yavari, G.R.; Mehrara, M.; Yazdani, S. Estimating the recreational value of Rudkhan Castel Forest Park: Application of One and One-half Bound (OOHD) dichotomous choice contingent valuation. *Agric. Econ. Dev.* **2019**, *33*, 313–327.

26. Kim, J.H.; Yoo, S.H. South Koreans' perspective on assisting the power supply to North Korea: Evidence from a contingent valuation. *Energy Policy* **2020**, *139*, 111336. [CrossRef]

27. Segerson, K. Valuing environmental goods and services: An economic perspective. In *A Primer on Nonmarket Valuation*, 2nd ed.; Champ, P.A., Boyle, K.J., Brown, T.C., Eds.; Springer: Dordrecht, The Netherlands, 2017.

28. Smith, V.K. Fifty years of contingent valuation. In *Handbook on Contingent Valuation*; Alberini, A., Kahn, J.R., Eds.; Edward Elgar: Cheltenham, UK, 2006.

29. Carson, R.T. *Contingent Valuation: A Comprehensive Bibliography and History*; Edward Elgar: Cheltenham, UK, 2011.

30. Loomis, J.; González-Cabán, A.; Champ, J. Estimating the robustness of contingent valuation estimates of WTP to survey mode and treatment of protest responses. In *The International Handbook on Non-Market Environmental Evaluation*; Bennett, J., Ed.; Edward Elgar: Cheltenham, UK, 2011.

31. Freeman, A.M., III; Herriges, J.A.; Kling, C.L. *The Measurement of Environmental and Resource Values: Theory and Methods*, 3rd ed.; RFF Press: New York, NY, USA, 2014.

32. Flores, N.E. A Conceptual framework for nonmarket valuation. In *A Primer on Nonmarket Valuation*, 2nd ed.; Champ, P.A., Boyle, K.J., Brown, T.C., Eds.; Springer: Dordrecht, The Netherlands, 2017.

33. Bateman, I.J.; Carson, R.T.; Day, B.; Hanemann, M.; Hanley, N.; Hett, T.; Sugden, R. *Economic Valuation with Stated Preference Techniques: A Manual*; Edward Elgar: Cheltenham, UK, 2002.

34. Haab, T.; Lewis, L.; Whitehead, J. *State of the Art of Contingent Valuation*; Oxford Research Encyclopedia of Environmental Science; Oxford University Press: Oxford, UK, 2020.

35. Boyle, K.J. Contingent valuation in practice. In *A Primer on Nonmarket Valuation*, 2nd ed.; Champ, P.A., Boyle, K.J., Brown, T.C., Eds.; Springer: Dordrecht, The Netherlands, 2017.

36. Ahmed, S.U.; Gotoh, K. *Cost-Benefit Analysis of Environmental Goods by Applying the Contingent Valuation Method: Some Japanese Case Studies*; Springer: Tokyo, Japan, 2006.

37. Hoyos, D.; Mariel, P. Contingent valuation: Past, present and future. *Prague Econ. Pap.* **2010**, *4*, 329–343. [CrossRef]

38. Whitehead, J.C. A practitioner's primer on the contingent valuation method. In *Handbook on Contingent Valuation*; Alberini, A., Kahn, J.R., Eds.; Edward Elgar: Cheltenham, UK, 2006.

39. Johnston, R.J.; Boyle, K.J.; Adamowicz, W.; Bennett, J.; Brouwer, R.; Cameron, T.A.; Hanemann, W.M.; Hanley, N.; Ryan, M.; Scarpa, R.; et al. Contemporary guidance for stated preference studies. *J. Assoc. Environ. Resour. Econ.* **2017**, *4*, 319–405. [CrossRef]

40. Kriström, B. Spike models in contingent valuation. *Am. J. Agric. Econ.* **1997**, *79*, 1013–1023. [CrossRef]

41. Habb, T.C.; McConnell, K.E. *Valuing Environmental and Natural Resources*; Edward Elgar: Cheltenham, UK, 2002.

42. Yoo, S.H.; Kwak, S.J. Using a spike model to deal with zero response data from double bounded dichotomous contingent valuation survey. *Appl. Econ. Lett.* **2002**, *9*, 929–932. [CrossRef]

43. Hanemann, W.M. Welfare evaluations in contingent valuation experiments with discrete responses. *Am. J. Agric. Econ.* **1984**, *66*, 332–341. [CrossRef]

44. Carson, R.T.; Hanemann, W.M. Contingent valuation. In *Handbook of Environmental Economics*; Maler, K.G., Vincent, J.R., Eds.; North-Holland: Amsterdam, The Netherlands, 2005; Volume 2, pp. 821–936.

45. Carson, R.T.; Hanemann, W.M.; Whittington, D. The existence value of a distinctive native American culture: Survival of the Hopi reservation. *Environ. Resour. Econ.* **2020**, *75*, 931–951. [CrossRef]

46. Krinsky, I.; Robb, A.L. On approximating the statistical properties of elasticities. *Rev. Econ. Stat.* **1986**, *68*, 715–719. [CrossRef]

47. Statistics Korea. Available online: http://kosis.kr (accessed on 4 June 2020).

48. Korea Ministry of Oceans and Fisheries. *The Third Marine Debris Management Master Plan. (2019–2023)*; Ministry of Oceans and Fisheries: Sejong, Korea, 2019. (In Korean)

© 2020 by the authors. Licensee MDPI, Basel, Switzerland. This article is an open access article distributed under the terms and conditions of the Creative Commons Attribution (CC BY) license (http://creativecommons.org/licenses/by/4.0/).

Article

Green Roofs as Effective Tools for Improving the Indoor Comfort Levels of Buildings—An Application to a Case Study in Sicily

Laura Cirrincione [1,2,*], **Maria La Gennusa** [1] , **Giorgia Peri** [1], **Gianfranco Rizzo** [1], **Gianluca Scaccianoce** [1,3], **Giancarlo Sorrentino** [1] **and Simona Aprile** [4]

1   Department of Engineering, University of Palermo, Viale delle Scienze Bld. 9, 90128 Palermo, Italy;
    maria.lagennusa@unipa.it (M.L.G.); giorgia.peri@unipa.it (G.P.); gianfranco.rizzo@unipa.it (G.R.);
    gianluca.scaccianoce@unipa.it (G.S.); giancarlo.sorrentino@unipa.it (G.S.)
2   ERIN—Environmental Research & Innovation Department, Luxembourg Institute of Science and
    Technology (LIST), 41 rue du Brill, L-4422 Belvaux, Luxembourg
3   Institute of Biomedicine and Molecular Immunology, National Research Council of Italy, via Ugo La Malfa
    153, 90146 Palermo, Italy
4   Research Center for Plant protection and Certification, Council for Agricultural Research and Economics,
    Italy—SS 113; Km 245, 500-90011 Bagheria, Palermo, Italy; simona.aprile@crea.gov.it
*   Correspondence: laura.cirrincione@unipa.it

Received: 27 December 2019; Accepted: 22 January 2020; Published: 29 January 2020

Abstract: In the line of pursuing better energy efficiency in human activities that would result in a more sustainable utilization of resources, the building sector plays a relevant role, being responsible for almost 40% of both energy consumption and the release of pollutant substances in the atmosphere. For this purpose, techniques aimed at improving the energy performances of buildings' envelopes are of paramount importance. Among them, green roofs are becoming increasingly popular due to their capability of reducing the (electric) energy needs for (summer) climatization of buildings, hence also positively affecting the indoor comfort levels for the occupants. Clearly, reliable tools for the modelling of these envelope components are needed, requiring the availability of suitable field data. Starting with the results of a case study designed to estimate how the adoption of green roofs on a Sicilian building could positively affect its energy performance, this paper shows the impact of this technology on indoor comfort and energy consumption, as well as on the reduction of direct and indirect CO_2 emissions related to the climatization of the building. Specifically, the ceiling surface temperatures of some rooms located underneath six different types of green roofs were monitored. Subsequently, the obtained data were used as input for one of the most widely used simulation models, i.e., EnergyPlus, to evaluate the indoor comfort levels and the achievable energy demand savings of the building involved. From these field analyses, green roofs were shown to contribute to the mitigation of the indoor air temperatures, thus producing an improvement of the comfort conditions, especially in summer conditions, despite some worsening during transition periods seeming to arise.

Keywords: innovative envelope; building components; green roofs; indoor comfort; energy consumption; building modelling; simulation models

---

1. Introduction

The reduction of energy consumption and the related decrease of greenhouse gases emissions represent important aspects to which much attention has been paid at global, European and countries levels, especially with regard to the building sector.

Worldwide, energy consumption in the building sector is responsible for 36% of total energy use (corresponding to a 39% of energy-related CO_2 emissions) [1,2], while at the European level, the energy consumption in the same sector accounts for a share of the total energy comprised between 25% and 40% (corresponding to about 35% of CO_2 emissions throughout Europe) [3–5].

From this perspective, various strategies have been implemented. At the global level, the UN 2030 Agenda for Sustainable Development, along with the 17 Sustainable Development Goals (SDGs) [6], need to be mentioned.

At the European scale, the EU has been very committed to this issue by setting the well known ambitious targets for 2020 ("climate and energy package") [7], and even more so for 2030 ("climate and energy framework") [8,9] and 2050 ("long-term strategy") [10,11]. Other relevant goals have been set out in the seventh Environment Action Program (EAP) [12] aimed at decarbonizing and making more sustainable European cities. Among European standards and regulations issued on this matter, the EPBD Directive and its recast must be cited [13–15].

Italy's National Energy Strategy 2017 [16] lays down the actions to be achieved by 2030, in accordance with the long-term scenario drawn up in the EU Energy Roadmap 2050, which translate to a reduction of emissions by at least 80% from their 1990 levels.

However, despite these standards and regulations being in force, in recent years, the energy consumption in the building sector has increased, particularly in Italy [17]. That is why more effort in promoting actions and finding new strategies to improve energy savings and efficiency are necessary [18].

Generally speaking, apart from all the design strategies typical of the principles of bioclimatic architecture (such as, for instance, space organisation, wall-window-ratio, orientation, thermal mass, operation management [19,20]), more relevant energy savings achievable in buildings can be attributed to two main categories of components: technical plants (HVAC system) and the building envelope, which have a synergistic relationship. In fact, a reduction in energy consumption related to the HVAC consists in the use of active systems which entail further energy consumption. As regards the building envelope, passive systems (not energy depending) can be used, which allow to actually obtain a reduction of the energy consumption (and at the same time, to also save on the use and the size of the HVAC system). Clearly, the occupants' behaviours and attitudes might also significantly influence energy saving, as demonstrated, for instance, in [21–24]. Starting from the above considerations, in this work, it was decided to pay attention to the use of a passive system to be applied to the building envelope, that is, green roofs equipped with different vegetation types.

Among the passive systems, green roofs are becoming more and more popular due to their capability of reducing the energy needs for the climatization of buildings [25–27], especially for cooling purposes [28–30].

At the same time, vegetated roofs also have a positive impact on the outdoor urban environment in terms of regenerative sustainability, allowing to induce various environmental benefits [26,31], such as reducing air pollution [32,33], mitigating noise [34,35], improving the management of runoff water [32,36,37], easing the urban heat island (UHI) effects [38–40], and increasing the urban biodiversity [41,42]. Moreover, the European Union is evaluating the possibility of including criteria specifically referring to green coverings within the EU Ecolabel scheme for buildings [43].

In addition to experimental studies [44–46], the effect on the built environment of vegetated roofs in diverse climates has also been investigated from analytical [47–49] and modelling points of view [50–52] over the years. In particular, the relevant parameters for energy modelling of green roofs have been explored in the literature, particularly referring to the role played by leaves and solar radiation in the thermal exchanges between vegetated layers and the surrounding environment [53].

The reported literature indicates that green roofs represent very promising building components, also in the Mediterranean context, as demonstrates the incremental number of studies and analyses carried out in recent years concerning both the experimental [44,54] and the simulating approach [47,55].

Other studies have underlined that additional issues would probably need more attention regarding plants growing on the roof, especially their influence on the thermal performance of green roofs [56,57], and the influence of the evapotranspiration component on the green roof heat and mass transmission [50,58].

Taking into account the studies reported above, it is evident how green roofs can have a strong impact in attenuating the average radiant temperature on building roofs [44,59,60]. This capability of acting as thermal insulation positively influences the indoor comfort conditions for the occupants of the rooms sited under the roof [27,45,61,62]: this aspect has always been critical in the design phase of a building envelope.

Kuan-Teng Lei et al. [63], by means of a field experiment performed in a school building in Taipei, have developed a finite element analysis model for the improvement of indoor thermal comfort in the presence of extensive green roofs. The researchers found a decrease of the indoor temperature up to 4 °C, compared to bare roofs. Costa Junior et al. [64], through an experimental analysis conducted in the city of Recife, Pernambuco (Brazil), compared the performance of four roofs made up of chanana green roof (Turnera subulata), daisy green roof (Sphagneticola trilobata), parsley green roof (Ipomoea asarifolia), and fiber cement tile. Through the comparison, the index of discomfort (ID), effective temperature (ET) and the human comfort index (HCI) were calculated. The three vegetated options mitigated both the internal air temperature with a reduction of 0.71°C, 0.19°C and 0.35°C, respectively and the internal surface temperature with a reduction of 1.5°C, 0.8°C and 0.8°C, respectively, compared to the fiber cement tile-made roof. Di Giuseppe and Orazio [65] experimentally analysed the effect of cool and green roofs compared to traditional ones in a Nearly Zero Energy Building, on the internal comfort and the air temperatures of the surrounding environment. The outcomes, on one hand, confirmed the effectiveness of green and cool roofs for the mitigation of the Urban Heat Island effect, and on the other hand, indicated the little effectiveness of high-albedo materials on roofing systems with a very low U-value for internal comfort.

Furthermore, the impact of green walls on thermal comfort have been compared to that of green roofs. For instance, Malys et al. [66], using the SOLENE-microclimat tool, compared the effect caused by different 'greening strategies' on buildings' energy consumption and indoor comfort in the summer season. The outcomes of the investigation indicate that, while green roofs seemingly mainly affect the upper floor, green walls directly affect the indoor comfort throughout the entire building.

To help to provide a contribution to this important and often overlooked matter, the aim of this paper was to assess the influence that green roofs have on the indoor thermal comfort levels, particularly considering the indoor radiative heat exchanges. For this purpose, a case study was conducted to estimate how the adoption of the proposed interventions could impact the indoor thermal comfort and the energy consumption of a building and contribute to the reduction of the direct and indirect CO_2 emissions. In particular, the ceiling temperatures of some rooms located underneath six different types of green roofs were monitored. The choice of detecting this parameter resides on the circumstance that the ceiling internal surface's temperature is a relevant component of the mean radiant temperature of the room that, in turn, greatly affects the value of the indoor parameter PMV [67]. Subsequently, the obtained data were used as input data for one of the most widely used simulation models (EnergyPlus [68]) to evaluate the indoor comfort levels and the achievable energy demand savings of the involved building.

Three scenarios were adopted. Scenarios #1 and #2 refer to a building equipped with a green roof; Scenarios #3 refers to the pre-existing roof. The simulation of Scenario #2 is made by means of the Energy Plus code, through its resident routine; on the other hand, Scenario #1 is modelled by imposing, for the indoor temperatures of the ceiling, the experimental data detected on the site. The aim of this approach was to compare the PMVs obtained from Energy Plus with those calculated on the basis of the monitored experimental data, i.e., the indoor ceiling surface temperatures.

2. Materials and Methods

The presented study is part of a joint research project between the University of Palermo and the "Consiglio per la ricerca in agricoltura e l'analisi dell'economia agraria - CREA", operating in Sicily. To accomplish the task mentioned in the Introduction, a mixed approach, partly modelling and partly experimental, was used in the work.

At the same time, the impact that green roofs have on the energy consumption of a building was evaluated. In addition, the estimation of the achievable savings in direct and indirect CO_2 emissions due to the use of such building component is reported as well.

2.1. Description of the Experimental Site

The installation of the experimental green roof was settled by the CREA Research Center and the University of Palermo with the support of a building materials enterprise, on the roof of a one-storey detached house (Figure 1) owned by CREA and sited in Bagheria, a Sicilian town near Palermo (Southern Italy).

Figure 1. Site of the installation of the experimental field (source: Google-Earth).

To conduct the present case study, it was decided to install the green coverage on the pre-existing roof of the building, made of hollow bricks, with a surface of approximately 80 m^2.

As regards the weather conditions of the site, they were typical of the South of Italy, characterized by a temperate climate with warm summers and mild winters. Figures 2–4 show the trend of outdoor air temperature (T), relative humidity (RH) and solar radiation (IR), respectively, during the monitoring period of one year.

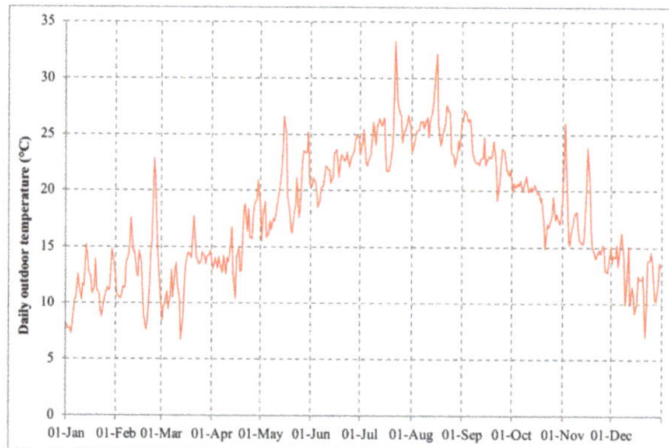

Figure 2. Trend of the outdoor air temperature during the monitoring period.

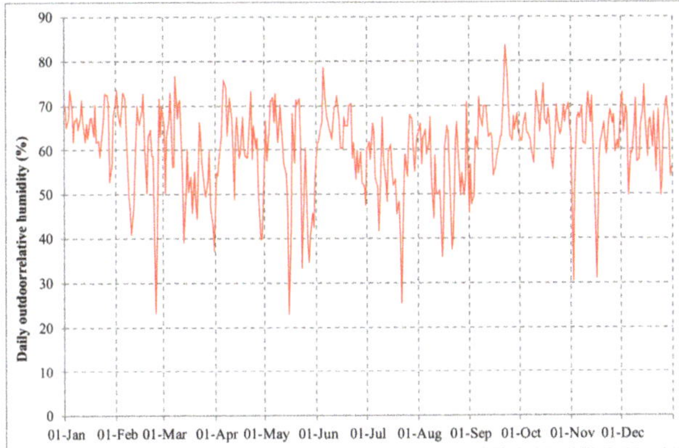

Figure 3. Trend of the outdoor air relative humidity during the monitoring period.

Figure 4. Trend of the solar radiation during the monitoring period.

2.2. Description of the Analysed Green Roof Installed in the Experimental Site

Going from the indoor to the outdoor sides of the building, as shown in Figure 5, the green roof compound is composed of the following layers: a root barrier (with a waterproofing membrane), a drainage layer (made of a polyethylene geo-net, hot-coupled with a non-woven geotextile with filtering functions), a water storage layer (constituted by cushions filled with expanded perlite), a filter fabric (composed by a geotextile felt, 100% polypropylene calendered), a growing medium (which is a mixture of peat, lapillus, pumice, zeolite and slow releasing fertilizers and is infesting weeds free) and the vegetation layer.

Figure 5. Sketch of the green-roof layers.

In order to also analyse the effects provided by different plant species and different thicknesses of the water storage layer on the green roof thermal behaviour, the roof was divided into six sectors, where three different Mediterranean autochthonous species (*Halimione Portulacoides, Rosmarinus Officinalis Prostratus* and *Crithmum Maritimum*) and two different thicknesses of the water storage layer (10 cm for plots 1, 2, 3 – P1, P2, P3 – and 15 cm for plots 4, 5, 6 – P4, P5, P6 –) were used according to the scheme reported in Figure 6.

Figure 6. Scheme of the plant species planted in the different sectors.

A brief structural description of each layer is reported in Table 1.

Table 1. Description of the layers constituting the green roof plot.

Layer	Element Type	Thickness [cm]	Plant Species
1	Structural support	20	*Hollow brick*
2	Waterproofing membrane and root barrier	-	*Bituminous paint*
3	Drainage layer	0.5	*Polyethylene geo-net, hot-coupled with a no woven geotextile*
4	Water storage layer	10 (P1, P2, P3) 15 (P4, P5, P6)	*Pillows filled with expanded perlite*
5	Filter fabric	-	*Geotextile felt, 100%polypropylene calendered*
6	Growing medium	15	*Pumice, lapillus and peat*
7	Vegetation layer	-	*Halimione Portulacoides (P1 and P6) Rosmarinus Officinalis Prostratus (P2 and P5) Crithmum Maritimum (P3 and P4)*

Table 2 reports, instead, the main physical parameters characterizing each of the green roofs' six plots. The data listed in Table 2 are the same as those used in [69].

Table 2. Description of the layers constituting the green roof plot.

Parameters	Plots					
	P1	**P2**	**P3**	**P4**	**P5**	**P6**
Water storage layer thickness (cm)	15	15	15	10	10	10
Height of Plants (m)	0.35	0.28	0.12	0.12	0.22	0.30
Leaf Area Index (-)	4.0	2.8	1.2	0.9	2.3	3.8
Leaf Reflectivity (-)	0.19	0.18	0.17	0.20	0.21	0.21
Substrate total thickness (m)	0.30	0.30	0.30	0.25	0.25	0.25
Thermal conductivity of dry soil (W/m·K)	0.0738	0.0738	0.0738	0.0816	0.0816	0.0816
Density of dry soil (kg/m^3)	530	530	530	446	446	446
Specific heat of dry soil (J/kg·K)	1050	1050	1050	1060	1060	1060

In addition to thermal conductivity, density and specific heat, which characterize the thermo-physical behaviour of the soil, some other properties typical of the specific plants, which have an important impact on the heat exchanges through the green-roof, were also considered.

Particularly, the "leaf reflectance" (dimensionless), that is, the ratio of the incoming light which is reflected by a leaf, and the "leaf area index"—LAI (m^2/m^2)—defined as the one-sided green leaf area per unit of ground surface area. The latter, in particular, which has a great influence on the shading and transpiration effects, has a positive effect, especially during summer seasons; in fact, the higher the LAI, the higher the cooling reduction [50,69,70].

As for the vegetation characteristics, Figure 7 shows how the *Halimione Portulacoides* (both P1 and P6) and the *Rosmarinus Officinalis Prostratus* (only P5) reached full coverage (100%) in less than 12 months while the *Rosmarinus Officinalis Prostratus* in P2 achieved a maxim coverage of about 85% in the same period and the *Crithmum Maritimum* (both P3 and P4) did not accomplish more than 40%–60%, showing the difficulty of establishing it in the considered environment [52].

Figure 7. The six-plots green coverages system.

2.3. Data Monitoring System Adopted

The field experimental part of the proposed approach essentially consisted in a monitoring campaign of the ceiling temperature values of the building.

Since the monitoring of the temperatures profiles of the ceiling was aimed at checking the effects of the presence of the green roof on the indoor conditions, particularly in terms of thermal comfort levels, measures were performed in the center of the rooms' ceiling, far from thermal bridges and lights fixtures, as shown in Figure 8.

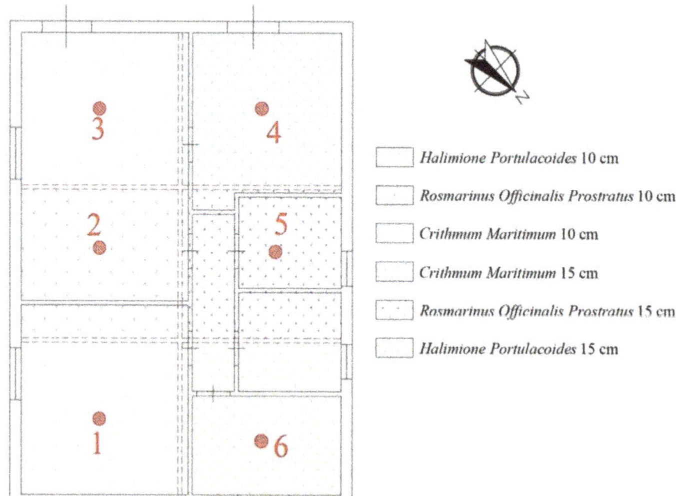

Figure 8. Layout of the building, green roof plots' arrangement and positions of probes for the temperature measurement.

Concerning the measuring method, temperatures were recorded with a sample rate of 10 minutes, during both the winter and the summer seasons, by means of insulated T type thermocouple probes.

The monitoring campaign lasted one year and started six months after the installation of the green roofs in order to have the green coverage well stabilized and to allow the testing of the acquisition system.

2.4. Simulations Performed in the Study

The modelling part of the proposed approach consisted in utilizing the very popular EnergyPlus simulation code to run the building's thermal calculations. For this purpose, different scenarios were implemented, specifically:

- Scenario #1, in which the monitored ceiling temperatures were utilized in the simulation as boundary fixed conditions for the ceiling of the investigated rooms. In this scenario, a detailed schedule for the HVAC was implemented, based on assumptions made of its "real" use according to the typical time of occupation of the building, considering a power capacity of 10000 Watt. This value was obtained from simulations previously conducted using the climatic design-day typical of the examined area, characterized by a temperature of 5.2°C (± 0) for winter conditions and 31°C (± 6) for summer conditions.

- Scenario #2, in which the simulation was carried out utilizing the green roof configuration provided by EnergyPlus (EP+GR), trying to simulate the previously described six plots as faithfully as possible by using the parameters reported in Tables 1 and 2. In fact, the EnergyPlus green roof simulation tool sets numerical limits for some parameters, which could not therefore have been set according to their real values.

- Scenario #3, in which the simulation was conducted by implementing a standard case (STD), that is, considering the original roof of the building without the presence of green coverage.

Regarding the HVAC schedule of Scenario #2 and Scenario #3, it was decided to use a simple on/off schedule, with the HVAC working between 7:00 and 17:00, considering the same power capacity as that used in Scenario #1.

The authors would like to underline here that Scenarios #1 and #2 are characterized by the presence of green roofs, while Scenario #3 refers to a standard roof. The difference consists in the fact that, while the simulation of Scenario #2 totally complies with the Energy Plus code (by utilizing its typical green roof simulation routine), Scenario #1 is modelled by forcing the Energy Plus code, that is, imposing the monitored indoor ceiling temperatures as boundary conditions. In this way, the model was driven with real data based on the presence of the experimental green roof, avoiding the actual simulation of the green roof element itself. Hence, this allowed us to compare the PMVs obtained from the Energy Plus green roof simulation tool (with its relative assumptions limits) with those calculated on the basis of the real monitored experimental data.

In order to assess the direct and indirect reduction of CO2 emissions, in this work, a value of 85 kgCO2/ha per year [71] was considered for the direct reduction of CO2 emission, based on the extension of green covering, while a value of 0.531 tCO2/MWh (the average emissions for the current electric Italian energy mix [34]) was used to estimate the indirect reduction of CO2 emission based on energy saving for climatization purposes and having set a COP value equal to 3 for the cooling season and 3.5 for the heating season.

3. Results and Discussions

In this section, the results relative to the monitoring campaign and to the energy performance simulations are reported.

3.1. Monitored Data

Table 3 shows the average temperatures measured on the ceiling of each room sited below the green roof's six plots, for summer (July) and winter (February) conditions, in periods during which the air-conditioning system was working.

Table 3. Monitoring results of the green roofs six plots.

Plots	Plant Species	Water Storage Layer Thickness (cm)	Maximum Green Coverage (%)	T_{avg} (°C) of the Ceiling February	July
P1	*Halimione Portulacoides*	10	100%	18.1 ± 2.1	26.6 ± 0.2
P2	*Rosmarinus Officinalis Prostratus*	10	85%	16.0 ± 1.8	27.5 ± 0.9
P3	*Crithmum Maritimum*	10	58%	15.8 ± 1.4	26.9 ± 1.1
P4	*Crithmum Maritimum*	15	38%	17.9 ± 1.2	30.8 ± 0.8
P5	*Rosmarinus Officinalis Prostratus*	15	100%	17.6 ± 1.2	28.5 ± 0.6
P6	*Halimione Portulacoides*	15	100%	19.0 ± 1.1	28.2 ± 0.3

The monitoring results point out a general tendency to attain lower temperatures when the green coverage is higher, i.e., P1 (*Halimione Portulacoides*). Indeed, this plot shows that ceiling temperatures were generally 1–3 °C lower with respect to the other plots in summer and 1–2 °C higher during winter, hence representing a benefit for both the summer and winter seasons.

Moreover, it must be noted that the LAI has a positive influence on the green-roof thermal behaviour; P1 and P6 have, in fact, higher LAI, unlike P3 and P4. In addition, another factor that could have influenced the obtained results is represented by the light colour of the plants' leaf surface, which enabless a higher amount of solar radiation to be reflected.

The results shown in Table 3 also highlight the influence of the different type of plants. In particular, *Halimione Portulacoides* (P1 and P6) reduces temperature peaks more consistently. Therefore, this type of plant seems to be more suitable for lowering the summer temperature values and increasing the winter temperature peaks.

Anyway, as reported in Table 3, it should be pointed out that during the summer season, a mean temperature of about 26.6 °C has been recorded by the thermocouples placed on the rooms' ceilings under P1 (*Halimione Portulacoides*), with maximum peaks of 27 °C, that lies within the suggested range for the indoor comfort in summer conditions [72]. The same cannot be stated for the other plots, where, even with a 100% green coverage, ceiling temperatures of about 27–28 °C were registered, with maximum peaks going beyond 30 °C.

The above-discussed outcomes demonstrate the importance of selecting a proper plant species during the green roof design phase.

Apart from these considerations, strictly related to the physical characteristics of the green roof, it is also necessary to take into consideration some aspects related to the building features that may have influenced the monitored ceiling temperatures, in particular:

- the room located underneath the plot P1, facing North, is almost always in the shade (and not often sunlit); therefore, it is likely that the indoor environment is characterized by an air temperature lower than that of the other rooms;
- the room sited below the plot P4, on the other hand, is subjected both to greater solar radiation levels on the west-faced external wall and to heat released by several refrigerators aimed at the storage of biomass; it is then possible that the temperature inside such a space is constantly higher than that of the other rooms;
- the sensors located under plots P2 and P3, despite being associated to two different plant species, are located within a single large environment, which could make the distinction of their readings quite difficult;
- the rooms where the sensors relative to plots P3 and P4 are placed in, border on the left with a small greenhouse that, reasonably, is characterized by a higher air temperature than the outdoor.

3.2. Outcomes of Energy Simulations

Since the main aim of this work was to assess how the use of green roofs can affect the indoor comfort and the energy consumption of a building, it was decided to report, in the first part of this section, a comparison between the simulation results of Scenario #1 and Scenario #3. In particular, considering that such estimation is affected by the temperature changes during the actual HVAC system operating periods and in light of the detailed schedule utilized to run the simulations, it was chosen to divide the resulting data into two five-days periods representative of winter (Figure 9, on the left) and summer (Figure 9, on the right) conditions. Specifically, in Figure 9, the green lines represent an average of the values obtained for the six plots (Scenario #1), with its relative ranges of variation and the blue lines represent the standard case (Scenario #3); on the other hand, the red lines indicate the HVAC system start-up intervals.

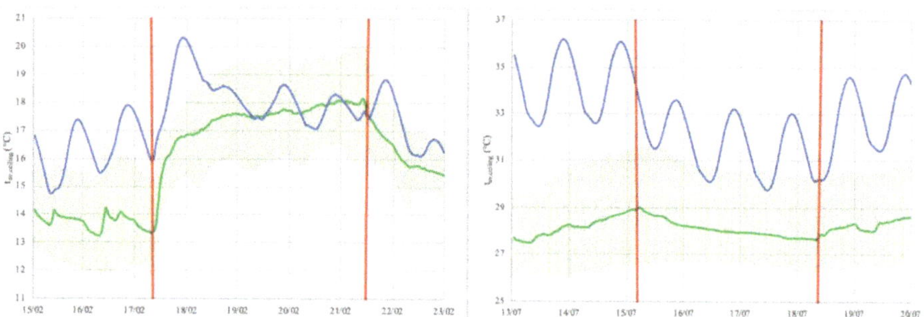

Figure 9. Comparison between Scenario #1 (green lines) and Scenario #3 (blue lines) for winter (left) and summer (right) conditions.

Looking at Figure 9, it can be noted how after an initial start-up phase of the HVAC system, the presence of a green roof during the winter season does not seem to improve the indoor thermal conditions, while during the summer season, it brings a noticeable improvement of the indoor comfort levels. It must be underlined here that the amplitude of the variation range relative to the green roofs' temperature values is due to the fact that as reported in Section 2.2, the six plots are characterized by different features and therefore, describing parameters; in particular, the type of species and its relative coverage percentage have a strong influence on the monitored temperatures.

The temperature differences noticed also had an impact on the energy consumption. Over the representative five-day periods considered, in fact, a 18% increase for heating needs (220.05 kWh for the standard roof against 259.59 kWh for the green roof) and a 44% saving for cooling needs (189.38 kWh for the standard roof against 106.37 kWh for the green roof) were observed.

As mentioned earlier, after this first comparison, the authors wondered what results would have been obtained by simulating a green roof similar to the real one using the green roof configuration tool provided by EnergyPlus. The second part of this section shows, therefore, a comparison between the results obtained from the Scenario # 1 and Scenario # 2 simulations.

Similarly to the previously reported Figure 9, Figure 10 contains the obtained results for the two five-days periods representative of winter (on the left) and summer (on the right) conditions. In particular, the green lines (Scenario #1) and the red lines (HVAC system start-up intervals) are the same as shown in Figure 9, while the black lines represent an average of the values obtained for the six plots, with the relative ranges of variation, using the Scenario #2 EnergyPlus settings.

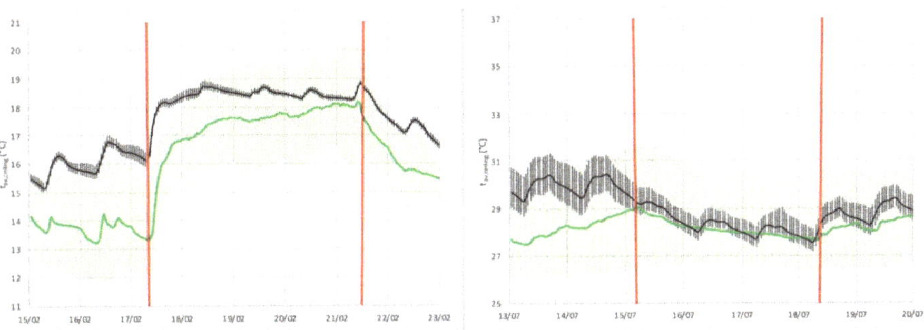

Figure 10. Comparison between Scenario #1 (green lines) and Scenario #2 (black lines) for winter (left) and summer (right) conditions.

By analysing Figure 10, it can be seen how, in the winter conditions, the green roof simulated according to Scenario #2 shows a very similar behaviour to that in Scenario #1, particularly during the air-conditioning working periods. In winter conditions, however, Scenario #2 allows to obtain higher temperatures than Scenario #1, corresponding to a further improvement in the indoor comfort levels. Furthermore, in summer conditions, contrarily to Scenario #1, Scenario #2 shows an evident very variable temperature trend between day and night, typical of a context highly influenced from solar radiation, which does not seem to reflect reality.

As for the fact that the changes of the Scenario #2 temperatures range are much narrower than those in Scenario #1, it must be observed that this is most likely due to the fact that, as previously highlighted, the model used by EnergyPlus does not allow to set all the parameters of the green roof freely but imposes some constraints to their numerical values. Due to this reason, in fact, the Scenario #2 results show no differences relating to the two different thicknesses of water storage used for each species, but only some small differences between the different species.

As for the energy consumption of Scenario #2, over the representative five-day periods considered, a 4% saving for heating needs (220.05 kWh for the standard roof against 212.34 kWh for the green roof) and a 41% saving for cooling needs (189.38 kWh for the standard roof against 112.54 kWh for the green roof) were observed.

Finally, in the last part of this section, it was decided to report a rough estimate, on a monthly basis, relative to both aspects of indoor comfort improvement and energy consumption savings. In this regard, it was chosen to compare the results deriving from the simulations of Scenarios #2 and #3.

This choice was suggested due to the fact that, given the intended use of the building (i.e., research laboratory) and its real use (i.e., infrequent), especially in terms of the HVAC system, it was assumed that the results of Scenario #1 were not considered actually representative for a long-term estimate.

Once again, in order to make the results visually more easily readable, an average of the results obtained for the green roof six plots was used to display the results of Scenario #2.

To compare Scenario #2 with Scenario #3 in relation to indoor comfort levels, it was decided to report, in Figure 11, a graph of the monthly temperatures. The graph indeed allows to show the deviations of the average values of the green roof ceiling temperatures ($\Delta t_{av,ceiling}$) compared to those of the standard roof, where the black bars represent the range of deviation from the average values.

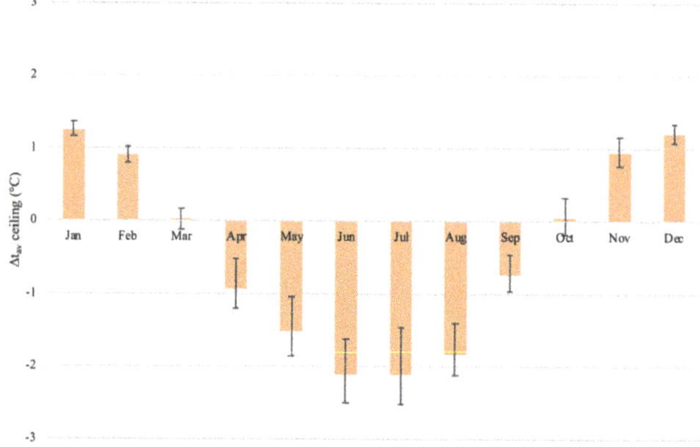

Figure 11. Deviations of the average values of the green roof ceiling temperatures compared to those of the standard roof.

Figure 11 highlights the positive effects due to the presence of the green roof, which, with respect to the standard roof, allows maintaining higher ceiling temperatures in winter and lower ceiling temperatures during summer.

Other than the temperature, another important indicator when assessing the indoor comfort levels is represented by the PMV (Predicted Mean Vote). For this reason, it was also decided to report, in Figure 12, a comparison between the monthly PMV average and peak values of Scenario #2 (GR) and Scenario #3 (ST).

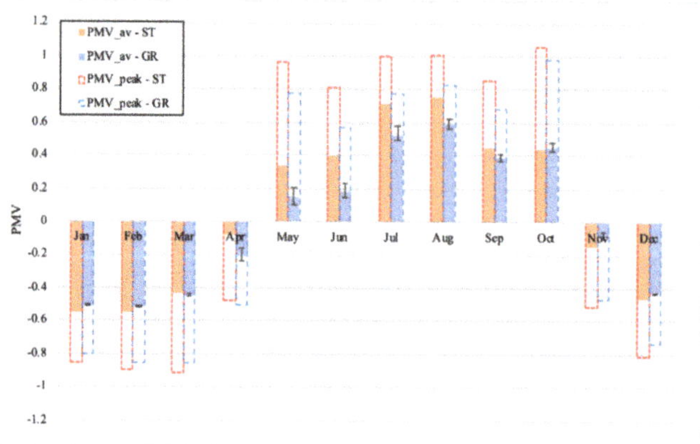

Figure 12. Monthly PMV average and peak values for Scenario #2 (GR) and Scenario #3 (ST).

By looking at the differences in the obtained PMV average values (Figure 12) with and without the presence of the green roof, especially those relative to July and August, it arises that the presence of the green roof reduces PMV average values from more than 0.7 to approximately 0.5. Hence, accordingly to the standard currently in force for the design of the indoor environment, i.e., EN 16798-1:2019 [72], the presence of the green roof contributes to shift the indoor thermal environmental conditions from Category III (acceptable, moderate level of expectations) to Category II (normal level of expectation). In other words, the presence of the green roof contributes to bring the building within comfort conditions (PMV = 0.5), starting from a slight warm condition (PMV = 0.7).

Moreover, by analysing Figure 12, it can also be seen how, although a general positive effect due to the presence of the green roof is evident, some critical issues emerged in the months of April and October (transition months), for which the standard roof seems to perform better than the green roof. This condition, which needs to be better investigated, is probably due to the additional thermal inertia that the presence of the green roof brings to the structure: this slows down the response of the green roof compound to the changes of climatic conditions occurring in the transition periods of the end of spring (April) and the beginning of winter (October).

For the sake of completeness, it was decided to report, in Figure 13, an annual plot where the average daily external temperatures (Outdoor) are compared with those of the ceiling in the presence of the green roof (GR_mean) and with those relative to the standard roof (ST).

Figure 13. Annual average daily temperatures (i.e. $T_{\text{outdoor air}}$, $T_{\text{ceiling with GR}}$ and $T_{\text{ceiling without GR}}$) trends.

Regarding the energy consumptions for heating and cooling needs, these are summarized in Table 4 by reporting the absolute values (kWh) obtained for the standard roof scenario and the correlated average percentage deviations (including the respective variation ranges) relative to the achievable savings due to the green roof presence.

Table 4. Monitoring results of the green roofs six plots.

		Standard Roof (kWh)	Green Roof (% Savings)
Jan	**Heating**	**655.1**	**17.9 ± 1.7%**
	Cooling	0	
Feb	**Heating**	**610.6**	**12.5 ± 1.8%**
	Cooling	0	
Mar	**Heating**	**409.9**	**2.2 ± 3.5%**
	Cooling	0	
Apr	Heating	15.9	−43.6 ± 34.3%
	Cooling	0	
May	Heating	4.4	2.1 ± 24.2%
	Cooling	0	
Jun	Heating	0	
	Cooling	**239.6**	**50.1 ± 9.8%**
Jul	Heating	0	
	Cooling	**534.2**	**28.7 ± 7.4%**
Aug	Heating	0	
	Cooling	**630.0**	**24.7 ± 4.9%**
Sep	Heating	0	
	Cooling	**323.1**	**20.4 ± 4.7%**
Oct	Heating	2.0	62.9 ± 6.7%
	Cooling	45.9	33.0 ± 10.7%
Nov	Heating	69.5	59.3 ± 7.0%
	Cooling	0	
Dec	**Heating**	**429.8**	**26.4 ± 2.5%**
	Cooling	0	

Specifically, for each month, the amount energy consumed for both heating and cooling with and without the presence of the green roof is listed. The use of the bold character is intended to show more easily the actual HVAC working periods, that is December–March for winter conditions and June–September for summer conditions. Therefore, in Table 4, data related to months when the HVAC is working have been highlighted using the colour black. The results confirm the advantage of using green roofs as a solution capable of achieving valuable energy savings.

Moreover, the mean indirect reduction of CO_2 emissions due to the green roof installation was 145.6 ± 13.8 tCO_2/year, while for the direct reduction, a value of only 56 gCO_2/year was observed.

Finally, some further considerations need reporting. When an energy restoration of a roof is in context, such as the one considered in this work, it is easier and safer to add a vegetated coverage to an existing roof than making it larger. That is why here, it was chosen to exclude a theoretical comparison between a green roof and a hypothetical alternative high massive one and to limit the analysis to a specific comparison between the behaviour of the pre-existent standard roof equipping the building and the improvements brought by the installation of the green coverage. For this purpose, the thermo-physical characteristics of the roof are those typical of the building habit of the considered geographical area.

In addition, the benefits of the presence of a green roof cannot be simply evaluated in terms of thermal insulation, since it generates other positive effects regarding the evaporative phenomena and the change of the albedo of the roof. Therefore, it was decided to exclude a comparison with insulated roofs too, in accordance with existing literature studies—such as that by Niachou et al. [73]—demonstrating that the presence of green roof on an insulated roof is practically irrelevant.

4. Conclusions

The capability of green roofs in reducing the electric energy needs for the climatization of buildings and their environmental benefits has been extensively demonstrated in the literature.

The idea behind the presented work derived from considerations regarding the possible influence of green roofs on the indoor thermal comfort levels; that is, instead, an aspect often overlooked when estimating the performance of such building components. Therefore, the analysis methodology approach utilized (partly modelling and partly experimental) was implemented in light of such considerations.

Accurate knowledge of the internal temperatures of the ceiling is indeed an important prerequisite for establishing both achievable indoor comfort conditions and the energy demand for the air conditioning of the building itself. A lowering of the indoor temperatures in summer, and a rise during winter, lead, in fact to an improvement in the comfort levels for the occupants, and consequently, a saving on the use of the HVAC system, which, in turn, translates into a reduction of polluting emissions.

For this purpose, the comparison between Scenario #1 (where the monitored ceiling temperatures were used as boundary conditions) and Scenario #3 (standard roof case), and that between Scenario #2 (where the green roof configuration provided by EnergyPlus was utilized) and Scenario #3, allowed to prove how the presence of the experimental green roof on the monitored building improved the indoor comfort levels during summer by moderating the ceiling temperatures (Figures 9–11 and 13), despite some worsening during winter periods seeming to occur. Moreover, the temperature differences noticed had also a positive impact on the building energy consumption and the CO_2 emissions.

On the other hand, by comparing Scenario #1 with Scenario #2, it was possible to highlight the possible limits of the code's ability in adequately simulating the green roof behaviour. These limits are mainly represented by the lack of flexibility that EnergyPlus allows in the setting-process of some of the green roof physical parameters and in the way in which the code takes into account the solar radiation components.

Another aspect which should be better investigated, regards the PMV results (Figure 12). In fact, even though, also, in this case, a general positive effect due to the presence of the green roof can be seen, some criticalities emerged during some transition periods, for which the standard roof seems to perform slightly better than the green roof.

In conclusion, the proposed analysis made it possible to highlight how it is possible to assess the impact that green roofs have specifically on the indoor comfort levels, other than on the energy consumption.

Furthermore, the availability of field data put into evidence the importance of adequate simulation tools to facilitate the green roof design and assessment processes.

Author Contributions: All authors contributed equally to the design, experimental analyses and editing of this research. Conceptualization, L.C., M.L.G., G.P., G.R., G.S. (Gianluca Scaccianoce), G.S. (Giancarlo Sorrentino) and S.A.; Data curation, L.C., M.L.G., G.P., G.R., G.S. (Gianluca Scaccianoce), G.S. (Giancarlo Sorrentino) and S.A.; Methodology, L.C., M.L.G., G.P., G.R., G.S. (Gianluca Scaccianoce), G.S. (Giancarlo Sorrentino) and S.A.; Writing—original draft, L.C., M.L.G., G.P., G.R., G.S. (Gianluca Scaccianoce), G.S. (Giancarlo Sorrentino) and S.A.; Writing—review and editing, L.C., M.L.G., G.P., G.R., G.S. (Gianluca Scaccianoce), G.S. (Giancarlo Sorrentino) and S.A. All authors have read and agreed to the published version of the manuscript.

Funding: This research received no external funding.

Conflicts of Interest: The authors declare no conflict of interest.

References

1. International Energy Agency (IEA). *2018 Global Status Report—Towards a Zero-Emission, Efficient and Resilient Buildings and Construction Sector Global Alliance for Buildings and Construction—UN Environment*; International Energy Agency (IEA): Paris, France, 2018.
2. International Energy Agency. *CO₂ Emissions from Fuel Combustion Highlights*; International Energy Agency (IEA): Paris, France, 2017.
3. Tsemekidi-Tzeiranaki, S.; Bertoldi, P.; Labanca, N.; Castellazzi, L.; Serrenho, T.; Economidou, M.; Zangheri, P. *Energy Consumption and Energy Efficiency Trends in the EU-28 for the Period 2000–2016*; JRC Science for Policy Report; Joint Research Centre (JRC): Brussels, Belgium, 2018.

4. Energy Performance of Buildings. Statistics from European Commission on Energy Performance of Buildings. Available online: https://ec.europa.eu/energy/en/topics/energyefficiency/energy-performance-of-buildings (accessed on 4 September 2019).

5. European Union. *EU Energy in Figures Statistical Pocketbook*; European Commission: Brussels, Belgium, 2017.

6. Sustainable Development Goals. Available online: https://www.un.org/sustainabledevelopment/sustainable-development-goals/ (accessed on 4 September 2019).

7. European Commission. *Communication from the Commission Europe 2020. A strategy for Smart, Sustainable and Inclusive Growth*; 3 March 2010 COM(2010)2020; European Commission: Brussels, Belgium, 2010.

8. United Nations. *Transforming our World: The 2030 Agenda for Sustainable Development*; A/RES/70/1; General Assembly. Distr.: General 21 October 2015; United Nations: New York, NY, USA, 2015.

9. European Commission. *Communication from the Commission to the European Parliament, the Council, the European Economic and Social Committee and the Committee of the Regions. A Policy Framework for Climate and Energy in the Period from 2020 to 2030*; 22 January 2014 COM(2014) 15 final; European Commission: Brussels, Belgium, 2014.

10. European Commission. *A Roadmap for Moving to a Competitive Low Carbon Economy in 2050*; COM(2011) 112 final; European Commission: Brussels, Belgium, 2011.

11. European Commission. *Communication from the Commission to the European Parliament, the European Council, the Council, the European Economic and Social Committee, the Committee of the Regions and the European Investment Bank. A Clean Planet for All a European Strategic Long-Term Vision for a Prosperous, Modern, Competitive and Climate Neutral Economy*; 28 November 2018 COM(2018) 773 final; European Commission: Brussels, Belgium, 2018.

12. The European Parliament and the Council. Decision No 1386/2013/EU of the European Parliament and of the Council of 20 November 2013 on a general Union environment action programme to 2020 'living well, within the limits of our planet'. *Off. J. Eur. Union* **2013**, L 354/171.

13. The European Parliament and the Council. Directive 2002/91/EC of the European parliament and of the council of 16 December 2002 on the energy performance of buildings. 4.1. *Off. J. Eur. Communities* **2003**, L 1/65.

14. The European Parliament and the Council. Directive 2010/31/EU of the European Parliament and of the Council of 19 May 2010 on the energy performance of buildings (recast) 18.6. *Off. J. Eur. Union* **2010**, L 153/13.

15. European Commission. Commission Delegated Regulation (EU) No 244/2012 of 16 January 2012 supplementing Directive 2010/31/EU on the energy performance of buildings. *Off. J. Eur. Union* **2012**, L 081.

16. Ministero dello Sviluppo Economico; Ministero dell'Ambiente e della tutela del Territorio e del Mare. *Italy's National Energy Strategy (2017)*; Ministero dello Sviluppo Economico: Rome, Italy; Ministero dell'Ambiente e della tutela del Territorio e del Mare: Rome, Italy, 2017.

17. Agenzia Nazionale per le Nuove Tecnologie, l'Energia e lo Sviluppo Economico Sostenibile (ENEA). *Rapporto Annuale Efficienza Energetica (RAEE) 2019*; Agenzia Nazionale per le Nuove Tecnologie, l'Energia e lo Sviluppo Economico Sostenibile (ENEA): Rome, Italy, 2019; ISBN 978-88-8286-382-1.

18. Bisegna, F.; Cirrincione, L.; Casto, B.M.L.; Peri, G.; Rizzo, G.; Scaccianoce, G.; Sorrentino, G. Fostering the energy efficiency through the energy savings: The case of the University of Palermo. In Proceedings of the 2019 IEEE International Conference on Environment and Electrical Engineering and 2019 IEEE Industrial and Commercial Power Systems Europe, EEEIC/I and CPS Europe 2019, Palermo, Italy, 11–14 June 2019.

19. Hens, H. *Applied Building Physics. Boundary Conditions, Building Performance and Material Properties*; Ernst and Sohn: Berlin, Germany, 2012; ISBN 978-3-433-02962-6.

20. Underwood, C.P.; Yik, H.F.W. *Modelling Methods for Energy in Buildings*; Blackwell Science: Hoboken, NY, USA, 2004; ISBN 0-63205936-2.

21. Caniato, M.; Gasparella, A. Discriminating people's attitude towards building physical features in sustainable and conventional buildings. *Energies* **2019**, *12*, 1429. [CrossRef]

22. Diaz Lozano Patino, E.; Siegel, J.A. Indoor environmental quality in social housing: A literature review. *Build. Environ.* **2018**, *131*, 231–241. [CrossRef]

23. Castaldo, V.L.; Pigliautile, I.; Rosso, F.; Cotana, F.; de Giorgio, F.; Pisello, A.L. How subjective and non-physical parameters affect occupants' environmental comfort perception. *Energy Build.* **2018**, *178*, 107–129. [CrossRef]

24. Sant'Anna, D.O.; Dos Santos, P.H.; Vianna, N.S.; Romero, M.A. Indoor environmental quality perception and users' satisfaction of conventional and green buildings in Brazil. *Sustain. Cities Soc.* **2018**, *43*, 95–110. [CrossRef]

25. Silva, C.M.; Gomes, M.G.M.; Silva, M. Green roofs energy performance in Mediterranean climate. *Energy Build.* **2016**, *116*, 318–325. [CrossRef]

26. Shafique, M.; Kima, R.; Rafiq, M. Green roof benefits, opportunities and challenges—A review. *Renew. Sustain. Energy Rev.* **2018**, *90*, 757–773. [CrossRef]

27. La Roche, P.; Berardi, U. Comfort and energy savings with active green roofs. *Energy Build.* **2014**, *82*, 492–504. [CrossRef]

28. Colmenar-Santos, A.; de Lober, L.N.T.; Borge-Diez, D.; Castro-Gil, M. Solutions to reduce energy consumption in the management of large buildings. *Energy Build.* **2013**, *56*, 66–77. [CrossRef]

29. Ascione, F.; Bianco, N.; de' Rossi, F.; Turni, G.; Vanoli, G.P. Green roofs in European climates. Are effective solutions for the energy savings in air-conditioning? *Appl. Energy* **2013**, *104*, 845–859. [CrossRef]

30. Cao, J.J.; Hu, S.; Dong, Q.; Liu, L.J.; Wang, Z.L. Green roof cooling contributed by plant species with different photosynthetic strategies. *Energy Build.* **2019**, *195*, 45–50. [CrossRef]

31. Di Lorenzo, D.; Lupo, V.; Peri, G.; Rizzo, G.; Scaccianoce, G. A simple methodology for comparing cost-benefit of traditional, green and cool roofs. In Proceedings of the 13th REHVA World Congress CLIMA 2019, Bucharest, Romania, 26–29 May 2019.

32. Zhang, Q.; Miao, L.; Wang, X.; Liu, D.; Zhu, L.; Zhou, B.; Sun, J.; Liu, J. The capacity of greening roof to reduce stormwater runoff and pollution. *Landsc. Urban Plan.* **2015**, *144*, 142–150. [CrossRef]

33. Abhijith, K.V.; Kumar, P.; Gallagher, J.; McNabola, A.; Baldauf, R.; Pilla, F.; Broderick, B.; Di Sabatino, S.; Pulvirenti, B. Air pollution abatement performances of green infrastructure in open road and built-up street canyon environments—A review. *Atmos. Environ.* **2017**, *162*, 71–86. [CrossRef]

34. Van Renterghem, T. Green roofs for acoustic insulation and noise reduction. In *Nature Based Strategies for Urban and Building Sustainability*; Pérez, G., Perini, K., Eds.; Butterworth-Heinemann: Oxford, UK, 2018; Chapter 3.8; pp. 167–179. ISBN 9780128121504.

35. Liu, C.; Hornikx, M. Effect of water content on noise attenuation over vegetated roofs: Results from two field studies. *Build. Environ.* **2018**, *146*, 1–11. [CrossRef]

36. Soulis, K.X.; Ntoulas, N.; Nektarios, P.A.; Kargas, G. Runoff reduction from extensive green roofs having different substrate depth and plant cover. *Ecol. Eng.* **2017**, *102*, 80–89. [CrossRef]

37. Vijayaraghavan, K.; Reddy, D.H.K.; Yun, Y.-S. Improving the quality of runoff from green roofs through synergistic biosorption and phytoremediation techniques: A review. *Sustain. Cities Soc.* **2019**, *46*, 101381. [CrossRef]

38. Yang, J.; llamathy, D.; Kumar, M.; Pyrgou, A.; Chong, A.; Santamouris, M.; Kolokotsa, D.; Lee, S.E. Green and cool roofs' urban heat island mitigation potential in tropical climate. *Sol. Energy* **2018**, *173*, 597–609. [CrossRef]

39. Peri, G.; Rizzo, G.; Scaccianoce, G.; Sorrentino, G. Role of green coverings in mitigating heat island effects: An analysis of physical models. *Appl. Mech. Mater.* **2013**, *261–262*, 251–256. [CrossRef]

40. Bevilacqua, P.; Mazzeo, D.; Bruno, R.; Arcuri, N. Surface temperature analysis of an extensive green roof for the mitigation of urban heat island in southern Mediterranean climate. *Energy Build.* **2017**, *150*, 318–327. [CrossRef]

41. Köhler, M.; Ksiazek-Mikenas, K. Green roofs as habitats for biodiversity. In *Nature Based Strategies for Urban and Building Sustainability*; Pérez, G., Perini, K., Eds.; Butterworth-Heinemann: Oxford, UK, 2018; Chapter 3.14; pp. 239–249. ISBN 9780128121504.

42. Francis, L.F.M.; Jensen, M.B. Benefits of green roofs: A systematic review of the evidence for three ecosystem services. *Urban For. Urban Green.* **2017**, *28*, 167–176. [CrossRef]

43. Peri, G.; Rizzo, G. The overall classification of residential buildings: Possible role of tourist EU Ecolabel award scheme. *Build. Environ.* **2012**, *56*, 151–161. [CrossRef]

44. Porcaro, M.; Ruiz de Adana, M.; Comino, F.; Peña, A.; Martín-Consuegra, E.; Vanwalleghem, T. Long term experimental analysis of thermal performance of extensive green roofs with different substrates in Mediterranean climate. *Energy Build.* **2019**, *197*, 18–33. [CrossRef]

45. Tang, M.; Zheng, X. Experimental study of the thermal performance of an extensive green roof on sunny summer days. *Appl. Energy* **2019**, *242*, 1010–1021. [CrossRef]

46. Ferrante, P.; La Gennusa, M.; Peri, G.; Rizzo, G.; Scaccianoce, G. Vegetation growth parameters and leaf temperature: Experimental results from a six plots green roofs' system. *Energy* **2016**, *115*, 1723–1732. [CrossRef]

47. Brunetti, G.; Porti, M.; Piro, P. Multi-Level numerical and statistical analysis of the hygrothermal behavior of a non-vegetated green roof in a Mediterranean climate. *Appl. Energy* **2018**, *221*, 204–219. [CrossRef]

48. Morakinyo, T.E.; Dahanayake, K.W.D.K.C.; Ng, E.; Chow, C.L. Temperature and cooling demand reduction by green-roof types in different climates and urban densities: A co-simulation parametric study. *Energy Build.* **2017**, *145*, 226–237. [CrossRef]

49. Ferrante, P.; La Gennusa, M.; Peri, G.; Scaccianoce, G.; Sorrentino, G. Comparison between conventional and vegetated roof by means of a dynamic simulation. *Energy Procedia* **2015**, *78*, 2917–2922. [CrossRef]

50. Li, S.-X.; Qin, H.-P.; Peng, Y.-N.; Khu, S.T. Modelling the combined effects of runoff reduction and increase in evapotranspiration for green roofs with a storage layer. *Ecol. Eng.* **2019**, *127*, 302–311. [CrossRef]

51. Heusinger, J.; Sailor, D.J.; Weber, S. Modeling the reduction of urban excess heat by green roofs with respect to different irrigation scenarios. *Build. Environ.* **2018**, *131*, 174–183. [CrossRef]

52. Zirkelbach, D.; Mehra, S.R.; Sedlbauer, K.P.; Künzel, H.M.; Stöckl, B. A hygrothermal green roof model to simulate moisture and energyperformance of building components. *Energy Build.* **2017**, *145*, 79–91. [CrossRef]

53. Peri, G.; Rizzo, G.; Scaccianoce, G.; La Gennusa, M.; Jones, P. Vegetation and soil—Related parameters for computing solar radiation exchanges within green roofs: Are the available values adequate for an easy modeling of their thermal behavior? *Energy Build.* **2016**, *129*, 535–548. [CrossRef]

54. Saiz Alcazar, S.; Olivieri, F.; Neila, J. Green roofs: Experimental and analytical study of its potential for urban microclimate regulation in Mediterranean-continental climates. *Urban Clim.* **2016**, *17*, 304–317. [CrossRef]

55. Karteris, M.; Theodoridou, I.; Mallinis, G.; Tsiros, E.; Karteris, A. Towards a green sustainable strategy for Mediterranean cities: Assessing the benefits of large-scale green roofs implementation in Thessaloniki, Northern Greece, using environmental modelling, GIS and very high spatialr esolution remote sensing data. *Renew. Sustain. Energy Rev.* **2016**, *58*, 510–525. [CrossRef]

56. Vaz Monteiro, M.; Blanusa, T.; Verhoef, A.; Richardson, M.; Hadley, P.; Cameron, R.W.F. Functional green roofs: Importance of plant choice in maximizing summertime environmental cooling and substrate insulation potential. *Energy Build.* **2017**, *141*, 56–68. [CrossRef]

57. Zhao, M.; Tabares-Velasco, P.C.; Srebric, J.; Komarneni, S.; Berghage, R. Effects of plant and substrate selection on thermal performance of green roofs during the summer. *Build. Environ.* **2014**, *78*, 199–211. [CrossRef]

58. Marasco, D.E.; Culligan, P.J.; McGillis, W.R. Evaluation of common evapotranspiration models based on measurements from two extensive green roofs in New York City. *Ecol. Eng.* **2015**, *84*, 451–462. [CrossRef]

59. Di Lorenzo, D.; Maini Lo Casto, B.; Peri, G.; Rizzo, G.; Scaccianoce, G.; Tambani, C. Enhancing values of roofs albedo for lowering cities air temperature and electric demand of buildings: A simple economic evaluation. In Proceedings of the 10th International Conference on Indoor Air Quality, Ventilation and Energy Conservation in Buildings (IAQVEC 2019), Bari, Italy, 5–7 September 2019.

60. Yin, H.; Kong, F.; Dronova, I.; Middel, A.; James, P. Investigation of extensive green roof outdoor spatio-temporal thermal performance during summer in a subtropical monsoon climate. *Sci. Total Environ.* **2019**, *696*, 133976. [CrossRef]

61. Calvino, F.; La Gennusa, M.; Nucara, A.; Rizzo, G.; Scaccianoce, G. Evaluating human body area factors from digital images: A measurement tool for a better evaluation of the ergonomics of working places. *Occup. Ergon.* **2005**, *5*, 173–185.

62. Marino, C.; Nucara, A.; Peri, G.; Pietrafesa, M.; Pudano, A.; Rizzo, G. An MAS-based subjective model for indoor adaptive thermal comfort. *Sci. Technol. Built Environ.* **2015**, *21*, 114–125. [CrossRef]

63. Lei, K.T.; Tang, J.S.; Chen, P.H. Numerical simulation and experiments with green roofs for increasing indoor thermal comfort. *J. Chin. Inst. Eng.* **2019**, *42*, 346–356. [CrossRef]

64. Costa, C.R., Jr.; Cordeiro, J.J.F., Jr.; Omar, A.J.S.; Guiselini, C.; Loges, V.; Silva, G.R., Jr.; Pandorfi, H. Thermal comfort in rural buildings with green roofs. *Acta Hortic.* **2018**, *1215*, 291–294. [CrossRef]

65. Di Giuseppe, E.; D'Orazio, M. Assessment of the effectiveness of cool and green roofs for the mitigation of the Heat Island effect and for the improvement of thermal comfort in Nearly Zero Energy Building. *Archit. Sci. Rev.* **2015**, *58*, 134–143. [CrossRef]

66. Musy, M.; Malys, L.; Inard, C. Direct and indirect impacts of vegetation on building comfort: A comparative study of lawns, greenwalls and green roofs. *Procedia Environ. Sci.* **2017**, *38*, 603–610. [CrossRef]

67. Marino, C.; Nucara, A.; Peri, G.; Pietrafesa, M.; Rizzo, G. A generalized model of human body radiative heat exchanges for optimal design of indoor thermal comfort conditions. *Sol. Energy* **2018**, *176*, 556–571. [CrossRef]

68. EnergyPlus. Available online: https://energyplus.net (accessed on 1 November 2019).

69. La Gennusa, M.; Peri, G.; Scaccianoce, G.; Sorrentino, G.; Aprile, S. A case-study of green roof monitoring: The building of council for agricultural research and economics in Bagheria, (Italy). In Proceedings of the 2018 IEEE International Conference on Environment and Electrical Engineering and 2018 IEEE Industrial and Commercial Power Systems, Palermo, Italy, 12–15 June 2018.

70. Kotsiris, G.; Androutsopoulos, A.; Polychroni, E.; Nektarios, P.A. Dynamic U-value estimation and energy simulation for green roofs. *Energy Build.* **2012**, *45*, 240–249. [CrossRef]

71. Cascone, S.; Catania, F.; Gagliano, A.; Sciuto, G. A comprehensive study on green roof performance for retrofitting existing buildings. *Build. Environ.* **2018**, *136*, 227–239. [CrossRef]

72. European Committee for Standardization. *EN 16798-1:2019—Energy Performance of Buildings—Ventilation for Buildings—Part 1: Indoor Environmental Input Parameters for Design and Assessment of Energy Performance of Buildings Addressing Indoor Air Quality, Thermal Environment, Lighting and Acoustics—Module M1-6—CEN;* European Committee for Standardization: Brussels, Belgium, 2019.

73. Niachou, A.; Papakonstantinou, K.; Santamouris, M.; Tsangrassoulis, A.; Mihalakakou, G. Analysisi of green roof thermal properties and investigation of its energy performance. *Energy Build.* **2001**, *33*, 719–729. [CrossRef]

 © 2020 by the authors. Licensee MDPI, Basel, Switzerland. This article is an open access article distributed under the terms and conditions of the Creative Commons Attribution (CC BY) license (http://creativecommons.org/licenses/by/4.0/).

Article

Towards Nearly Zero Energy and Environmentally Sustainable Agritourisms: The Effectiveness of the Application of the European Ecolabel Brand

Laura Cirrincione [1,*] , **Maria La Gennusa** [1] , **Giorgia Peri** [1] , **Gianfranco Rizzo** [1] and **Gianluca Scaccianoce** [1,2]

1. Department of Engineering, University of Palermo, Viale delle Scienze Bld. 9, 90128 Palermo, Italy; maria.lagennusa@unipa.it (M.L.G.); giorgia.peri@unipa.it (G.P.); gianfranco.rizzo@unipa.it (G.R.); gianluca.scaccianoce@unipa.it (G.S.)
2. National Research Council of Italy, Institute of Biomedicine and Molecular Immunology, via Ugo La Malfa 153, 90146 Palermo, Italy
* Correspondence: laura.cirrincione@unipa.it

Received: 28 July 2020; Accepted: 17 August 2020; Published: 19 August 2020

Abstract: Tourism represents an important economic driver in Italy, being responsible for approximately 13.2% of the total GDP (a value higher than the reference European average) and for nearly 10% of the regional GDP. Among the touristic sectors, the agritourist ones show a persistent growth, experiencing in 2019 a 6.7 point percentage improvement compared to the 2017 figures. Given this situation, the transition towards a low-carbon path, affecting the building sector for some time, should also involve agritourist buildings, through the release of EU directives, member state laws, and technical rules. On the other hand, agritourism sites could be awarded the Community EU Ecolabel. Unfortunately, awarding the EU environmental excellence brand implies the availability of several data on building energy behavior that should then be managed by complex evaluation tools. To overcome this issue, the use of the simplified ARERA (Italian Regulatory Authority for Energy Networks and Environment) technical datasheets, issued to assess environmental improvements consequent to energy efficiency interventions in the urban residential building stock, is proposed. The application of this tool totally avoids using building computer-based simulation models, thus facilitating the preparation of the EU Ecolabel request documentation by agritourism owners. Being awarded the Community EU Ecolabel also implies approaching a net zero energy condition because of a lower energy consumption and a minor recourse to fossil fuels. For this purpose, an application of an easy graphical method, previously developed for residential and commercial buildings, which visually represents improvements achievable by a given agritourism when implementing energy efficiency measures, is presented.

Keywords: building energy efficiency; European environmental brands; tourism sector; agritourism; nearly zero energy buildings (nZEB)

---

1. Introduction

Tourism, and the activities connected to it, represent an important sector of the economic system. According to recent statistics the tourism industry represents about 10% of total global gross domestic product (GDP) and 7% of global trade [1,2], accounting for approximately 11% of the world's employment, with an expected positive economic growth trend [3,4]. Tourism constitutes a significant contributor to energy consumption, both at a global and European scale [5–7], which translates to a significant impact on the environment and ecosystem; it is in fact responsible for about 5% of the global CO_2 emitted by human activities [1,8].

Accommodation (thus the building), in particular, is the third energy consuming item (after travel and transport), much of which is consumed in space heating or air conditioning (up to 50% in some cases), followed by hot water, and cooking [9,10]. Moreover, a study conducted by the World Tourism Organization and the United Nations Environment Programme [11] estimates that accommodation generates 21% of tourism's total greenhouse gas (GHG) emissions. Accordingly, the number of papers analyzing tourism significance, in terms of energy consumption [12,13] and impacts on emissions [14,15], has been increasing lately.

Consequently, in recent years much attention has been paid to the concept of sustainable tourism, which in accordance with the United Nations Environment Programme (UNEP) and the United Nations World Tourism Organization (UNTWO) is defined as "development of tourism activities with a suitable balance between the dimensions of environmental, economic, and sociocultural aspects to guarantee its long-term sustainability". Hence, the challenge of sustainable tourism is to mitigate its negative impacts, consisting mainly in: (i) high energy consumption, (ii) increasing GHG emissions in the atmosphere, and (iii) the contribution to climate change [16].

Therefore, taking into consideration global [17,18], European [19–22], and national [23] policies, the UNWTO recommended three central actions on which the tourism sector should concentrate in order to contribute in achieving a more sustainable development [1,24], which are resource efficiency, environmental protection, and climate change (linked to sustainable development goals (SDGs) 6, 7, 8, 11, 12, 13, 14, and 15) [25].

At the European scale, the European Commission set the basis for the best environmental management practice in the tourism sector in accordance with Article 46 of the Eco-Management and Audit Scheme (EMAS) regulation [26,27]. Furthermore, by means of the "Guide on EU funding for the tourism sector 2014–2020" [28] the EU states that effective governance, policies, frameworks, and tools need to be implemented in order to properly guide and support (also from an economic point of view) the development and promotion of sustainable tourism practices.

Tools like these are indeed important because they encourage the owners and/or managers of the accommodation facilities to use practices and systems that allow both energy savings and pollutant emissions, by favoring the visibility of these structures in terms of environmental sustainability, which represent an added value, given that tourists are becoming increasingly more attentive to this issue.

In this regard one of the first initiatives undertaken by the European Community has been the releasing of the EU Ecolabel for tourist accommodation services [29], created to improve the environmental performance of hotels, campsites, hostels, agritourisms, holiday homes, and bed & breakfasts, by providing efficient guidelines on the action to be implemented in order to lower their environmental impact; and which still remains one of the most implemented initiatives.

The promotion of sustainable tourism is also the basis of the nearly zero-energy hotels (neZEH) project, launched by the Intelligent Energy Europe Programme of the European Commission, with the intent of supporting European hotels in complying with the nZEB (nearly-Zero Energy Buildings) regulations [1]. On this subject, various studies have been conducted aimed at analyzing the achievable energy saving measures [30–32] and proposing suitable strategies and policies to be adopted [33,34].

Looking at the national scenario, the tourism issue is particularly relevant, considering that 16.5% of EU accommodation facilities are located in Italy [35], and since in the last two years Italy resulted to be amongst the top five most visited European tourist destinations (for accommodation in hospitality facilities), with a 13.4% share of the total of the EU-28 [36,37].

According to some recent statistics, the Italian tourism sector represents 13.2% of the national GDP (for a total contribution of around 230 billion euros), higher than both the world and European figures (which stand at around 10%). The economic impact of tourism is significantly reflected in the job market, accounting for 14.9% of the country's total employment [38]. Tourism is in fact one of the fastest growing industrys in Italy, and both public and private business organizations are strongly interested in its economic and environmental impact, both at national and regional level [39,40].

Thus, from the collaboration between such organizations and the national and regional governments, different initiatives have been undertaken from an environmental sustainability point of view in recent years. These include the creation of a set of national environmental quality certifications (besides the previously cited EU Ecolabel), including the "Green Key" [41], "Bandiera Blu" [42], and "Spighe Verdi" [43], born from the collaboration between the Italian Foundation for Environmental Education—FEE Italia (whose actions are supported by ONU, UNEP, UNWTO, and UNESCO) and national authorities dealing with environmental policies [44].

Furthermore, other economic initiatives have been implemented to encourage the use of sustainable energy solutions through financial incentives. The "Tax Credit Alberghi—Bonus alberghi e agriturismi" (bonus for hotels and agritourism), a tax facility that encourages various upgrading activities, including those aimed at improving energy efficiency, has recently been introduced, specifically for accommodation facilities [45].

Agritourism, or rural tourism, has been promoted as a practice able to encourage the use of green practices, making farms sustainable and also maintaining the local historical and natural settings [8,46].

Thanks to this, according to recent statistics in Italy, the agritourism sector continues to record a growing trend, both in the number of structures, and in the presence of customers and its economic value. Agriculture economic reports make it possible to measure the economic dimension of the agritourism sector, which is equal to 1.36 billion euros, up 6.7% compared to the previous year. In particular, 60% of agritourisms are located in the regions of central and southern Italy, where Sicily prevails with more than 600 farms [47].

The growing interest in the agritourism sector is also reflected in the academic world, where studies concerning both the economic and social benefits of various tourist activities in the rural area, including agritourism [48], and the environmental performance of agritourism companies in terms of energy performance [49,50], can be found.

In the present work we verified whether the simplified ARERA (Italian Regulatory Authority for Energy Networks and Environment) technical datasheets [51], issued for the urban residential building stock, can be easily applied to estimate the increase in energy efficiency (or the corresponding decrease in the release of polluting substances) consequent to the adoption of some improvements to a building or plant, planned for the issuance of the EU Ecolabel brand for accommodation facilities [29]. The convenience in the use of these technical datasheets lies in the fact that they allow the estimation of the energy demand reductions without necessarily going through the building simulation. For this purpose, a case study has been conducted to estimate what advantages agritourism owners could gain in adopting a well-known brand such as the EU Ecolabel [29], with particular reference to the actions aimed at saving energy and reducing emissions of pollutants, from the perspective of a possible "nearly Zero Energy Agritourism (nZEA)", in parallel with the previously cited nZEB and neZEH projects.

The idea at the base of this work stems from the numerical consistency of agritourisms in Sicily [47] and their conceivable growth trend, which is a consequence of the increased interest in the rural landscape of the territory and in the products of the land that are strongly orienting tourism, directing it not only towards the urban context. The adoption of an environmental certificate like the EU Ecolabel [29] can therefore represent an advantage both for agritourism owners and for the entire territory.

Furthermore, the owners of agritourism in Sicily can apply for subsidized loans and financial funding [28,52] in the regional area and beyond. However, such requests must be supported by information concerning the consumption and energy efficiency of the agritourism and, in line with the new European directives on sustainability [17–19], by information on the environmental performance of the buildings themselves (premises).

Normally this information is of a complex nature and tends to imply the use of sophisticated simulation models, the use of which is not always the prerogative of (or available to) the managers of the holiday farms. The same problem can be found by analyzing the work of the decision makers who have to assess the adequacy of the requests for funding.

Essentially, the availability of simple but reliable tools for evaluating these premises is of paramount importance for the orientation of this important tourism sector towards a sustainable path.

Hence, as previously mentioned, in order to provide a contribution to this important issue we assessed the reliability of a scheme of simple computational methods provided by the Italian Regulatory Authority for Energy Networks and Environment—ARERA [51], specifically for the residential and tertiary building stock. The advantage in the use of this computational scheme lies indeed in the fact that it is based on excel spreadsheets (technical datasheets) which, as already mentioned, allow the estimation of the energy demand reductions without the need of simulating the building behavior.

2. Materials and Methods

The proposed methodology aims at considering together in an easy and accessible way two aspects of the sustainability, which are energy efficiency and environmental safety, in order to help agritourism owners, and/or managers, to make decisions that are more favorable to them and consistent with the European policies in force. Specifically, according to the presented approach, the selection of energy efficiency interventions is based on a combination of the ARERA technical datasheets and the EU Ecolabel criteria, hence taking into account the environmental sustainability aspects, and also in view of achieving a possible nearly zero energy condition (nZEA). Therefore, two Sicilian agritourisms have been selected to show how the application of the proposed methodology actually works.

The considered approach can also be seen as a simple diagnosis method aimed at facilitating the social appropriation of knowledge and technology, so that the owners of agritourism facilities can confidently check their level of eco-efficiency. Moreover, the method can be utilized in order to choose between addressing actions concerning the energy performance of the structure or interventions regarding the installation of new (renewable) energy plants.

2.1. Agritourism Definition

The Italian national legislation [53], and the regional Sicilian one [54,55], define as 'agritourism' activities, those reception and hospitality activities exercised by agricultural entrepreneurs, through the use of their own company connected with the activities of cultivation of the land, forestry, and animal breeding. Thus, agritourism activities include:

- providing accommodation;
- administering meals and beverages consisting mainly in products of their own production and products from farms in the local area;
- organizing recreational, cultural, educational, sports, and excursion activities aimed at promoting and supporting the territory and the rural heritage.

2.2. The ARERA Technical Datasheets

In the present work, the use of ARERA technical datasheets [51] was not an arbitrary (random) choice, but it was decided to turn to these methods since, although simplified, they constitute an official reference at the Italian national level.

The Italian Regulatory Authority for Energy Networks and Environment—ARERA is indeed an independent body, established with the task of protecting consumers' interests and promoting competition, efficiency, and the spread of services, and having adequate quality levels, through regulation and control activities. The action of ARERA concerns the sectors of electricity and natural gas [56], water services [57], district heating and district cooling [58], and the waste cycle [59].

One of the main tasks of ARERA is to promote the rational use of energy, with particular reference to the promotion and diffusion of end use energy efficiency and/or energy saving actions, and the adoption of measures for sustainable development. Among the feasible actions, there are both active measures, which involve the installation of high efficiency equipment, or the insertion of regulation devices for a more efficient use of energy, and passive interventions such as the modification of buildings' envelope in order to reduce losses.

In this regard, the technical datasheets proposed by ARERA establish the guidelines for the preparation, execution, and final evaluation of specific actions, aimed at increasing energy efficiency (or promoting energy saving), providing reduced rates of primary energy consumption actually achieved (expressed in toe—Tons of oil equivalent), and also for the purpose of issuing energy efficiency certificates. Table A1 in Appendix A reports a comprehensive list of the current standardized and analytical ARERA technical datasheets.

2.3. The EU Ecolabel Brand

Established in 1992 (by Regulation n. 880/92 [60], now disciplined by Regulation (EC) n. 66/2010 [61] in force in the EU-28) and recognized across Europe and worldwide (Figure 1), the EU Ecolabel is a voluntary environmental performance certificate that is awarded to products and services meeting high environmental standards. The EU Ecolabel encourages companies to develop products and provide services that consume less energy, and generate less waste and CO_2 emissions. As of March 2019, an increase by 88%, with respect to 2016, of the number of EU Ecolabelled products/services has been registered. Leading countries for number of products/services are: Spain, Italy, Germany, Belgium, and France [62].

Figure 1. Official EU Ecolabel logo [62].

The EU Ecolabel provides exigent criteria, and relative guidelines, depending on the type of product and/or service, in order to reduce their overall environmental impact. Such criteria are established at a European scale with a wide participation of interested parties, including both public authorities, and consumer and environmental associations [63].

In particular, the EU Ecolabel for tourist accommodation services [29] was created specifically for hotels, campsites, hostels, agritourisms, holiday homes, and bed & breakfasts, in order to improve their environmental performance, by providing a set of criteria on the action to be implemented in order to lower their impact. Such criteria are divided into mandatory and optional, and focus on the five categories; general management, energy, water, waste and wastewater, and other, as shown in Table A2 in Appendix A.

In order to be awarded the EU Ecolabel a tourist accommodation service, other than falling within the product group "tourist accommodation" according to the legal obligations of the country in which the accommodation is located, must comply with all the mandatory criteria (if applicable), and receive at least twenty points under the optional criteria [29].

An added value, in terms of visibility, for tourist accommodation owners lies in the fact that the EU Ecolabel is recognized by the majority of travelers as a way of legitimizing the accommodation's claims that it is making real efforts to reduce its impact on the environment in its operational activities.

2.4. Analysis Methodology

2.4.1. Merging the ARERA Technical Datasheets and the EU Ecolabel Criteria

As mentioned in the introduction section, the aim of the present work is to verify whether the simplified ARERA technical datasheets can be applied to estimate the increase in energy efficiency consequent to the adoption of some actions planned for the issuance of a EU Ecolabel for Tourist Accommodation Services, without necessarily going through the building simulation. Obviously, an increase in the energy efficiency implies a corresponding decrease in the release of polluting substances.

Therefore, since this work is mainly focused on the energy criteria, starting from the assumption that the considered agritourism meets all the mandatory criteria, it was decided to analyze possible "environmental action packages", consisting of different combinations of the actions established by the optional energy criteria, which are better suited to a scenario such as agritourism, and which allow the obtaining of the weight of the energy category on the twenty points minimum limit set by the regulation, which corresponds to 7.34 points.

To this purpose, only the ARERA datasheets regarding the actions related to the improvement of the structure energy efficiency that could be transferred and applied to agritourism structures, according to the EU Ecolabel for Tourist Accommodation energy criteria, have been considered, as reported in Table 1, where the correspondent energy consumption categories have also been reported.

Table 1. Correspondence between the Italian Regulatory Authority for Energy Networks and Environment (ARERA) datasheets and the EU Ecolabel energy criteria.

ARERA Technical Data Sheet N.	EU Ecolabel Criterion N.	EU Ecolabel Achievable Points	Energy Consumption Category [1]
5	33	4	HVAC
7	39, 40, 41	3.5	RES
8T	6	2	DHW
15T	6, 7	1.5	HVAC
19T	7	3.5	HVAC
27T	6	1.5	DHW
6	-	-	HVAC
20T	-	-	

[1] HVAC—Heating, Ventilation, and Air Conditioning; DHW—Domestic Hot Water; RES—Renewable Energy Source.

As can be seen in Table 1, under the dotted line, the two ARERA datasheets 6T and 20T, additional to those that can be associated to the EU Ecolabel, have also been taken into account. In fact, even though these two intervention typologies are not foreseen by the current Ecolabel scheme, they represent actions that can actually be applied to an agritourism structure in the perspective of a possible "nearly Zero Energy Agritourism (nZEA)" as a parallel with the well-known nZEB concept; and also, in view of a possible future improvement of the Ecolabel scheme.

The selected datasheets (Table 1), have been then put into the form of appropriate excel spreadsheets.

The equations relative to the ARERA calculation procedures, for each considered technical datasheet, are given in Appendix B.

One aspect that must be highlighted here regards the fact that while datasheets 5, 6, 15T, 19T, 20T, and 27T enable obtaining savings of consumed energy (energy saving measures, ESM), datasheets 7 and 8T allow, instead, the production of energy from renewable sources (renewable energy sources, RES).

2.4.2. Methodology Application Feasibility

With the aim of assessing the potential energy savings, with reference to a real context, it was decided to select two agritourisms situated in the Sicilian province of Palermo, considered as representative of the whole regional agritourism context regarding the size, the provided services, and more importantly for the purpose of the proposed methodology object of the present work, in terms of energy consumption. Apart from these physical and energy features, both agritourisms were selected thanks to their wide offer of services, which are representative of these kind of farms, and due to the fact that they operate in the two climatic zones where agritourisms are mainly sited in

Sicily. Specifically, the two agritourism are *Villa Dafne*, sited in Alia and belonging to climatic zone D, and *Bergi*, located in Castelbuono and classified as climatic zone C. Both agritourisms fall into solar belt 3. Table 2 describes the main general characteristics of the two structures.

Table 2. General characteristics of the two selected agritourisms.

Characteristic	*Villa Dafne*	*Bergi*
Covered surface (m^2)	1000	1400
Glazed surface (m^2)	236	305
Opaque surface (m^2)	2768.5	2791.75
Surface/Volume ratio (-)	0.62	0.5
N. of seats in the dining area	150	180
N. of rooms	35	34

In Table 3 is reported the information relative to the energy characteristics of interest for the conducted study, which were obtained by on field surveys and interviews with the owners of the two businesses, thanks to which it was possible to reconstruct the energy consumption relative to an entire year of operation of the structures. In particular, the data regarding the energy consumption were distributed between four main categories and accordingly broken down into percentages, and corresponding toe/year, also with reference to the corresponding energy sources. As for the energy sources' average costs, the following values were used:

- 0.19 €/kWh for electricity;
- 1.17 €/Sm^3 for natural gas;
- 0.90 €/lt. for diesel oil.

Table 3. Energy sources and energy consumption breakdown for the two selected agritourisms.

Category	Source	*Villa Dafne*		*Bergi*	
		%	Toe/Year	%	Toe/Year
Domestic Hot Water (DHW)	natural gas/diesel oil	0	0.00	16	5.09
	electricity	22	9.72	6	1.91
Lighting	electricity	15	6.63	15	4.77
Heating, Ventilation and Air Conditioning (HVAC)	natural gas/diesel oil	25	11.05	15	4.77
	electricity	14	6.19	24	7.64
Other	natural gas/diesel oil	0	0.00	0	0.00
	electricity	24	10.60	24	7.64
Total			**44.18**		**31.82**

Subsequently, it was hence possible to obtain the achievable energy savings (AES), in terms of percentage of electricity consumption covered by the datasheets proposed interventions on an annual basis, by comparing the values obtained from Equations (A1) to (A8), and the total energy consumption (Table 3), by means of the following equation:

$$AES_i = \frac{R_i}{Tot.\ cons._j}\ [\%] \tag{1}$$

where:

- R_i represent the energy savings obtained from Equations (A1) to (A8);
- *Tot. cons.$_j$* is the total figure reported in Table 3;
- *i* and *j* represent the selected intervention and the considered agritourism, respectively.

Regarding the pollutant emissions, an assessment of the CO_2 emissions' reduction was conducted assuming for the considered climatic context an emission factor equal to 2.30 tCO$_2$eq/toe for the electrical supply [64,65], while for natural gas and diesel oil an emission factor of 3.08 tCO$_2$eq/toe and 2.34 tCO$_2$eq/toe, respectively [66].

In order to single out the most convenient aforementioned "environmental actions packages", an economic estimation relative to the interventions suggested by the ARERA technical datasheets was also performed. To this purpose, the information relative to the costs of supply and installation for the materials, used to calculate the proposed interventions costs, were obtained from the current regional price list [67] and from local market surveys, as reported in Table 4.

Table 4. ARERA technical datasheets proposed interventions costs.

Datasheet N°	Proposed Intervention	Cost
5	Replacement of simple glazing with double glazing	407.13 €/m^2
7	Use of photovoltaic systems with an electrical power of less than 20 kW	1898.42 €/kW$_p$
8T	Installation of solar collectors for the production of domestic hot water	578.73 €/m^2
15T	Installation of outdoor air electric heat pumps instead of boilers in newly built or renovated residential buildings	4901.323 €/UFR *
19T	Installation of high efficiency outdoor air conditioners with cooling capacity lower than 12 kW$_f$	490.13 €/kW
27T	Installation of electric heat pump for domestic hot water production in new and existing plants	570.65 €/UFR *
6	Wall and roof insulation	29.32 €/m^2
20T	Thermal insulation of walls and roofs for summer cooling in domestic and service sectors	29.32 €/m^2

* UFR—Reference physical unit.

Successively, the economic savings, in terms of saved €/year, were obtained by multiplying the energy savings with the energy sources' average costs, according to the considered categories breakdown (Table 1). Furthermore, in order to select the optimal "environmental actions packages", for these the pay-back periods (not discounted) were also calculated and expressed in years.

3. Results

In this section the outcomes of the application of analysis methodology are reported.

Regarding the input parameters used in the equations relative to the ARERA calculation procedures, for each considered technical datasheets, these are given in Table A3 in Appendix B.

The following Figures 2 and 3 show the achievable energy savings (AES) on the total annual consumption, relative to the application of the intervention proposed by each considered ARERA datasheet, to the two agritourisms. On the right side of the graphs, the EU Ecolabel points corresponding to each datasheet are also reported.

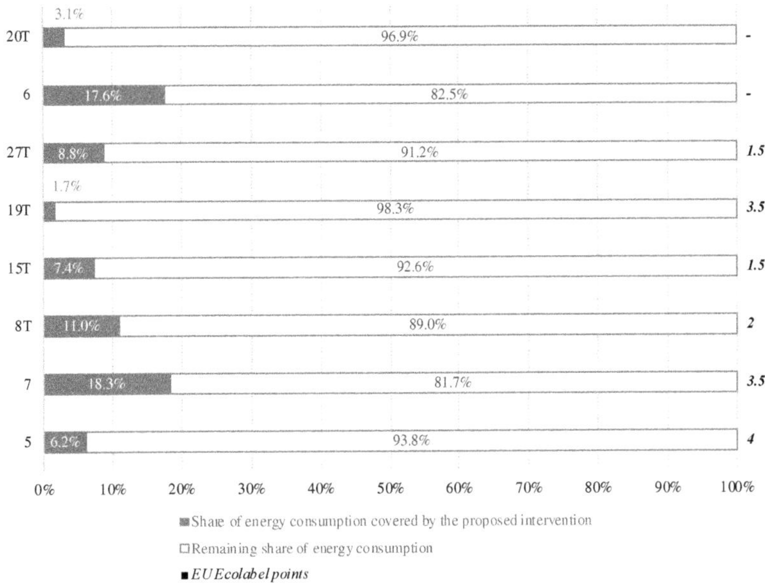

Figure 2. Achievable energy savings (AES), on an annual basis, and EU Ecolabel points relative to each considered ARERA datasheet for *Villa Dafne* agritourism.

Figure 3. Achievable energy savings (AES), on an annual basis, and EU Ecolabel points relative to each considered ARERA datasheet for *Bergi* agritourism.

Figures 2 and 3 show how, amongst the ARERA proposed interventions enabling the obtaining of an EU Ecolabel score, the implementation of a photovoltaic (PV) system (datasheet 7) would be the one allowing the gain of a greater advantage in terms of energy consumption. Concerning the savings related to the domestic hot water (DHW) category, by comparing datasheets 8T and 27T, which are alternatives to each other, it can be observed how 8T would be the more convenient choice. Regarding the heating, ventilation, and air conditioning (HVAC) category, the savings achievable through the application of datasheets 15T and 19T should, instead, be considered jointly (15T + 19T) as they can be attributed to the same improvement intervention. As for datasheet 5, although it represents one of the easiest measures to implement, it does not seem to bring the benefits that would have been expected.

With respect to datasheets 6 and 20T, also in this case the consideration that the attainable benefits must be considered together (6 + 20T) is valid. As already explained these two datasheets fall out of the Ecolabel scoring scheme, nevertheless they represent the second-best intervention that allows the highest energy savings, after datasheet 7.

The overall obtained results for the two considered agritourisms are reported in Tables 5 and 6.

Table 5. Proposed interventions costs and environmental benefits for *Villa Dafne* agritourism.

ARERA Datasheet N.	EU Ecolabel Points	Proposed Intervention Cost (€)	Energy Savings (*AES*)	CO_2 Emissions Reduction (tCO_2eq/Year)	Economic Savings (€/Year)
5	4	123,971.09	6.20%	7.7	2831.21
7	3.5	37,588.70	18.26%	18.6	8053.63
8T	2	13,392.91	11.00%	11.2	4851.56
15T	1.5	98,026.40	7.37%	10.0	3429.31
19T	3.5		1.67%	1.7	738.75
27T	1.5	20,930.15	8.80%	9.0	3881.25
6	-	81,172.42	17.55%	23.9	738.74
20T	-		3.13%	3.2	1381.90

Table 6. Proposed interventions costs and environmental benefits for *Bergi* agritourism.

ARERA Datasheet N.	EU Ecolabel Points	Proposed Intervention Cost (€)	Energy Savings (*AES*)	CO_2 Emissions Reduction (tCO_2eq/Year)	Economic Savings (€/Year)
5	4	95,980.90	3.70%	2.73	1350.79
7	3.5	37,588.70	25.35%	18.59	8053.63
8T	2	14,608.10	11.00%	8.18	4595.95
15T	1.5	73,519.80	6.29%	4.69	2767.13
19T	3.5		1.74%	1.28	554.06
27T	1.5	10,182.10	6.00%	4.40	1905.96
6	-	81,854.11	13.16%	9.81	1212.67
20T	-		4.39%	3.22	1393.51

As can be observed, according to what has been previously pointed out, a single intervention cost was given to datasheets 15T and 19T as the proposed intervention corresponds to the same type of system, i.e., the same system allows operation for both heating and cooling. The same consideration can be made for datasheets 6 and 20T in relation to the insulation of the building.

Looking at the economic savings column it can be noticed how, from this point of view greater advantages can be associated with datasheets 7 and 8T, followed by 15T + 19T, 27T and lastly 5. Considering the whole set of interventions, instead, the (6 + 20T) option would also result second in this case.

Referring to CO_2 emissions reduction, the obtained results are obviously in line with what was seen beforehand (Figures 2 and 3) and commented on with the energy savings.

4. Discussion

The application of the ARERA data sheets to agritourism raises a question concerning the suitability of these simplified forms to the energy performances of agritourisms sites, being originally developed for residential and commercial buildings.

On the other hand, a possible improvement of the energy features of an agritourism, due to the actions referred to in the ARERA datasheets, should be evaluated on the base of its effectiveness in addressing a given site, towards a nearly-zero energy path, as required by the current international standards [20,21,28].

Both issues are briefly discussed in the following.

4.1. Effectiveness of the Proposed Actions

The obtained results could seem not too encouraging in terms of energy savings. In fact, the reduction of the energy demand following the proposed actions accounts for about one third of the annual energy consumption for both the considered agritourisms. However, this is not surprising; the fact that the ARERA technical datasheets proposed interventions have been designed for the residential sector, in fact, place some limits on their application in a wider context, such as the agritourism one. Specifically, the limitations set on the reference physical units (UFRs) sizes might have made the outcomes much lower than the actually achievable results.

For instance, concerning datasheet 7 a maximum kW_p of 20 kW is reductive for an agritourism, which could employ PV better having wide areas available to install such systems. Supporting this observation, during the survey of the agritourisms, it arose that both currently have a 100-kW PV undergoing design phase. In this context it would be more sensible to impose a limit on the maximum percentage of yearly energy consumption to be covered with the proposed intervention.

The latter consideration also applies to datasheet 8T.

Regarding, instead, datasheet 15T the application problem is mainly related to the residential standard apartment size (80–90 m^2), which is difficult to translate into an agritourism setting. In the conducted analysis, for instance, in order to comply with such a parameter, three to four rooms were grouped and assumed equal to 1.5 standard apartments, but it could be a questionable criterion.

As for datasheets 19T, it would be more reasonable to install a centralized system rather than considering the replacement of the single air conditioning units (the same goes for the heat pumps proposed by datasheets 15T).

Nevertheless, since one of the aims of this work was that of singling the most convenient EU Ecolabel "environmental actions packages", based on the comparison of the results reported in Tables 5 and 6 it was decided to tentatively choose three alternative options, both for *Villa Dafne* (*VD-n*) and *Bergi* (*B-n*), as follows:

- options *VD-1* and *B-1*, constituted by datasheets number 5 and 7;
- options *VD-2* and *B-2*, constituted by datasheets number 7, 8T and (15T + 19T), the latter two must be considered together for the reasons indicated at the end of Section 2.4.2.;
- options *VD-3* and *B-3*, constituted by datasheets number 7, (15T + 19T) and 27T.

Table 7 summarizes the obtained results relative to the selected "environmental actions packages".

By analyzing the data reported in Table 7 it was, therefore decided to consider as optimal options *VD-2* for *Villa Dafne* and *B-2* for *Bergi*. These two options allow, in fact, the obtaining of greater economic and energy savings and, correspondingly, higher CO$_2$ emissions reductions. Moreover, they are characterized by the lower pay back periods.

It must be observed that the availability of effective and reliable methods for evaluating the energy actions involving agritourism is of paramount importance for suitable planning of this important sector. Therefore, the ARERA technical data sheets should be properly reconsidered in order to render them more complicit with the energy features of agritourism buildings and dwellings.

Table 7. Summarized results for the two agritourisms.

Agrit.	Environm. Actions Package	EU Ecolabel Points	Environmental Actions Package Cost (€)	Energy Savings (AES)	CO$_2$ Emissions Reduction (tCO$_2$eq/Year)	Economic Savings (€/Year)	Pay Back Period—Not Discounted (Years)
Villa Dafne	*VD-1*	7.5	161,559.79	24.46%	26.3	10,884.84	14.8
	VD-2	10.5	149,008.01	36.63%	39.8	16,334.51	9.1
	VD-3	10	156,545.25	34.43%	37.6	15,364.20	10.2
Bergi	*B-1*	7.5	133,569.60	29.05%	21.3	9,404.42	14.2
	B-2	10.5	125,716.60	42.64%	31.5	15,416.72	8.2
	B-3	10	121,290.60	37.64%	27.7	12,726.73	9.5

4.2. Towards a Nearly Zero Energy Agritourism

As already mentioned, another intention of this work concerned the possibility of applying some actions to the agritourism structures, additional to those envisioned by the EU Ecolabel, in order to move towards a potential "nearly Zero Energy Agritourism (nZEA)". For this purpose, the results relative to ARERA datasheets 6 and 20T were added to the selected optimal options *VD-2* and *B-2* for *Villa Dafne* and *Bergi*, respectively; the outcomes of such combinations are reported in Figures 4 and 5.

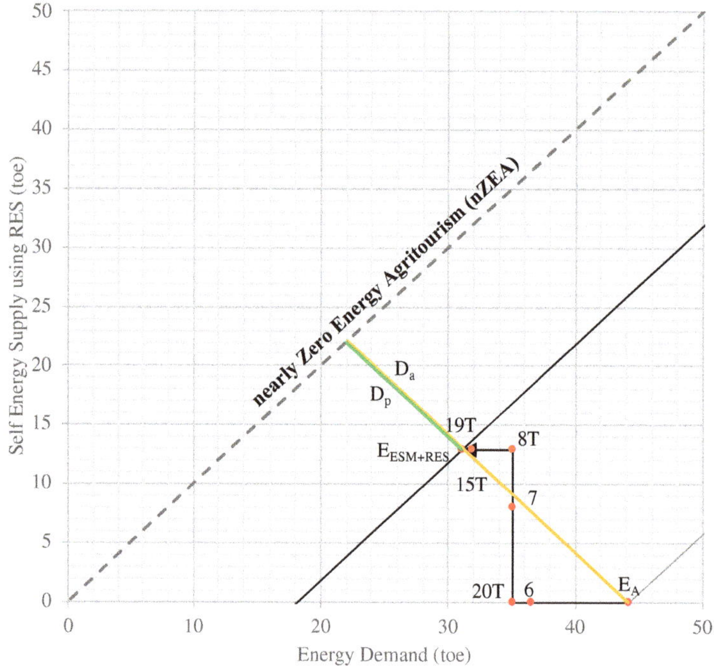

Figure 4. Path towards a nearly zero energy condition (nZEA) for *Villa Dafne*.

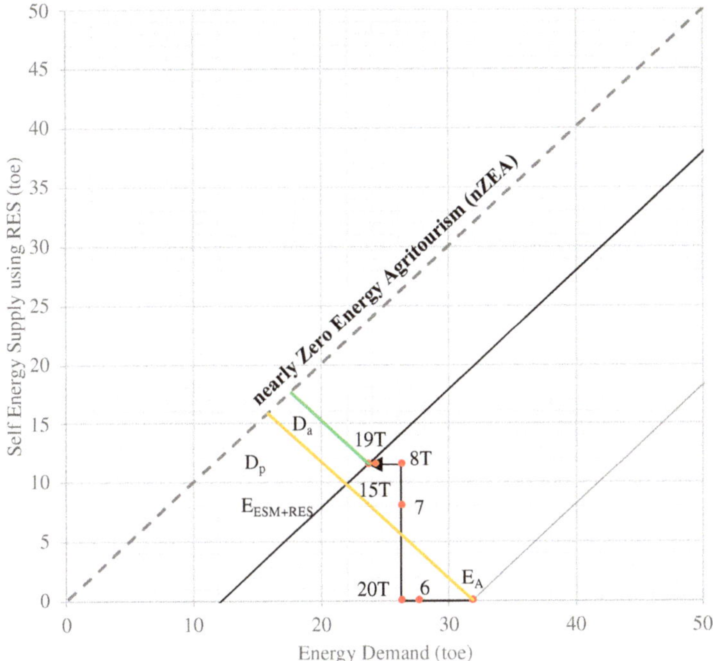

Figure 5. Path towards a nearly zero energy condition (nZEA) for *Bergi*.

According to such approaches and visual representations, already used in the literature [68–70], the nearly zero energy condition (nZEA) is reached when the energy demand (reported on the x axis) is completely covered by the self-energy supply from renewable sources (reported on the y axis).

Therefore, the effectiveness of the ARERA proposed interventions in moving agritourism towards a sustainable path, nZEA, is given as a simple summation of the effects provided by the energy saving measures—E_{ESM} (datasheets 6, 15T, 19T, and 20T) and those attributable to the renewable energy sources—E_{RES} (datasheets 7 and 8T). E_A represents, instead, the current energy consumption and, D_a and D_p the current (ante operam) and achievable (post operam) minimum distances (hence the perpendicularity) from the nZEA condition, respectively.

For the sake of simplicity, this assumption does not take into account the synergetic effects that are likely induced by the contemporary adoption of different energy actions on a given agritourism site.

Consequently, the results reported in Figures 4 and 5 show that the selected combinations of interventions allow an improvement of 59% for *Villa Dafne* and a 62% for *Bergi*, in terms of approaching the nZEA condition with respect to the current conditions.

Regardless of the obtained results, the proposed methodology can be seen as a simplified scheme for analyzing and ranking the "environmental actions packages" applicable to agritourisms, and could be usefully adopted by local administrations to define the impact of different scenarios in order to better define environmental policies concerning the agritourism sector.

The proposed assessment/estimation methodology could, therefore, also represent important information for the design of the rural tourism sector, and of the/a regional energy plan by stakeholders and decision makers [71,72].

4.3. On the Correspondence between the EU Ecolabel Criteria and the ARERA Technical Datasheets

The application of the ARERA technical datasheets and the EU Ecolabel criteria to two different agritourisms in Sicily (here considered representative of the whole regional agritourism context) enabled us to better understand the level of compliance between two such schemes. The level of correspondence cannot totally match, since the ARERA methodology has been designed specifically for residential buildings and, on the other hand, the EU Ecolabel for Tourist Accommodation Services applies expressly to tourism facilities, with features that could not be perfectly applicable to agritourisms. These latter, in fact, are generally characterized by the presence of cultivated soils and the production of agrifarm foods and products.

Nevertheless, the comparison exerted on the two sites has shown that some useful correspondences can be assessed. In fact, by means of the combination of the ARERA technical datasheets and the EU Ecolabel energy optional criteria, it is likely possible to identify some "environmental actions packages", suitable to the agritourism context. Such packages are allowed to obtain a 7.34 points minimum limit for the energy category, set by European regulation. In particular, it emerged that the combination of datasheets 5 and 7 (options *VD-1* and *B-1*) allowed obtaining 7.5 EU Ecolabel points, while by adding datasheets 7, 8T and 15T + 19T (options *VD-2* and *B-2*) it is possible to achieve 10.5 points, and from the union of datasheets 7, 15T + 19T, and 27T (options *VD-3* and *B-3*) a total of 10 points can be reached.

Therefore, a suitable implementation of the ARERA technical datasheets (that, apart from other things, permits an easy computation of the energy performances of various building and system components) is recommended to be ancillary utilized with the EU criteria in order to assess a unique scheme for the application of the EU Ecolabel brand.

In addition, the above verified correspondence, allowed us to introduce a criterion for ranking the effectiveness of the proposed measures within the framework of the nearly Zero Energy Buildings approach (nearly Zero Energy Agritourism, in this case). In other words, once the ARERA datasheets have provided useful energy saving results, achieved thanks to the implementation of the proposed interventions, it is easy to report such results in terms of closeness to a zero energy situation for a given agritourism.

5. Conclusions

Agritourisms represent an important reality in the Italian tourism sector, specifically in Sicily due to their numerical consistency and constantly growing trend. The idea at the base of the presented work stems from some considerations regarding the use of a simple method, based on the ARERA technical datasheets (which constitute an official Italian reference), to assess the energy, environmental, and economic benefits related to the implementation of some energy efficiency measures on a given agritourism, specifically aimed at achieving the EU Ecolabel environmental excellence brand, in the perspective of approaching a potential nearly Zero Energy condition.

The results of the conducted analysis put in evidence some discrepancies regarding the application of the ARERA calculation methods, devised for the residential sector, in a wider context, like that of agritourism. Such an outcome was foreseeable, but it has probably been highlighted even more by the fact that the datasheets results are outdated, having not been updated in the last few years.

Nevertheless, the adoption of the proposed efficiency interventions, despite not being specifically defined for the agritourism context, contributed in addressing both structures toward a nearly Zero Energy path, hence, improving their performances in terms of sustainability.

Apart from the interventions proposed by the ARERA, clearly agritourism sites can be interested in further renewable technologies in order to promote their energy sustainability. In fact, the application of solutions like micro wind turbines, biomass, and high efficiency cogeneration for such purposes has been demonstrated [73]. Similar and/or recently available technologies could represent a driver for implementing new ARERA technical datasheets, in order to render them more compliant with the agritourism context, and the EU targets for energy efficiency and emissions reductions in the civil sector.

In conclusion, it arose that, although it was possible to combine the ARERA technical datasheets with the EU Ecolabel criteria, in order to apply the proposed analysis methodology to the agritourism context in a more efficient way, the existing ARERA technical datasheets should be suitably updated and/or replaced by other more effective tools, expressly planned for the accommodation and catering business sector.

Author Contributions: All authors contributed equally to the design, experimental analyses and editing of this research. Conceptualization, L.C., M.L.G., G.P., G.R., G.S.; Data curation, L.C., M.L.G., G.P., G.R., G.S.; Methodology, L.C., M.L.G., G.P., G.R., G.S.; Writing—original draft, L.C., M.L.G., G.P., G.R., G.S.; Writing—review and editing, L.C., M.L.G., G.P., G.R., G.S. All authors have read and agreed to the published version of the manuscript.

Funding: This work was carried out within the research funds provided by the XXXIII Cycle Doctoral Course in Energy and Information Technologies of the University of Palermo.

Conflicts of Interest: The authors declare no conflict of interest.

Appendix A

Table A1. ARERA technical datasheets.

Datasheet N°	Proposed Action
1-tris	Installation of high-quality compact fluorescent lamps, not exceeding 15 W power
2	Replacement of electric water heater with gas water heater with sealed chamber and piezoelectric ignition
3	New installation of 4-star single-family efficiency boiler fueled with natural gas
4	Replacement of gas water heater (with open chamber and pilot flame) with gas water heater (with sealed chamber and piezoelectric ignition)
5	Replacement of simple glazing with double glazing
6	Wall and roof insulation
7	Use of photovoltaic systems with an electrical power of less than 20 kW
8T	Installation of solar collectors for the production of domestic hot water
9T	Installation of electronic frequency regulation systems (inverters) in electric motors operating on pumping systems with power lower than 22 kW
10T	Electricity recovery from natural gas decompression
11T	Installation of engines with higher efficiency
13a-bis	Installation, in residential environments, of water saving kits consisting of low-flow aerators and low-flow shower heads
13b-bis	Installation of low flow shower dispensers in hotels and guest houses
13c-bis	Installation of low flow shower dispensers in sports facilities
15T	Installation of outdoor air electric heat pumps instead of boilers in newly built or renovated residential buildings
16T	Installation of electronic frequency regulation systems (inverters) in electric motors operating on pumping systems with power greater than or equal to 22 kW
17T	Installation of luminous flux regulators for mercury vapor lamps and high-pressure sodium vapor lamps in outdoor lighting systems
18T	Replacement of mercury vapor lamps with high pressure sodium vapor lamps in public lighting systems
19T	Installation of high efficiency outdoor air conditioners with cooling capacity lower than 12 kW$_f$
20T	Thermal insulation of walls and roofs for summer cooling in domestic and service sectors
21T	Application in the civil sector of small cogeneration systems for winter and summer air-conditioning of rooms and the production of domestic hot water

Table A1. *Cont.*

Datasheet N°	Proposed Action
22T	Application in the civil sector of district heating systems for room air conditioning and domestic hot water production
23T	Replacement of incandescent traffic lights with LED traffic lights
24T	Replacement of incandescent votive lamps with votive LED lamps
25Ta	Installation of devices for automatically switching off equipment in standby mode in the residential sector
25Tb	Installation of devices for automatically switching off equipment in stand-by mode in the hotel sector
26T	Installation of centralized systems for winter and/or summer air conditioning in civil use buildings
27T	Installation of electric heat pump for domestic hot water production in new and existing plants
28T	Realization of high efficiency systems for the illumination of main motorway and extra-urban tunnels
29Ta	Implementation of new high-efficiency lighting systems for roads destined to motorized traffic
29Tb	Installation of high efficiency lighting fixtures in existing lighting systems for roads destined to motorized traffic

Table A2. EU Ecolabel for Tourist Accommodation Services criteria.

	Mandatory		Optional
General management criteria			
1	Basis of an Environmental Management System	23	EMAS registration, ISO certification of the tourist accommodation (up to 5 points)
2	Staff training	24	EMAS registration or ISO certification of suppliers (up to 5 points)
3	Information to guests	25	Ecolabelled services (up to 4 points)
4	General maintenance	26	Environmental and social communication and education (up to 2 points)
5	Consumption monitoring	27	Consumption monitoring: Energy and water sub-metering (up to 2 points)
Energy criteria			
6	Energy efficient space heating and water heating appliances	28	Energy efficient space heating and water heating appliances (up to 3 points)
7	Energy efficient air conditioning and air-based heat pumps appliances	29	Energy efficient air conditioning and air-based heat pumps appliances (up to 3.5 points)
8	Energy efficient lighting	30	Air-based heat pumps up to 100 kW heat output (3 points)
9	Thermoregulation	31	Energy efficient household appliances and lighting (up to 4 points)
10	Automatic switching off of HVAC and lighting	32	Heat recovery (up to 3 points)

Table A2. *Cont.*

	Mandatory		Optional
11	Outside heating and air conditioning appliances	33	Thermoregulation and window insulation (up to 4 points)
12	Procurement of electricity from a renewable electricity supplier	34	Automatic switch off appliances/devices (up to 4.5 points)
13	Coal and heating oils	35	District heating/cooling and cooling from cogeneration (up to 4 points)
		36	Electric hand driers with proximity sensor (1 point)
		37	Space Heater emissions (1.5 points)
		38	Procurement of electricity from a renewable electricity supplier (up to 4 points)
		39	On site self-generation of electricity through renewable energy sources (up to 5 points)
		40	Heating energy from renewable energy sources (up to 3.5 points)
		41	Swimming pool heating (up to 1.5 points)
	Water criteria		
14	Efficient water fittings: Bathroom taps and showers	42	Efficient water fittings: Bathroom taps and showers (up to 4 points)
15	Efficient water fittings: Toilets and urinals	43	Efficient water fittings: Toilets and urinals (up to 4.5 points)
16	Reduction in laundry achieved through reuse of towels and bedclothes	44	Dishwasher water consumption (2.5 points)
		45	Washing machine water consumption (3 points)
		46	Indications on water hardness (up to 1.5 points)
		47	Optimised pool management (up to 2.5 points)
		48	Rainwater and grey water recycling (up to 3 points)
		49	Efficient irrigation (1.5 points)
		50	Native or non-invasive alien species used in outdoor planting (up to 2 points)
	Waste and wastewater criteria		
17	Waste prevention: Food service waste reduction plan	51	Paper Products (up to 2 points)
18	Waste prevention: Disposable items	52	Durable goods (up to 4 points)
19	Waste sorting and sending for recycling	53	Beverages provision (2 points)
		54	Detergents and toiletries procurement (up to 2 points)
		55	Minimisation of the use of cleaning products (1.5 point)
		56	De-icing (1 point)
		57	Used textiles and furniture (up to 2 points)
		58	Composting (up to 2 points)
		59	Waste water treatment (up to 3 points)

Table A2. *Cont.*

	Mandatory		Optional
		Other criteria	
20	No smoking in common areas	60	No smoking in rooms (1 point)
21	Promotion of environmentally preferable means of transport	61	Social policy (up to 2 points)
22	Information appearing on the EU Ecolabel	62	Maintenance vehicles (1 point)
		63	Environmentally preferable means of transport offer (up to 2.5 points)
		64	Unsealed surfaces (1 point)
		65	Local and organic products (up to 4 points)
		66	Pesticide avoidance (2 points)
		67	Additional environmental and social actions (up to 3 points)

Appendix B

In the following the equations relating to the ARERA calculation procedures for each considered technical datasheet are given, in order to define the parameters reported in the following, Table A3.

Datasheet N° 5, "Replacement of simple glazing with double glazing", allows obtaining the gross primary energy savings (RL) achievable per individual building:

$$RL = RSL \times S_{window} \ [toe/year/building] \tag{A1}$$

where:

- RSL is the specific gross primary energy savings per m^2 of replaced glass surface, dependent on the climatic zone and the buildings intended use (residential, office, school, hospital, etc.), expressed in toe/year/m^2;
- S_{window} is the replaced glass surface, expressed in m^2.

Datasheet N° 7, "Use of photovoltaic systems with an electrical power of less than 20 kW", allows obtaining the achievable specific gross primary energy savings (RSL) for each reference physical unit (UFR), represented by a photovoltaic system with electrical power <20 kW:

$$RSL = kW_p \times h_{eq} \times k_1 \times 0.22 \cdot 10^{-3} \ [toe/year] \tag{A2}$$

where:

- kW_p is the peak power of the system, expressed in kW;
- h_{eq} is a coefficient dependent on the solar belt of the considered province, expressed in h/year;
- k_1 is a dimensionless coefficient that varies in function of the inclination (β) of the photovoltaic modules on to the horizontal plane, that is $k_1 = 0.70$ for $\beta > 70°$, otherwise $k_1 = 1$;

Datasheet N° 8T, "Installation of solar collectors for the production of domestic hot water", allows obtaining the annual shares of primary net energy savings (RN_C) for each reference physical unit (UFR), represented by the opening surface (m^2) of the installed collectors:

$$RN_C = RSN \times UFR \ [toe/year] \tag{A3}$$

where:

- *RSN* is the net specific primary energy savings achievable per m^2 of *UFR*, based on the system typology and on the solar belt to which the site belongs, expressed in toe/year/m^2;

Datasheet N° 15T, "Installation of outdoor air electric heat pumps instead of boilers in newly built or renovated residential buildings", allows obtaining the annual shares of primary net energy savings (RN_C) for each reference physical unit (*UFR*), represented by a standard apartment, which in terms of square meters of heated surface corresponds to about 80–90 m^2:

$$RN_C = a \times RSL \times UFR \ [toe/year] \tag{A4}$$

where:

- *a* is the additionality coefficient (dimensionless);
- *RSL* is the specific gross primary energy savings per single *UFR*, based on the COP of the heat pump typology, the surface/volume (S/V) ratio of the heated environment and the climatic zone (c.z.), expressed in toe/year/m^2.

Datasheet N° 19T, "Installation of high efficiency outdoor air conditioners with cooling capacity lower than 12 kW$_f$", allows obtaining the annual shares of primary net energy savings (RN_C) for each reference physical unit (*UFR*), represented by 1 kW cooling capacity of the air conditioning system at nominal conditions (expressed in actual installed cooling capacity):

$$RN_C = a \times RSL \times UFR \ [toe/year] \tag{A5}$$

where:

- *a* is the additionality coefficient (dimensionless);
- *RSL* is the specific gross primary energy savings per *UFR*, dependent on the solar belt of the considered province, expressed in toe/year/m^2.

Datasheet N° 27T, "Installation of electric heat pump for domestic hot water production in new and existing plants", allows obtaining the annual shares of primary net energy savings (RN_C) for each reference physical unit (*UFR*), represented by an electric heat pump water heater for the production of domestic hot water (expressed in number of units):

$$RN_C = a \times RSL \times UFR \ [toe/year] \tag{A6}$$

where:

- *a* is the additionality coefficient (dimensionless);
- *RSL* is the specific gross primary energy savings per single *UFR*, based on the COP of the heat pump typology and on the climatic zone, expressed in toe/year/m^2.

Datasheet N° 6, "Wall and roof insulation", allows obtaining the gross primary energy savings (*RL*) achievable per insulated surface unit (m^2):

$$RL = RSL \times S_{wall-roof} \ [toe/year/building] \tag{A7}$$

where:

- *RSL* is the specific gross primary energy savings per m^2 of insulated surface, dependent on the climatic zone and the building intended use (residential, office, school, hospital, etc.), expressed in toe/year/m^2;
- $S_{wall-roof}$ is the insulated surface of walls and/or roof, expressed in m^2.

Datasheet N° 20T, "Thermal insulation of walls and roofs for summer cooling in domestic and service sectors", allows obtaining the annual shares of primary net energy savings (RN_C) achievable per m^2 of insulated surface unit (UFR):

$$RN_C = a \times RSL \times UFR \ [toe/year] \tag{A8}$$

where:

- a is the additionality coefficient (dimensionless);
- RSL is the specific gross primary energy savings per m^2 of insulated surface, based on the thermal transmittance K (W/m^2/K) of the structure (walls and/or) before the intervention, expressed in toe/year/m^2.

Table A3. Equations (A1) to (A8) input data parameters for the two considered agritourisms.

Data-Sheet N. (Eq)	*Villa Dafne*		*Bergi*	
	Equation Input Data	Equation Result	Equation Input Data	Equation Result
5 (1)	RSL = 0.009 toe/year/m^2	RL = 2.7 toe/year	RSL = 0.005 toe/year/m^2	RL = 1.2 toe/year
	S_{window} = 305 m^2		S_{window} = 236 m^2	
7 (2)	kWp = 19.9 kW	RSL = 8.1 toe/year	kWp = 19.9 kW	RSL = 8.1 toe/year
	h_{eq} = 1852 h/year		h_{eq} = 1852 h/year	
	k_1 = 1		k_1 = 1	
8T (3)	RSN$_{electrical}$ = 0.210 toe/year/m^2	RN$_C$ = 4.9 toe/year	RSN$_{electrical}$ = 0.210 toe/year/m^2	RN$_C$ = 3.5 toe/year
	RSN$_{gas}$ = 0.123 toe/year/m^2		RSN$_{gas}$ = 0.123 toe/year/m^2	
	UFR$_{electrical}$ = 23 m^2		UFR$_{electrical}$ = 5 m^2	
	UFR$_{gas}$ = 0		UFR$_{gas}$ = 21 m^2	
15T (4)	a = 1	RN$_C$ = 3.3 toe/year	a = 1	RN$_C$ = 2.0 toe/year
	RSL = 0.181 toe/year/UFR		RSL = 0.143 toe/year/UFR	
	UFR = 18		UFR = 14	
19T (5)	a = 1	RN$_C$ = 0.7 toe/year	a = 1	RN$_C$ = 0.6 toe/year
	RSL = 0.0037 toe/year/UFR		RSL = 0.0037 toe/year/UFR	
	UFR = 200 kWf		UFR = 150 kWf	
27T (6)	a = 1	RN$_C$ = 3.9 toe/year	a = 1	RN$_C$ = 1.9 toe/year
	RSL = 0.106 toe/year/UFR		RSL = 0.107 toe/year/UFR	
	UFR = 37		UFR = 18	
6 (7)	RSL = 0.0028 toe/year/m^2	RL = 7.8 toe/year	RSL = 0.0015 toe/year/m^2	RL = 4.2 toe/year
	$S_{wall\text{-}roof}$ = 2768.5 m^2		$S_{wall\text{-}roof}$ = 2791.75 m^2	
20T (8)	a = 1	RN$_C$ = 1.4 toe/year	a = 1	RN$_C$ = 1.4 toe/year
	RSL = 0.0005 toe/year/UFR		RSL = 0.0005 toe/year/UFR	
	UFR = 2768.5 m^2		UFR = 2791.75 m^2	

References

1. *UNWTO Annual Report 2017*; United Nations World Tourism Organization: Badri, Spain, 2018.
2. *WTTC Report 2017*; World Travel and Tourism Council: London, UK, 2018.
3. Silva, F.B.; Herrera, M.A.M.; Rosina, K.; Barranco, R.R.; Freire, S.; Schiavina, M. Analysing spatiotemporal patterns of tourism in Europe at high-resolution with conventional and big data sources. *Tour. Manag.* **2018**, *68*, 101–115. [CrossRef]
4. Isik, C.; Dogan, E.; Ongan, S. Analyzing the Tourism-Energy-Growth Nexus for the Top 10 Most-Visited Countries. *Economies* **2017**, *5*, 40. [CrossRef]
5. Nižić, M.K.; Grdić, Z.Š.; Hustić, A. The Importance of Energy for the Tourism Sector. *Acad. Tur.* **2016**, *9*, 77–84.

6. Tsemekidi-Tzeiranaki, S.; Bertoldi, P.; Labanca, N.; Castellazzi, L.; Serrenho, T.; Economidou, M.; Zangheri, P. *Energy Consumption and Energy Efficiency Trends in the EU-28 for the Period 2000–2016*; JRC Science for Policy Report; Joint Research Centre (JRC): Brussels, Belgium, 2018.

7. *World Energy Balances 2018*; International Energy Agency (IEA): Paris, France, 2018.

8. UNEP. *Tourism Green Economy and Trade: Trends, Challenges and Opportunities*; United Nations Environment Programme: Nairobi, Kenya, 2013; pp. 260–291.

9. Beccali, M.; Gennusa, M.L.; Coco, L.L.; Rizzo, G. An empirical approach for ranking environmental and energy saving measures in the hotel sector. *Renew. Energy* **2009**, *34*, 82–90. [CrossRef]

10. Pablo-Romero, M.P.; Sánchez-Braza, A.; Sánchez-Rivas, J. Relationships between Hotel and Restaurant Electricity Consumption and Tourism in 11 European Union Countries. *Sustainability* **2017**, *9*, 2109. [CrossRef]

11. Aissa, S.B.; Aissa, S.B.; Goaied, M.; Goaied, M. Determinants of tourism hotel market efficiency. *Int. J. Cult. Tour. Hosp. Res.* **2016**, *10*, 173–190. [CrossRef]

12. Paramati, S.R.; Shahbaz, M.; Alam, M.S. Does tourism degrade environmental quality? A comparative study of Eastern and Western European Union. *Transp. Res. Part D Transp. Environ.* **2017**, *50*, 1–13. [CrossRef]

13. Tang, C.; Zhong, L.; Jiang, Q. Energy efficiency and carbon efficiency of tourism industry in destination. *Energy Effic.* **2018**, *11*, 539–558. [CrossRef]

14. Paramati, S.R.; Alam, M.S.; Chen, C.F. The effects of tourism on economic growth and CO_2 emissions: A comparison between developed and developing economies. *J. Travel Res.* **2017**, *56*, 712–724. [CrossRef]

15. Isik, C.; Kasımatı, E.; Ongan, S. Analyzing the causalities between economic growth, financial development, international trade, tourism expenditure and/on the CO_2 emissions in Greece. *Energy Sources Part B Econ. Plan. Policy* **2017**, *12*, 665–673. [CrossRef]

16. Pan, S.Y.; Gao, M.; Kim, H.; Shah, K.J.; Pei, S.L.; Chiang, P.C. Advances and challenges in sustainable tourism toward a green economy. *Sci. Total. Environ.* **2018**, *635*, 452–469. [CrossRef] [PubMed]

17. United Nations. *Transforming Our World: The 2030 Agenda for Sustainable Development*; United Nations: New York, NY, USA, 2015.

18. Sustainable Development Goals. Available online: https://www.un.org/sustainabledevelopment/sustainable-development-goals/ (accessed on 4 March 2020).

19. European Commission. *Communication from the Commission Europe 2020. A Strategy for Smart, Sustainable and Inclusive Growth*; European Commission: Brussels, Belgium, 2010.

20. European Commission. *Communication from the Commission to the European Parliament, the Council, the European Economic and Social Committee and the Committee of the Regions. A Policy Framework for Climate and Energy in the Period from 2020 to 2030*; European Commission: Brussels, Belgium, 2014.

21. European Commission. *A Roadmap for Moving to a Competitive Low Carbon Economy in 2050*; European Commission: Brussels, Belgium, 2011.

22. European Commission. *Communication from the Commission to the European Parliament, the European Council, the Council, the European Economic and Social Committee, the Committee of the Regions and the European Investment Bank. A Clean Planet for All a European Strategic Long-Term Vision for a Prosperous, Modern, Competitive and Climate Neutral Economy*; European Commission: Brussels, Belgium, 2018.

23. Ministero dello Sviluppo Economico; Ministero dell'Ambiente e della tutela del Territorio e del Mare. *Italy's National Energy Strategy (2017)*; Ministero dello Sviluppo Economico: Rome, Italy; Ministero dell'Ambiente e della tutela del Territorio e del Mare: Rome, Italy, 2017.

24. UNWTO. *A Roadmap for Celebrating Togerther*; World Tourism Organization: Madrid, Spain, 2016.

25. Saarinen, J.; Rogerson, C.M. Tourism and the millennium development goals: Perspectives beyond 2015. *Tour. Geogr.* **2013**, *16*, 23–30. [CrossRef]

26. Styles, H.; Schönberger, J.L. *European Commission JRC Scientific and Policy Report on Best Environmental Management Practice in the Tourism Sector D*; Galvez Martos European Union: Brussels, Belgium, 2013.

27. *Regulation (EC) No 1221/2009 of the European Parliament and of the Council of 25 November 2009 on the Voluntary Participation by Organisations in a Community Eco-Management and Audit Scheme (EMAS)*; Official Journal of the European Union, European Commission: Brussels, Belgium, 2009.

28. European Commission. *Guide on EU Funding for the Tourism Sector 2014–2020*; European Union: Brussels, Belgium, 2016; ISBN 978-92-79-58401-5.

29. *Commission Decision (EU) 2017/175 of 25 January 2017 on Establishing EU Ecolabel Criteria for Tourist Accommodation*; European Union, European Commission: Brussels, Belgium, 2017.

30. González, J.P.; Yousif, C. Prioritising energy efficiency measures to achieve a zero net-energy hotel on the island of Gozo in the central Mediterranean. *Energy Procedia* **2015**, *83*, 50–59. [CrossRef]

31. Tsoutsos, T.; Tournaki, S.; De Santos, C.A.; Vercellotti, R. Nearly Zero Energy Buildings Application in Mediterranean hotels. *Energy Procedia* **2013**, *42*, 230–238. [CrossRef]

32. Buso, T.; Becchio, C.; Corgnati, S.P. NZEB, cost- and comfort-optimal retrofit solutions for an Italian Reference Hotel. *Energy Procedia* **2017**, *140*, 217–230. [CrossRef]

33. Parpairi, K. Sustainability and Energy Use in Small Scale Greek Hotels: Energy Saving Strategies and Environmental Policies. *Procedia Environ. Sci.* **2017**, *38*, 169–177. [CrossRef]

34. Bianco, V.; Scarpa, F.; Tagliafico, L.A. Modeling energy consumption and efficiency measures in the Italian hotel sector. *Energy Build.* **2017**, *149*, 329–338. [CrossRef]

35. Datatur Report 2018. Available online: http://www.federalberghi.it/rapporti/rapporti.aspx?IDEL=160#.WCM66S3hDIU (accessed on 4 March 2020).

36. Eurostat Report 2018. Available online: http://ec.europa.eu/eurostat/web/tourism/data/database (accessed on 4 March 2020).

37. ISTAT Report 2018, Tourist flow in Italy Year 2017. Available online: https://www.istat.it/it/files//2018/11/EN_Tourism_2017.pdf (accessed on 4 March 2020).

38. ONT Report 2019. *Turismo in Cifre 2018/2019 (ONT, Italian National Tourism Observatory, Elaboration on WTTC, World Travel & Tourism Council Data)*; Agenzia Nazionale Turismo Italia, Ufficio Studi ENIT: Rome, Italy, 2019.

39. Turismo in Cifre. Available online: http://www.ontit.it/opencms/export/sites/default/ont/it/documenti/files/ONT_2019-07-30_03066.pdf (accessed on 4 March 2020).

40. Ali, Y.; Ciaschini, M.; Pretaroli, R.; Severini, F.; Socci, C. *Economic Relevance of Tourism Industry: The Italian Case*; Quaderno di Dipartimento n. 72 Università degli Studi di Macerata, Dipartimento di Economia e Diritto: Roma, Italy, 2014.

41. ISPRA Report 2018. Available online: https://annuario.isprambiente.it/ada/macro/28 (accessed on 4 March 2020).

42. Green Key. Available online: https://www.greenkey.global (accessed on 4 March 2020).

43. Bandiera Blu. Available online: http://www.bandierablu.org/common/blueflag.asp?anno=2019&tipo=bb (accessed on 4 March 2020).

44. Spighe Verdi. Available online: http://www.spigheverdi.net/il-programma/ (accessed on 4 March 2020).

45. Fee-Italia. Available online: http://www.feeitalia.org/fee/index.aspx (accessed on 4 March 2020).

46. Ministero delle Politiche Agricole Alimentari Forestali e del Turismo. Available online: https://www.politicheagricole.it/flex/cm/pages/ServeBLOB.php/L/IT/IDPagina/202 (accessed on 4 March 2020).

47. D'Alessandro, F. Green Building for a Green Tourism. A New Model of Eco-Friendly Agritourism. *Agric. Agric. Sci. Procedia* **2016**, *8*, 201–210. [CrossRef]

48. ISTAT Report 26 settembre 2018, "Anno 2017—LE AZIENDE AGRITURISTICHE IN ITALIA". Available online: https://www.istat.it/it/files//2018/09/ReportAGR_2017.pdf (accessed on 4 March 2020).

49. Mastronardi, L.; Giaccio, V.; Giannelli, A.; Scardera, A. Primi risultati di un'analisi aziendale, Agriregionieuropa. *Agritur. Sostenibilità Ambient. anno 11 n° 40* **2015**.

50. Peri, G.; Sanyé-Mengual, E.; Rieradevall, J.; Ciulla, G.; Rizzo, G. Proposal of a New Operative Brand for Environmentally Labelling Agritourism: Embodying Tourism, Buildings and Transportation Requirements. *World Appl. Sci. J.* **2014**, *32*, 1764–1774.

51. Peri, G.; Rizzo, G. The overall classification of residential buildings: Possible role of tourist EU Ecolabel award scheme. *Build. Environ.* **2012**, *56*, 151–161. [CrossRef]

52. Technical Datasheets ARERA—Autorità di Regolazione per Energia Reti e Ambiente. Available online: https://www.arera.it/it/ (accessed on 4 March 2020).

53. *P.S.R Sicilia 2014-2020—Programma di Sviluppo Rurale—Regione Sicilia*; Regione Siciliana: Sicily, Italy, 2013.

54. Disciplina dell'agriturismo. *Gazz. Uff. 16 marzo 2006, n. 63* **2006**.

55. Regione Sicilia—Legge Regionale. *Norme Sull'Agriturismo*; Regione Siciliana: Sicily, Italy, 1997.

56. Disciplina dell'agriturismo in Sicilia. *Gazz. Uff. Reg. Sicil. 1 Mar 2010 Parte I* **2010**.

57. LEGGE. *Norme per la Concorrenza e la Regolazione dei Servizi di Pubblica Utilità. Istituzione delle Autorità di Regolazione dei Servizi di Pubblica Utilità*. GU Serie Generale n.270 del 18-11-1995 - Suppl. Ordinario n. 136. 1995. Available online: https://www.federalismi.it/nv14/articolo-documento.cfm?Artid=26481 (accessed on 4 March 2020).

58. Legge. *Recante Disposizioni Urgenti per la Crescita, L'equità e il Consolidamento dei Conti Pubblici.* GU Serie Generale n.284 del 06-12-2011 - Suppl. Ordinario n. 251. 2011. Available online: http://www.senato.it/service/PDF/PDFServer/BGT/00737571.pdf (accessed on 4 March 2020).

59. Decreto Legislativo. *Attuazione della Direttiva 2012/27/UE Sull'Efficienza Energetica, che Modifica le Direttive 2009/125/CE e 2010/30/UE e Abroga le Direttive 2004/8/CE e 2006/32/CE.* GU Serie Generale n.165 del 18-07-2014. 2014. Available online: http://documenti.camera.it/leg17/dossier/pdf/AP0020.pdf (accessed on 4 March 2020).

60. LEGGE. *Bilancio di Previsione dello Stato per L'anno Finanziario 2018 e Bilancio Pluriennale per il Triennio 2018–2020.* GU Serie Generale n.302 del 29-12-2017 - Suppl. Ordinario n. 62. 2017. Available online: https://www.tuttoambiente.it/leggi/bilancio-2018/ (accessed on 4 March 2020).

61. *Council Regulation (EEC) No 880/92 of 23 March 1992 on a Community Eco-Label Award Scheme*; European Commission: Brussels, Belgium, 1992.

62. *Regulation (Ec) No 66/2010 of the European Parliament and of the Council of 25 November 2009 on the EU Ecolabel*; European Commission: Brussels, Belgium, 2010.

63. EU Ecolabel. Available online: https://ec.europa.eu/environment/ecolabel/ (accessed on 4 March 2020).

64. Ecolabel UE. Available online: http://certificazioni.isprambiente.it/valcer/node/3 (accessed on 4 March 2020).

65. Moro, A.; Lonza, L. Electricity carbon instensity in European Member States: Impacts on GHG emissions of electric vehicles. *Transp. Res. Part D* **2018**, *64*, 5–14. [CrossRef]

66. Vandepaer, L.; Treyer, K.; Mutel, C.; Bauer, C.; Amor, B. The integration of long-term marginal electricity supply mixes in the ecoinvent consequential database version 3.4 and examination of modeling choices. *Int. J. Life Cycle Assess.* **2019**, *24*, 1409–1428. [CrossRef]

67. *ISPRA Rapporti 317/2020, Fattori di Emissione Atmosferica di Gas a Effetto Serra nel Settore Elettrico Nazionale e nei Principali Paesi Europei*; ISPRA: Roma, Italy, 2020; ISBN 978-88-448-0992-8.

68. Supplemento ordinario alla GAZZETTA UFFICIALE DELLA REGIONE SICILIANA (p. I) n. 5 del 1 febbraio **2019** (n. 8)—Prezzario unico regionale per i lavori pubblici anno 2019. Available online: http://www.gurs.regione.sicilia.it/Gazzette/g19-05o/g19-05o.pdf (accessed on 4 March 2020).

69. Guerrieri, M.; Gennusa, M.L.; Peri, G.; Rizzo, G.; Scaccianoce, G. University campuses as small-scale models of cities: Quantitative assessment of a low carbon transition path. *Renew. Sustain. Energy Rev.* **2019**, *113*, 109263. [CrossRef]

70. Yoshida, Y.; Shimoda, Y.; Ohashi, T. Strategies for a sustainable campus in Osaka University. *Energ. Build.* **2017**, *147*, 1–8. [CrossRef]

71. Bisegna, F.; Cirrincione, L.; Maini, B.L.C.; Peri, G.; Rizzo, G.; Scaccianoce, G.; Sorrentino, G. Fostering the energy efficiency through the energy savings: The case of the University of Palermo. In Proceedings of the 2019 IEEE International Conference on Environment and Electrical Engineering and 2019 IEEE Industrial and Commercial Power Systems Europe, Palermo, Italy, 11–14 June 2019.

72. Giaccone, A.; Lascari, G.; Peri, G.; Rizzo, G. An ex post criticism, based on stakeholders' preferences, of a residential sector's energy master plan: The case study of the Sicilian region. *Energy Effic.* **2017**, *10*, 129–149. [CrossRef]

73. Vourdoubas, J. The Nexus between Agriculture and Renewable Energy Sources in the Island of Crete, Greece. *Eur. J. Appl. Sci.* **2020**, *8*, 101–110.

© 2020 by the authors. Licensee MDPI, Basel, Switzerland. This article is an open access article distributed under the terms and conditions of the Creative Commons Attribution (CC BY) license (http://creativecommons.org/licenses/by/4.0/).

Article

Barriers of Consumer Behavior for the Development of the Circular Economy: Empirical Evidence from Russia

Svetlana Ratner [1,2], Inna Lazanyuk [1,*], Svetlana Revinova [1] and Konstantin Gomonov [1]

[1] Department of Economic and Mathematical Modelling, Peoples' Friendship University of Russia (RUDN University), 6 Miklukho-Maklaya Street, 117198 Moscow, Russia; ratner-sv@rudn.ru (S.R.); revinova-syu@rudn.ru (S.R.); gomonov-kg@rudn.ru (K.G.)

[2] Economic Dynamics and Innovation Management Laboratory, V.A. Trapeznikov Institute of Control Sciences, Russian Academy of Sciences, 65 Profsoyuznaya Street, 117997 Moscow, Russia

* Correspondence: lazanyuk-iv@rudn.ru; Tel.: +7-495-433-4065

Abstract: This paper contributes to the literature on sustainable consumption by in-depth analysis of the factors affecting the probability of 57 different practices of proenvironmental behavior (PEBs) in Russia. The set of studied PEBs includes not only popular energy-saving and waste-management practices but also more circular patterns of plastic consumption, shopping, and city mobility. To study real and potential barriers to greening consumer behavior models, we conducted a survey of 623 respondents using a questionnaire developed based on a comparative analysis of similar studies conducted in other countries. The processing of the survey results was carried out using nonparametric statistics due to the absence of a normal distribution of the sample for most of the studied characteristics. The results of the study revealed that the main barriers to sustainable consumption in Russia are the lack of appropriate infrastructure as well as a lack of knowledge. Infrastructural barriers in some situations makes sustainable consumer behavior impossible or inconvenient (in this case, preference is given to other types of consumption), or in some cases necessitates spending additional time and money (then sustainable consumer behavior is not completely denied but practiced less often).

Keywords: proecological behavior; circular economy; environmental management; survey; nonparametric methods

Citation: Ratner, S.; Lazanyuk, I.; Revinova, S.; Gomonov, K. Barriers of Consumer Behavior for the Development of the Circular Economy: Empirical Evidence from Russia. *Appl. Sci.* **2021**, *11*, 46. https://dx.doi.org/10.3390/app11010046

Received: 11 November 2020
Accepted: 21 December 2020
Published: 23 December 2020

Publisher's Note: MDPI stays neutral with regard to jurisdictional claims in published maps and institutional affiliations.

Copyright: © 2020 by the authors. Licensee MDPI, Basel, Switzerland. This article is an open access article distributed under the terms and conditions of the Creative Commons Attribution (CC BY) license (https://creativecommons.org/licenses/by/4.0/).

1. Introduction

Circular Economy (CE) describes an economic system based on business models for reusing, recycling, and recovering materials in the production and consumption of goods, works, and services. The transition to the circular economy is necessary to maximize each process's performance in the life cycle of goods or services. The concept of CE opens new radical changes in consumption patterns and lifestyles. Implementing the idea of a circular economy requires rethinking the value chain, a conscious consumer approach, and using new business models such as sharing platforms to extend the life of a product [1]. This principle of improving the economy implies that the global production system is designed to have a no waste concept. All useful elements taken from the environment can be reused and waste from one production chain is the starting material for building another one. In contrast to the economy of the linear type, this stimulates the consumer to change items of consumption always to replace them with newer ones. When the economy transitions according to the circular type, it is essential to extend the life cycle of products as long as possible by the following cycles: (1) reuse, (2) remanufacturing, (3) recycling, (4) disposal [2]. At the same time, the more a product is in the first cycles, the cheaper it is to produce in general than when the product goes for disposal immediately after use (the traditional linear model of the production system).

The European Development Plan for a Circular Economy is one of the best-developed practical guidelines for the transition in today's economic environment to the new, more sustainable patterns of production and consumption. The European plan considers the formation of sustainable consumption as one of the priority areas of the circular economy [3].

The transition to the concept of a circular economy requires not only the restructuring of production chains but also the reformatting of many logistic, informational, and managerial links, as well as a change in consumer behavior models [4,5]. The lack of technology and infrastructure readiness to support consumer behavior patterns can inhibit the introduction of proecological consumer behavior patterns. The sustainability of the ecological infrastructure ("eco" -infrastructure) and the living environment as socioecological subsystems necessary for a person is achieved by the ability to adapt in a changing world. By "eco" -infrastructure, we mean a complex of natural, natural-anthropogenic and artificial objects and systems that provide conditions for preserving the human environment. The "eco" infrastructure includes elements of traditional infrastructure (natural resources, all mining systems, waste disposal systems, energy, transport, etc.). Many elements of traditional infrastructure must be environmentally stable in order to be included as components of the "eco" infrastructure that a person must carry out in accordance with ethics and environmental regulation. However, modern world literature in the field of sustainable development shows that it is cultural and economic barriers that have the most significant impact on which of the practices of proecological behavior are spread and which are not [6–8].

Consumer demand is a powerful incentive to create new and transformative industries to increase environmental and social responsibility. However, consumers' readiness for such a radical change in traditional patterns should be formed gradually with the transition from simple resource conservation and waste management practices to more complex ones.

Information on the barriers of proenvironmental conduct in a different cultural, economic, and professional context is scattered across the body of literature [9–12]. However, as far as we know, no previous research has investigated the problem of the barriers of proenvironmental behavior in the context of circular economy development in Russia. This country is only in the beginning of its way to the circular economy but has a comparatively high level of environmental awareness in some groups of consumers and strong cultural traditions in sustainable consumption of food and clothes [13].

Most scholars recognize that environmental awareness is the first important step in the adoption of proenvironmental behavior [14,15]. At the same time, some authors argue that there is still a gap between environmental awareness and proenvironmental behavior and rely on behavioral cost concept in explaining the obstacles of sustainable consumption [14,16]. Since there are no consistent government policies in supporting sustainable consumption in Russia, it is possible to assume that behavioral cost of proenvironmental behavior is high in this country. This makes Russia an interesting case to study both from empirical and theoretical points of view.

The purpose of this paper is to assess barriers of proenvironmental behavior associated with the transition towards a more circular daily practice of energy and water consumption, waste management, city mobility, and shopping in Russia. The study contributes to the literature by bringing new empirical evidence of consumers' attitude toward different forms of proenvironmental behavior in Russia and the cultural influences on it.

To achieve the study's objectives, we used a face-to-face questionnaire survey. The survey group mostly included students from the Peoples' Friendship University of Russia (Moscow) and the Kuban State University (Krasnodar), who studies in several training areas, whose educational programs include courses in environmental management, environmental engineering, etc. This choice of respondents is explained by the fact that this category of consumers is the most informed and the most flexible in forming consumer behavior patterns.

2. Literature Review

Experts realize that the circular economy is characterized by a restorative and closed nature. In the literature, there are three key features inherent in a circular economy: first, enhanced control over natural resources and maintaining a sustainable balance of renewable resources to conserve and maintain natural capital at an inexhaustible level; second, optimization of consumption processes by developing and distributing products, components, and materials that meet the highest level of reuse; third, the identification and prevention of negative externalities of current production activities in order to improve the efficiency of economic and ecological systems [17].

For the functioning of the circular economy model, the Ellen MacArthur Foundation identifies several essential components of the transition and functioning of the circular economy of groups of activities [17]: (a) Regenerate, (b) Share, (c) Optimize, (d) Loop, (e) Exchange.

Despite the relatively recent interest in circular economics, recycled materials are being returned to manufacturing processes at much lower levels. If this system could be improved, loss of value, dependence on volatile product markets, reduced resource productivity, and externalities in the form of environmental pollution could be avoided [18]. Germany's waste management system is one of the most advanced globally; waste disposal is carried out safely for people and the environment. However, only about 14% of the raw materials used in the industry come from processing; the rest still come from raw materials. The introduction of a circular economy is primarily an information problem.

Promoting proecological behavior in people's daily lives is critical in the circular economy's direction and the industrial and commercial sectors' efforts. Therefore, this issue is relatively well studied in modern scientific literature. As evidenced by the results of numerous studies, awareness of the existence of environmental problems concern about the state of the environment is characteristic of almost all social strategy of society, regardless of gender, age, education, type of activity, etc. [19–21]. However, people's environmental activity—their ability to abandon their habitual consumer behavior to take the path of more responsible behavior in everyday practices—is in its infancy [6,22,23]. There are many internal and external barriers to taking real measures to prevent them or reduce the negative impact and consequences if people are well aware of environmental problems [24]. These barriers can be due to various factors such as traditional values, lifestyles, policies, infrastructure, environmental circumstances, various cultures, and countries [5,25].

A significant amount of research addresses selecting factors that determine the choice in favor of proecological behavior. In studies on the role of the individual behavioral aspect in sustainable development, several primary determinants are highlighted that can influence an individual's choice in favor of proecological actions. Thus, the article [26] states that a combination of social and personal moral standards positively impacts the efficient use of electricity, water, other resources, ecological consumerism, and the desire to recycle materials. A person's attitudes and beliefs, reflecting their awareness of problems and a desire to change the situation also stimulate ecofriendly behavior, although in a less significant way. In papers [27,28], it was noted that in addition to these factors, there are several more: the so-called behavioral control, which means the ease with which an individual can put his intentions into action (the presence of monetary funds, time, opportunity); the intrinsic usefulness of environmental good and the willingness to pay for its offer. First of all, the last two parameters affect the desire to purchase environmentally friendly goods instead of conventional ones. Sustainability is based on a balanced relationship of the triple bottom line—people, profit, and planet [29].

The article [30] describes a study of proecological behavior in Canada and the United States. As a result of this study, the authors found a positive relationship between life satisfaction and proecological behavior. Life satisfaction as well as proecological behavior was predicted by more social behavior.

There was high interest in the study, which was conducted by the students of Malaysian universities. The authors of the study [31] carried out a personal assessment of proecological behavior. They assessed such qualities of a student's character as openness, benevolence, and conscientiousness. It was found that conscientiousness was the only trait that most influenced the proenvironmental behavior (PEB). The multilevel analysis [32] showed that the relationship between self-transcendence/self-development values–proecological behavior was weaker among societies with higher cultural and socioecological constraints (for example, with lower values of self-expression and economic development). These results clarify when values can promote or inhibit proecological behavior and highlight the need to consider human interaction and context in understanding how personal factors translate into proecological behavior. The article [33] examined the impact of education on proecological behavior in Thailand's educational institutions. The authors concluded that the higher the level of education, the more often the respondents are ready to participate in environmental actions. However, education does not have any impact on willingness to pay environmental taxes. The study [34] estimated that people who received environmental training would demonstrate 4.7 times more voluntary proenvironmental behavior than those who did not receive any training. The article [35] examined personal factors in proecological tourism behavior from a gender perspective. This study is based on a sample of 347 golfers from 16 European countries. The results confirmed the relationship between ecological habits, personal ability, and attitude to the environment. It is shown that only the interaction between personal capabilities and externally-oriented habits influence environmental attitudes.

In studies [36,37], they used the European countries' example. The authors showed that, despite evidence of a positive influence of education on individuals' proecological behavior, clear conclusions are impossible due to an ambiguous causal relationship. Moreover, it may be overlooked factors that force individuals to get more education and care for the environment.

The article [38] examines students' knowledge and behavior in ecological civilization. Based on 13,404 questionnaires from 152 universities and using a polynomial logit regression model, the authors investigated the relationship between Chinese university students' cognition and behavior. This article divides students' ecological civilization knowledge into two parts. One of which is common sense knowledge that comes from long-term public environmental and environmental education in China. The other is the knowledge of the ecological civilization's national strategy, which comes from the recent intense and comprehensive political advertising. This article concludes that university students with the latest knowledge are more willing to implement ecological civilizational behavior. This result illustrates the role of political propaganda in China for ecological civilization to influence its citizens' behavior and inspires other countries to promote environmental policy and promote public environmental protection through the media.

The paper [39] examines the role of stakeholder engagement in establishing and strengthening the sustainability culture in a company transitioning toward a circular economy. The authors also emphasize that the government and universities play an active role in promoting sustainability culture.

In addition to education, energy conservation, for example, is influenced by other aspects. Thus, authors from China [40] considered factors affecting consumer behavior with energy-saving. The authors concluded that urban inhabitants are more economical in terms of energy consumption than rural inhabitants. The authors confirmed a more responsible attitude towards the environment due to improved knowledge in energy conservation and more developed consumption habits. Scientists from Sweden [41] carried out a comprehensive analysis of energy conservation factors. The authors identified the three most significant factors: age of people, type of house, and income—with the decisive factor being the type of housing. People living in private houses are more likely to save than people living in apartment buildings. Socioeconomic factors influence savings on heating more than attitude to the environment.

An interesting study was conducted [42] on procrastination on household energy-saving behavior. The author concluded that environmental awareness increases energy savings but only affects actions that do not require additional costs (lowering the house's temperature during the absence).

One of the most common methods for studying consumer proecological behavior barriers is to conduct surveys [8]. Although this method has its significant drawbacks, which are also actively discussed in the literature (see, for example, works [7,43,44], it remains a priority tool in primary research. The goal is to obtain an overall picture of the prevalence of a particular consumer behavior model and identify the reasons for the nonproliferation of other models.

3. Materials and Methods

3.1. Procedure and Participants

To study real and potential barriers to greening consumer behavior models, we conducted a survey of 623 respondents using a questionnaire developed based on a comparative analysis of similar studies conducted in other countries [6,25]. The study was launched in early 2019 on a pilot group of students (N = 100) from Kuban State University. On the stage of pilot research, a face-to-face survey method was used. A personal survey in the case of small- and medium-scale gives an opportunity to study of phenomena in a real-life context [45]. After processing the results of the pilot study and finding interesting consistent patterns, the study was extended to different age groups and regions and continued until the May of 2020.

On the second stage of survey a method of responded-driven sampling was used [46], in which one interviewer from the invited group (students) interviews several (up to 10) people at a time. It allowed us to increase the speed of the study and the size of the sample.

The procedure of responded-driven sampling led us to the following distribution of respondents by age: the majority of respondents belong to the age category from 20 to 29 years old (55%), besides 8% of respondents belongs to the category from 16 to 20 years old. This makes it possible to test the hypothesis about whether the educational process affects the respondent's values and their attitude to the practices of proecological behavior. In addition, the sample has a reasonably well-represented age group from 30 to 39 (11%), from 40 to 49 (11%) and 50 to 59 (9%). Age categories over 60 are less represented (6%).

The distribution of respondents by involvement in the educational process is the following: 62% of respondents are high school or college students, as well as undergraduate and graduate university students. All other respondents (38%) form the category, which was called "not a student" in order to highlight that they are not involved in the education process at the time.

The distribution of respondents by economic activity is the following: 93% of respondents are economically active and only 7% are not. All working respondents are classified as economically active, regardless of whether they are employed (be it self-employed, sole proprietor, an employee at the enterprise) or students. The economically inactive group includes pensioners, housewives, persons on parental leave, or persons with disabilities. This division of respondents into groups was performed to test the hypothesis of how the availability of free time (it is believed that economically active people have less free time) affects the frequency of using various practices of proenvironmental behavior.

The distribution of respondents by region of residence is uneven. The majority live in Moscow (69%). A significant proportion of respondents live in Krasnodar (17%) and the Krasnodar region (10%). Other regions of Russia are slightly represented in the sample, which is a consequence of the research technique—the main groups were formed precisely in Moscow and Krasnodar based on two large universities. Nevertheless, 2% of respondents are representatives of other regions of the Southern Federal District (except for the Krasnodar region), 1% of respondents live in St. Petersburg, and another 1% live in other regions (Samara Oblast, Kamchatka Krai, Yuzhno-Sakhalin Oblast, etc.).

The level of education of the interviewed respondents in general can be defined as high. The majority of respondents (62%) have higher education (31%—bachelor's level and 41%—specialist or master's level). Another 6% of respondents have postgraduate education—

postgraduate, doctoral, or residency). Sixteen percent of respondents have secondary education (6%—general and 10%—professional) and another 16% are university students.

Regarding the distribution of respondents by income level, the majority of respondents (60%) determined their income level using the following description "In general ok, but sometimes I have to save". The share of those who defined their financial situation with the following phrase "I can satisfy all my needs and the needs of my family" is also quite large in the sample (23%). Only 1% of the respondents defined their financial situation as "I live in poverty", 11% chose the phrase "I have to save" to describe their financial situation. Five percent of respondents found it difficult to determine their financial situation or refused to answer.

The distribution of respondents by the level of participation in environmental activities is the following: 39% of respondents participated in environmental campaigns to plant trees, clean garbage, collect wastepaper, glass containers, etc. at least once. Almost the same number of respondents did not participate in any environmental activities (38%). Six percent of respondents filed complaints about any environmental pollution, participated in collecting signatures for appeals to the authorities with demands to improve the environmental situation and another 6% made donations for environmental activities. Only 3% of respondents took part in environmental protests and only 2% participated in two or more environmental events.

3.2. Measure of Attitude to Proenvironmental Behavior

In this study, we use the concept of ecological behavior proposed by Stern [14]. He understands behavior as an interactive product of a personal value system and a set of contextual factors. Accordingly, ecological behavior is "a product of the interaction of personality characteristics (internal factors) and contextual (external) factors" [14]. Internal factors are associated with citizens' characteristics and include environmental knowledge, environmental values, motivation, locus of control, and personal responsibility. External factors can be divided into several categories: political, economic, social, and technological. Political factors are related to the level and speed of adoption of environmental laws and standards for sustainable development. The economic factors include the state of the country's economy as a whole, the volume and structure of environmental investments, the use of natural resources, measures to restore natural resources, and the demand for environmentally friendly products. Social factors influence environmental competence through social cultural values and traditions, the propensity to accept innovations, and change consumer behavior. External technological factors include the level of development of technologies for the treatment of discharges and emissions, disposal and recycling of waste, secondary use of raw materials, environmental research, research in the field of alternative energy, etc.

As noted above, the questionnaire consisted of three parts. The first part of the questionnaire has fixed data internal factors of the respondents.

The second part of the questionnaire aimed to determine the respondents' attitudes on environmental responsibility issues (general environmental self-awareness). We asked the respondents to answer whether they believe that they can improve their city's environment and indicate the types of environmental measures they took part in at least once. The third part of the questionnaire aimed to assess the frequency and reasons for applying (or not applying) practices of proenvironmental behavior in energy conservation, water conservation, waste management, and reducing the use of disposable products and mobility.

In compiling a list of proenvironmental behaviors, we used a variation of the questionnaire proposed by Lee, Kurisu and Hanaki [6]. We chose this study as a prototype for the following reasons: firstly, it contains a large list of possible practices of proenvironmental behavior (includes 58 practices); secondly, it has already been used by other scientists as a basis for international comparisons [25], which allows us to further (when conducting a more extensive study) compare the results obtained for Russia with the results of other countries. It should be noted that from the entire list of 58 practices of proenvironmental behavior, some are almost unknown in modern Russian society. Nevertheless, they were included in the study to identify possible patterns of behavior transmitted from generation to generation as a way of lean housekeeping. Several practices were excluded from the list due to Russia's inability to use a contradiction to cultural or legal norms. Instead, several new energy-saving practices have been added. In general, our set of PEBs is 90% similar to the set of Lee, Kurisu, and Hanaki [6].

The selected 57 patterns in our study were divided into six groups: (1) patterns in the field of energy conservation; (2) patterns in the field of water conservation; (3) patterns in the field of waste management; (4) patterns that can be arbitrarily called "against plastic" (reducing the use of disposable tableware, packaging, etc.); (5) shopping patterns; (6) urban mobility patterns. This division is explained by the fact that some groups of patterns (for example, in the field of energy efficiency) have been actively promoting at the state level for more than ten years, while others (for example, deciding in favor of purchasing more environmentally friendly goods) have not yet been stimulated. The list of patterns, divided into groups, is presented in Table 1

When answering the question of how often respondents practice each of the 57 models of proenvironmental behavior, one could choose the answer options "never", "rarely", "often", or "always", which, when processed, were translated into a scale from 1 to 4. Besides, respondents were asked to choose the reasons for the application or nonapplication of these practices. Among the possible reasons for applying the practice were "Habit", "Laziness", "Money saving", "Sense of duty", "Fashion" and among the reasons for nonapplication—"Laziness", "Time consuming", "No consideration", "Forget", "Nobody doing", "Costly", "There are no conditions for application", "I did not know that it was so necessary." The choice of each of the reasons was encoded as a Boolean variable (0 or 1).

Table 1. Proenvironmental patterns of consumers' behavior.

Group of PEBs	Description of PEBs
Energy saving	P1 Avoiding overloading the refrigerator P2 Reducing opening and closing the door of the refrigerator P3 Using a lower setting in the refrigerator compartment P4 Putting hot food into the refrigerator after cooling P5 Using stairs instead of elevators P6 Cleaning filter of the air conditioner or cleaner P7 Adjusting the temperature of the air conditioner P8 Turning off lights in empty rooms P9 Unplugging appliances not in use P10 Turning off the TV when people are not watching P11 Using energy-saving mode or turning off when not in use P12 Doing ironing collectively P13 Setting a lower shower temperature P14 Adjusting the temperature of the radiator P29 Avoiding over-volume cooking P30 Water heating of the required volume in an electric kettle P31 Covering the pan with a lid when cooking or boiling water P40 Buying energy efficient appliances P54 Using LED lamp instead of a fluorescent lamp P56 Flame adjustment for cooking P57 Use of residual heat when cooking on an electric stove
Water saving	P15 Using toothbrush cup P16 Turning off the water when washing face or brushing teeth P17 Taking short showers P18 Washing dishes using jugged water P19 Reducing detergent P20 Cutting down on the frequency of washing clothes P55 Using dishwasher

Table 1. *Cont.*

Group of PEBs	Description of PEBs
Waste management	P21 Avoiding throwing away waste cooking oil P22 Following garbage rules P23 Garbage separation P24 Giving used clothes to other people or using a recycle box P25 Collection and delivery of glass containers to appropriate collection points P26 Wastepaper collection and delivery to appropriate collection points P27 Collection and delivery of used batteries, light bulbs to appropriate collection points P32 Composting kitchen garbage P33 Throwing away kitchen garbage after it has dried P46 Using both sides of the paper
No plastic	P28 Using own cup P34 Using receptacle instead of plastic bag P35 Using own bag when going shopping P36 Reducing use of disposable products P37 Not buying over-packaged products
Shopping	P38 Buying organic products P39 Buying recycled goods P41 Buying ecomark-appliances P42 Choosing goods with their CO_2 emission in mind (carbon footprint) P43 Not buying unnecessary products P44 Trying to repair things before buying replacements P45 Using refill goods
Mobility	P47 Using bicycle or walking P48 Using public transportation P49 Joining the one day without car program P50 Doing car checks regularly P51 Avoiding overloading the car P52 Reducing idling of the car P53 Maintaining air pressure of the tire

Further, the investigation for answers to research questions was carried out using descriptive and nonparametric statistics. Nonparametric tests were used in case the studied variables are not distributed normally and are measured on an ordinal scale. Research questions and methods of verification are summarized in Table 2.

Table 2. Research questions and methods.

Research Questions	Method of Study
Q1: What are the most popular/unpopular PEBs?	Descriptive statistics
Q2: Which groups of PEBs are the most popular/unpopular	Descriptive statistics
Q3: The most coherent and discordant PEBs (single, groups)	Descriptive statistics
Q4: Does the probability of PEBs depends on income level?	Kruskal–Wallis Test
Q5: Does the probability of PEBs depends on the level of education?	Mann–Whitney Test
Q6: Does the probability of PEBs depends on the degree of involvement in the educational process?	Mann–Whitney Test
Q7: Does the probability of PEBs depends on the level of environmental responsibility (theoretical, real)?	Mann–Whitney Test
Q8: What are the most common reasons for not practicing PEBs?	Descriptive statistics
Q9: What is the degree of population involvement in ecopractices (waste separation, wastepaper collection, etc., adherence to waste management standards, etc.)?	Descriptive statistics

Table 2. *Cont.*

Research Questions	Method of Study
Q10: Does the probability of PEBs depends on age (what is the influence of old patterns of behavior)	Correlation of Spearman (R) and Tau-Kendall (K)
Q11: Does the probability of PEBs differ by region of residence?	Kruskal–Wallis Test
Q12: Does the probability of PEBs depends on the level of economic activity of the respondent?	Mann–Whitney Test

4. Results

The survey data showed that most Russians are interested in the environment (73%). More than half (57%) have taken part in nature conservation activity at least once. Every third respondent spoke in the affirmative on personal participation to improve the city's environmental situation. The relatively widespread opinion among respondents about the need to develop initiatives to protect the environment and the degree of responsibility for the environment indicates the lack of necessary changes. From the point of view of the locus of control the respondents' opinions were divided in groups, which are quite close to such a factor as responsibility and determines the measure of a person's moral attitude towards other people and the world around him. Citizens with an internal locus of control who recognize each resident's possibility and ability to influence the current environmental situation were 54% of respondents. On the contrary, 42% of respondents believe that an individual cannot influence the situation with the environment, thereby shifting their responsibility to the relevant social institutions, namely, the federal or local authorities. Despite these differences in answers, Russians are interested in environmental consumption and behavior issues and economic and social challenges. The study's main focus was on determining the statistical relationships between age, involvement in the educational process, economic activity, region of residence, education level, income level, and the degree of participation in environmental activities on the frequency of application of proenvironmental behavior practices.

4.1. Statistical Relationship between the Age and Gender of the Respondent and the Frequency of Application of the Practices of Proecological Behavior

Since some age categories of respondents were less represented in the sample, we used nonparametric statistical methods for processing research results that allow working with small samples [47,48]. Calculation of nonparametric correlation coefficients Spearman (R) and tau-Kendall (K) revealed the presence of some statistically significant relationships between the age of the respondent and his attitude to various practices of proecological behavior (Table 3). As shown from the nonparametric correlation results, the older generation more often practices more than 50% of the main patterns of proecological behavior. The exception is practice P5—using stairs instead of elevators, for which there is a negative correlation with the respondent's age, which is understandable. The highest values of nonparametric correlation coefficients were obtained in practices: P56—Flame adjustment for cooking, P53—Maintaining air pressure of the tire, P43—Not buying unnecessary products, P51—Avoiding overloading the car, P52—Reducing idling of the car, P50—Doing car checks regularly, P25—Collection and delivery of glass containers to appropriate collection points.

Statistically significant differences in the frequency PEBs between men and women were identified with Mann–Whitney test for practices P4 "Putting hot food into refrigerator after cooling", P50 "Doing car checks regularly", P21 "Avoiding throwing away waste cooking oil", P22 "Following garbage rules", P33 "Throwing away kitchen garbage after it has dried", P35 "Using own bag when going shopping", P37 "Not buying over-packaged products", P47 "Using bicycle or walking" and P49 "Joining the one day without car program". All these PEBs are practiced more often by men than by women.

Table 3. Correlations (at the p = 0.05 level) between the respondent's age and the frequency of using the practices of proecological behavior.

Designation of Practice	Description of PEBs	R	K
P1	Avoiding overloading the refrigerator	0.110939	0.093358
P4	Putting hot food into the refrigerator after cooling	0.092132	0.078012
P5	Using stairs instead of elevators	−0.086402	−0.072836
P8	Turning off lights in empty rooms	0.108368	0.092159
P9	Unplugging appliances not in use	0.098662	0.082177
P14	Adjusting the temperature of the radiator	0.129772	0.109882
P16	Turning off the water when washing face or brushing teeth	0.095665	0.080276
P24	Giving used clothes to other people or using a recycle box	0.114770	0.097202
P28	Using own cup	0.153335	0.131443
P30	Water heating of the required volume in an electric kettle	0.129507	0.109186
P31	Covering the pan with a lid when cooking or boiling water	0.143026	0.124615
P32	Composting kitchen garbage	0.104058	0.089667
P35	Using own bag when going shopping	0.082377	0.069287
P36	Reducing use of disposable products	0.090911	0.076527
P37	Not buying over-packaged products	0.098018	0.082512
P38	Buying organic products	0.147661	0.124152
P40	Buying energy efficient appliances	0.104952	0.088397
P43	Not buying unnecessary products	0.177359	0.150145
P44	Trying to repair things before buying replacements	0.125854	0.106487
P45	Using refill goods	0.091414	0.077797
P46	Using both sides of the paper	0.093672	0.079129
P50	Doing car checks regularly	0.155318	0.132766
P51	Avoiding overloading the car	0.170529	0.147143
P52	Reducing idling of the car	0.161546	0.138124
P53	Maintaining air pressure of the tire	0.192696	0.162217
P54	Using LED lamp instead of a fluorescent lamp	0.126195	0.107098
P55	Using dishwasher	0.113048	0.093301
P56	Flame adjustment for cooking	0.195606	0.162289
P10	Turning off the TV when people are not watching	is not significant	0.065431
P17	Taking short showers	is not significant	0.066025
P19	Reducing detergent	is not significant	0.059806

4.2. Influence of Involvement in the Educational Process on the Frequency of Application of Practices of Proecological Behavior

We tested whether education can significantly impact proenvironmental behavior, involving more efficient use of natural resources, recycling materials, and green consumerism. Conducting a series of nonparametric Mann–Whitney tests made it possible to reveal statistically significant differences: in the use of specific practices of proenvironmental behavior between respondents involved in the learning process (schoolchildren, students) and those who completed their education (Table 4).

Table 4. Statistically significant (at the $p < 0.05$ level) results of the Mann–Whitney test with a grouping variable—involvement in the educational process.

Designation of Practice	Description of PEBs	Average in a Student Group	Average in the Nonstudent Group
P8	Turning off lights in empty rooms	2.85	3.05
P14	Adjusting the temperature of the radiator	2.39	2.62
P24	Giving used clothes to other people or using a recycle box	2.28	2.49
P28	Using own cup	2.51	2.78

Table 4. *Cont.*

Designation of Practice	Description of PEBs	Average in a Student Group	Average in the Nonstudent Group
P30	Water heating of the required volume in an electric kettle	2.43	2.67
P31	Covering the pan with a lid when cooking or boiling water	2.58	2.85
P32	Composting kitchen garbage	2.00	2.22
P40	Buying energy efficient appliances	2.45	2.65
P41	Buying ecomark-appliances	2.18	2.39
P43	Not buying unnecessary products	2.51	2.68
P44	Trying to repair things before buying replacements	2.58	2.78
P50	Doing car checks regularly	1.97	2.43
P51	Avoiding overloading the car	1.95	2.35
P52	Reducing idling of the car	1.87	2.24
P53	Maintaining air pressure of the tire	1.45	2.10
P54	Using LED lamp instead of a fluorescent lamp	2.33	2.61
P56	Flame adjustment for cooking	1.85	2.27

Concerning the revealed, statistically significant differences between the groups of respondents receiving education and have completed their education, the second group of respondents demonstrate more conscious proecological behavior in absolutely all aspects.

4.3. The Statistical Relationship between the Grouping Variable—Economic Activity and Frequency of Application Practices Proenvironmental Behavior

Conducting a series of nonparametric Mann–Whitney tests made it possible to identify statistically significant differences in the use of some proecological behavior practices between economically active and economically inactive respondents (Table 5).

Table 5. Statistically significant (at the $p < 0.05$ level) results of the Mann–Whitney test with a grouping variable—economic activity.

Designation of Practice	Description of PEBs	The Average in the Group of Economically Active	Average in the Group of Economically Inactive
P5	Using stairs instead of elevators	2.61	2.18
P6	Cleaning filter of the air conditioner or cleaner	2.40	1.95
P8	Turning off lights in empty rooms	2.95	3.32
P11	Using energy-saving mode or turning off when not in use	2.80	2.28
P19	Reducing detergent	2.47	2.85
P25	Collection and delivery of glass containers to appropriate collection points	2.12	1.51
P26	Wastepaper collection and delivery to appropriate collection points	2.15	1.61
P27	Collection and delivery of used batteries, light bulbs to appropriate collection points	2.11	1.76
P31	Covering the pan with a lid when cooking or boiling water	2.71	3.15
P33	Throwing away kitchen garbage after it has dried	2.10	1.63
P34	Using receptacle instead of a plastic bag	2.14	1.68
P43	Not buying unnecessary products	2.58	3.12
P54	Using LED lamp instead of a fluorescent lamp	2.48	3.00
P56	Flame adjustment for cooking	2.03	3.37

The revealed differences in the frequency of the use of some practices by economically active and economically inactive respondents do not confirm the original author's hypothesis that having more free time in the absence of a permanent job can contribute to the more careful handling of all resources and the use of more labor-intensive waste management practices. On the contrary, economically active respondents demonstrate greater environmental awareness. They are more likely to practice collecting and handing over recycling wastepaper, glass containers, e-waste, drying food waste, and reusable shopping bags. In addition, economically active respondents are more competent in handling household appliances to reduce their energy consumption. At the same time, nonworking respondents more often use those practices associated with saving, first of all, financial resources. Besides, they more often use traditional household practices in cooking, which are aimed not so much at saving resources but simply correspond to the usual established order of things.

4.4. The Impact of the Region of Residence on the Frequency of Use of Practices of Proenvironmental Behavior

Given the strong uneven distribution of respondents by region of residence, to process the results of statistical testing hypotheses related to regions of residence, nonparametric statistics were also used. In addition, we tried to interpret the results as carefully as possible. The influence of the respondent's region of residence on the frequency of application of proecological behavior practices was investigated using a series of nonparametric Kruskal–Wallis tests. Statistically significant values of the χ^2-statistic are given in Table 6.

Table 6. The results of calculating the Kruskal–Wallis statistics test the hypothesis about the influence of the region of residence on the frequency of using proecological behavior practices.

Practice.	H-Statistics	*p*-Value
P1	12.85229	0.0248
P2	23.98897	0.0002
P4	39.28119	0.0000
P6	15.12456	0.0098
P7	19.66623	0.0014
P8	25.955667	0.0001
P10	14.93873	0.0106
P12	15.00023	0.0104
P13	12.89271	0.0244
P15	46.54626	0.0000
P17	16.29421	0.0061
P18	33.89423	0.0000
P20	16.76744	0.0050
P21	43.66621	0.0000
P22	15.42830	0.0087
P23	47.3464	0.0000
P25	49.1394	0.0000
P26	39. 5141	0.0000
P27	33.7303	0.0000
P28	16.1706	0.0064
P30	19.0606	0.0019
P31	12.5835	0.0276
P32	36.8298	0.0000
P33	54.84471	0.0000
P34	47.85524	0.0000
P36	14.10583	0.0150
P38	17.90442	0.0031
P39	34.38250	0.0000
P41	11.75316	0.0383

Table 6. *Cont.*

Practice.	H-Statistics	*p*-Value
P42	38.94659	0.0000
P43	21.59409	0.0006
P44	23.25259	0.0003
P50	50.10451	0.0000
P51	35.87414	0.0000
P52	18.91370	0.0020
P53	56.74829	0.0000
P54	26.04608	0.0001
P55	17.85102	0.0031
P56	70.39943	0.0000

Residents of two regions of Moscow and Krasnodar took part in the survey, and the range of answers was quite comprehensive.

4.5. Effect of Educational Level on the Frequency of Usage Practices Proenvironmental Behavior

To test the hypothesis about the influence of the respondent's education level on his attitude to various proecological practices, we calculated the nonparametric Spearman (R) and tau-Kendall (K) correlation coefficients. As a result, a weak positive statistically significant result was obtained at the $p = 0.05$ level for some nontrivial practices. It should be noted that Practices P23 (Garbage separation), P25 (Collection and delivery of glass containers to appropriate collection points), P27 (Collection and delivery of used batteries, light bulbs to appropriate collection points), P33 (Throwing away kitchen garbage after it has dried) from the Waste Management category, Practices P28 (Using own cup), P34 (Using receptacle instead of plastic bag) from the category "Rejection of Plastic" and Practices P38 (Buying organic products), P42 (Choosing goods with their CO2 emission in mind (carbon footprint)) from the Purchase category are relatively new for the Russian consumer and/or labor-intensive. Thus, we can talk about the positive influence of the respondent's education level on forming a proecological consumer behavior (Table 7).

Table 7. Correlations (at the $p = 0.05$ level) between the respondent's educational level and the frequency of using proenvironmental practices.

Designation of Practice.	Description of PEBs	R	K
P4	Putting hot food into the refrigerator after cooling	−0.092740	−0.078767
P6	Cleaning filter of the air conditioner or cleaner	0.1010145	0.082851
P25	Collection and delivery of glass containers to appropriate collection points	0.081862	0.068227
P27	Collection and delivery of used batteries, light bulbs to appropriate collection points	0.104634	0.087570
P28	Using own cup	0.086698	0.072704
P33	Throwing away kitchen garbage after it has dried	0.108604	0.091953
P34	Using receptacle instead of plastic bag	0.116350	0.097265
P39	Buying recycled goods	0.084708	0.070035
P42	Choosing goods with their CO_2 emission in mind (carbon footprint)	0.091716	0.075898
P3	Using a lower setting in the refrigerator compartment	is not significant	−0.057050
P23	Garbage separation	is not significant	0.052729
P44	Trying to repair things before buying replacements	is not significant	−0.056678

4.6. The Impact of Income on the Frequency of Use of Practices of Proenvironmental Behavior

As a result of testing the hypothesis about the influence of the respondent's income level on the frequency of his use of various practices of proecological behavior by calculating the Spearman and Kendall nonparametric correlations, negative correlations were revealed between the income level and the use of his own bags for shopping (P35), as well as the refusal to use personal vehicles in urban areas (P47, P48). A weak positive correlation was also found between the respondent's income level and his use of such energy saving practices as cleaning the air conditioner filter (Table 8).

Table 8. Correlations (at the *p* = 0.05 level) between the respondent's income level and the frequency of using proenvironmental practices.

Designation of Practice	Description of PEBs	R	K
P6	Cleaning filter of the air conditioner or cleaner	is not significant	0.054452
P35	Using own bag when going shopping	−0.084402	−0.072595
P47	Using bicycle or walking	−0.100478	−0.087474
P48	Using public transportation	−0.087542	−0.075391

Thus, we can say that the level of income has practically no effect on the frequency of using proecological patterns of consumer behavior. The exception is urban mobility practices, where there is a negative impact of income on proecological mobility practices.

4.7. The Influence of the Level of Involvement in Environmental Activities on the Frequency of Use of Practices of Proenvironmental Behavior

The influence of the level of involvement in environmental activities (which can be interpreted as the level of environmental awareness) on the frequency of applying proenvironmental behavior was also investigated using a series of nonparametric Kruskal–Wallis tests. Statistically significant values χ^2-statistics are given in Table 9.

Table 9. The results of calculating the Kruskal–Wallis statistics to test the hypothesis about the influence of the respondent's level of involvement in environmental activities on the frequency of using proenvironmental practices.

Practice	H-Statistics	*p*-Value
P4	15.64615	0.0158
P11	14.51215	0.0244
P16	14.22839	0.0272
P18	17.12929	0.0088
P23	13.33903	0.0380
P28	20.03792	0.0027
P29	13.21269	0.0398
P31	14.12257	0.0283
P34	12.65754	0.0488
P37	23.60005	0.0006
P38	15.38802	0.0174
P47	16.87749	0.0097
P53	20.70417	0.0021
P54	13.48528	0.0359
P56	23.78589	0.0006

4.8. Results of the Analysis of Descriptive Statistics on Proenvironmental Behavior Practices

The calculation of descriptive statistics on the respondents' assessments of the frequency of using the proposed patterns of proecological consumer behavior shows that the most popular practices across all 57 were P1 "Avoiding overloading the refrigerator", P2 "Reducing the opening and closing of the refrigerator door", "P4 "Putting hot food into refrigerator after cooling", P7 "Adjusting the temperature of the air conditioner", P8 "Turning off lights in empty rooms", P10 "Turning off the TV when people are not watching", P11 "Using energy saving mode or turning off when not in use", P28 "Using own cup", P31 "Covering the pan with a lid when cooking or boiling water" и P44 "Attempting to fix things before buying a replacement", they have a median score of 2.7 to 3.

The least popular practices, with a median score of 1.5 to 2.1, are: P21 "Avoiding throwing away waste cooking oil", P25 "Collection and delivery of glass containers to appropriate collection points", P27 "Collection and delivery of used batteries, light bulbs to appropriate collection points", P33 "Throwing away kitchen garbage after it has dried", P34 "Using receptacle instead of plastic bag", P42 "Choosing goods with their CO_2 emission in mind (carbon footprint)", P49 "Joining the one day without car program", P52 "Reducing idling of the car", P53 "Maintaining the air pressure in the tire", P55 "Using the dishwasher" и P57 "Using residual heat when cooking on an electric stove" (Figure 1).

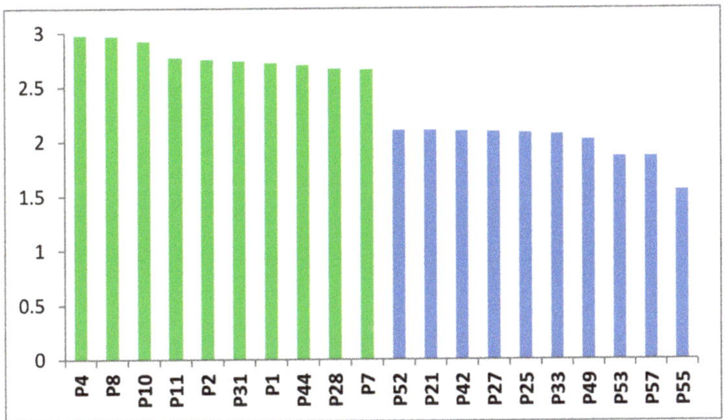

Figure 1. Average ratings of 10 most and 10 least popular practices of proecological behavior in Russia.

The most consistent respondents from the regions rated the practices: P1 "Avoiding overloading the refrigerator", P2 "Reducing opening and closing the door of the refrigerator", P3 "Using a lower setting in the refrigerator compartment", P17 "Taking short showers", P36 "Reducing use of disposable products", P41 "Buying ecomark-appliances" (the standard deviation is slightly more than 1). The largest scatter of assessments is observed across practices: P57 "Use of residual heat when cooking on an electric stove", P16 "Turning off the water when washing face or brushing teeth", P51 "Avoiding overloading the car", P53 "Maintaining air pressure of tire". Moscow's respondents did not show consistency in applying practices (standard deviation is on average more than 1.0). The most considerable variation among respondents is observed for practices: P53 "Maintaining air pressure of tire", P55 "Using the dishwasher", P56 "Flame control when cooking on a gas stove", P57 "Using residual heat when cooking on an electric stove" (standard deviation is on average more than 1.5). Among the groups of practices, the most popular were energy saving practices (median 2.8, average 2.6), the least popular practices for waste management (median 1.9, average 2.21) (Figure 2).

One of the central principles of ecological behavior is saving resources in everyday life. Research data showed that a significant proportion of citizens (46%) try to save electricity, water, and gas. Regarding people's behavior, which can be called eco-oriented (saving resources in everyday life, separating garbage, following the norms of waste management), we can conclude that Russians rarely adhere to several pro-eco practices simultaneously, but they are used quite often separately. The most frequently chosen reason for nonapplication of proenvironmental practices of handling household waste was the answer option "There are no necessary conditions for application", which suggests that the overwhelming majority of respondents could make them at least more commonly used with appropriate collection points for glass containers, wastepaper, old clothing, used batteries and light bulbs. However, also popular answers about the reasons for nonapplication were "I see no need", "laziness", "waste of time", "forgetting", which indicates an insufficient level of environmental self-awareness even in one of the most informed and flexible in household skills group of respondents. The identified differences are most likely explained not so much by the respondent's income level as by other factors that may be indirectly related to the income level: cultural traditions and living conditions. More research is needed to understand better the impact of income on the frequency of using a particular behavior model.

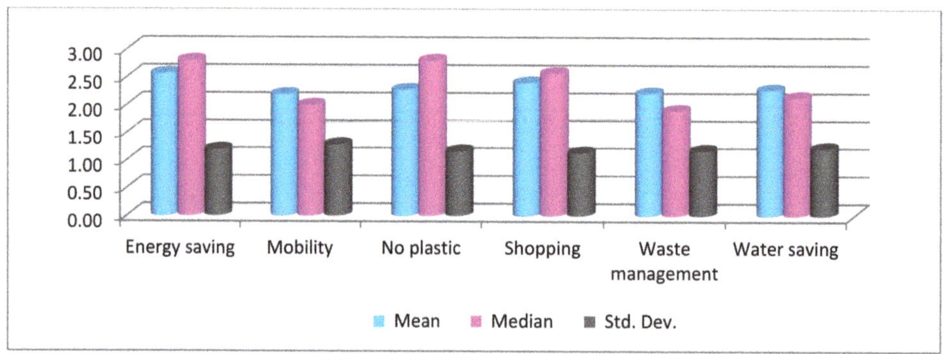

Figure 2. Most popular groups of practices.

4.9. Box and Whisker Chart Analysis Results for Proenvironmental Behavior Practices

The revealed differences were analyzed by constructing box and whisker diagrams, reflecting the median, and quartiles of 25%–75% and the maximum and minimum values of each group's respondents' ratings. Differences due to the discrepancy between only the quartiles of 25–75% and the discrepancy between the median estimates of only groups 4, 5, and 6 (the smallest, cannot be generated randomly) were excluded from the meaningful interpretation. In those cases, when the median values of the most numerous groups of respondents (1, 2, and 3) differed among themselves, we attempted to analyze and substantively explain the statistically significant differences revealed. In Figures 3–16 are box and whisker diagrams with the most noticeable differences in the respondents' assessments between groups 1—Moscow, 2—Krasnodar, 3—Krasnodar region.

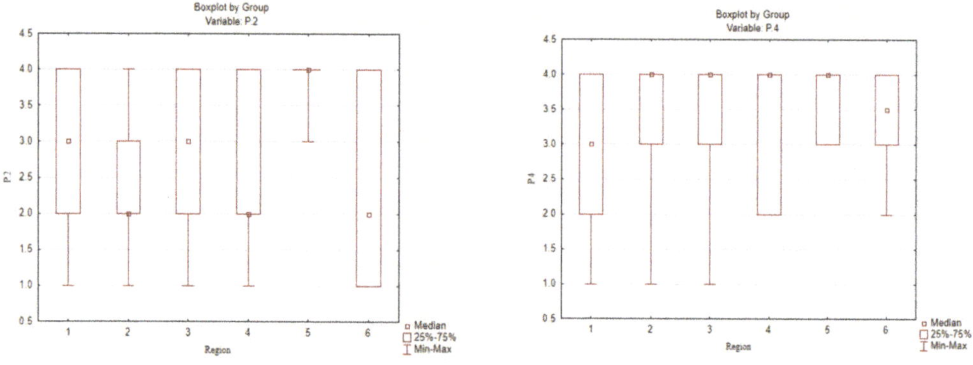

Figure 3. Box and whisker diagrams for practices P2 (Reducing opening and closing the door of the refrigerator) and P4 (Putting hot food into the refrigerator after cooling) from the category "energy saving".

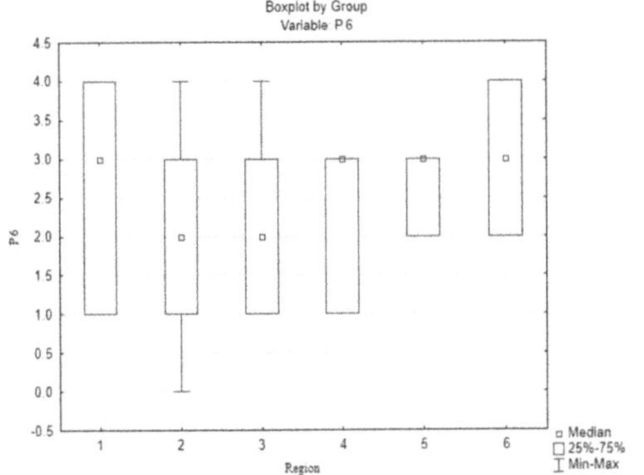
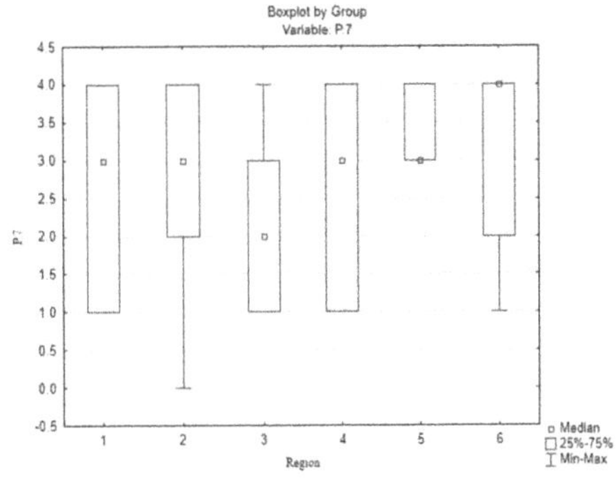

Figure 4. Box and whisker diagrams for practice P6 (Cleaning filter of the air conditioner or cleaner) and P7 (Adjusting the temperature of the air conditioner), the category "energy saving".

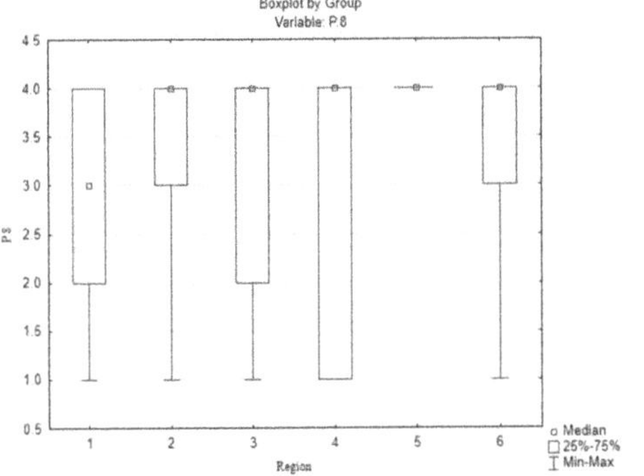
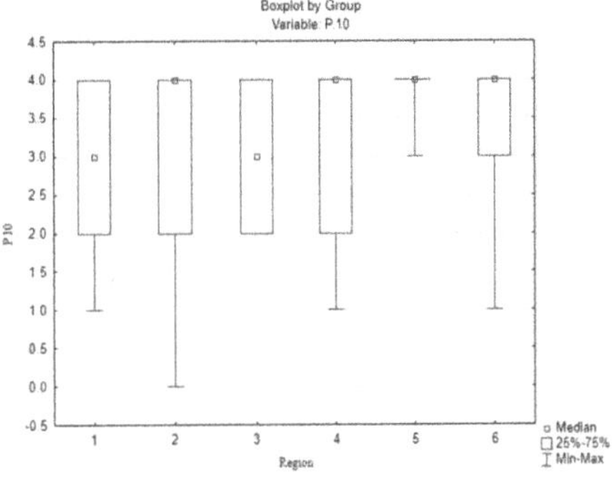

Figure 5. Box and whisker diagrams for practice P8 (Turning off lights in empty rooms) and P10 (Turning off the TV when people are not watching), the category "energy saving".

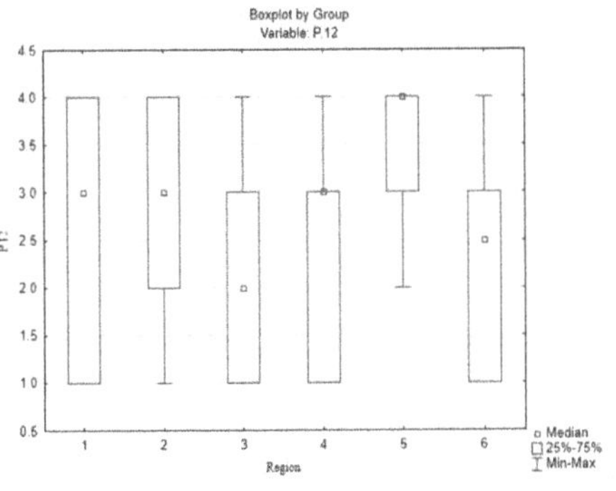
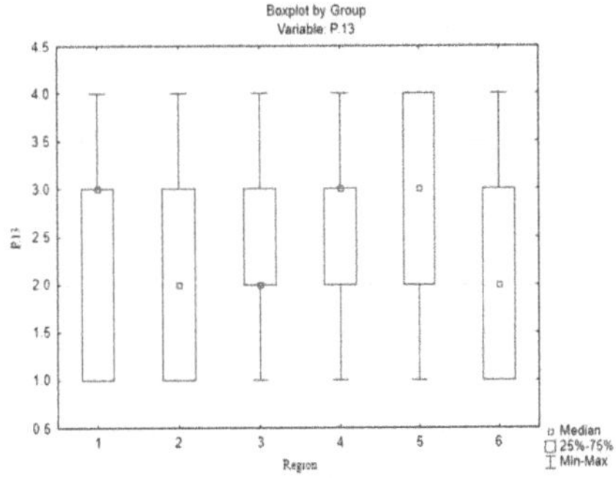

Figure 6. Box and whisker diagrams for practice P12 (Doing ironing collectively) and P13 (Setting a lower shower temperature), the category "energy saving".

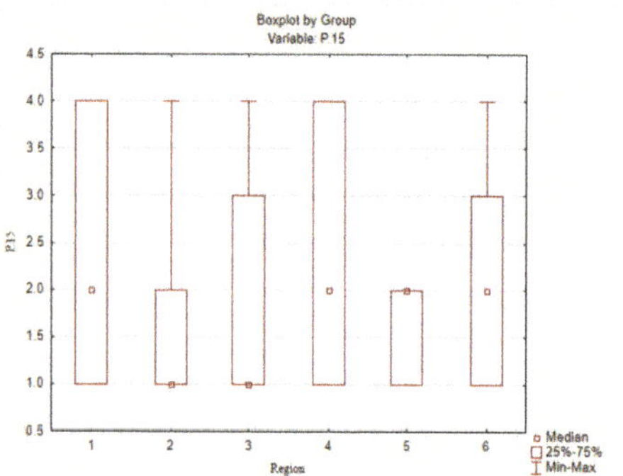
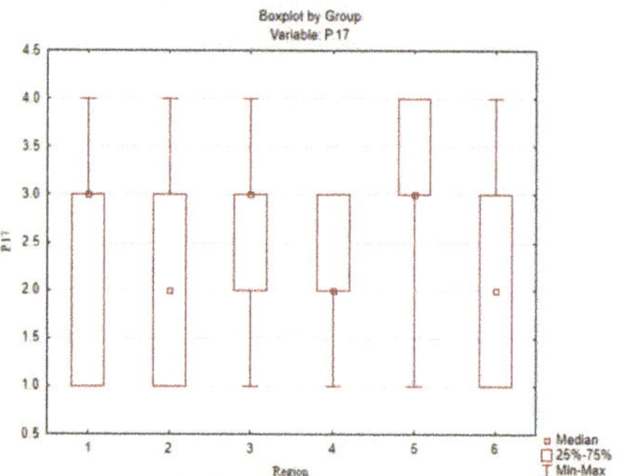

Figure 7. Box and whisker diagrams for practice P15 (Using toothbrush cup) and P17 (Taking short showers), category—"Water saving".

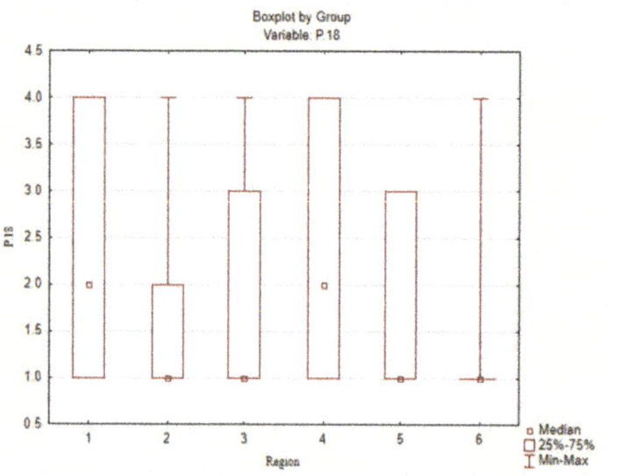
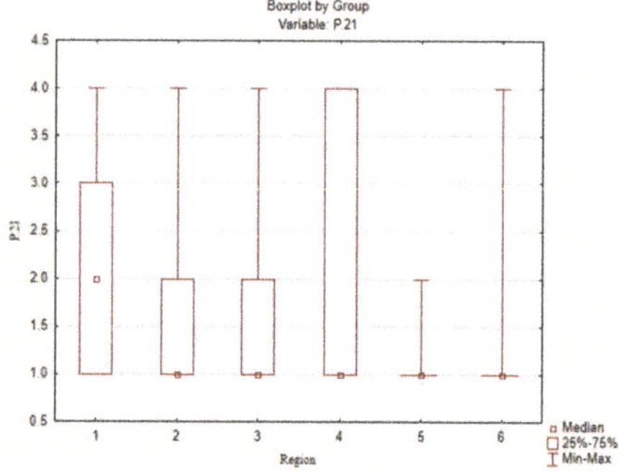

Figure 8. Box and whisker diagrams for practice P18 (category "Water saving") and P21 (category "Waste management").

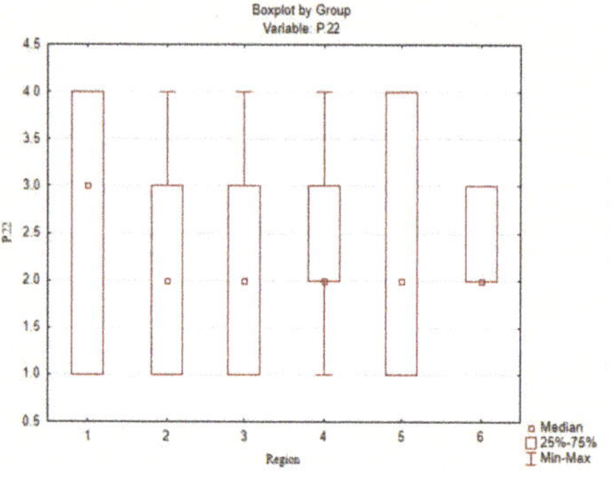
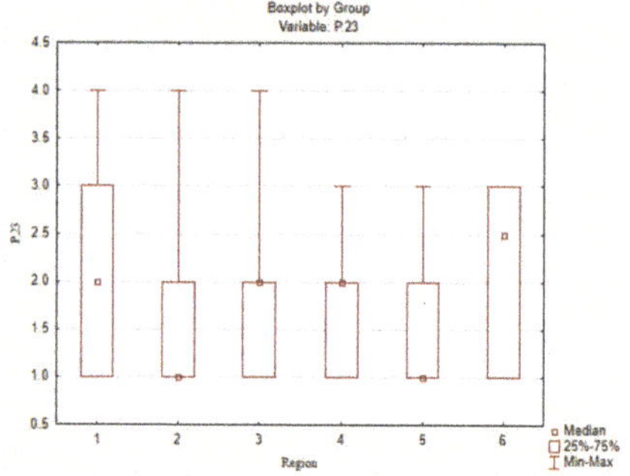

Figure 9. Box and whisker diagrams for practices P22 (Following garbage rules) and P23 (Garbage separation), the category "Waste management".

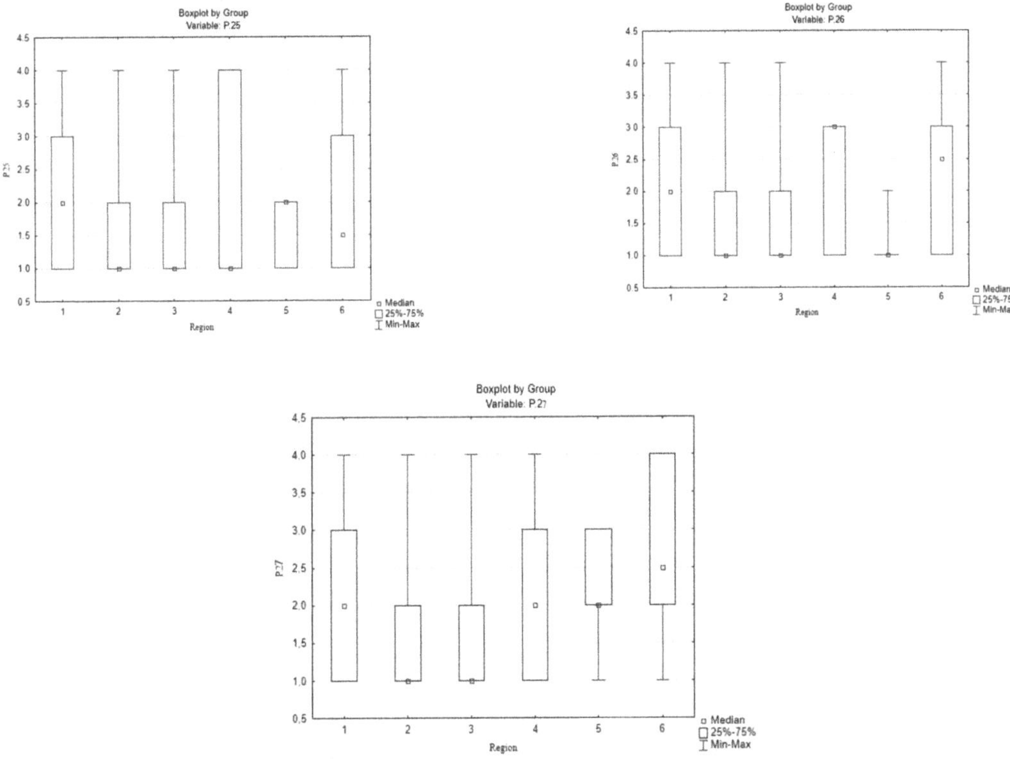

Figure 10. Box and whisker diagrams for practices P25 (Collection and delivery of glass containers to appropriate collection points), P26 (wastepaper collection and delivery to appropriate collection points) and P27 (Collection and delivery of used batteries, light bulbs to appropriate collection points) from the category "Waste management".

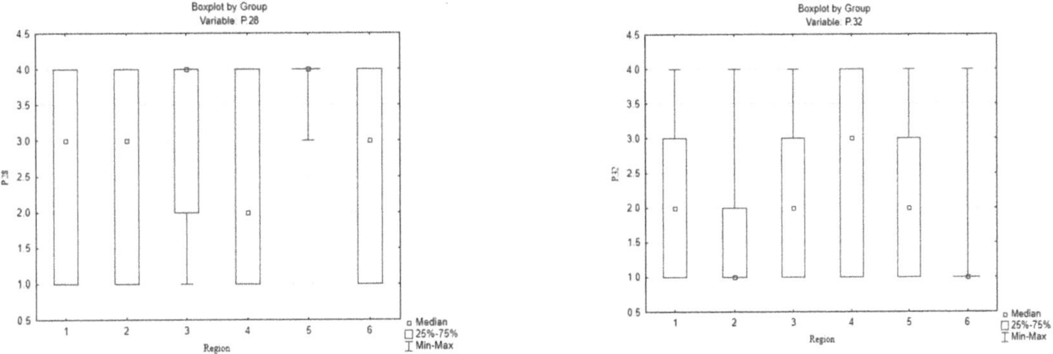

Figure 11. Box and whisker diagrams by practices P28 (Using own cup), P32 (Composting kitchen garbage), from the categories "No plastic" and "Waste management".

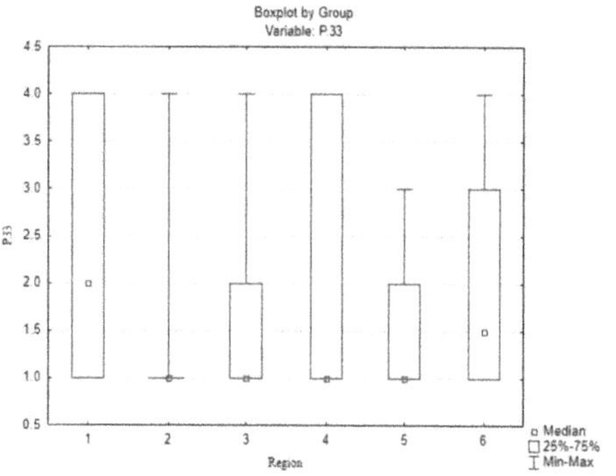
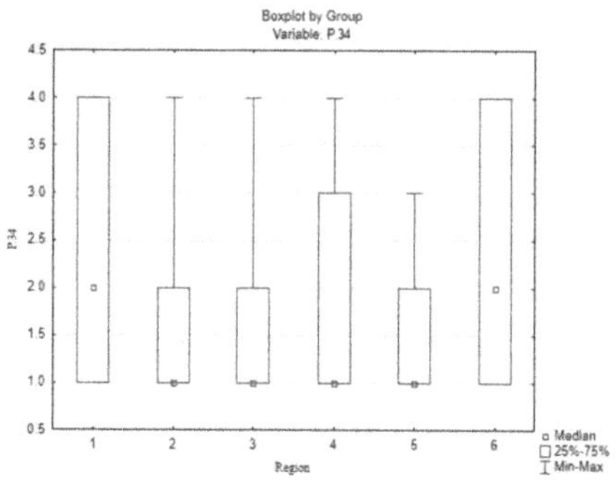

Figure 12. Box and whisker diagrams by practices P33 (Throwing away kitchen garbage after it has dried), P34 (Using receptacle instead of plastic bag), from the categories "No plastic" and "Waste management".

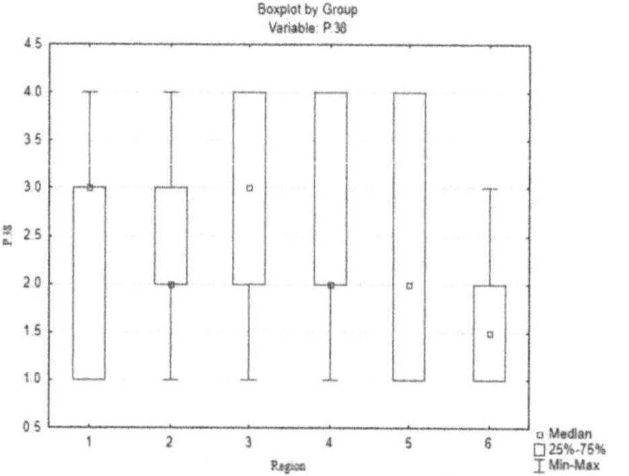
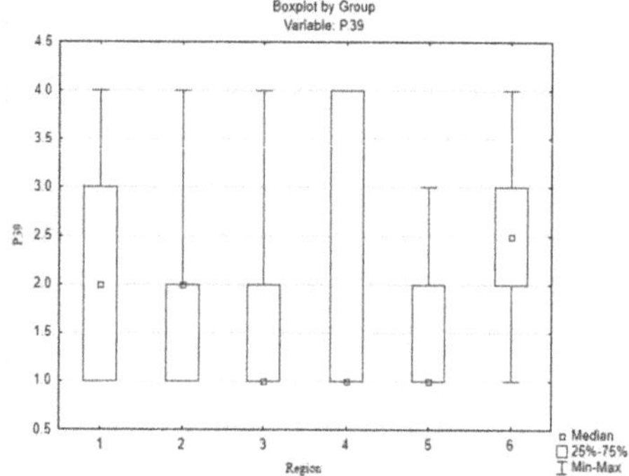

Figure 13. Box and whisker charts by practices P38 (Buying organic products), P39 (Buying recycled goods), from category "Shopping".

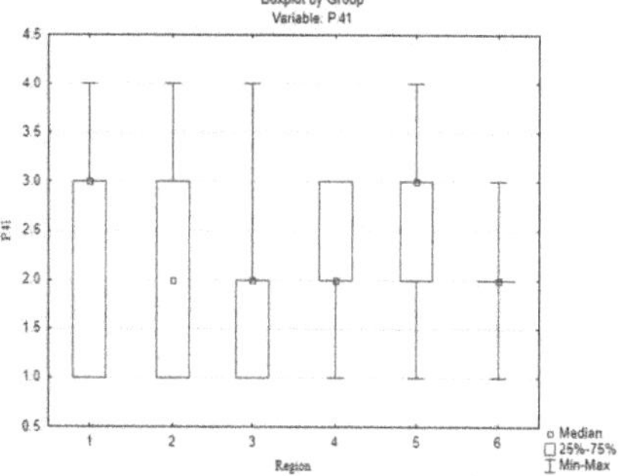
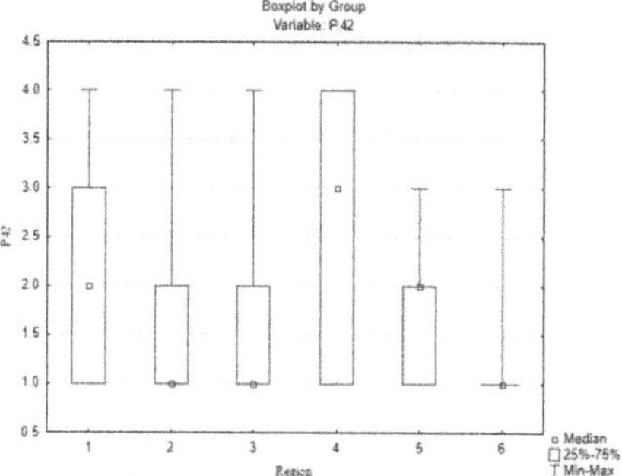

Figure 14. Box and whisker charts by practices P41 (Buying ecomark-appliances), P42 (Choosing goods with their CO_2 emission in mind (carbon footprint)), from category "Shopping".

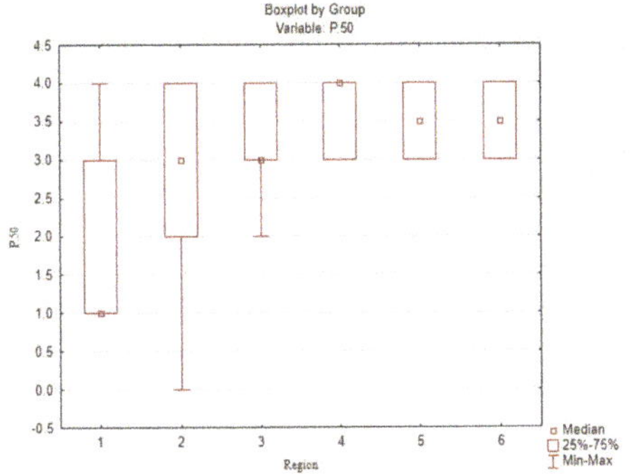
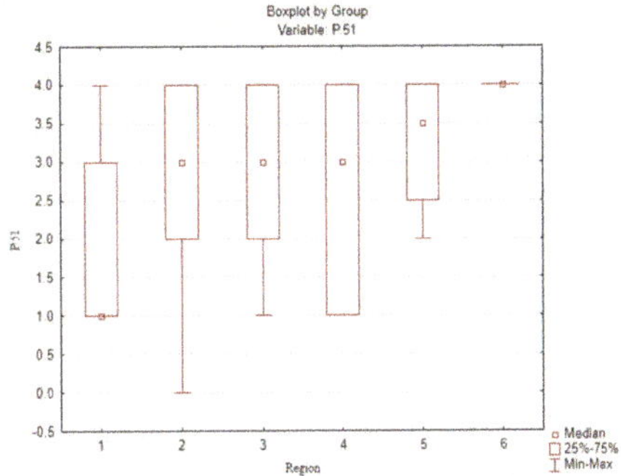

Figure 15. Box and whisker diagrams by practices P50 (Doing car checks regularly), P51 (Avoiding overloading the car), category "Mobility".

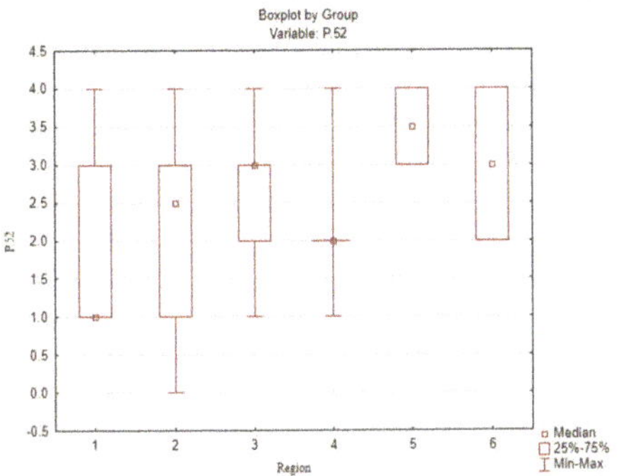
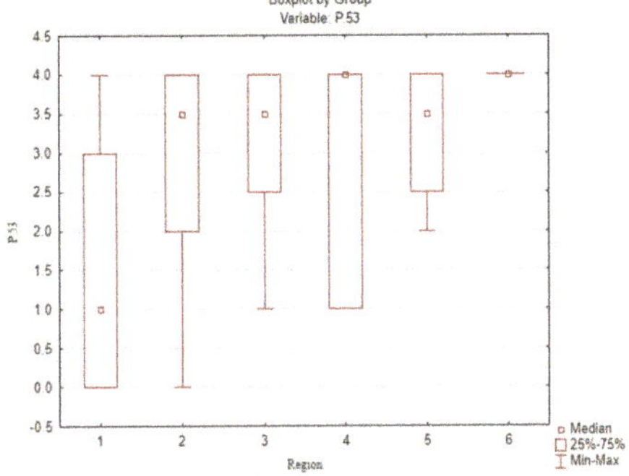

Figure 16. Box and whisker diagrams by practices P52 (Reducing idling of the car), P53 (Maintaining air pressure of the tire), category "Mobility".

Control over opening the refrigerator (P2): in Krasnodar, this practice is used less frequently (median 2.0, which corresponds to the answer "rarely") than in Moscow and small towns of the Krasnodar Territory (median 3.0, which corresponds to the answer "often"). Cooling hot food before placing it in the refrigerator (P4): in Krasnodar and Krasnodar region it is used more often (median 4, which corresponds to the answer "always") than in Moscow (median—3, which corresponds to the answer "often") (Figure 3). Air conditioner filter cleaning (P6) is used much more frequently in Moscow (median 3, which corresponds to the answer "often") than in Krasnodar and Krasnodar region (median 2—"rarely"). Air conditioner temperature control (P7) is more often used in large cities—Moscow and Krasnodar (median 3—"often"), and in Krasnodar region-less often (median 2—"rarely") (Figure 4).

Lighting off in empty rooms (P8) is more commonly used in Krasnodar and Krasnodar region (median 4—"always") than in Moscow (median 3—"often"). Switching off the TV after watching (P10) is most often used in Krasnodar (median 4—"always"), while in Moscow and Krasnodar region—less often (median 3—"often") (Figure 5). Selective ironing (P12) is often practiced in large cities—Moscow and Krasnodar (median 3—"often"). In small towns and rural areas of the Krasnodar region, this practice of energy saving is rarely used (median 2—"rarely). Lowering the water temperature when taking a shower/washing/washing dishes (P13) is more often practiced in Moscow (median 3—"often"), while in Krasnodar and Krasnodar region this practice of energy saving is used less often (median 2—"rarely") (Figure 6).

Using a water cup when brushing teeth (P15) is one of Russia's least popular resource-saving practices. In Moscow (as well as in the regions that are represented by a small number of respondents), it is still used more often (median 2—"rarely") than in Krasnodar and Krasnodar region (median 1—"never"). Shortening of shower time (P17) as a water saving practice is more often used in Moscow and small towns of the Krasnodar region (median 3—"often") than in Krasnodar (median 2—"rarely") (Figure 7).

Washing dishes in a basin (P18) is more commonly used in Moscow (median 2—"rarely") than in Krasnodar and Krasnodar region (median 1—"never"). A similar situation is observed in the practice of recycling vegetable oil (P21). This situation is rather strange since it is logical to assume that such practices are more accessible to apply in the case of living in your own house and/or running a subsidiary farm, which is much more common in small towns of the Krasnodar region than in such a metropolis as Moscow (Figure 8).

Compliance with waste management standards (P22) is much more common in Moscow (median 3—"rare") than in Krasnodar and Krasnodar region (median 2—"rare"). Judging by the assessments of respondents from regions with a low representation, it can be expected that the same situation of ignoring waste management standards is observed in other regions of Russia. The practice of sorting waste (P23) is more often used in Moscow and small towns of the Krasnodar region (median 2—"rarely") than in Krasnodar (median 1—"never") (Figure 9).

Collection and delivery of wastepaper, glass containers, and electronic waste occurs (P25) more often in Moscow (median 2—"rarely") than in Krasnodar and other settlements of the Krasnodar region (median 1—"never") (Figure 10).

The use of one's tableware instead of disposable ones is equally often practiced in Moscow and Krasnodar (median 3—"often") but even more often in small towns of Krasnodar region (median 4—"always"). Composting of kitchen waste is sometimes practiced in Moscow and small towns of the Krasnodar region (median 2—"rare"), in Krasnodar it does not occur at all (median 1—"never") (Figure 11).

Such a practice of handling household waste as throwing away organic waste only after drying (P33) is sometimes observed in Moscow (median 2—"rarely"), in Krasnodar and other points of the Krasnodar region do not occur at all (median 1—"never"). The use of a container instead of a plastic bag for storing food (P34) is also more often used in Moscow (median 2—"rarely") than in the Krasnodar region and Krasnodar (median 1—"never") (Figure 12).

The purchase of organic products (P38) is more common in Moscow and small towns of the Krasnodar region (median 3—"often") than in Krasnodar (median 2—"rare"). The purchase of processed goods (P39) is sometimes found in large cities—Moscow and Krasnodar but not at all in small towns of the Krasnodar region (median 1—"never") (Figure 13).

Purchases of goods subject to ecolabeling (P41) are more likely to occur in Moscow (median 3—"often") than in Krasnodar and Krasnodar region (median 2—"rare"). The purchase of goods taking into account the carbon footprint (P42) is still found only in Moscow (median 2—"rare"), in the Krasnodar region not found at all (Figure 14).

As for the practices that help reduce the negative impact of the car on the environment, the practice of regular technical inspection and avoidance of overloading the car (P50) is sometimes used in Krasnodar and Krasnodar region (median 2—"rarely"), in Moscow—never (median 1) (Figure 15). A similar picture is observed with a decrease in the idling of a car and tire pressure maintenance (P52). (Figure 16). In the Krasnodar region and Krasnodar, they are often used (median 2.5-3), in Moscow—never (median 1).

The respondents' distribution by the level of participation in environmental events held by local communities, regional and federal authorities, and their initiative is shown in Figure 16. Of the most interesting differences between the behavior of respondents, the following can be noted: respondents who show their environmental activity in the form of participation in protest actions are less likely than other groups (including respondents who show their activity in other forms and respondents who do not participate in any what environmental actions) use in practice the patterns P4 (Putting hot food into the refrigerator after cooling), P11 (Using energy-saving mode or turning off when not in use) and P29 (Avoiding overvolume cooking) from the category "energy saving", P16 (Turning off the water when washing face or brushing teeth) from the category "Water use", P23 (Garbage separation) from the category "Waste management", P28 (Using own cup) from the category "Refusal from plastic", P37 (Not buying over-packaged products) and P38 (Buying organic products) from the category "Shopping", P47 (Using bicycle or walking) and P53 (Maintaining air pressure of the tire) from the category "Mobility". For all other practices, where differences in respondents' behavior from different groups are revealed, this group does not differ from most others. Judging by the identified features of consumer behavior, this group of respondents cannot be classified as environmentally conscious. Their participation in protest actions is most likely motivated not by environmental considerations as by a generally negative attitude towards the authorities' actions.

The most active respondents who took part in more than one type of environmental activities demonstrate a more frequent use of practices P4 (Putting hot food into the refrigerator after cooling) and P56 (Flame adjustment for cooking) from the category "energy saving", P16 (Turning off the water when washing face or brushing teeth) from the category "Water use", P28 (Using own cup) and P37 (Not buying over-packaged products) from in the category "Refusal of plastic", P38 (Buying organic products) from the category "Purchases", P53 (Maintaining air pressure of the tire) from the category "Mobility". At the same time, according to other practices, such as P18 (Washing dishes using jugged water) from the category "Water use", P23 (Garbage separation) from the category "Waste management", P34 (Using receptacle instead of plastic bag) from the category "Avoiding plastic", this group of respondents has the lowest or one of the lowest rates of the frequency of use. Such a spread in the frequency of using various practices of the most active respondents' proecological behavior is most likely evidence that not all of the studied practices are realized as essential and useful. Any striking differences in the behavior of groups of respondents who have never taken part in environmental actions at all, who filed complaints or collected signatures under appeals to the authorities and who participated in tree planting, garbage collection, collection of wastepaper, glass containers, etc., was not identified. This may indicate that the respondents took part in environmental campaigns more out of solidarity, communication, charity, etc., rather than based on clear ideas about the severity of environmental problems. Another interpretation of the results obtained can be offered: it is much easier for even environmentally conscious people to take part in a one-time action than change the usual patterns of their daily consumer behavior.

5. Discussion

In the extensive literature on various proecological behavior issues, we can highlight several articles that consider various factors and barriers that affect environmental responsibility, which implies a more efficient use of natural resources. Thus, the article [27,49] noted that in addition to environmental awareness, the availability of financial resources, time, and opportunities affect the desire to purchase environmental goods. Studies [50–52] demonstrate that time preferences influence the valuation of PEBs even in very different cultural contexts (US and India), and people who have free time are more engaged in energy-saving proenvironmental behavior. Unlike previous research, our study did not reveal the differences in the frequency of some PEBs between the groups of economically active and inactive respondents, which can support our original hypothesis that having more free time in the absence of a permanent job can contribute to more careful handling of all resources and the use of more labor-intensive waste management practices. The exception is the pattern P8 "turning off lights in empty rooms".

On the contrary, economically active respondents demonstrate greater environmental awareness. In a study [37] on the example of European countries, the author showed evidence of a positive influence of education on the proecological behavior of individuals. The same kind of positive correlation was recently found in [53] on the example of Peru for the PEBs, connected with plastic consumption. Our results also showed that involvement in the educational process has a positive effect on the respondent's values and attitude towards practices. As for the revealed statistically significant differences between the groups of respondents who are receiving education and have completed their education, the second group of respondents demonstrates more conscious proecological behavior in absolutely all aspects.

Statistically significant differences in the frequency of PEBs for the variable "gender" were found only for nine PEBs, of which three PEBs belong to the mobility group and three PEBs belong to the waste-management group. This result is rather unexpected, since in other countries, for example, in Spain [54], France [55], and China [56], women demonstrate more active proecological behavior.

In the study of authors from China [40], it is shown that other aspects affect energy conservation in addition to education. The authors believe that urban residents are more economical in terms of energy consumption than rural residents. In this, the results of our study are quite consistent with the results of surveys of Chinese scientists conducted in large urban agglomerations and presented in the study [11]. Our results showed that concerning interregional differences, we could say that Moscow is a regional leader in infrastructure development and information support for more "advanced" practices of proenvironmental behavior. The application of these practices requires a certain level of environmental knowledge and specific technical equipment. In small towns and rural areas, traditional patterns of proenvironmental behavior are more often used. However, more detailed conclusions about regional differences can be made only after additional research and expansion of other regions of Russia in the sample.

Despite the fact that the obtained results cannot be directly compared with the parallel results of surveys in other countries presented in the literature, since they were obtained in different macro and micro contexts [56], it can be noted that composting kitchen waste is unpopular in Russia as much as in cities like Tokyo, Bangkok, and Seoul [25]. On the contrary, energy-saving practices are the most popular both in Russia and in all three Asian cities examined with the same methodology. In an article [41,57], the authors identified the most significant factors: age, income, and housing type. They showed that people living in private houses are more likely to save than residents of apartment buildings. The results of the study confirmed the effect of age on the frequency of practice. Namely, the older generation more often practices more than 50% of proecological behavior's main patterns. They are more conscious about their consumption, which can be seen in many practices. However, this behavior is not associated with a decrease in economic activity and a decrease in income. On the contrary, economically active respondents are more likely to practice proenvironmental behavior than economically inactive, especially in cases where proenvironmental behavior is more costly than irresponsible consumption. The main reasons for not applying several consumer behavior patterns are either a lack of understanding of their significance or a complete lack of information about the possibility of such consumer behavior.

The selection of methods for studying barriers to consumer prosustainable behavior confirmed significant shortcomings of the questionnaire used in this study. This method is actively discussed in the literature [7,8,43]. However, when conducting primary research to obtain a general picture of the spread of a particular behavior model and identifying the reasons for the nonproliferation of other models, this method, in our opinion, remains a priority tool. It is possible to identify the main reasons for the weak operation of CE in reality. Central to the effective circular economy design is the observation that there is still a problem with information. For example, potential consumers do not know how many years they can use it when buying a supported product. If the customer can receive such information on time, this may affect his choice. There is also a distorted perception among potential customers that recycled materials are usually considered substandard. In primary markets, external effects (pollution of the environment and air) are not taken into account, which leads to unjustified price advantages of primary materials. It is necessary to increase the literacy of citizens. As the key factors that hinder environmentally sound behavior, the respondents highlight the lack of their initiative and specific environmental measures on the part of the government.

6. Conclusions

This paper contributes to the literature on sustainable consumption by in-depth analysis of the factors affecting the probability of 57 different PEBs in Russia. The advantages of our research are as follows: the set of studied PEBs includes not only popular energy-saving and waste-management practices but also more circular patterns of plastic consumption, shopping, and city mobility.

The study has revealed that involvement in the educational process (being a student) has a positive effect on the respondent's attitude towards the most energy-saving practices. Besides, the respondents who completed their education demonstrate more conscious proenvironmental behavior in all studied areas: energy and water saving, plastic consumption, food consumption, waste management, and sustainable mobility. At the same time, the probability of PEBs practically not related to the level of income and, therefore, is not the result of intention to save money. Economically active respondents demonstrate greater environmental awareness and are more likely to practice waste collection and recycling and reusable shopping bags. In addition, economically active respondents are more competent in handling household appliances to reduce their energy consumption.

Thus, it is possible to confirm the positive externality of education-higher quality proecological behavior. Given the high cost of recycling modern waste, investment in education can be a preventive measure for improving the environment. In the short term, the effect is hardly possible, but for the regions of Russia, raising awareness and education of the population in the long term can significantly affect the level of environmental awareness.

The main reasons for not applying some sustainable behavior patterns are either a lack of understanding of their significance or a complete lack of information about the possibility of such consumer behavior. Waste management practices are the least popular. We can explain that the unpopularity of waste management practice by the underdevelopment of "eco" infrastructure, which is a barrier to the formation of proecological behavior. As a rule, respondents understand which of the patterns is correct and realize its importance, but the lack of infrastructure at the proper level does not allow using this pattern. The study showed that there are inter-regional differences. We can say that Moscow is a leader in infrastructure development and information support for more "advanced" practices of proenvironmental behavior, the use of which requires a certain level of environmental knowledge and specific technical equipment. In small towns and rural areas, traditional patterns of proecological behavior are more commonly used. However, more detailed conclusions about regional differences can be made only after additional research and expansion of other regions of Russia in the sample. The findings can be used in the educational process and social work with young people.

We realize that the obtained results cannot be directly compared with results of surveys in other countries presented in the literature, since they were obtained in different macro and micro contexts. It should be noted that the factors influencing PEB are still not clearly understood outside a high-income country. However, our major findings do not contradict with conclusions of other authors, arguing that the level of education is one of the most important factors for practicing PEBs.

As for the limitations of our study, we can point out a moderate sample size. This has prevented us from studying the regional differences in behaviors in a more detailed way. Besides the sample size limitation, our study has also neglected the role of cultural values and social norms in stimulating proenvironmental patterns of consumer's behavior.

In our questionnaire design, we have provided questions aimed at identifying the main reasons as to why respondents do not practice PEBs. The answers to these questions should have led us to estimates of behavioral cost. However, in reality, the main reason for not using PEBs was the lack of information about its environmental importance. We will continue our study in trying to redesign the questionnaire, collecting more even samples, and using advanced methods, such as structural modelling, for processing the result of survey.

Author Contributions: Conceptualization, S.R. (Svetlana Ratner); Data curation, S.R. (Svetlana Ratner); Formal analysis, I.L.; Funding acquisition, I.L.; Investigation, S.R. (Svetlana Ratner); Methodology, S.R. (Svetlana Ratner); Project administration, I.L.; Resources, K.G., S.R. (Svetlana Revinova); Software, S.R. (Svetlana Ratner); Supervision, S.R. (Svetlana Ratner); Validation, S.R. (Svetlana Ratner), K.G. and S.R. (Svetlana Revinova); Visualization, K.G.; Writing—original draft, I.L.; Writing—review and editing, S.R. (Svetlana Ratner), K.G., S.R. (Svetlana Revinova) and I.L. All authors have read and agreed to the published version of the manuscript.

Funding: The publication has been prepared with the support of the "RUDN University Program 5-100".

Institutional Review Board Statement: Not applicable.

Informed Consent Statement: Informed consent was obtained from all subjects involved in the study.

Conflicts of Interest: The authors declare no conflict of interest.

References

1. Revinova, S.; Ratner, S.; Lazanyuk, I.; Gomonov, K. Sharing Economy in Russia: Current Status, Barriers, Prospects and Role of Universities. *Sustainability* **2020**, *12*, 4855. [CrossRef]
2. Korhonen, J.; Honkasalo, A.; Seppälä, J. Circular Economy: The Concept and its Limitations. *Ecol. Econ.* **2018**, *143*, 37–46. [CrossRef]
3. Ratner, S. European experience in transition to circular economy. *Econ. Anal. Theory Pract.* **2020**, *19*, 598–617. [CrossRef]
4. Haas, W.; Krausmann, F.; Wiedenhofer, D.; Heinz, M. How Circular is the Global Economy? An Assessment of Material Flows, Waste Production, and Recycling in the European Union and the World in 2005. *J. Ind. Ecol.* **2015**, *19*, 765–777. [CrossRef]
5. Kirchherr, J.; Piscicelli, L.; Bour, R.; Kostense-Smit, E.; Muller, J.; Huibrechtse-Truijens, A.; Hekkert, M. Barriers to the Circular Economy: Evidence From the European Union (EU). *Ecol. Econ.* **2018**, *150*, 264–272. [CrossRef]
6. Lee, H.; Kurisu, K.; Hanaki, K. Influential Factors on Pro-Environmental Behaviors—A Case Study in Tokyo and Seoul. *Low Carbon Econ.* **2013**, *4*, 104–116. [CrossRef]
7. Kormos, C.; Gifford, R. The validity of self-report measures of proenvironmental behavior: A meta-analytic review. *J. Environ. Psychol.* **2014**, *40*, 359–371. [CrossRef]
8. Lange, F.; Dewitte, S. Measuring pro-environmental behavior: Review and recommendations. *J. Environ. Psychol.* **2019**, *63*, 92–100. [CrossRef]
9. Yuriev, A.; Boiral, O.; Francoeur, V.; Paillé, P. Overcoming the barriers to pro-environmental behaviors in the workplace: A systematic review. *J. Clean. Prod.* **2018**, *182*, 379–394. [CrossRef]
10. Fu, L.; Sun, Z.; Zha, L.; Liu, F.; He, L.; Sun, X.; Jing, X. Environmental Awareness and Pro-environmental Behavior within China's Road Freight Transportation Industry: Moderating Role of Perceived Policy Effectiveness. *J. Clean. Prod.* **2019**, 119796. [CrossRef]
11. Huang, L.; Wen, Y.; Gao, J. What ultimately prevents the pro-environmental behavior? An in-depth and extensive study of the behavioral costs. *Resour. Conserv. Recycl.* **2020**, *158*, 104747. [CrossRef]
12. Chwialkowska, A.; Bhatti, W.A.; Glowik, M. The influence of cultural values on pro-environmental behavior. *J. Clean. Prod.* **2020**, 122305. [CrossRef]
13. Ali, E.B.; Anufriev, V.P. Towards environmental sustainability in Russia: Evidence from green universities. *Heliyon* **2020**, *6*, e04719. [CrossRef] [PubMed]
14. Stern, P. New Environmental Theories: Toward a Coherent Theory of Environmentally Significant Behavior. *J. Soc. Issues* **2000**, *56*, 407–424. [CrossRef]
15. Andersson, L.; Shivarajan, S.; Blau, G. Enacting ecological sustainability in the MNC: A test of an adapted value-belief-norm framework. *J. Bus. Ethics* **2005**, *59*, 295–305. [CrossRef]
16. Kaiser, F.G.; Wolfing, S.; Fuhrer, U. Environmental attitude and ecological behavior. *J. Environ. Psychol.* **1999**, *19*, 1–19. [CrossRef]
17. Ellen MacArthur Foundation: Towards a Circular Economy: Business Rationale for an Accelerated Transition. Available online: https://www.ellenmacarthurfoundation.org/assets/downloads/TCE_Ellen-MacArthur-Foundation_9-Dec-2015.pdf (accessed on 15 October 2020).
18. Wilts, H. The Digital Circular Economy: Can the Digital Transformation Pave the Way for Resource-Efficient Materials Cycles? *Int. J. Environ. Sci. Nat. Resour.* **2017**, *7*. [CrossRef]
19. Ando, K.; Ohnuma, S.; Chang, E. Comparing normative influences as determinants of environmentally conscious behaviours between the USA and Japan. *Asian J. Soc. Psychol.* **2007**, *10*, 171–178. [CrossRef]
20. Ando, K.; Ohnuma, S.; Anke, B.; Matthies, E.; Sugiura, J. Determinants of Individual and Collective Pro-Environmental Behaviors: Comparing Germany and Japan. *J. Environ. Inf. Sci.* **2010**, *38*, 21–32.
21. Schultz, P.W. Knowledge, Information, and Household Recycling: Examining the Knowledge-Deficit Model of Behavior Change. In *New Tools for Environmental Protection: Education, Information, and Voluntary Measures*; The National Academic Press: Washington, DC, USA, 2012; pp. 67–82.
22. Aoki, E.; Kurisu, K.H.; Nakatani, J.; Hanaki, K. Current State and Interregional Comparison of Citizen's Environmental Behavior by 47 Prefectures. *Jpn. J. Jsce* **2010**, *38*, 17–26.

23. Cordano, M.; Welcomer, S.; Scherer, R.; Pradenas, L.; Parada, V. A Cross-Cultural Assessment of Three Theories of Pro-Environmental Behavior. *Environ. Behav.* **2010**, *43*, 634–657. [CrossRef]
24. Li, D.; Zhao, L.; Ma, S.; Shao, S.; Zhang, L. What influences an individual's pro-environmental behavior? A literature review. *Resour. Conserv. Recycl.* **2019**, *146*, 28–34. [CrossRef]
25. Phuphisith, S.; Kurisu, K.; Hanaki, K. A comparison of the practices and influential factors of pro-environmental behaviors in three Asian megacities: Bangkok, Tokyo, and Seoul. *J. Clean. Prod.* **2020**, *253*, 119882. [CrossRef]
26. Moon, M.A.; Habib, M.D.; Attiq, S. Analyzing the Sustainable Behavioral Intentions: Role of Norms, Beliefs and Values on Behavioral Intentions. *Pak. J. Commer. Soc. Sci.* **2015**, *9*, 524–539.
27. Yadav, R.; Pathak, G. Determinants of Consumers' Green Purchase Behavior in a Developing Nation: Applying and Extending the Theory of Planned Behavior. *Ecol. Econ.* **2017**, *134*, 114–122. [CrossRef]
28. Setyawan, A.; Noermijati, N.; Sunaryo, S.; Aisjah, S. Green product buying intentions among young consumers: Extending the application of theory of planned behavior. *Probl. Perspect. Manag.* **2018**, *16*, 145–154. [CrossRef]
29. Gallagher, V.; Hrivnak, M.; Valcea, S.; Mahoney, C.; LaWong, D. A comprehensive three-dimensional sustainability measure: The 'missing P' of 'people'—A vital stakeholder in sustainable development. *Corp. Soc. Responsib. Environ. Manag.* **2018**, *25*, 772–787. [CrossRef]
30. Schmitt, M.; Aknin, L.; Axsen, J.; Shwom, R. Unpacking the Relationships Between Pro-environmental Behavior, Life Satisfaction, and Perceived Ecological Threat. *Ecol. Econ.* **2018**, *143*, 130–140. [CrossRef]
31. Dalvi-Esfahani, M.; Alaedini, Z.; Nilashi, M.; Samad, S.; Asadi, S.; Mohammadi, M. Students' green information technology behavior: Beliefs and personality traits. *J. Clean. Prod.* **2020**, *257*, 120406. [CrossRef]
32. Chan, H. When do values promote pro-environmental behaviors? Multilevel evidence on the self-expression hypothesis. *J. Environ. Psychol.* **2020**, *71*, 101361. [CrossRef]
33. Chankrajang, T.; Muttarak, R. Green Returns to Education: Does Schooling Contribute to Pro-Environmental Behaviours? Evidence from Thailand. *SSRN Electron. J.* **2015**. [CrossRef]
34. Donmez-Turan, A.; Kiliclar, I. The analysis of pro-environmental behaviour based on ecological worldviews, environmental training/knowledge and goal frames. *J. Clean. Prod.* **2020**, *279*, 123518. [CrossRef]
35. López-Bonilla, J.; Reyes-Rodríguez, M.; López-Bonilla, L. Interactions and Relationships between Personal Factors in Pro-Environmental Golf Tourist Behaviour: A Gender Analysis. *Sustainability* **2019**, *12*, 332. [CrossRef]
36. Gaspar, R. Understanding the Reasons for Behavioral Failure: A Process View of Psychosocial Barriers and Constraints to Pro-Ecological Behavior. *Sustainability* **2013**, *5*, 2960–2975. [CrossRef]
37. Meyer, A. Does education increase pro-environmental behavior? Evidence from Europe. *Ecol. Econ.* **2015**, *116*, 108–121. [CrossRef]
38. Wang, R.; Qi, R.; Cheng, J.; Zhu, Y.; Lu, P. The behavior and cognition of ecological civilization among Chinese university students. *J. Clean. Prod.* **2020**, *243*, 118464. [CrossRef]
39. Salvioni, D.; Almici, A. Transitioning Toward a Circular Economy: The Impact of Stakeholder Engagement on Sustainability Culture. *Sustainability* **2020**, *12*, 8641. [CrossRef]
40. Ding, Z.; Wang, G.; Liu, Z.; Long, R. Research on differences in the factors influencing the energy-saving behavior of urban and rural residents in China–A case study of Jiangsu Province. *Energy Policy* **2017**, *100*, 252–259. [CrossRef]
41. Martinsson, J.; Lundqvist, L.; Sundström, A. Energy saving in Swedish households. The (relative) importance of environmental attitudes. *Energy Policy* **2011**, *39*, 5182–5191. [CrossRef]
42. Lillemo, S. Measuring the effect of procrastination and environmental awareness on households' energy-saving behaviours: An empirical approach. *Energy Policy* **2014**, *66*, 249–256. [CrossRef]
43. Kaiser, F.; Merten, M.; Wetzel, E. How do we know we are measuring environmental attitude? Specific objectivity as the formal validation criterion for measures of latent attributes. *J. Environ. Psychol.* **2018**, *55*, 139–146. [CrossRef]
44. Ratner, S.; Iosifov, V. Eco-management and eco-standardization in Russia: The perspectives and barriers for development. *J. Environ. Manag. Tour.* **2017**, *8*, 247–258. [CrossRef]
45. Yin, R. *Case Study Research and Applications: Design and Methods*; Sage Publications: Thousand Oaks, CA, USA, 2018.
46. Heckathorn, D. Respondent-driven sampling: A new approach to the study of hidden populations. *Soc. Probl.* **1997**, *44*, 174–199. [CrossRef]
47. Ratner, S.; Zaretskaya, M. Forecasting the Ecology Effects of Electric Cars Deployment in Krasnodar Region (Russia): Learning Curves Approach. *J. Environ. Manag. Tour.* **2018**, *9*, 82–94. [CrossRef]
48. Bol'shakov, A.A.; Karimov, R.N. *Metody Obrabotki Mnogomernyh Dannyh i Vremennyh Ryadov: Ucheb. Posobie dlya Vuzov. M.: Goryachaya Liniya—Telekom*; Hotline —Telekom: Moscow, Russia, 2007.
49. Kadic-Maglajlic, S.; Arslanagic-Kalajdzic, M.; Micevski, M.; Dlacic, J.; Zabkar, V. Being engaged is a good thing: Understanding sustainable consumption behavior among young adults. *J. Bus. Res.* **2019**, *104*, 644–654. [CrossRef]
50. Qiu, Y.; Colson, G.; Grebitus, C. Risk preferences and purchase of energy-efficient technologies in the residential sector. *Ecol. Econ.* **2014**, *107*, 216–229. [CrossRef]
51. Newell, R.G.; Siikamki, J. Individual time preferences and energy efficiency. *Am. Econ. Rev. Pap. Proc.* **2015**, *105*, 196–200. [CrossRef]
52. Fuerst, F.; Singh, R. How present bias forestalls energy efficiency upgrades: A study of household appliance purchases in India. *J. Clean. Prod.* **2018**, *186*, 558–569. [CrossRef]

53. Fuhrmann-Riebel, H.; D'Exelle, B.; Verschoor, A. The role of preferences for pro-environmental behaviour among urban middle class households in Peru. *Ecol. Econ.* **2021**, *180*, 106850. [CrossRef]
54. Casaló, L.; Escario, J.; Rodriguez-Sanchez, C. Analyzing differences between different types of pro-environmental behaviors: Do attitude intensity and type of knowledge matter? *Resour. Conserv. Recycl.* **2019**, *149*, 56–64. [CrossRef]
55. Bradley, G.L.; Babutsidze, Z.; Chai, A.; Reser, J.P. The role of climate change risk perception, response efficacy, and psychological adaptation in pro-environmental behavior: A two nation study. *J. Environ. Psychol.* **2020**, *68*, 101410. [CrossRef]
56. Ling, M.; Xu, L. Relationships between personal values, micro-contextual factors and residents' pro-environmental behaviors: An explorative study. *Resour. Conserv. Recycl.* **2020**, *156*, 104697. [CrossRef]
57. Binder, M.; Blankenberg, A.; Guardiola, J. Does it have to be a sacrifice? Different notions of the good life, pro-environmental behavior and their heterogeneous impact on well-being. *Ecol. Econ.* **2020**, *167*, 106448. [CrossRef]

Article

Which Influencing Factors Could Reduce Ecological Consumption? Evidence from 90 Countries for the Time Period 1996–2015

Shuai Zhang [1],*, Dajian Zhu [2], Jiaping Zhang [2] and Lilian Li [2]

[1]   College of Design and Innovation, Tongji University, Shanghai 200092, China
[2]   School of Economics and Management, Tongji University, Shanghai 200092, China; dajianzhu@263.net (D.Z.); jz789@scarletmail.rutgers.edu (J.Z.); lll_pyh123@163.com (L.L.)
*   Correspondence: zhangshuaiboshi@tongji.edu.cn

Received: 18 November 2019; Accepted: 14 January 2020; Published: 18 January 2020

Abstract: In the "full world" and Anthropocene, global ecological consumption is beyond natural capital's regenerative and absorptive abilities, and ecological consumption of humanity has to be reduced to have an ecologically sustainable future. To achieve the goal of ecological sustainability, influencing factors that could reduce ecological consumption need to be explored. Based on three panel datasets for the time period 1996–2015, this paper estimates the impacts of urbanization, renewable energy consumption, service industries, and internet usage on ecological consumption for all 90 sample countries, the 42 developed countries, and the 48 developing countries. Education and income are taken as control variables in the panel regressions. As a consumption-side indicator, the ecological footprint is selected to measure ecological consumption. The estimations find that (1) urbanization has negative impacts for all sample countries and the developed countries, and it is insignificant for the developing countries, (2) renewable energy consumption and service industries have negative impacts for all of the three samples, and (3) internet usage has lagged negative impacts for all sample countries, and it is an independent and significant force of reducing ecological consumption in the developing countries rather than the developed countries. It is found that there is a positive linear relationship, an inversed U-shaped relationship, and a U-shaped relationship between ecological consumption and income in all sample countries, the developed countries, and the developing countries, respectively. The estimated results provide guidance for evidence-based policymaking on reducing ecological consumption.

Keywords: ecological consumption; influencing factors; panel regressions; ecological footprint

1. Introduction

In the increasingly "full world" and Anthropocene, and according to the paradigm of strong sustainability, the limiting factor to well-being improvement switched from manmade capital to natural capital [1]. The epochal and fundamental change in the pattern of scarcity warns us that the physical stock of natural capital has to be kept constant, only then enabling an ecologically sustainable future. However, the undisputed fact is that ecological consumption of humanity is beyond natural capital's regenerative and absorptive abilities and that we are living off the "principal" of natural capital [2,3]. Humanity stepped over at least four planetary boundaries, i.e., climate change, rate of biodiversity loss, nitrogen cycle, and change in land use [4,5]. Natural capital is being gradually liquidated and degraded, which could cause declines in provision of ecosystem goods and services and have negative impacts on ecosystem stability and resilience [6].

Therefore, to have an ecologically sustainable future, ecological consumption of humanity has to be reduced until it is kept below natural capital's regenerative and absorptive abilities [7,8]. To achieve the goal of ecological sustainability, influencing factors that could reduce ecological consumption need to be explored in detail, which would provide specific guidance for sound and evidence-based policymaking on reducing ecological consumption. Inspired by previous literature, urbanization, renewable energy consumption, service industries, and internet usage are treated as the latent influencing factors. This paper explores the impacts of the influencing factors on ecological consumption at the global level by selecting representative sample countries covering all of the world. More importantly, this paper explores whether the impacts of the influencing factors on ecological consumption are different or not for countries at different development stages, i.e., developed countries and developing countries. Lessons could be drawn and policy implications could be obtained from the possible estimation differences between developed countries and developing countries.

Because exploring how to reduce ecological consumption from the consumption side is more related to individual lifestyles and consumption habits [9], the ecological footprint (EF) is employed as the proxy of ecological consumption. Regressions based on panel datasets are used to estimate the impacts of the influencing factors on ecological consumption, which could minimize estimation bias caused by omitted explanatory variables. In comparison with time series and cross-section regressions, panel regressions are more capable of controlling econometric problems such as serial correlation and heterogeneity [10]. Two important economic–social factors, education and income, are employed as the control variables in the panel regressions. To some extent, the unobserved and unmentioned influencing factors of ecological consumption could be controlled for by education and income. By incorporating income into the panel regressions, whether the "environmental Kuznets curve" (EKC) hypothesis is valid or not could also be estimated.

The developed countries and the developing countries selected as the samples are those with a population larger than one million and 10 million in 2018, respectively, which to some extent guarantees that their empirical data are relatively reliable [11] (classification of developed countries and developing countries is based on the M49 Standard made by the United Nations, which can be seen from the web page of https://unstats.un.org/unsd/methodology/m49/, accessed on 5 October 2019). Furthermore, the development patterns of developed countries with a population of less than one million and developing countries with a population of less than 10 million are more likely to be unstable and more prone to distortion, which may make the empirical estimations biased and unrepresentative. After sorting out all of the data obtained from public sources, a panel dataset of 90 sample countries for the time period 1996–2015 was available for empirical estimations. The EF data of the 90 sample countries in 2016 were also obtained to estimate the lagged impacts of the influencing factors on ecological consumption, which could reduce estimation bias caused by endogeneity and serve as robust checks for the estimation results. In 2018, the population of the 90 sample countries accounted for 83.85% of the total population, which demonstrates that the sample is quite representative of the whole world and that the empirical findings could provide general guidance on reducing ecological consumption. Among the 90 sample countries, there are 42 developed countries and 48 developing countries.

The remaining of this paper is organized as follows: Section 2 introduces the EF and depicts global and national ecological consumption. Section 3 discusses why the four latent influencing factors are chosen and conducts a literature review. Section 4 presents the regression variables, data sources, and econometric framework. Regression estimations of the impacts of the influencing factors and control variables on ecological consumption for all 90 sample countries, the 42 developed countries, and the 48 developing countries are conducted successively in Section 5. Finally, a discussion and conclusions are presented in Section 6.

2. Ecological Footprint and Levels of Ecological Consumption

Despite some criticisms of its rationale and methodology, the EF is one of the most popular and inclusive indicators of ecological consumption [12,13]. Some authors argued that the EF is now the most widely used indicator in sustainable development research [14]. The EF measures ecological consumption by calculating the area of biologically productive and mutually exclusive land and water that is required to provide the resources a population demands and to absorb the corresponding wastes in a given year [8]. The EF consists of grazing land footprint (providing animal-based food and other animal products), cropland footprint (providing plant-based food and fiber products), fish product footprint (providing fish-based food products), forest product footprint (providing timber and other forest products), carbon footprint (providing carbon uptake land for absorption of anthropogenic carbon dioxide emissions), and built-up land (representing ecological productivity lost due to occupation of physical space for shelter and other infrastructure) [15].

The EF measures humanity's final demand on a wide range of ecological resources and services from the consumption side ($EF_{consumption} = EF_{production} + EF_{imports} - EF_{exports}$) [16]. Therefore, the land and water to be calculated are not only within national borders but also outside national borders. Because EF values vary greatly with consumption behaviors and habits, it is not difficult to understand global ecological impacts of individual daily lives with the use of the EF [17,18]. The EF is a biophysical rather than monetary accounting approach to measuring ecological consumption. The measuring unit of the EF is global hectares (gha) per capita. A global hectare represents an ecologically productive hectare with global average biological productivity.

Another prominent advantage of the EF is that it has a counterpart, i.e., the biocapacity (BIO), which measures the theoretical maximum capabilities of ecological systems to meet humanity's demands for ecological consumption. The measuring unit of the BIO is also gha per capita. Comparing national EF to globally available BIO provides a quantitative criterion to assess whether national ecological consumption exceeds globally average ecological capacities. For countries, if their EF values are higher than globally available BIO values, they are countries with an ecological deficit; otherwise, they are countries with an ecological surplus.

Data of the EF and BIO were obtained from the National Footprint Account results (2019 Edition) provided by Global Footprint Network (GFN). Figure 1 depicts temporal trends of the global EF and BIO from 1961 to 2016. The globally available BIO values declined continuously from 3.12 gha to 1.63 gha per capita. The global EF values increased from 2.28 gha per capita in 1961 to 2.87 gha per capita in 1973 and fluctuated between 2.54 gha and 2.87 gha per capita for the time period 1973–2016. Following 1970, the global EF values were larger than the globally available BIO values, which demonstrates that humanity's ecological consumption exceeded the regenerative and absorptive capacities of natural capital and that humanity is living in a state of an ecological deficit.

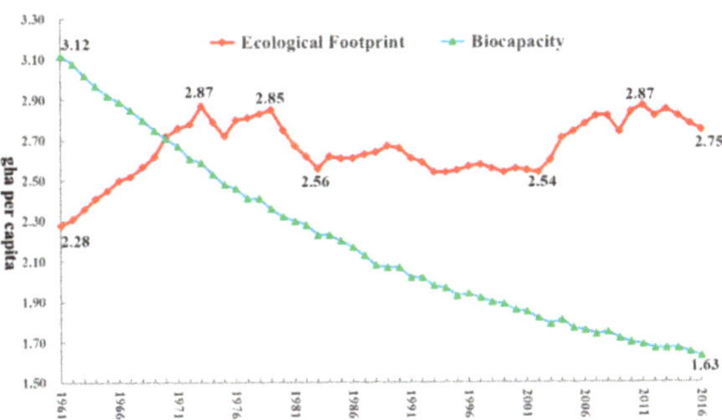

Figure 1. Temporal trends of global ecological footprint (EF) and biocapacity (BIO) (1961–2016). Data source: National Footprint Account results (2019 Edition) from the Global Footprint Network (GFN).

By calculating the ratio of global EF to globally available BIO, how many Earths are needed to support humanity to be ecologically sustainable could be obtained. Figure 2 depicts evolution of the "number of Earths" needed. In 1961, 1970, 1980, 1990, 2000, 2010, and 2016, 0.73, 1.00, 1.19, 1.29, 1.38, 1.67, and 1.69 Earths were needed to support humanity to be ecologically sustainable, respectively. Before 1970, one Earth was sufficient to support humanity to be ecologically sustainable. Since 1970, we needed more than one Earth and, in general, more Earths.

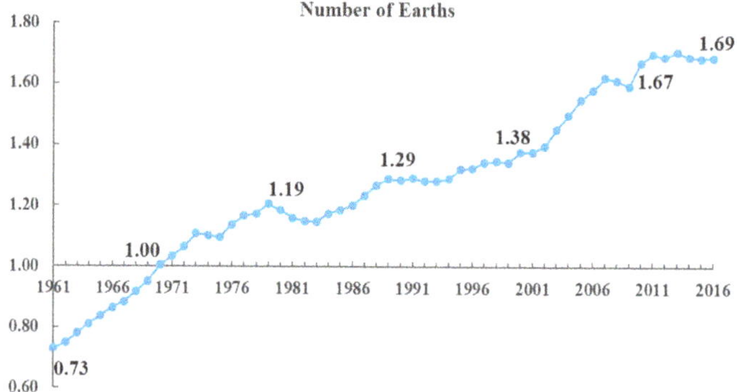

Figure 2. Evolution of "number of Earths" needed (1961–2016). Data source: National Footprint Account results (2019 Edition) from the GFN.

As the most influential countries, the G20 countries are used as examples to illustrate national ecological consumption. Table 1 lists the EF values and "number of Earths" of the G20 countries from 1990 to 2016 ("number of Earths" is the ratio of national EF to globally available BIO, which means that, if the level of ecological consumption of one certain country is universal, this is how many Earths would be needed to support humanity to be ecologically sustainable). For the time period 1990–2016, only the EF values of India were lower than the globally available BIO, and India was the only country with an ecological surplus; China and Indonesia transformed from countries with an ecological surplus into countries with an ecological deficit; the EF values of the United States and Canada were extremely large. In 2000, the EF value of the United States was even higher than 10 gha per capita and 5.52 Earths were needed to support the lifestyle of the United States.

Table 1. Ecological footprint (EF) values and "number of Earths" of G20 countries for the time period 1990–2016.

	1990		2000		2010		2016	
	EF	Earths	EF	Earths	EF	Earths	EF	Earths
Argentina	3.07	1.49	3.13	1.69	3.25	1.91	3.37	2.06
Australia	8.04	3.88	8.06	4.33	8.32	4.89	6.64	4.07
Brazil	2.89	1.40	3.08	1.66	3.00	1.76	2.81	1.73
Canada	8.94	4.32	9.10	4.90	8.34	4.90	7.74	4.75
China	1.53	0.74	1.92	1.03	3.36	1.98	3.62	2.22
France	5.59	2.70	5.54	2.98	5.25	3.09	4.45	2.73
Germany	6.90	3.34	5.51	2.96	5.39	3.17	4.84	2.97
India	0.78	0.38	0.86	0.46	1.07	0.63	1.17	0.72
Indonesia	1.20	0.58	1.35	0.73	1.51	0.89	1.69	1.04
Italy	5.18	2.51	5.60	3.01	5.29	3.11	4.44	2.72
Japan	5.46	2.64	5.29	2.84	4.69	2.76	4.49	2.76
Republic of Korea	3.74	1.81	5.06	2.72	5.88	3.46	6.00	3.68
Mexico	2.50	1.21	2.85	1.53	3.18	1.87	2.60	1.60
Russia	6.90	3.34	4.69	2.52	5.35	3.15	5.16	3.17
Saudi Arabia	2.13	1.03	3.77	2.03	5.66	3.33	6.23	3.83
South Africa	3.36	1.62	3.05	1.64	3.60	2.12	3.15	1.93
Turkey	2.58	1.25	2.92	1.57	3.21	1.89	3.36	2.06
United Kingdom	5.84	2.82	5.73	3.08	5.31	3.12	4.37	2.68
United States	9.87	4.77	10.25	5.52	8.94	5.26	8.10	4.97

Notes: The EF value of Russia of 1990 is not available and the data in the row is of the year 1992.

It is encouraging to find that the EF values of four developed countries (France, Germany, Japan, and the United Kingdom) show downward trends. The EF values of Germany decreased by the largest extent (29.84%). In 2016, the EF value of Germany was about 40% lower than the EF of the United States. The EF values of seven countries (Argentina, China, India, Indonesia, Turkey, Republic of Korea, and Saudi Arabia) show upward trends. The EF values of Republic of Korea, China, and Saudi Arabia increased by 60.63%, 136.72%, and 193.10%, respectively.

3. Influencing Factors and Control Variables

Why the four influencing factors and two control variables are selected is discussed in detail. The related and most recent literature is reviewed. In terms of the relationships between the four influencing factors and ecological consumption, four hypotheses are established.

3.1. Urbanization

As an important economic and social transformation, urbanization is treated as one of foremost influencing factors of ecological consumption. However, the impacts of urbanization on ecological consumption are controversial. On the one hand, because urbanization typically goes hand-in-hand with the industrial process, urbanization would increase ecological consumption due to more consumption of fossil fuels, construction land, cars, electric appliances, and so on. On the other hand, urbanization goes in parallel with a high population density and more ecologically oriented institutions, policies, plans, and technologies, which would permit more efficient use of ecological consumption and, thus, reduce ecological consumption.

By employing the EF as the proxy of ecological consumption and based on panel datasets, Danish and Wang confirmed a positive relationship between urbanization and ecological consumption in 11 emerging countries during 1971–2014 [19], Baloch et al. confirmed a positive relationship in 59 Belt and Road countries for the time period 1990–2016 [20], and Wang and Dong confirmed a positive relationship in 14 sub-Saharan Africa countries for the time period 1990–2014 [21].

By employing the EF as the proxy of ecological consumption, some literature confirmed a negative relationship between urbanization and ecological consumption. Based on time series datasets, Nathaniel et al. confirmed a negative relationship in South Africa for the time period 1965–2014 [22], Ahmed and Wang confirmed a negative relationship in India for the time period 1971–2014 [23], and Dogan et al. confirmed a negative relationship in Nigeria for the time period 1971–2013 [24]. Based on a panel dataset during 1990–2013, Balsalobre-Lorente et al. confirmed a negative relationship in MINT countries (Mexico, Indonesia, Nigeria, and Turkey) [25]. Based on three panel datasets for the time period 1975–2007, Charfeddine and Mrabet found that urbanization would reduce ecological consumption in 15 Middle East and North African (MENA) countries, eight oil-exporting MENA countries and seven non-oil-exporting MENA countries [26].

It is expected that the negative impacts of urbanization on ecological consumption are stronger than the positive impacts and would dominate the relationships between urbanization and ecological consumption. High-density, compact, and modern urban lifestyles are more likely to bring lower levels of ecological consumption. Therefore, the following hypothesis is established:

Hypothesis 1 (H1). *Urbanization would reduce ecological consumption.*

3.2. Renewable Energy Consumption

By employing the EF to measure ecological consumption, the estimated relationship between renewable energy consumption and ecological consumption is consistent. Based on two time series datasets, Dogan et al. found a negative relationship between renewable energy consumption and ecological consumption in Nigeria and Turkey for the time period 1971–2013 [24]. Based on panel datasets, Alola et al. confirmed a negative relationship in 16 European Union (EU) countries for the time period 1997–2014 [27], Shujah-ur-Rahman et al. confirmed a negative relationship in 16 Central and Eastern European Countries for the time period 1991–2014 [28], Wang and Dong confirmed a negative relationship in 14 sub-Saharan African countries for the time period 1990–2014 [21], Balsalobre-Lorente et al. confirmed a negative relationship in the MINT countries for the time period 1990–2013 [25], and Olanipekun et al. confirmed a negative relationship in eleven Central and West African countries for the time period 1996–2015 [29].

Renewable energy consumption would reduce greenhouse gas emissions and other pollutants, which constitute major parts of ecological consumption [30,31]. Renewable energy consumption is considered as a key option for reduction in fossil-fuel consumption [32,33]. Whether the EKC hypothesis is valid or not is determined by the significance of renewable energy consumption [10], and that increasing the role of renewable energy consumption is a fundamental strategy in decreasing environmental pressures. In addition, renewable energy consumption may make individuals conscious of ecologically friendly behaviors and lifestyles. Therefore, the following hypothesis is established:

Hypothesis 2 (H2). *Renewable energy consumption would reduce ecological consumption.*

3.3. Service Industries

Relative to agricultural and industrial industries, service industries are generally less material- and energy-intensive [29,34]. More importantly, service industries could improve technical efficiencies in using ecological consumption [35]. If service industries account for larger proportions of economic output, national ecological consumption is more likely to be reduced. A paradigm shift from material-intensive and energy-intensive industries to service-centered industries is urgently needed to mitigate negative impacts of ecological overshoot and crisis [36]. Therefore, the following hypothesis is established:

Hypothesis 3 (H3). *Service industries would reduce ecological consumption.*

3.4. Internet Usage

Internet changes traditional ways of consumption and production and creates a new economic form, i.e., the internet economy. The internet economy has the potential to reduce ecological consumption in two ways. Firstly, the internet economy is much less material- and energy-intensive than traditional industries; secondly and more importantly, the internet economy could improve the efficiency of every sector of the economy in transforming ecological consumption into economic output [37]. In addition, the internet promotes and facilitates the collaborative economy or the sharing economy, which has the potential to reduce ecological consumption [38,39]. Therefore, the following hypothesis is established:

Hypothesis 4 (H4). *Internet usage would reduce ecological consumption.*

3.5. Control Variables: Education and Income

Because education embodies a lot of information on economic and social progress such as scientific and technological progress and human capital accumulation, it is an important and essential control variable. Education may reduce ecological consumption by stimulating ecological awareness and increasing pro-ecological practices. Higher education levels would enable individuals to have more access to various scientific information and knowledge to understand complicated environmental issues and identify causes and consequences of ecological crisis [29]. Furthermore, higher levels of education would increase individual willingness to live an ecologically sustainable life such as installing more renewable energy equipment and participating more in recycling activities [23].

The negative impacts of education may be insignificant because of the attitude–behavior gap [40]. In practice, higher education levels and enough information and comprehension of the ecological crisis may not be transformed into ecologically friendly lifestyles and consumption habits. Moreover, individuals with higher levels of education tend to have high levels of living standards, which often demand higher levels of ecological consumption. Therefore, it is unclear whether education would reduce ecological consumption or not.

Because the EKC hypothesis is quite well known and the EF is a consumption-side proxy of ecological consumption, income is the most important control variable. The EKC hypothesis implies that there is an inversed U-shaped relationship between ecological consumption and income. At low levels of income, levels of ecological consumption and income tend to increase simultaneously. When income reaches a threshold point, levels of ecological consumption would decrease along with further increases in income levels.

Based on a panel dataset of 16 Central and Eastern European Countries during 1991–2014, Shujah-ur-Rahman et al. validated an N-shaped relationship between income and the EF [28]. Based on a panel dataset of 26 EU countries for the time period 1990–2013 and employing the sub-footprints of the EF as the indicators of ecological consumption, Aydin et al. revealed that the EKC hypothesis is not valid [17]. However, based on a panel dataset of 16 EU countries for the time period 1997–2014, Alola et al. found that a 1% increase in real GDP would reduce total EF by 0.81%, which supports the EKC [27]. The above literature shows that the estimations of the EKC hypothesis for developed countries are inconsistent.

By employing the EF to measure ecological consumption, estimations of the EKC hypothesis for developing countries are also inconclusive. Based on time series datasets, Ahmed and Wang confirmed the EKC in India for the time period 1971–2014 [23], and Dogan et al. confirmed the EKC in each of the MINT countries during 1971–2013 [24]. Based on panel datasets, Balsalobre-Lorente et al. validated the EKC in the MINT countries for the time period 1990–2013 [25], but Wang and Dong indicated that economic growth and ecological consumption were positively related in 14 sub-Saharan Africa countries for the time period 1990–2014 [21]. Based on three panel datasets during 1975–2007, Charfeddine and Mrabet showed that the EKC hypothesis was valid in 15 MENA countries and eight oil-exporting MENA countries, and that the relationship between economic growth and ecological consumption was U-shaped in seven non-oil-exporting MENA countries [26].

4. Regression Variables, Data Sources, and Econometric Framework

To conduct empirical estimations, "urban population (% of total population)" (URB) is employed to measure urbanization, "renewable energy consumption (% of total final energy consumption)" (REN) is employed to measure renewable energy consumption, "services, value added (% of GDP)" (SER) is employed to measure service industries, and "individuals using the internet (% of population)" (INT) is employed to measure internet usage. For the control variables, "mean years of schooling (years)" (MYS) is employed to measure education, and "gross national income per capita, PPP (current international $)" (GNIPC) is employed to measure income. Abbreviations of all of the regression variables are listed in Table 2.

Table 2. List of abbreviations of the regression variables.

Abbreviation	Variable
EF	Ecological footprint
URB	Urban population (% of total population)
REN	Renewable energy consumption (% of total final energy consumption)
SER	Services, value added (% of GDP)
INT	Individuals using the internet (% of population)
MYS	Mean years of schooling
GNIPC	Gross national income per capita

Data of URB, REN, SER, INT, and GNIPC were obtained from the World Bank Indicators. Data of MYS were obtained from the Human Development Reports of the United Nations Development Program. All of the data used were obtained from public and reliable data sources, which could guarantee that the regression estimations are repeatable and testable.

After sorting out all of the data of the seven variables and based on the criteria of selecting the sample countries, we finally obtained three panel datasets, i.e., all sample countries (90), the developed countries

(42), and the developing countries (48), for the time period 1996–2015. Lists of the developed countries and the developing countries are presented in Tables 3 and 4, respectively. Statistical descriptions of the seven variables for all sample countries, the developed countries, and the developing countries are presented in Tables 5–7, respectively. Furthermore, to estimate the lagged impacts of the independent variables on the EF (lagged by one year) and have as many observations as possible, data of the EF in 2016 for all sample countries were obtained.

Table 3. List of 42 developed countries.

Developed Countries		
Albania	France	North Macedonia
Australia	Germany	Norway
Austria	Greece	Poland
Belarus	Hungary	Portugal
Belgium	Ireland	Romania
Bosnia and Herzegovina	Israel	Russian Federation
Bulgaria	Italy	Serbia
Canada	Japan	Slovak Republic
Croatia	Republic of Korea	Slovenia
Cyprus	Latvia	Spain
Czech	Lithuania	Sweden
Denmark	Moldova	Switzerland
Estonia	Netherlands	United Kingdom
Finland	New Zealand	United States

Table 4. List of 48 developing countries.

Developing Countries		
Angola	Ecuador	Niger
Argentina	Egypt	Pakistan
Bangladesh	Ethiopia	Peru
Benin	Ghana	Philippines
Bolivia	Guatemala	Rwanda
Brazil	Guinea	Senegal
Burkina Faso	Haiti	South Africa
Burundi	India	Sri Lanka
Cambodia	Indonesia	Thailand
Cameroon	Kazakhstan	Tunisia
Chad	Madagascar	Turkey
Chile	Malawi	Uganda
China	Malaysia	Venezuela
Colombia	Mali	Vietnam
Cote d'Ivoire	Mexico	Zambia
Dominican Republic	Morocco	Zimbabwe

Table 5. Statistical descriptions of the variables for all sample countries (1996–2015). Obs—observations; Min—minimum; Max—maximum.

Variable	Obs	Mean	SD	Min	Max
EF	1790	3.40	2.22	0.50	10.48
URB	1800	57.13	22.22	7.41	97.88
REN	1796	33.77	29.10	0.61	98.09
SER	1786	53.12	9.86	17.99	77.02
INT	1778	27.40	28.85	0.00	96.81
MYS	1773	8.12	3.40	0.90	14.10
GNIPC	1794	14,562.42	13,688.90	450.00	68,100.00

Table 6. Statistical descriptions of the variables for the developed countries (1996–2015).

Variable	Obs	Mean	SD	Min	Max
EF	830	5.27	1.73	1.09	10.48
URB	840	70.41	13.27	39.47	97.88
REN	840	15.79	12.95	0.61	60.19
SER	838	59.22	7.96	35.70	76.92
INT	826	45.83	29.29	0.00	96.81
MYS	832	10.93	1.49	6.50	14.10
GNIPC	835	24,869.68	13,180.40	2180.00	68,100.00

Table 7. Statistical descriptions of the variables for the developing countries (1996–2015).

Variable	Obs	Mean	SD	Min	Max
EF	960	1.78	0.99	0.50	6.83
URB	960	45.50	21.97	7.41	91.50
REN	956	49.57	30.17	1.15	98.09
SER	948	47.73	8.08	17.99	77.02
INT	952	11.41	16.10	0.00	76.63
MYS	941	5.63	2.58	0.90	11.70
GNIPC	959	5587.91	5123.61	450.00	26,360.00

Based on panel datasets, the ordinary least square (OLS) was employed to conduct the estimations. We employed Equation (1) to estimate the impacts of the influencing factors and control variables on ecological consumption. Equation (2) is the specification of Equation (1).

$$EF = f(URB, REN, SER, INT, MYS, GNIPC). \tag{1}$$

$$EF_{i,t} = \alpha + \beta_1 URB_{i,t} + \beta_2 REN_{i,t} + \beta_3 SER_{i,t} + \beta_4 INT_{i,t} + \beta_5 MYS_{i,t}$$
$$+ \beta_6 Ln(GNIPC)_{i,t} + \beta_7 Ln(GNIPC)^2{}_{i,t} + \varepsilon_{i,t} \tag{2}$$

EF was the dependent variable and URB, REN, SER, INT, MYS, $Ln(GNIPC)$, and $Ln(GNIPC)^2$ were the independent variables. $Ln(GNIPC)$ is the natural log form of $GNIPC$. $Ln(GNIPC)^2$ is the square of $Ln(GNIPC)$. Because marginal impacts of income on ecological consumption are supposed to be diminished, $Ln(GNIPC)$ rather than $GNIPC$ is used in Equation (2). Relative to $GNIPC$, $Ln(GNIPC)$ could minimize the potential estimation bias caused by extreme income values. α represents the intercept term. $\beta_1, \beta_2, \beta_3, \beta_4, \beta_5, \beta_6$, and β_7 represent the slope coefficients of URB, REN, SER, INT, MYS, $Ln(GNIPC)$, and $Ln(GNIPC)^2$, respectively. i represents the sample countries (cross-section), which indicates the country-specific effects. t denotes the time period (years), which indicates the time series effects. $\varepsilon_{i,t}$ is the stochastic error term, which captures the impacts of all unobserved variables on EF.

According to the hypotheses in Section 3, $\beta_1, \beta_2, \beta_3$, and β_4 are expected to be negative. We still could not predict whether β_5 (the coefficient of MYS) is expected to be negative. For ecological consumption and income, there is a monotonically increasing linear relationship if $\beta_6 > 0$ and $\beta_7 = 0$, there is a monotonically decreasing linear relationship if $\beta_6 < 0$ and $\beta_7 = 0$, there is an inversed U-shaped relationship if $\beta_6 > 0$ and $\beta_7 < 0$, which validates the EKC hypothesis, and there is a U-shaped relationship if $\beta_6 < 0$ and $\beta_7 > 0$, which is contrary to the EKC hypothesis. For the inversed U-shaped or U-shaped relationship, it is easy to be calculated that the turning point values of income are exp $(-\beta_6/2\beta_7)$ (the marginal impacts of $Ln(GNIPC)$ on EF are equal to $\frac{dEF}{dLn(GNIPC)} = \beta_6 + 2\beta_7 Ln(GNIPC)$. At the turning points, the marginal impacts are zero, and the corresponding values of $Ln(GNIPC)$ are $-\beta_6/2\beta_7$. Therefore, the corresponding values of $GNIPC$ are exp $(-\beta_6/2\beta_7)$).

Because values of $Ln(GNIPC)$ are above zero, an inversed U-shaped or U-shaped relationship between ecological consumption and income cannot be validated if $\beta_6 = 0$ and $\beta_7 \neq 0$. Furthermore, it could not be concluded that income is not a significant influencing factor of ecological consumption if $\beta_6 = 0$ and $\beta_7 = 0$. Under the circumstances, a liner relationship rather than an inversed U-shaped or

U-shaped relationship between ecological consumption and income should be explored and estimated. Therefore, we needed to revise Equation (2) and another estimation equation was proposed. The new estimation equation is as follows:

$$EF_{i,t} = \alpha + \beta_1 URB_{i,t} + \beta_2 REN_{i,t} + \beta_3 SER_{i,t} + \beta_4 INT_{i,t} + \beta_5 MYS_{i,t}$$
$$+\beta_6 Ln(GNIPC)_{i,t} + \varepsilon_{i,t} \tag{3}$$

In order to explore the lagged impacts of the independent variables on *EF* (lagged by one year), the following two equations were estimated:

$$EF_{i,t} = \alpha + \beta_1 URB_{i,t-1} + \beta_2 REN_{i,t-1} + \beta_3 SER_{i,t-1} + \beta_4 INT_{i,t-1} + \beta_5 MYS_{i,t-1}$$
$$+\beta_6 Ln(GNIPC)_{i,t-1} + \beta_7 Ln(GNIPC)^2_{i,t-1} + \varepsilon_{i,t-1} \tag{4}$$

$$EF_{i,t} = \alpha + \beta_1 URB_{i,t-1} + \beta_2 REN_{i,t-1} + \beta_3 SER_{i,t-1} + \beta_4 INT_{i,t-1} + \beta_5 MYS_{i,t-1}$$
$$+\beta_6 Ln(GNIPC)_{i,t-1} + \varepsilon_{i,t-1} \tag{5}$$

We could present the regression results by mainly estimating Equations (2) and (4). If an inversed U-shaped or U-shaped relationship between ecological consumption and income could not be validated, Equations (3) and (5) were estimated instead to explore the linear relationship. Based on the three panel datasets, this paper follows the subsequent seven regression procedures:

I.   This paper estimates Equation (2) by employing the country random effects model. The estimation is called Model (1);

II.   This paper estimates Equation (2) by employing the country fixed effects model. The estimation is called Model (2);

III.   Between the country random effects model and the country fixed effects model, this paper selects an appropriate model by employing the Hausman test;

IV.   To minimize the potential estimation bias caused by heteroscedasticity, this paper estimates Equation (2) by employing the selected appropriate model and robust standard errors. Robust standard errors are clustered on the country. The estimation is called Model (3);

V.   To explore the lagged impacts of the independent variables on EF (lagged by one year), this paper estimates Equation (4) by employing the selected appropriate model and robust standard errors. Robust standard errors are clustered on the country. The estimation is called Model (4);

VI.   If there is not an inversed U-shaped or U-shaped relationship between ecological consumption and income, this paper estimates Equation (3) by employing the selected appropriate model and robust standard errors. Robust standard errors are clustered on the country. The estimation is called Model (5);

VII.   If there is not an inversed U-shaped or U-shaped relationship between ecological consumption and lagged income (lagged by one year), this paper estimates Equation (5) by employing the selected appropriate model and robust standard errors. Robust standard errors are clustered on the country. The estimation is called Model (6).

5. Regression Estimation Results

Estimation results of the impacts of the influencing factors and control variables on ecological consumption for the three samples are presented. For all sample countries, Models (1)–(6) are presented, and Models (5) and (6) should be used to describe the impacts. For the developed countries and the developing countries, Models (1)–(4) are presented, and Models (3) and (4) should be used to describe the impacts.

5.1. All Sample Countries (90)

Estimation results of the impacts of the influencing factors and control variables on ecological consumption for all sample countries are presented in Table 8. As can be seen from Models (1) and (2), the estimation results based on the country random effects model and the country fixed effects model were different, especially for the impacts of URB and MYS. Therefore, the Hausman test was conducted to select an appropriate model. The result of the Hausman test (the chi-square statistic was significant at the 1% level) shows that the null hypothesis, i.e., the country random effects model is appropriate, was rejected. Therefore, the country fixed effects model was selected.

Table 8. Estimations of impacts of influencing factors and control variables for all sample countries.

	Model (1)	Model (2)	Model (3)	Model (4)	Model (5)	Model (6)
	Coefficient (Prob.)	Coefficient (Prob.)	Coefficient (RSE)	Coefficient (RSE)	Coefficient (RSE)	Coefficient (RSE)
URB	−0.003 (0.53)	−0.033 *** (0.00)	−0.033 *** (0.01)	−0.029 ** (0.01)	−0.034 *** (0.01)	−0.030 ** (0.01)
REN	−0.028 *** (0.00)	−0.031 *** (0.00)	−0.031 *** (0.01)	−0.026 *** (0.01)	−0.027 *** (0.01)	−0.025 *** (0.01)
SER	−0.014 *** (0.00)	−0.018 *** (0.00)	−0.018 *** (0.00)	−0.018 *** (0.01)	−0.019 *** (0.01)	−0.019 *** (0.01)
INT	−0.012 *** (0.00)	−0.005 *** (0.00)	−0.005 * (0.00)	−0.005 * (0.00)	−0.002 (0.00)	−0.004 * (0.00)
MYS	0.075 *** (0.00)	0.013 (0.58)	0.013 (0.05)	0.015 (0.05)	0.001 (0.05)	0.010 (0.05)
$Ln(GNIPC)$	−2.958 *** (0.00)	−0.956 *** (0.01)	−0.956 (0.80)	−0.078 (0.89)	0.592 *** (0.18)	0.591 *** (0.20)
$Ln(GNIPC)^2$	0.215 *** (0.00)	0.093 *** (0.00)	0.093 * (0.05)	0.040 (0.06)		
Constant	13.858 *** (0.00)	8.228 *** (0.00)	8.228 *** (3.16)	4.353 (3.48)	2.031 (1.29)	1.673 (1.40)
Prob > F-statistic		0.00	0.00	0.00	0.00	0.00
Prob > chi^2	0.00					
R-squared	0.68	0.18	0.18	0.15	0.17	0.15
Obs	1728	1728	1728	1729	1728	1729
Groups	90	90	90	90	90	90

Notes: *, **, and *** denote significance at the 10%, 5%, and 1% levels, respectively; for Models (1) and (2), probability values (Prob.) are reported in parentheses; for Models (3–6), robust standard errors (RSE) are reported in parentheses.

In Model (3), $Ln(GNIPC)$ was not statistically significant. In Model (4), neither $Ln(GNIPC)$ nor $Ln(GNIPC)^2$ was significant. According to the arguments in Section 4, an inversed U-shaped or U-shaped relationship between ecological consumption and income could not be statistically validated. Therefore, a linear relationship was explored instead. The scatter plot of EF and GNIPC (Figure 3) further demonstrates that a linear relationship was more appropriate. We ought to interpret the estimated relationships between the independent variables and EF based on the Models (5) and (6).

In Models (5) and (6), the estimated coefficients and significant extents of URB, REN, SER, MYS, and $Ln(GNIPC)$ showed small differences. URB, REN, and SER had statistically significant and negative impacts on EF. $Ln(GNIPC)$ had statistically significant and positive impacts on EF. MYS was statistically insignificant. The fact that INT was statistically significant in Model (6) and insignificant in Model (5) demonstrates that INT only had significant lagged impacts on EF. Relative to URB, REN, and SER, the lagged impacts of INT were weaker. The variance inflation factor (VIF) values of MYS and INT were 4.69 and 3.22, respectively, which demonstrates that the estimations of MYS and INT were not likely affected by the problems with multicollinearity.

To sum up, for all sample countries, urbanization, renewable energy consumption, and service industries were the significant influencing factors of reducing ecological consumption; internet usage only had lagged negative impacts on ecological consumption; education had no significant impacts on ecological consumption; ecological consumption and income were positively related. The relationship was linear rather than inversely U-shaped, and the EKC hypothesis was not supported.

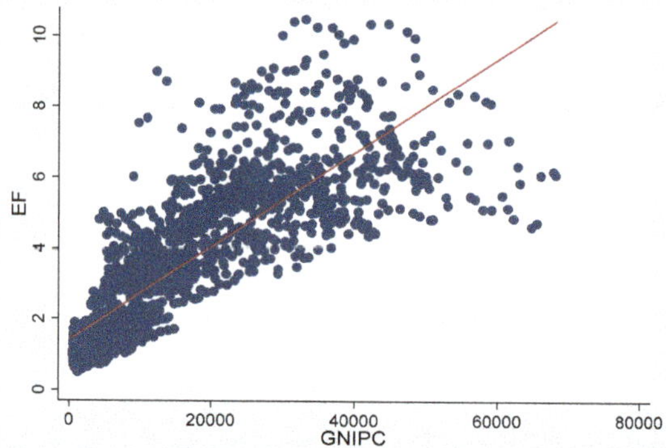

Figure 3. Scatter plot of EF and gross national income per capita (GNIPC) of all sample countries.

5.2. Developed Countries (42)

Estimation results of the impacts of the influencing factors and control variables on ecological consumption for the developed countries are presented in Table 9. As can be seen from Models (1) and (2), the estimation results based on the country random effects model and the country fixed effects model were different, especially for the impacts of URB and INT. Therefore, the Hausman test was conducted to select an appropriate model. The result of the Hausman test shows that the country fixed effects model should be selected.

Table 9. Estimations of impacts of influencing factors and control variables for the developed countries.

	Model (1)	Model (2)	Model (3)	Model (4)
	Coefficient (Prob.)	Coefficient (Prob.)	Coefficient (RSE)	Coefficient (RSE)
URB	0.004	−0.046 ***	−0.046 **	−0.046 **
	(0.64)	(0.00)	(0.02)	(0.02)
REN	−0.051 ***	−0.058 ***	−0.058 ***	−0.046 ***
	(0.00)	(0.00)	(0.01)	(0.01)
SER	−0.060 ***	−0.077 ***	−0.077 ***	−0.069 ***
	(0.00)	(0.00)	(0.01)	(0.02)
INT	−0.005 ***	0.003	0.003	0.003
	(0.01)	(0.18)	(0.00)	(0.00)
MYS	0.122 ***	0.119 ***	0.119	0.091
	(0.00)	(0.00)	(0.07)	(0.08)
Ln(GNIPC)	4.498 ***	6.017 ***	6.017 ***	8.407 ***
	(0.00)	(0.00)	(2.01)	(2.32)
Ln(GNIPC)^2	−0.188 ***	−0.285 ***	−0.285 **	−0.423 ***
	(0.00)	(0.00)	(0.11)	(0.13)
Constant	−17.801 ***	−18.954 ***	−18.954 **	−29.425 ***
	(0.00)	(0.00)	(8.83)	(10.28)
Prob > F-statistic		0.00	0.00	0.00
Prob > chi^2	0.00			
R-squared	0.12	0.32	0.32	0.28
Obs	810	810	810	811
Groups	42	42	42	42

Notes: *, **, and *** denote significance at the 10%, 5%, and 1% levels, respectively; for Models (1) and (2), probability values (Prob.) are reported in parentheses; for Models (3) and (4), robust standard errors (RSE) are reported in parentheses.

In Models (3) and (4), the estimated coefficients and significant extents of URB, REN, SER, INT, MYS, Ln(GNIPC), and Ln(GNIPC)^2 showed small differences. URB, REN, and SER had statistically significant and negative impacts on EF. Ln(GNIPC) and Ln(GNIPC)^2 had significantly positive and negative impacts on EF, respectively, which validated an inversed U-shaped relationship between ecological consumption and income. According to the estimated coefficients of Ln(GNIPC) and Ln(GNIPC)^2 in Model (3), the turning point value of GNIPC of the inversed U-shaped relationship was 39,014. The scatter plot of EF and GNIPC (Figure 4) further verifies the turning point. For the sample, there were about 15% of the observations with GNIPC values higher than the turning point. INT and MYS were statistically insignificant. The VIF values of INT and MYS were 3.09 and 2.02, respectively, which demonstrates that the estimations of INT and MYS were not likely affected by the problems with multicollinearity.

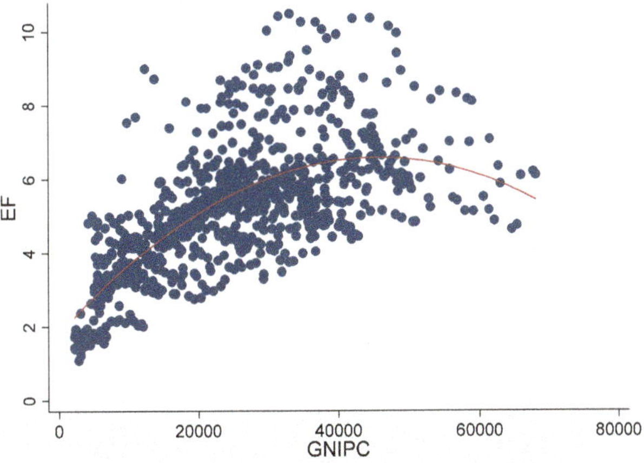

Figure 4. Scatter plot of EF and GNIPC of the developed countries.

To sum up, for the developed countries, urbanization, renewable energy consumption, and service industries were the significant influencing factors of reducing ecological consumption; internet usage and education had no significant impacts on ecological consumption; the relationship between ecological consumption and income was inversely U-shaped, and the EKC was supported.

5.3. Developing Countries (48)

Estimation results of the impacts of the influencing factors and control variables on ecological consumption for the developing countries are presented in Table 10. As can be seen from Models (1) and (2), the estimation results based on the country random effects model and the country fixed effects model showed some differences. The Hausman test was conducted to select an appropriate model. The result of the Hausman Test shows that the country fixed effects model was more appropriate and should be selected.

In Models (3) and (4), the estimated coefficients and significant extents of URB, REN, SER, INT, MYS, Ln(GNIPC), and Ln(GNIPC)^2 showed small differences. REN, SER, and INT had statistically significant and negative impacts on EF. Ln(GNIPC) and Ln(GNIPC)^2 had significantly negative and positive impacts on EF, respectively, which validated a U-shaped relationship between ecological consumption and income. According to the estimated coefficients of Ln(GNIPC) and Ln(GNIPC)^2 in Model (3), the turning point value of GNIPC of the U-shaped relationship was 706. The scatter plot of EF and GNIPC (Figure 5) further certifies the U-shaped relationship and the turning point. For the sample, there were about 95% of the observations with GNIPC values higher than the turning point.

URB and MYS were statistically insignificant. The VIF values of URB and MYS were 3.24 and 2.82, respectively, which demonstrates that the estimations of URB and MYS were not likely affected by the problems with multicollinearity.

Table 10. Estimations of impacts of influencing factors and control variables for the developing countries.

	Model (1)	Model (2)	Model (3)	Model (4)
	Coefficient (Prob.)	Coefficient (Prob.)	Coefficient (RSE)	Coefficient (RSE)
URB	−0.005 *	−0.014 ***	-0.014	−0.014
	(0.08)	(0.00)	(0.01)	(0.01)
REN	−0.008 ***	−0.008 ***	−0.008 **	−0.006 *
	(0.00)	(0.00)	(0.00)	(0.00)
SER	−0.004 **	−0.004 **	−0.004 *	−0.006 **
	(0.02)	(0.02)	(0.00)	(0.00)
INT	−0.005 ***	−0.004 ***	−0.004 *	−0.006 *
	(0.00)	(0.00)	(0.00)	(0.00)
MYS	−0.042 **	−0.061 ***	−0.061	−0.042
	(0.02)	(0.00)	(0.06)	(0.04)
Ln(GNIPC)	−3.090 ***	−2.759 ***	−2.759 ***	−2.941 ***
	(0.00)	(0.00)	(0.91)	(1.13)
Ln(GNIPC)^2	0.228 ***	0.210 ***	0.210 ***	0.225 ***
	(0.00)	(0.00)	(0.07)	(0.09)
Constant	12.724 ***	11.683 ***	11.683 ***	12.137 ***
	(0.00)	(0.00)	(3.31)	(4.01)
Prob > F-statistic		0.00	0.00	0.00
Prob > chi^2	0.00			
R-squared	0.63	0.39	0.39	0.39
Obs	918	918	918	918
Groups	48	48	48	48

Notes: *, **, and *** denote significance at the 10%, 5%, and 1% levels, respectively; for Models (1) and (2), probability values (Prob.) are reported in parentheses; for Models (3) and (4), robust standard errors (RSE) are reported in parentheses.

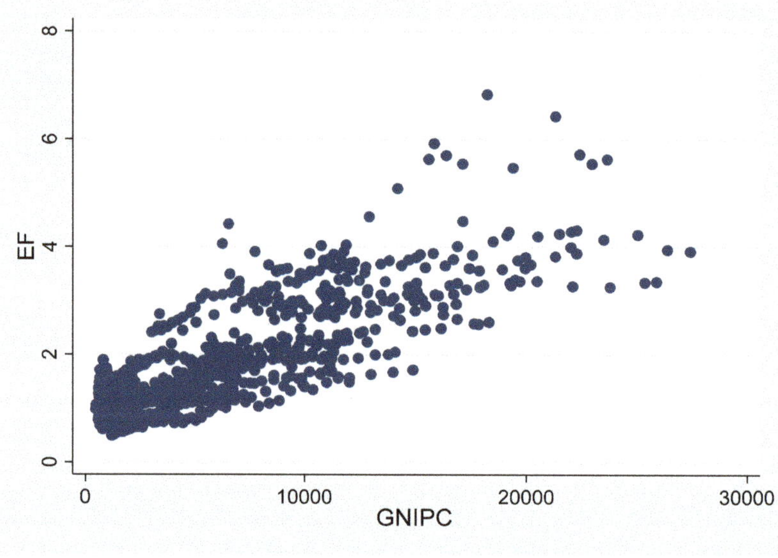

Figure 5. Scatter plot of EF and GNIPC of the developing countries.

To sum up, for the developing countries, renewable energy consumption, service industries, and internet usage were the significant influencing factors of reducing ecological consumption; urbanization and education had no significant impacts on ecological consumption; the relationship between ecological consumption and income was U-shaped, and the EKC hypothesis was not supported.

6. Discussion and Conclusions

Humanity as a whole is living with an ecological deficit, which is widening in general. Reducing ecological consumption is a basic prerequisite and necessary condition of achieving global ecological sustainability. More and more literature explored the influencing factors of ecological consumption from multiple research perspectives. Based on three panel datasets for the time period 1996–2015, this paper adds to the literature by exploring the impacts of urbanization, renewable energy consumption, service industries, and internet usage on ecological consumption and taking education and income as the control variables.

The main contributions of this paper are as follows: (1) this paper employed the EF as the indicator of ecological consumption. The EF tracks ecological consumption in dimensions of both sources and sinks from the consumption side, which enlarges the discussion on ecological sustainability beyond a certain specific domain and is more illuminating for the demand-side policymaking; (2) by selecting representative sample countries covering all of the world, this paper estimated the impacts of the influencing factors on ecological consumption at the global level, which is helpful of summarizing universal laws of reducing ecological consumption; (3) this paper divided all 90 sample countries into 42 developed countries and 48 developing countries and found the different impacts of the influencing factors for the developed countries and the developing countries, especially the different impacts of urbanization and internet usage.

Urbanization would reduce ecological consumption in all sample countries and the developed countries. However, urbanization was not an independent and significant influencing factor of ecological consumption in the developing countries. Based on a panel dataset for the time period 1980–2015, Adams and Acheampong also demonstrated that urbanization had indeterminate and insignificant impacts on carbon emissions in 46 sub-Saharan Africa countries [30]. As discussed in Section 3.1, the negative impacts do not dominate the potential relationships between urbanization and ecological consumption in the developing countries. To further strengthen the negative impacts and weaken the positive impacts of urbanization on ecological consumption, the urban areas should be more ecologically planed (more vertical and compact rather than horizontal and sprawled), and the urban residents ought to have easier and more convenient access to ecologically efficient technologies and public infrastructure such as consumer durables and mass transit. Unplanned, scattered, and disordered urban sprawl in developing countries must be controlled through joint governance at various levels and by different public departments [41].

For all three samples, renewable energy consumption and service industries would reduce ecological consumption. Developing new, reliable, and affordable green and clean technologies to further promote renewable energy consumption and accelerating economic structure transition (less agricultural and industrial industries and more service industries) are useful and effective ways of reducing ecological consumption. By comparing Tables 9 and 10, it could be found that the negative impacts of both renewable energy consumption and service industries in the developed countries were much stronger than those in the developing countries. The much stronger negative impacts could help explain the inversed U-shaped and U-shaped relationships between ecological consumption and income in the developed countries and the developing countries, respectively [10].

Internet usage was an independent and significant force of reducing ecological consumption in the developing countries. Internet not only brings new information, knowledge, technologies, and goods and services markets, but also more ecologically sustainable lifestyles. However, internet usage was not a significant influencing factor of ecological consumption in the developed countries. By employing the same proxy of internet usage and based on a panel dataset, Salahuddin et al.

also found that internet usage could not reduce CO_2 emissions in 31 OECD (Organization for Economic Cooperation and Development) countries for the time period 1991–2012 [42]. Therefore, "internet for ecological sustainability" should be a new guidance principle when developed countries design internet businesses and industries.

Education was not an independent and significant influencing factor of ecological consumption for all three samples. The attitude–behavior gap still exists and prevents ecological awareness from being turned into real and specific ecologically friendly actions. The estimated results remind us that the ecological education we need in the "full world" and Anthropocene does not only have to do with "knowing" but also with "doing". It is hoped that individuals with higher education levels would become the pioneers on living and promoting ecologically sustainable lives.

By employing income as the most important control variable, this paper enriches the long-lasting discussions on the EKC hypothesis and validates three kinds of relationships between ecological consumption and income. For all sample countries, income was a significant factor of increasing ecological consumption, and higher income was not the solution to ecological degradation. For the developing countries, the turning point value of GNIPC of the U-shaped curve was 706, and 95% of the observations were on the right side of the U-shaped curve, which means that, after a very early development stage, ecological consumption began increasing along with further increases in income levels. To our delight, the EKC hypothesis was valid in the developed countries. A positive relationship between ecological consumption and income was significantly reversed when GNIPC values approximately reached 39,014. For the developed countries, because only 15% of the observations were on the right side of the inversed U-shaped curve, more timely and effective measures should be taken to decouple income from ecological consumption and accelerate the arrival of the turning point of the inversed U-shaped curve, which would stimulate developing countries to make corresponding changes and help reduce global ecological consumption by much larger extents.

Much more work needs to be conducted to improve our empirical analysis and further deepen the research context. Because the variables in this paper are defined generally, future research should explore different definitions of the variables, which would give us different perspectives of exploring the influencing factors of ecological consumption. The data quality should also be improved. For example, cultural differences may affect the uniformity of the data in different countries, especially in developed and developing countries. A major drawback of the regression estimations is that we did not find instrumental variables of the influencing factors, especially the instrumental variables of urbanization and internet usage, to conduct sensitivity analyses of the estimation results. Although panel regressions were used to conduct the estimations and lagged impacts were estimated to serve as robust checks for the regression results, the causal relationships between the influencing factors and ecological consumption could not be validated. Additionally, because the regression estimations were on a global level, it was difficult to provide specific measures to reduce ecological consumption of individual countries. Future research can try to combine estimations of global and national levels.

More control variables, for example, political institutional quality and inequality, could be included in the panel regressions in order to refine the estimations. More importantly, the estimated results need further explanation. For example, why the negative impacts of renewable energy consumption and service industries are much stronger in developed countries than in developing countries needs to be explored. The reasons and evidence would provide more specific and scientific policy advice with regard to reducing ecological consumption. Finally, to have an ecologically sustainable future, humanity also has to find solutions to reverse the downward trend of globally available BIO and improve global ecological capacity, which provides more research directions in the field of ecological sustainability.

Author Contributions: Conceptualization, S.Z. and D.Z.; methodology, S.Z.; software, L.L.; validation, S.Z. and J.Z.; formal analysis, S.Z.; investigation, S.Z.; resources, D.Z.; data curation, J.Z.; writing—original draft preparation, S.Z.; writing—review and editing, S.Z.; visualization, J.Z.; supervision, D.Z.; project administration, D.Z.; funding acquisition, S.Z. All authors have read and agreed to the published version of the manuscript.

Funding: This work was supported by the Shanghai "Chen-Guang Project" (18CG20) and Shanghai Summit Discipline in Design (DA19102).

Conflicts of Interest: The authors declare no conflicts of interest.

Appendix A

According to the advice of reviewers, whether the regression results were robust to two different sub-samples (GNIPC values above/below median) was tested. The median value of GNIPC was 9990. The total 1728 observations were divided into 877 observations (GNIPC values above median), which covered 60 countries, and 851 observations (GNIPC values below median), which covered 62 countries. We conducted the tests by employing the same methods and procedures. The regression results of the sub-sample (GNIPC values above median) demonstrated that (1) urbanization, renewable energy consumption, and service industries had significant and negative impacts on ecological consumption, (2) internet usage and education were statistically insignificant, and (3) there was an inversed U-shaped relationship between ecological consumption and income (the EKC hypothesis was supported). As can be seen, the regression results were almost the same as those of the 42 developed countries. The regression results of the sub-sample (GNIPC values below median) demonstrated that (1) renewable energy consumption and service industries had significant and negative impacts on ecological consumption, (2) urbanization, internet usage, and education were statistically insignificant, and (3) there was a U-shaped relationship between ecological consumption and income (the EKC hypothesis was not supported). As can be seen, except for internet usage, the regression results of the variables were almost the same as those of the 48 developing countries. In conclusion, the regression results in this paper were robust to the two sub-samples (GNIPC values above/below median) generally. The details can be seen in Table A1.

Table A1. Estimations of impacts of influencing factors and control variables for the two sub-samples (GNIPC values above/below median).

	Sub-Sample (GNIPC Values above Median)		Sub-Sample (GNIPC Values below Median)	
	Model (3)	Model (4)	Model (3)	Model (4)
	Coefficient (RSE)	Coefficient (RSE)	Coefficient (RSE)	Coefficient (RSE)
URB	−0.036 *	−0.032 *	−0.000	0.001
	(0.02)	(0.02)	(0.01)	(0.01)
REN	−0.063 ***	−0.048 ***	−0.011 ***	−0.009 ***
	(0.01)	(0.01)	(0.00)	(0.00)
SER	−0.053 **	−0.059 ***	−0.004 *	−0.004 **
	(0.02)	(0.02)	(0.00)	(0.00)
INT	0.001	0.003	0.001	−0.000
	(0.00)	(0.00)	(0.00)	(0.00)
MYS	0.065	0.049	-0.037	−0.023
	(0.07)	(0.07)	(0.03)	(0.03)
$Ln(GNIPC)$	10.505 ***	12.745 ***	−1.584 ***	−1.812 ***
	(3.74)	(3.44)	(0.55)	(0.60)
$Ln(GNIPC)^2$	−0.500 ***	−0.638 ***	0.117 ***	0.132 ***
	(0.19)	(0.18)	(0.04)	(0.04)
Constant	−43.766 **	−52.587 ***	7.733 ***	8.345 ***
	(18.09)	(16.70)	(2.07)	(2.27)
Prob > F-statistic	0.00	0.00	0.00	0.00
R-squared	0.29	0.27	0.29	0.31
Obs	877	907	851	822
Groups	60	60	62	61

Notes: *, **, and *** denote significance at the 10%, 5%, and 1% levels, respectively; for Models (3) and (4), robust standard errors (RSE) are reported in parentheses.

References

1. Daly, H.E. Some overlaps between the first and second thirty years of ecological economics. *Ecol. Econ.* **2019**, *164*, 106372. [CrossRef]
2. Polasky, S.; Kling, C.L.; Levin, S.A. Role of economics in analyzing the environment and sustainable development. *Proc. Nati. Acad. Sci. USA* **2019**, *116*, 5233–5238. [CrossRef] [PubMed]
3. Sol, J. Economics in the anthropocene: Species extinction or steady state economics. *Ecol. Econ.* **2019**, *165*, 106392. [CrossRef]
4. Rockström, J.; Steffen, W.; Noone, K.; Persson, Å.; Chapin, F.S., III; Lambin, E.F.; Lenton, T.M.; Scheffer, M.; Folke, C.; Schellnhuber, H.J.; et al. A safe operating space for humanity. *Nature* **2009**, *461*, 472–475. [CrossRef]
5. Steffen, W.; Richardson, K.; Rockström, J. Planetary boundaries: Guiding human development on a changing planet. *Science* **2015**, *347*, 1259855. [CrossRef]
6. Freeman, R. A theory on the future of the rebound effect in a resource-constrained world. *Front. Energy Res.* **2018**, *6*, 81. [CrossRef]
7. Galli, A.; Iha, K.; Halle, M.; Bilali, H.E.; Bottalico, F. Mediterranean countries' food consumption and sourcing patterns: An Ecological Footprint viewpoint. *Sci. Total Environ.* **2017**, *578*, 383–391. [CrossRef]
8. Wackernagel, M.; Hanscom, L.; Lin, D. Making the Sustainable Development Goals consistent with sustainability. *Front. Energy Res.* **2017**, *5*, 18. [CrossRef]
9. Sahin, E.S.; Bayram, I.S.; Koc, M. Demand side management opportunities, framework, and implications for sustainable development in resource-rich countries: Case study Qatar. *J. Clean. Prod.* **2019**, *241*, 118332. [CrossRef]
10. Al-Mulali, U.; Ozturk, I.; Solarin, S.A. Investigating the environmental Kuznets curve hypothesis in seven regions: The role of renewable energy. *Ecol. Indic.* **2016**, *67*, 267–282. [CrossRef]
11. Klugman, J.; Rodríguez, F.; Choi, H.J. The HDI 2010: New controversies, old critiques. *J. Econ. Inequal.* **2011**, *9*, 249–288. [CrossRef]
12. Collins, A.; Galli, A.; Patrizi, N. Learning and teaching sustainability: The contribution of Ecological Footprint calculators. *J. Clean. Prod.* **2018**, *174*, 1000–1010. [CrossRef]
13. Mancini, M.S.; Galli, A.; Coscieme, L. Exploring ecosystem services assessment through Ecological Footprint accounting. *Ecosyst. Serv.* **2018**, *30*, 228–235. [CrossRef]
14. Jóhannesson, S.E.; Davíesdóttir, B.; Heinonen, J.T. Standard Ecological Footprint method for small, highly specialized economies. *Ecol. Econ.* **2018**, *146*, 370–380. [CrossRef]
15. Aşıcı, A.A.; Acar, S. Does income growth relocate ecological footprint? *Ecol. Indic.* **2016**, *61*, 707–714. [CrossRef]
16. Claborn, K.A.; Brooks, J.S. Can we consume less and gain more? Environmental efficiency of well-being at the individual level. *Ecol. Econ.* **2019**, *156*, 110–120. [CrossRef]
17. Aydin, C.; Esen, Ö.; Aydin, R. Is the ecological footprint related to the Kuznets curve a real process or rationalizing the ecological consequences of the affluence? Evidence from PSTR approach. *Ecol. Indic.* **2019**, *98*, 543–555. [CrossRef]
18. O'Neill, D.W.; Fanning, A.L.; Lamb, W.F.; Steinberger, J.K. A good life for all within planetary boundaries. *Nat. Sustain.* **2018**, *1*, 88–95. [CrossRef]
19. Danish; Wang, Z.H. Investigation of the ecological footprint's driving factors: What we learn from the experience of emerging economies. *Sustain. Cities Soc.* **2019**, *49*, 101626. [CrossRef]
20. Baloch, M.A.; Zhang, J.; Iqbal, K.; Iqbal, Z. The effect of financial development on ecological footprint in BRI countries: Evidence from panel data estimation. *Environ. Sci. Pollut. Res.* **2019**, *26*, 6199–6208. [CrossRef]
21. Wang, J.; Dong, K. What drives environmental degradation? Evidence from 14 Sub-Saharan African countries. *Sci. Total Environ.* **2019**, *656*, 165–173. [CrossRef] [PubMed]
22. Nathaniel, S.; Nwodo, O.; Adediran, A.; Sharma, G.; Shah, M.; Adeleye, N. Ecological footprint, urbanization, and energy consumption in South Africa: Including the excluded. *Environ. Sci. Pollut. Res.* **2019**, *26*, 27168–27179. [CrossRef] [PubMed]
23. Ahmed, Z.; Wang, Z. Investigating the impact of human capital on the ecological footprint in India: An empirical analysis. *Environ. Sci. Pollut. Res.* **2019**, *26*, 26782–26796. [CrossRef] [PubMed]
24. Dogan, E.; Taspinar, N.; Gokmenoglu, K.K. Determinants of ecological footprint in MINT countries. *Energy Environ.* **2019**, *30*, 1065–1086. [CrossRef]

25. Balsalobre-Lorente, D.; Gokmenoglu, K.K.; Taspinar, N.; Cantos-Cantos, J.M. An approach to the pollution haven and pollution halo hypotheses in MINT countries. *Environ. Sci. Pollut. Res.* **2019**, *26*, 23010–23026. [CrossRef]

26. Charfeddine, L.; Mrabet, Z. The impact of economic development and social-political factors on ecological footprint: A panel data analysis for 15 MENA countries. *Renew. Sustain. Energy Rev.* **2017**, *76*, 138–154. [CrossRef]

27. Alola, A.A.; Bekun, F.V.; Sarkodie, S.A. Dynamic impact of trade policy, economic growth, fertility rate, renewable and non-renewable energy consumption on ecological footprint in Europe. *Sci. Total Environ.* **2019**, *685*, 702–709. [CrossRef]

28. Shujah-ur-Rahman Chen, S.; Saud, S.; Saleem, N.; Bari, M.W. Nexus between financial development, energy consumption, income level, and ecological footprint in CEE countries: Do human capital and biocapacity matter? *Environ. Sci. Pollut. Res.* **2019**. [CrossRef]

29. Olanipekun, I.O.; Olasehinde-Williams, G.O.; Alao, R.O. Agriculture and environmental degradation in Africa: The role of income. *Sci. Total Environ.* **2019**, *692*, 60–67. [CrossRef]

30. Adams, S.; Acheampong, A.O. Reducing carbon emissions: The role of renewable energy and democracy. *J. Clean. Prod.* **2019**, *240*, 118245. [CrossRef]

31. Bekun, F.V.; Alola, A.A.; Sarkodie, S.A. Toward a sustainable environment: Nexus between CO_2 emissions, resource rent, renewable and nonrenewable energy in 16-EU countries. *Sci. Total Environ.* **2019**, *657*, 1023–1029. [CrossRef] [PubMed]

32. Paramati, S.R.; Apergis, N.; Ummalla, M. Dynamics of renewable energy consumption and economic activities across the agriculture, industry, and service sectors: Evidence in the perspective of sustainable development. *Environ. Sci. Pollut. Res.* **2018**, *25*, 1375–1387. [CrossRef] [PubMed]

33. Wang, Q.; Zhan, L. Assessing the sustainability of renewable energy: An empirical analysis of selected 18 European countries. *Sci. Total Environ.* **2019**, *692*, 529–545. [CrossRef] [PubMed]

34. Fourcroy, C.; Gallouj, F.; Decellas, F. Energy consumption in service industries: Challenging the myth of non-materiality. *Ecol. Econ.* **2012**, *81*, 155–164. [CrossRef]

35. Martínez, C.I.P.; Silveira, S. Analysis of energy use and CO_2 emission in service industries: Evidence from Sweden. *Renew. Sustain. Energy Rev.* **2012**, *16*, 5285–5294. [CrossRef]

36. Sarkodie, S.A.; Adams, S. Renewable energy, nuclear energy, and environmental pollution: Accounting for political institutional quality in South Africa. *Sci. Total Environ.* **2018**, *643*, 1590–1601. [CrossRef]

37. Romm, J. The internet and the new energy economy. *Resour. Conserv. Recycl.* **2002**, *36*, 197–210. [CrossRef]

38. Jian, H.; Xu, M.; Zhou, L. Collaborative collection effort strategies based on the Internet + recycling business model. *J. Clean. Prod.* **2019**, *241*, 118120. [CrossRef]

39. Vita, G.; Lundström, J.R.; Hertwich, E.G.; Quist, J.; Ivanova, D.; Stadler, K.; Wood, R. The environmental impact of green consumption and sufficiency lifestyles scenarios in Europe: Connecting local sustainability visions to global consequences. *Ecol. Econ.* **2019**, *164*, 106322. [CrossRef]

40. Farjam, M.; Nikolaychuk, O.; Bravo, G. Experimental evidence of an environmental attitude-behavior gap in high-cost situations. *Ecol. Econ.* **2019**, *166*, 106434. [CrossRef]

41. Feng, Y.; Wang, X.; Du, W.; Liu, J.; Li, Y. Spatiotemporal characteristics and driving forces of urban sprawl in China during 2003–2017. *J. Clean. Prod.* **2019**, *241*, 118061. [CrossRef]

42. Salahuddin, M.; Alam, K.; Ozturk, I. The effects of Internet usage and economic growth on CO_2 emissions in OECD countries: A panel investigation. *Renew. Sustain. Energy Rev.* **2016**, *62*, 1226–1235. [CrossRef]

© 2020 by the authors. Licensee MDPI, Basel, Switzerland. This article is an open access article distributed under the terms and conditions of the Creative Commons Attribution (CC BY) license (http://creativecommons.org/licenses/by/4.0/).

applied sciences

Article

The Impact of Virtual Water on Sustainable Development in Gansu Province

Weixuan Wang [1], Jan F. Adamowski [2], Chunfang Liu [3,4], Yujia Liu [1], Yongkai Zhang [5], Xueyan Wang [1], Haohai Su [1] and Jianjun Cao [1,*]

[1] College of Geography and Environmental Science, Northwest Normal University, Lanzhou 730070, China; 2019222456@nwnu.edu.cn (W.W.); 201675050111@nwnu.edu.cn (Y.L.); 2017222281@nwnu.edu.cn (X.W.); 2019222458@nwnu.edu.cn (H.S.)

[2] Department of Bioresource Engineering, Faculty of Agricultural and Environmental Sciences, McGill University, Sainte Anne de Bellevue, QC H9X 3V9, Canada; jan.adamowski@mcgill.ca

[3] College of Social Development and Public Administration, Northwest Normal University, Lanzhou 730070, China; liuchunfang@nwnu.edu.cn

[4] Gansu Engineering Research Center of Land Utilization and Comprehension Consolidation, Lanzhou 730070, China

[5] Department of Resources and Environment, Lanzhou University of Finance and Economics, Lanzhou 730070, China; zhangyk@lzufe.edu.cn

* Correspondence: caojj@nwnu.edu.cn; Tel.: +86-0931-7971565

Received: 14 November 2019; Accepted: 10 January 2020; Published: 13 January 2020

Abstract: The concept of virtual water, as a new approach for addressing water shortage and safety issues, can be applied to support sustainable development in water-scarce regions. Using the input-output method, the direct and the complete water use coefficients of industries categorized as primary, secondary, or tertiary, and the spatial flow patterns of the inter-provincial trade in the Gansu province region of China, were explored. The results show that in 2007, 2010, and 2012 the direct and complete water use coefficients of the primary industries were the greatest among the three industry categories, with direct water use coefficients of 1545.58, 882.28, and 762.16, respectively, and complete water use coefficients of 1692.22, 1005.38, and 873.44, respectively; whereas, the direct and complete water use coefficient values of the tertiary industry category were the lowest, with direct water use coefficients of 16.65, 7.74, and 66.89 for 2007, 2010, and 2012, respectively, and complete water use coefficients of 65.46, 66.89, and 72.81 for 2007, 2010, and 2012, respectively. In addition, study results suggest that the volume of virtual water supplied to Gasnu province's local industries has decreased annually, while virtual water exports from the province have increased annually, with the primary industry accounting for 95% of virtual water output. Overall, the virtual water of Gansu province in 2010 showed a net output trend, with a total output of 0.506 billion m^3, while in 2007 and 2012 it showed a net input trend with a total input of 0.104 and 1.235 billion m^3, respectively. Beijing, Shanghai, Guangdong, Ningxia and other water-scarce areas were the main input, or import source for Gansu's virtual water; during the years studied, these provinces imported more than 50 million m^3 individually. Based on these results, it is clear that under the current structure, virtual water is mainly exported to the well-developed coastal areas and their adjacent provinces or other water-abundant regions. Therefore, Gansu province should (1) adjust the industrial structure and develop water-saving and high-tech industries; (2) adjust the current trade pattern to reduce virtual water output while increasing its input to achieve balanced economic development and water resource security.

Keywords: water resources; virtual water trade; input-output method; Gansu province

1. Introduction

The concept of virtual water was first proposed by Tony Allan in 1993 [1] to refer to the amount of water used in the production of goods and services [2]. Since 2002, the concept has received extensive attention around the world [3,4]. In China, virtual water has been forwarded as a potential approach to safeguard water resources, especially in water-scarce areas such as the northwest [5]. This interest has prompted many empirical studies on the quantification of virtual water in global crop and livestock products (2352 kg of water in 1 kg of crop, and 6333 kg of water in 1 kg of livestock product, respectively), in Netherlands's and Canada's grain products (1000–2000 kg of water in 1 kg of grain), in Canada's beef products (16,000 kg of water in 1 kg of beef), in American computer chip products (16 kg of water in 1 g of a 32-megabyte computer chip) [6–10], and in Brazil's, Chile's and American paper products (1052.8 kg, 1227.3 kg, 3345.8 kg of water in 1 kg of paper, respectively) [11]. Virtual water studies have also been completed for services, for example, the assessment of virtual water in the Spanish tourism industry (9.7 kg of water per € 1 of tourism income in 2004) [12]. Virtual water trade, which refers to the practice of transporting this hidden water from one country to another, has also been studied. For example, studies have examined the amount and direction of virtual water in agri-food products transported from Italy to China [13], and the factors influencing its characteristics [14]. When employed appropriately, virtual water trade can be used to alleviate water shortages in water-scarce regions and to ensure local water security through the import of water-rich products from water-abundant regions [15,16]. Researchers have investigated the links between virtual water, food security [17], and water footprints [18,19], which is different compared to carbon footprints [20,21]. The former means the cumulative virtual water of all goods and services consumed by one individual or one country [22], while the latter means a measure of the total amount of greenhouse gas emission directly and indirectly produced by an activity or accumulated over the life stages of a product [23,24]. As there is a mutual feedback mechanism between water and carbon footprint, an understanding of water footprint can contribute to reflecting carbon footprint and proposing policies on the utilization of energy resources, and to achieving sustainable development [25–27].

A review of the existing empirical studies on virtual water shows that most are large-scale in scope, and focused on economically developed areas. There is a need for studies at the provincial scale, particularly for the less developed, arid inland provinces of China. Gansu province is located in the central region of northwest China and covers a total area of 425,900 km^2. It is a typical arid and semi-arid water-deficient area with a dry climate, sparse rainfall and severe water resource shortages. In addition, the unequitable distribution of industrial water use, low utilization efficiency, and over-exploitation of groundwater have resulted in a decrease in river water supply and an increase in desertification, with significant negative consequences for regional social and economic development [28].

The current paper analyzes virtual water content, the primary virtual water export industries and their export destinations, and the spatial patterns of virtual water in Gansu province between 2007 and 2012. The results of this study can be employed to support the sustainable use of water resources and the optimization of industry and trade structures in Gansu province.

2. Statistics and Methods

2.1. Inter-Regional Input-Output Model

In 1936, American economist W. Leontief proposed the input-output method to describe the relationships between inputs and outputs in all sectors of an economic system [29,30]. To inform the rational use of water resources, researches have applied the regional input-output method to the analysis of water consumption [31,32]. By compiling the resulting input-output tables and establishing mathematical models, researchers can identify the amount of virtual water in a system, the current direction of its movement [33], and calculate the ultimate water consumption and environmental emissions involved in product production [8,34].

The inter-regional input-output model (IO model) was first proposed by Isard [35] to reflect product trade between regions in a more systematic and comprehensive way compared to the single regional input-output model. The calculation of virtual water trade volumes in a certain area (Figure 1, specific meaning of each formula is presented in Table A1) is based on the assumption that an inter-regional input-output model contains n regions, and that each region has m sectors, so that the mathematical structure of the inter-regional input-output model contains m × n linear equations [36].

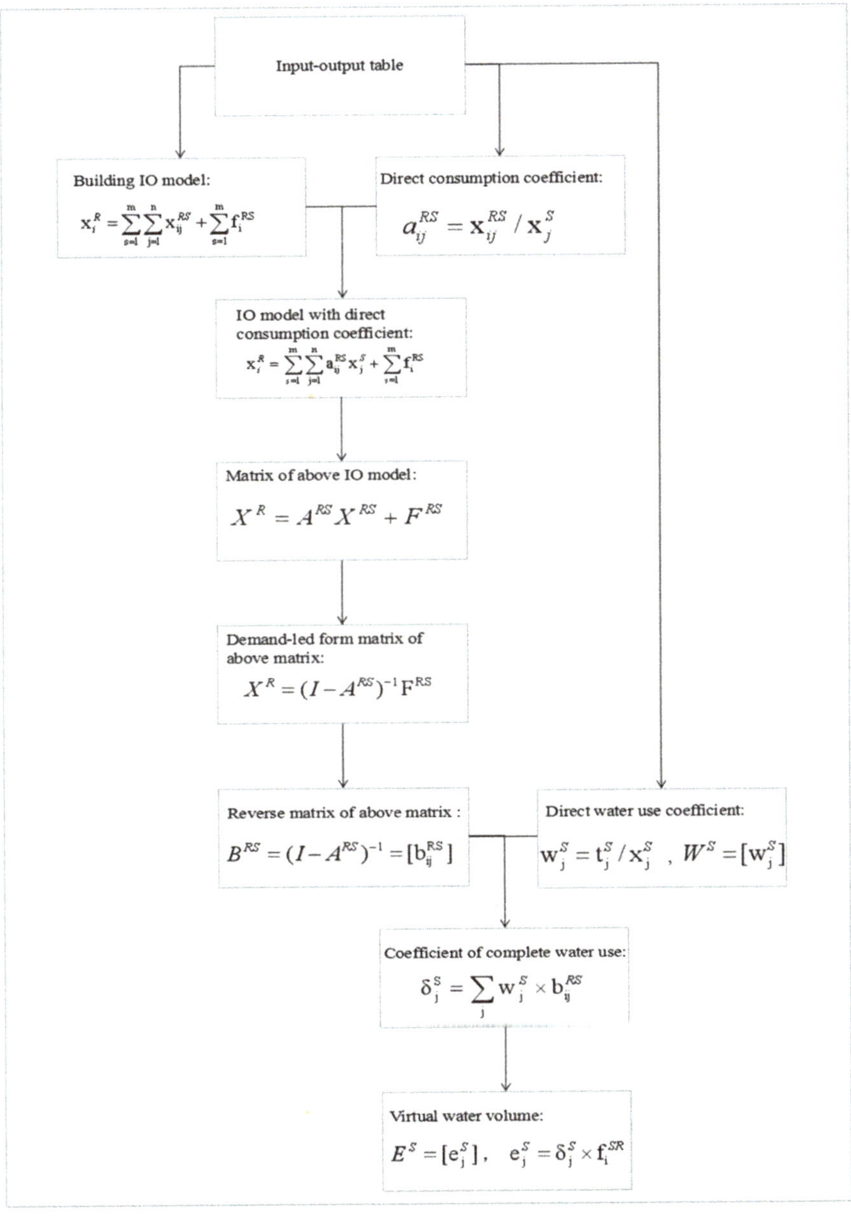

Figure 1. Flow chart of virtual water calculation [36].

2.2. Data Sources and Processing

Due to technical limitations and other difficulties associated with data acquisition, compiling the inter-regional input-output tables is demanding in terms of time and resources. These challenges can contribute to delays and irregularities in input-output table publication schedules. In China, for example, the initial data acquisition process for the regional virtual water input-output assessment was initiated in 1987 and has since been repeated at 5-year intervals. Within this process, input-output tables are launched every 3-years, however, these are usually presented 2–6 years after data acquisition for the cycle. The current study utilizes the input-output tables for 2007, 2010 and 2012.

Water consumption data for all industries in Gansu province were derived from the Gansu Water Resources Bulletin in 2007, 2010, and 2012 [37–39]. Forty-two industry sectors were classified into primary, secondary and tertiary industries according to the Industry Classification Regulations [40]. Primary industries include agriculture, forestry, animal husbandry and fishery; secondary industries consist of those in the manufacturing and construction category, while the remaining industries were classified as tertiary (Table A2 presents industry specific classifications). The complete and direct water use coefficients, with the former referring to the water demand of the entire economic system from the perspective of product life cycles, and the latter referring to water used in the production of intermediate products [41,42], were calculated and the volume of virtual water exported from Gansu to other provinces was inferred.

3. Results and Discussion

3.1. Industrial Virtual Water Consumption in Gansu Province between 2007 and 2012

3.1.1. Water Consumption Coefficients of Various Industries

Across all three industry categories, the complete water use coefficients are higher than the corresponding direct water use coefficients; this is expected as the concept of complete water use includes both direct and indirect water use (Table 1). In comparison to the secondary and tertiary industries, the complete water use coefficient of the primary industry category is much higher, and the difference between the complete water use coefficient and that of its direct water use coefficient is relatively insignificant, because large quantities of water are directly consumed by many of the industries in the primary industry and their production processes requiring few resources from the secondary and tertiary industries [40]. The secondary and tertiary industries complete water use is significantly higher than direct water use due to the high indirect water use of many of the composite industries that require considerable indirect water use [43,44]. The catering and lighting industries, for example, consume large amounts of agricultural and electrical products, respectively, resulting in large amounts of indirect water consumption during production. Consequently, these industries are labeled the "invisible water-consuming industries" [45,46]. In comparison with the tertiary industry class, the direct water use coefficient of the secondary industry category is relatively large. Secondary industries like the steel and electricity industries consume large amounts of direct water for cooling and rinsing [47,48]. These results suggest that improving water use efficiency in primary industries, and decreasing the use of primary industry products in secondary and tertiary industries, be employed to effectively reduce water consumption across all three industry classes [49]. From 2007 to 2012, the complete and direct water use coefficients of both the primary and secondary industries showed a trend of continuous decline. The decline in the complete water use coefficient in the primary industry is mainly related to the decline in the direct water use coefficient, while for the secondary industry the complete water use coefficient was influenced significantly by both direct water use and the use of water-abundant products of the primary industry [40,50]. The requirements of increasing water use efficiency and water recycling in Gansu province by the 11th Five-Year Plan (Outline of the Eleventh year plan for National Economy and Social Development in the People's Republic of China) from 2006 to 2010 with "six necessaries" principles, including maintaining steady and rapid economic development, accelerating transformation of economic growth modalities, improving self-directed innovation capabilities, promoting the coordinated development of urban and rural areas, strengthening the construction of a harmonious society, and deepening reforms openings [51], also contributed to the decline of direct water use in the primary and secondary industries. It is estimated that about 43 provincial-level pilot projects were designed to promote water-saving, through optimized and upgraded agricultural and industrial sectors [52]. These measures had significant results. Primarily, Gansu focused on promoting water-saving agricultural techniques specific to local conditions [53]. These techniques contributed to a marked decrease in the agricultural irrigation quota, an increase in grain output, and a significant reduction in the direct water use coefficient of primary industry [54]. The success of promoting appropriate local techniques should be noted - since the impacts of technical and infrastructure changes often vary from region to region, it is suggested that each city in Gansu establish context specific measures towards the promotion of coordinated economic and ecological development province-wide.

Table 1. Direct and complete water consumption coefficients of primary, secondary, and tertiary industries in Gansu province from 2007 to 2012.

	Water Use Coefficient in 2007		Water Use Coefficient in 2010		Water Use Coefficient in 2012	
	Direct	Complete	Direct	Complete	Direct	Complete
Primary Industry	1545.58	1692.22	882.28	1005.38	762.16	873.44
Secondary Industry	32.69	450.32	20.21	371.90	17.08	309.87
Tertiary Industry	16.65	65.46	7.74	66.89	5.09	72.81

Unit: m^3/million yuan.

During the study period, the complete water use coefficient of the tertiary industry class exhibited an increasing trend, while the direct water use coefficient decreased with time. This increase in the complete water use coefficient can be attributed primarily to infrastructure development in the western region of Gasnu Province in response to the 'develop-the-west' strategy under the 11th Five-Year Plan, in which tertiary industry consumed a large number of water-consuming products [48].

3.1.2. Virtual Water Flow and Water Resource Use among Primary, Secondary, and Tertiary Industries

Between 2007 and 2012, the majority of virtual water in Gansu province flowed towards tertiary industry and the smallest portion flowed towards primary industry (Figure 2). Under the 'develop-the-west' strategy of the 11th Five-Year Plan, Gansu focused on the development of infrastructure, basic industries and the tourism belt along the Silk Road. The "One Belt and One Road" has been one of the most important parts of China's strategy of domestic economic and social development, as well as an important part of China's foreign strategy. The "One Belt" refers to the Silk Road Economic Belt, and the "One Road" refers to the 21st Century Maritime Silk Road [55,56]. During this project, the transportation component of the tourism industry has consumed a large amount of water resources to this day [57].

The total volume of water resources demanded by the region is expressed in terms of gross, or total virtual water; this includes locally produced virtual water and that imported from other regions (Table 2) [58]. Compared to 2007, Gansu's primary industry virtual water use decreased by 16.63% in 2010. This reduction is attributed primarily to the emphasis on "grain for green" (conversion of farmland to forests) in the 'develop-the-west' strategy [59], which not only supports reduced virtual water consumption by primary industry, but also affects consumption in the other industry classes that consume raw materials from primary industry. In 2012, the total amount of virtual water increased significantly; this can be ascribed to the continuous development of the national economy which has brought about significant changes in the income level and consumption structure of Gansu residents, and thus contributes to an increase in water resource consumption [48,60]. In addition, the emergence and advancement of the production and service industries has contributed to increased virtual water use across all three industry categories themselves [61].

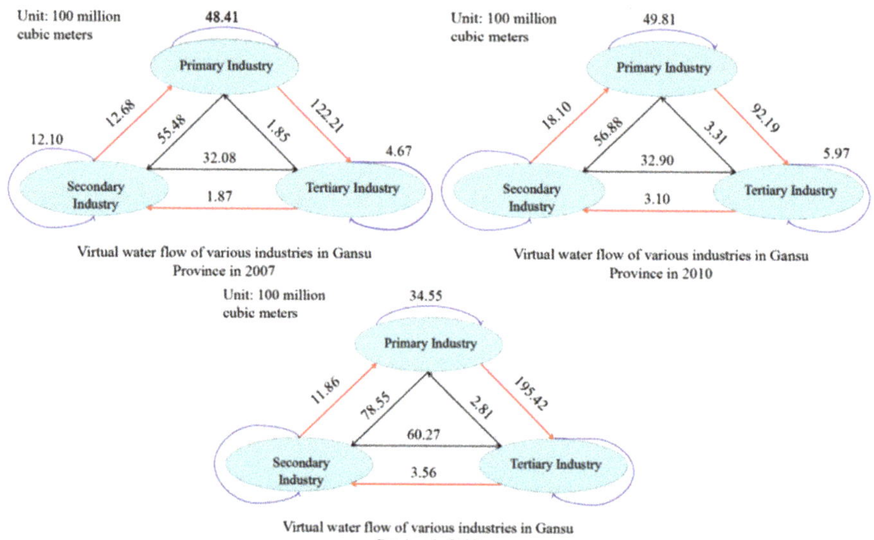

Figure 2. Virtual water flow momentum of primary, secondary, and tertiary industries in Gansu province from 2007 to 2012.

Table 2. Total virtual water consumption of primary, secondary, and tertiary industries in Gansu province from 2007 to 2012.

	Total Virtual Water in 2007		Total Virtual Water in 2010		Total Virtual Water in 2012	
	Local Production	Field Input	Local Production	Field Input	Local Production	Field Input
Primary Industry	210.92	1.11	175.84	1.11	226.64	1.61
Secondary Industry	56.13	12.94	65.05	19.68	80.40	57.40
Tertiary Industry	8.16	2.07	11.70	3.73	18.89	27.65
Total	291.33		277.11		412.59	

Unit: 100 million m^3.

3.1.3. Benefit Analysis of Virtual Water in Gansu Province

The results of the current study show a large gap in virtual water use between the three industry categories in Gansu between 2007 and 2012, but the total combined water use did not change significantly during this period. These findings indicate that, although Gansu province has made advances in various industries, inequitable distribution of water resources among the industries is still an issue. Upon examination of both gross production value and virtual water distribution, it appears that water consumption remains stable for each individual industry class, but that gross production value increases, suggesting an improvement in utilization efficiency. The primary industry class consumed the highest proportion of virtual water but exhibited the lowest production value, while the secondary and tertiary industry classes showed the opposite trend (Figure 3).

The added value of primary industries in Gansu province between 2007 and 2012 (17.063, 19.412 and 20.079 billion yuan, respectively) was lower than the added value of the national primary industry class (27.075, 37.895 and 49.391 billion yuan, respectively) [62–67]. This trend is related to grain output and prices; Gansu province's grain output and grain prices were lower than the national level, further indicating that the virtual water distribution in Gansu province in the primary industry is not appropriate. It is suggested that, to remedy this result, Gansu should adjust the distribution of water according to specific conditions (e.g., spatial distribution of water resources within Gansu province) through trade within the province, encourage the development of water-saving and profitable industries, and reform those industries that are water-consuming and less profitable [68,69].

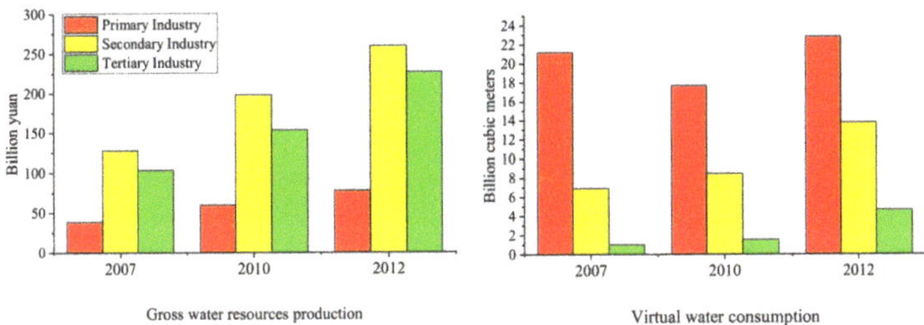

Figure 3. Comparison of the total water production and virtual water consumption in Gansu province from 2007 to 2012.

3.2. Virtual Water Trade in Gansu Province from 2007 to 2012

3.2.1. Spatial Patterns of Virtual Water Flow in and around Gansu Province

Patterns of virtual water supply and demand between Gansu province and other regions are important to consider. The results of the current study show that, in comparison with 2007, the use of locally produced virtual water in Gansu decreased by 2.262 billion m^3 in 2010 (Table 2). According to the Gansu Provincial Water Resources Bulletin in 2007 and 2010, in 2010 the water-saving irrigated area reached 3 million mu (1/15 ha) more than that in 2007. In 2012, the total amount of virtual water contained in products was 41.259 billion m^3, and 32.593 billion m^3 of virtual water was provided for local use, accounting for 79.00% of the total. On the basis of 2007 and 2010 statistics, the use of locally produced virtual water increased by 50.72 and 7.334 billion m^3, respectively in 2012 (Table 2); this is related to the increase in virtual water consumption across all three industry categories. The proportion of virtual water employed locally verses total virtual water decreased annually from 94.8%, 89.52% to 81.43%, while exports have been mounting, from 5.2%, 10.48%, to 18.57% (Table 3). This trend suggests that the increase in total virtual water may be due to a rise in the import of products with high virtual water content, rather than to the decline of virtual water exports. The industries with the highest direct water use are the dominant exporters, such as those in the primary industry category (Table 4). Therefore, Gansu should focus on improving water use efficiency and minimizing the development of industries with high direct water use to alleviate the pressure on water resources.

Table 3. The proportion of local usage and output of virtual water in Gansu province from 2007 to 2012.

	2007	2010	2012
Total amount of virtual water	100%	100%	100%
Local usage ratio (%)	94.80%	89.52%	81.43%
Output ratio (%)	5.2%	10.48%	18.57%

Table 4. Export of virtual water from Gansu and import of virtual water into Gasnu across primary, secondary, and tertiary industries.

	2007		2010		2012	
	Output	Input	Output	Input	Output	Input
Primary Industry	14.62	1.11	28.62	1.11	72.15	1.61
Secondary Industry	0.32	12.94	0.70	19.68	1.67	57.40
Tertiary Industry	0.16	2.09	0.26	3.73	0.49	27.65
Total	15.10	16.14	29.58	24.52	74.31	86.66

Unit: 100 million m^3.

Between 2007 and 2012, virtual water in Gansu mainly flowed towards developed coastal cities (Figure 4) such as Beijing, Shanghai, Guangdong and Tianjin. This is partially due to the encouragement of advances and reforms in emerging and traditional industries, and the implementation of virtual water strategies in these centers [70,71]. Over 80 million m^3 of virtual water flowed to the provinces of Jiangsu, Zhejiang and Hebei in 2007, 2010, and 2012, respectively. Gansu, as a water-scarce province, consistently exported large volumes of virtual water to water-rich areas from 2007 to 2012. Such a trade pattern no doubt increases the pressure on water resources and affects economic development in Gansu [72,73]. The spatial pattern of virtual water trade in Gansu province was developed by ranking the output volume of virtual water flowing from Gansu to different export regions. The export area included China's eastern coast along with the central, western, and northeast regions. Virtual water output volume was divided into three ranges: less than 80 million m^3, 40 million to 80 million m^3, and higher than 40 million m^3.

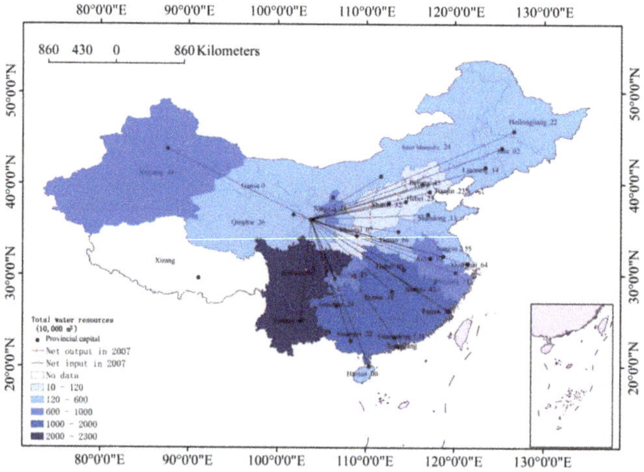

(**a**) Spatial pattern of interprovincial flow of virtual water from Gansu province in 2007

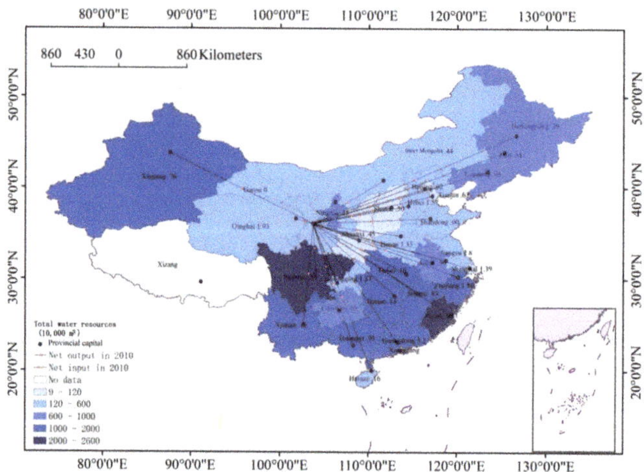

(**b**) Spatial pattern of interprovincial flow of virtual water from Gansu province in 2010

Figure 4. *Cont.*

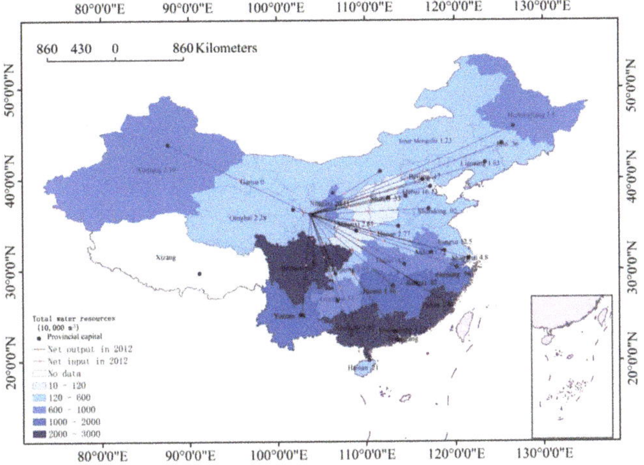

(c) Spatial pattern of interprovincial flow of virtual water from Gansu province from 2007 to 2012

Figure 4. Spatial pattern of interprovincial flow of virtual water from Gansu province from 2007 to 2012.

3.2.2. Industrial Structure of Virtual Water Inter-Provincial Flow Direction from Gansu Province

The inter-provincial trade of virtual water is carried out using industrial and agricultural products as carriers [74]. From 2007 to 2012, the largest virtual water exporter was primary industry, exporting 1.462 billion m^3, 2.862 billion m^3 and 7.215 billion m^3 of virtual water in 2007, 2010, and 2012, respectively, accounting for 96.81%, 96.78%, and 97.13% of the total output (Table 4). As such, the virtual water flow of the primary industry determines the trend of the virtual water current in Gansu province as a whole [75]. Because the agricultural sector dominates in Gansu province, it is the main contributor to virtual water exports. From 2007 to 2012, virtual water from primary industries mainly flowed to Beijing, Shanghai, Guangdong, Tianjin, Jiangsu, Zhejiang and Hebei, and the total output volume was over 80 million m^3, consistently increasing over time. These large export destinations are also the principal virtual water sinks, further supporting the concept that the basic pattern of virtual water output in Gansu province is affected by the trade flow of primary industry products [76]. However, most virtual water produced in Gansu province was consumed there, as agriculture, the primary industry of the province, suffers from low utilization efficiency in terms of irrigation water and rainfall [52,54].

In Gansu province, the secondary industry imported the highest virtual water volume from 2007 to 2012 (Table 4), at 1.294, 1.968 and 5.740 billion m^3, accounting for 80.17%, 80.26% and 66.24% of the total virtual water import, respectively. This virtual water stream flowed from Jiangsu, Zhejiang, and Henan, each of which retained more than 80 million m^3 during the study time period. This trend increased consistently, as these regions, in the midst of industrialization, consumed large amounts of water for industrial processes [77,78].

3.2.3. Variations and Rationality Analysis of Inter-Provincial Virtual Water Trade in Gansu Province

A net import trend was observed for all virtual water trade in Gansu province in 2007 and 2012; however, in 2010 virtual water displayed a net export trend (Table 4). Gansu province received 104 million m^3 of virtual water in 2007 because the import volume in the secondary and tertiary industry was higher than the export volume of the primary industry. The 'develop-the-west' strategy required the mid-eastern regions to provide facilities, resources, and techniques for the west during that period [55]. The volume of virtual water imported from the primary industry in 2010 was similar to that of 2007, yet the import volume from the secondary and the tertiary industries increased by 52.09% and 78.47%, respectively.

The export volume of virtual water for the three industry categories combined was two-times greater in 2010 than in 2007; net output of virtual water in 2010 was 506 million m^3. There are two primary reasons for this; first, although the number of export destinations were reduced, the export volume is extremely large, and second, Gansu increased intermediate inputs and final product exports to other provinces.

Net imports of virtual water were observed in 2012, with a virtual water import volume of 1.235 billion m^3. The output of the three industry classes combined was twice that of 2010, with the import of the second and tertiary industries, respectively, reaching 2.92 and 7.41 times that of their 2010 counterparts. This is likely related to the policy orientation, industry distribution and state of development of Gansu province, along with the important position of Gansu province in the construction of the "One Belt and One Road" [55].

The distribution of water resources in China is extremely unbalanced. Water resources are more plentiful, and the average annual precipitation is much higher, in coastal areas than in inland areas [79]. The main receivers of Gansu's virtual water exports include Zhejiang, Guangdong and other water-abundant areas, as well as Inner Mongolia, Chongqing, Shaanxi; all regions with more water storage than Gansu [72,80]. Conversely, Ningxia, Beijing, Shanghai and other regions with low water reserve capacity have been exporting water resources to Gansu province. These patterns reflect the regional failure to match the inter-provincial flow of virtual water trade with water source storage.

4. Conclusions and Recommendation

The current paper systematically explored virtual water flow among industries in Gansu province as well as the inter-provincial virtual water trade patterns. The results of this research indicate that the dynamic variations in virtual water patterns in Gansu province were mediated primarily by the primary and secondary industries from 2007 to 2012. During this period, virtual water for local use decreased, while virtual water exports to developed and water-abundant coastal areas, increased. In addition, it was found that a significant imbalance in virtual water distribution exists across primary, secondary, and tertiary industries, with the majority of virtual water flowing towards tertiary industries, and the smallest volumes flowing towards primary industries.

In view of the imbalance between virtual water distribution across the three industry categories, and the mismatch of inter-provincial virtual water trade with their water resources storage, it is suggested that Gansu emphasizes the promotion of water-saving agricultural techniques, adjusting the structure of the industrial sector by encouraging water-saving and highly-profitable industries, and reducing virtual water export while broadening their import paths, especially from water-abundant areas such as the import of virtual water from adjacent water-abundant areas (Shanxi province and Inner Mongolia), through the establishment of long-term cooperative relationships using appropriate economic policies, to reduce the proportion of water-reliant productions in all industries.

Author Contributions: Conceptualization, J.C., W.W. and J.F.A.; Methodology, J.C.; Validation, J.C.; Formal Analysis, W.W.; Investigation, W.W., C.L., Y.L., Y.Z., X.W. and H.S.; Data Curation, W.W.; Writing—Original Draft Preparation, W.W.; Writing—Review and Editing, J.C. and J.F.A.; Visualization, W.W.; Funding Acquisition, J.C. All authors have read and agreed to the published version of the manuscript.

Funding: This research was funded by the Major Program of the Natural Science Foundation of Gansu province, China (18JR4RA002), the Key Laboratory of Ecohydrology of Inland River Basins, Chinese Academy of Science (KLERB-ZS-16-01), the Open Fund for the Key Laboratory of Land Surface Processes and Climate Change in the Cold and Arid Region of the Chinese Academy of Sciences (LPCC2018008), and the Innovation Improvement Project for colleges and universities in Gansu Province in 2019 (2019B-033).

Conflicts of Interest: The authors declare no conflict of interest.

Appendix A Calculation of the Direct and Complete Water Use Coefficient and Virtual Water Use

Table A1. Calculation of the direct and complete water use coefficient and virtual water use [35,36].

Formula	Explanation
$x_i^R = \sum_{s=1}^{m} \sum_{j=1}^{n} x_{ij}^{RS} + \sum_{s=1}^{m} f_i^{RS}$	x_i^R is the total output of the R area i department; x_{ij}^{RS} is the intermediate input of the R area i department to the S area j department; f_i^{RS} is the R area i department's input to the final demand of the S area.
$a_{ij}^{RS} = x_{ij}^{RS} / x_j^S$	The direct input coefficient a_{ij}^{RS} indicates the direct input of the i-sector products of the R region when the unit j department produces the unit products.
$x_i^R = \sum_{s=1}^{m} \sum_{j=1}^{n} a_{ij}^{RS} x_j^S + \sum_{s=1}^{m} f_i^{RS}$ $X^R = A^{RS} X^R + F^{RS}$	A variation of $x_i^R = \sum_{s=1}^{m} \sum_{j=1}^{n} x_{ij}^{RS} + \sum_{s=1}^{m} f_i^{RS}$ containing a_{ij}^{RS} and its matrix representation, where X^R, A^{RS} and F^{RS} are respectively the output matrix, the direct input coefficient matrix, and the final demand matrix.
$x^R = \left(I - A^{RS}\right)^{-1} F^{RS}$ $B^{RS} = \left(I - A^{RS}\right)^{-1} = \left[b_{ij}^{RS}\right]$	$\left(I - A^{RS}\right)^{-1}$ is the inverse of the Leontief matrix; its matrix element b_{ij}^{RS} indicates the amount of input to the R-region i department that is needed to meet the final demand of the j-sector in a unit S region. Through the Leontief inverse coefficient matrix, the relationship between demand and output is finally established.
$E^B = \left[e_j^B\right]$ $e_j^B = w_j^B / x_j^B$	In order to further establish the output relationship between water consumption and input, it is necessary to determine the direct water use and complete water use coefficient. Among them, E^B is the direct water use coefficient matrix of B area; e_j^B is the direct water use coefficient of department j of B area; w_j^B is the direct water consumption of department j of B area; x_j^B is the total output of department j of B area.
$\delta = \sum_j e_j^B \times b_{ij}^{BR}$	The complete water use coefficient δ_j^B can be obtained by multiplying the direct water use coefficient by the Leontief inverse coefficient matrix, that is, the water consumption of the product in the B area by adding one unit of the final demand product. Among them, R indicates other areas outside the B area of the study area, and b_{ij}^{BR} indicates the complete water use coefficient of the R area B in other areas.
$T^B = t_j^B; t_j^B = \delta_j^B \times f_i^{BR}$	The virtual water volume is calculated from the complete water use coefficient. T^B is the virtual water matrix of the province B output; t_j^B is the virtual water of the j department, f_i^{BR} is the final use amount of the R area to the B area.

Table A2. Industry Specific Classification.

Primary Industry	**Agriculture, Forestry, Animal Husbandry, Fishery Services.**
Secondary Industry	Coal mining, Petroleum and gas, Metal mining, Nonmetal mining, Food processing and tobaccos, Textile, Clothing, leather, fur, etc. Wood processing and furnishing, Paper making, printing, stationery, etc. Petroleum refining, coking, etc. Chemical industry, Nonmetal products, Metallurgy, Metal products, General and specialist machinery, Transport equipment, Electrical equipment, Electronic equipment, Instrument and meter, Other manufacturing, Electricity and hot water production and supply, Gas and water production and supply, Construction.
Tertiary Industry	Transport and storage, Wholesale and retailing, Hotel and restaurant, Leasing and commercial services, Scientific research, Other services.

References

1. Allan, J.A. Fortunately There Are Substitutes for Water: Otherwise Our Hydropolitical Futures Would be Impossible. In *Priorities for Water Resources Allocation and Management*; ODA: London, UK, 1993; pp. 13–26.
2. Hoekstra, A.Y. *Perspectives on Water: An Integrated Model-Based Exploration of the Future*; International Books: Dublin, Ireland, 1998.
3. Dietzenbacher, E.; Velazquez, E. Analyzing Andalusian Virtual Water Trade in An Input-Output Framework. *Reg. Stud.* **2007**, *41*, 185–196. [CrossRef]
4. Guan, D.; Hubacek, K. Assessment of regional trade and virtual water flows in China. *Ecol. Econ.* **2007**, *61*, 159–170. [CrossRef]
5. Fang, S.; Pei, H. Water Resources Assessment and Regional Virtual Water Potential in the Turpan Basin, China. *Water Resour. Manag.* **2010**, *24*, 3321–3332. [CrossRef]
6. Mekonnen, M.M.; Hoekstra, A.Y. The green, blue and grey water footprint of crops and derived crop products. *Hydrol. Earth Syst. Sc.* **2011**, *15*, 1577–1600. [CrossRef]
7. Mekonnen, M.M.; Hoekstra, A.Y. A Global Assessment of the Water Footprint of Farm Animal Products. *Ecosystems* **2012**, *15*, 401–415. [CrossRef]
8. Chen, W.; Wu, S. Virtual water export and import in china's foreign trade: A quantification using input-output tables of China from 2000 to 2012. *Resour. Conserv. Recycl.* **2018**, *132*, 278–290. [CrossRef]
9. Chapagain, A.K.; Hoekstra, A.Y. Virtual water trade: A quantification of virtual water flows between nations in relation to international trade of livestock and livestock products. In *Value of Water Research Report Series*; UNESCO-IHE: Delft, The Netherlands, 2002; p. 12.
10. Williams, E.D.; Ayres, R.U. The 1.7 kilogram microchip: Energy and material use in the production of semiconductor devices. *Environ. Sci. Technol.* **2002**, *36*, 5504–5510. [CrossRef]
11. Alessandro, M.; Ren, J. Integration of water footprint accounting and costs for optimal chemical pulp supply mix in paper industry. *J. Clean. Prod.* **2014**, *72*, 167–173.
12. Cazcarro, I.; Hoekstra, A.Y. The water footprint of tourism in Spain. *Tour. Manag.* **2014**, *40*, 90–101. [CrossRef]
13. Lamastra, L.; Miglietta, P.P. Virtual water trade of agri-food products: Evidence from italian-chinese relations. *Sci. Total Environ.* **2017**, *599*, 474–482. [CrossRef]
14. Fracasso, A. A gravity model of virtual water trade. *Ecol. Econ.* **2014**, *108*, 215–228. [CrossRef]
15. Hoekstra, A.Y. Virtual water trade: A quantification of virtual water flows between nations in relation to international crop trade. In *Value of Water Research Report Series*; UNESCO-IHE: Delft, The Netherlands, 2002; p. 11.
16. Hoekstra, A.Y. The water footprint of India. In *Value of Water Research Report Series*; UNESCO-IHE: Delft, The Netherlands, 2008; p. 32.
17. Dalin, C.; Qiu, H. Balancing water resource conservation and food security in China. *Proc. Natl. Acad. Sci. USA* **2015**, *112*, 4588–4593. [CrossRef] [PubMed]
18. Hoekstra, A.Y.; Chapagain, A.K. Water footprints of nations: Water use by people as a function of their consumption pattern. *Water Resour.* **2007**, *21*, 35–48.
19. Cao, X.; Ren, J. Assessing agricultural water use effect of China based on water footprint framework. *Trans. CSAE* **2018**, *34*, 1–8.
20. Vasilis, K.; Anastasia, P. Allocating the cost of the CF produced along a supply chain, among the stakeholders involved. *J. Water Clim. Chang.* **2014**, *5*, 556–568.
21. Vasilis, K. Three alternative ways to allocate the CF produced in a water supply and distribution system. *Desalin. Water Treat.* **2015**, *54*, 2212–2222.
22. Hoekstra, A.Y. Virtual water: An introduction. Virtual water trade. In *Values of Water Research Report Series*; UNESCO-IHE: Delft, The Netherlands, 2003; p. 12.
23. Wiedmann, T.; Minx, J. A definition of carbon footprint. *Ecol. Econ. Res. Trends* **2008**, *1*, 1–11.
24. Wiedmann, T.; Minx, J. Allocating ecological footprints to final consumption categories with input-output analysis. *Ecol. Econ.* **2006**, *56*, 28–48. [CrossRef]
25. Eloise, M.B.; Eleanor, B. Sustainable development and the water-energy-food nexus: A perspective on livelihoods. *Environ. Sci. Policy* **2015**, *54*, 389–397.
26. Vasilis, K.; Stavroula, T. Integrating the Carbon and Water Footprints costs in the WFD Full Water Cost Recovery Concept: Basic principles towards their socially just allocation. *Water* **2012**, *4*, 45–62.

27. IWA. Task Force on Water Utility Efficiency in Low and Middle Income Countries (LAMIC TF). In *Increasing Energy Efficiency of Water Supply-How to Introduce Carbon Credits to the Water Sector?* Concept Note; IWA: London, UK, 2009.

28. Chao, C.; Sajjad, A. A dynamic model for exploring water-resource management scenarios in an inland arid area: Shanshan County, Northwestern China. *J. Mt. Sci. Engl.* **2017**, *14*, 1039–1057.

29. Leontief, W.W. Quantitative Input and Output Relations in the Economic Systems of the United States. *Rev. Econ. Stat.* **1936**, *18*, 105–125. [CrossRef]

30. Szyrmer, J.M. Input-output coefficients and multipliers from a total-flow perspective. *Environ. Plan A* **1992**, *24*, 921–937. [CrossRef]

31. Shao, L.; Guan, D. Multi-scale input-output analysis of consumption-based water resources: Method and application. *J. Clean. Prod.* **2017**, *164*, 338–346. [CrossRef]

32. Chen, W.; Wu, S. China's water footprint by province, and inter-provincial transfer of virtual water. *Ecol. Indic.* **2017**, *74*, 321–333. [CrossRef]

33. Boudhar, A.; Boudhar, S. An input-output framework for analysing relationships between economic sectors and water use and intersectoral water relationships in Morocco. *J. Econ. Struct.* **2017**, *6*, 1–25. [CrossRef]

34. Li, H.C. Situation Analysis of China's International Trade of Virtual Water. In *IOP Conference Series: Earth and Environmental Science*; IOP Publishing: Tokyo, Japan, 2018; Volume 199, pp. 32–49.

35. Isard, W. Interregional and Regional Input-Output Analysis: A Model of a Space-Economy. *Rev. Econ. Stat.* **1951**, *33*, 318–328. [CrossRef]

36. Zhao, X.; Yang, H. Applying the input-output method to account for water footprint and virtual water trade in the Haihe River basin in China. *Environ. Sci. Technol.* **2010**, *44*, 9150–9156. [CrossRef]

37. Department of National Accounts, National Bureau of Statistics. *2007 China Regional Input-Output Table*; Department of National Accounts, National Bureau of Statistics: Beijing, China, 2012.

38. Department of National Accounts, National Bureau of Statistics. *2010 China Regional Input-Output Table*; Department of National Accounts, National Bureau of Statistics: Beijing, China, 2015.

39. Department of National Accounts, National Bureau of Statistics. *2012 China Regional Input-Output Table*; Department of National Accounts, National Bureau of Statistics: Beijing, China, 2018.

40. Dong, H.; Geng, Y. Regional water footprint evaluation in China: A case of Liaoning. *Sci. Total Environ.* **2013**, *442*, 215–224. [CrossRef]

41. Feng, K.; Siu, L. Assessing regional virtual water flows and water footprints in the Yellow River Basin, China: A consumption based approach. *Appl. Geogr.* **2011**, *32*, 691–701. [CrossRef]

42. Mao, X.; Yang, Z. Ecological network analysis for virtual water trade system: A case study for the Baiyangdian Basin in Northern China. *Ecol. Inf.* **2012**, *10*, 17–24. [CrossRef]

43. Kleines, J.J.; Varbanov, P.S. Water footprint, water recycling and food-industry supply chains. In *Handbook of Waste Management and Co-Product Recovery in Food Processing*; Elsevier: Amsterdam, The Netherlands, 2009; pp. 134–168.

44. Miguel, Á.D.; Hoekstra, A.Y. Sustainability of the water footprint of the Spanish pork industry. *Ecol. Indic.* **2015**, *57*, 465–474. [CrossRef]

45. Ali, B.; Kumar, A. Development of life cycle water-demand coefficients for coal-based power generation technologies. *Energy Convers. Manag.* **2015**, *90*, 247–260. [CrossRef]

46. Wang, C.; Li, Y. Investigation of water-energy-emission nexus of air pollution control of the coal-fired power industry: A case study of Beijing-Tianjin-Hebei region, China. *Energy Policy* **2018**, *115*, 291–301. [CrossRef]

47. Denooyer, T.A.; Peschel, J.M. Integrating water resources and power generation: The energy–water nexus in Illinois. *Appl. Energy* **2016**, *162*, 363–371. [CrossRef]

48. Macknick, J.; Newmark, R. Operational water consumption and withdrawal factors for electricity generating technologies: A review of existing literature. *Environ. Res. Lett.* **2012**, *7*, 045802. [CrossRef]

49. Wang, Z.; Huang, K. An input–output approach to evaluate the water footprint and virtual water trade of Beijing, China. *J. Clean. Prod.* **2013**, *42*, 172–179. [CrossRef]

50. Hilda, R. Erratum to: Water Policy in Mexico: Economic, Institutional and Environmental Considerations. In *Water Policy in Mexico*; Springer: Berlin, Germany, 2019; pp. 89–113.

51. Notice of the State Council on the Implementation of the Main Objectives and Tasks of the Eleventh Five-Year Plan for the National Economic and Social Development of the People's Republic of China. Available online: http://www.gov.cn/zhengce/content/2008-03/28/content_1937.htm (accessed on 24 August 2006).

52. Akiyama, T.; Li, J. Assessment of water-saving society establishment program in China: A case study of Zhangye City, Gansu Province, China. In *EGU General Assembly Conference*; EGU Publishing: Vienna, Austria, 2014.

53. Fan, Y.; Wang, C. Comparative evaluation of crop water use efficiency, economic analysis and net household profit simulation in arid Northwest China. *Agric. Water Manag.* **2014**, *146*, 335–345. [CrossRef]

54. Chang, G.; Wang, L. Farmers' attitudes toward mandatory water-saving policies: A case study in two basins in northwest China. *J. Environ. Manag.* **2016**, *181*, 455–464. [CrossRef]

55. Wang, Y.; Shen, R. Conflicts Affecting Sustainable Development in West China Since the Start of China's Western Development Policy. *J. Resour. Ecol.* **2012**, *3*, 202–208.

56. Xia, L. The Development of the "One Belt and One Road" and Its Impacts on China-U.S. *Relat. Peace* **2015**, *2*, 17–20.

57. Dong, S.; Li, Z. Resources, Environment and Economic Patterns and Sustainable Development Modes of the Silk Road Economic Belt. *J. Resour. Ecol.* **2015**, *6*, 65–72.

58. Ahn, J.H.; Lee, J.G. Evaluation of Virtual Water Calculation Method in Korea. *J. Korea Water Resour. Assoc.* **2010**, *43*, 583–595. [CrossRef]

59. Zeng, Y.; Jin, W. Analysis and evaluation of cultivated land decrease in eastern part of Qinghai Plateau. *Trans. Soc. Agric. Eng.* **2013**, *29*, 214–222.

60. Bekchanov, M.; Bhaduri, A. The role of virtual water for sustainable economic restructuring: Evidence from Uzbekistan, Central Asia. Central Asia. *ZEF Discuss. Pap. Dev. Policy* **2012**, *167*, 34.

61. Li, W.; Shi, J. Grey Relational Analysis on the Effect of Income Growth of Rural Residents on Consumption Structure in Gansu Province. *Heilongjiang Agric. Sci.* **2015**, *7*, 134–138.

62. Gansu Provincial Bureau of Statistics. *2007 Gansu Development Yearbook*; Gansu Provincial Bureau of Statistics: Lanzhou, China, 2008.

63. Gansu Provincial Bureau of Statistics. *2010 Gansu Development Yearbook*; Gansu Provincial Bureau of Statistics: Lanzhou, China, 2010.

64. Gansu Provincial Bureau of Statistics. *2012 Gansu Development Yearbook*; Gansu Provincial Bureau of Statistics: Lanzhou, China, 2012.

65. National Bureau of Statistics. *2007 China Statistical Yearbook*; National Bureau of Statistics: Beijing, China, 2008.

66. National Bureau of Statistics. *2010 China Statistical Yearbook*; National Bureau of Statistics: Beijing, China, 2011.

67. National Bureau of Statistics. *2012 China Statistical Yearbook*; National Bureau of Statistics: Beijing, China, 2013.

68. Davijani, M.H.; Banihabib, M.E. Multi-Objective Optimization Model for the Allocation of Water Resources in Arid Regions Based on the Maximization of Socioeconomic Efficiency. *Water Resour. Manag.* **2016**, *30*, 927–946. [CrossRef]

69. Davijani, M.H.; Banihabib, M.E. Optimization model for the allocation of water resources based on the maximization of employment in the agriculture and industry sectors. *J. Hydrol.* **2016**, *533*, 430–438. [CrossRef]

70. Zhang, Z.; Shi, M. An Input—Output Analysis of Trends in Virtual Water Trade and The Impact on Water Resources and Uses in China. *Econ. Syst. Res.* **2011**, *23*, 431–446. [CrossRef]

71. Ye, Q.; Li, Y. Optimal allocation of physical water resources integrated with virtual water trade in water scarce regions: A case study for Beijing, China. *Water Res.* **2018**, *129*, 264–276. [CrossRef]

72. Zhang, C.; Anadon, L.D. A multi-regional input–output analysis of domestic virtual water trade and provincial water footprint in China. *Ecol. Econ.* **2014**, *100*, 159–172. [CrossRef]

73. Liu, J.; Sun, S. Inter-county virtual water flows of the Hetao irrigation district, China: A new perspective for water scarcity. *J. Arid Environ.* **2015**, *119*, 31–40. [CrossRef]

74. Antonelli, M.; Tamea, S. Intra-EU agricultural trade, virtual water flows and policy implications. *Sci. Total Environ.* **2017**, *587*, 439–448. [CrossRef] [PubMed]

75. Hassan, A.; Saari, M.Y. Virtual water trade in industrial products: Evidence from Malaysia. *Environ Dev Sustain.* **2017**, *19*, 877–894. [CrossRef]

76. Zhao, D.; Tang, Y. Water footprint of Jing-Jin-Ji urban agglomeration in China. *J. Clean. Prod.* **2017**, *167*, 919–928. [CrossRef]

77. Yue, Q.; Hou, L. Optimization of industrial structure based on water environmental carrying capacity in Tieling City. *Water Sci. Technol.* **2015**, *71*, 1255. [CrossRef] [PubMed]

78. Chen, L.; Xu, L. Optimization of urban industrial structure under the low-carbon goal and the water constraints: A case in Dalian, China. *J. Clean. Prod.* **2016**, *114*, 323–333. [CrossRef]

79. Li, P.; Qian, H. Water resources research to support a sustainable China. *Int. J. Water Resour. Dev.* **2018**, *34*, 327–336. [CrossRef]
80. Jiang, Y.; Cai, W. An index decomposition analysis of China's interregional embodied carbon flows. *J. Clean. Prod.* **2015**, *88*, 289–296. [CrossRef]

© 2020 by the authors. Licensee MDPI, Basel, Switzerland. This article is an open access article distributed under the terms and conditions of the Creative Commons Attribution (CC BY) license (http://creativecommons.org/licenses/by/4.0/).

Article

Optimal Manufacturing-Reconditioning Decisions in a Reverse Logistic System under Periodic Mandatory Carbon Regulation

Sadok Turki *, Soulayma Sahraoui, Christophe Sauvey and Nathalie Sauer

Laboratory of Informatics Engineering and Production of Metz (LGIPM), University of Lorraine, LGIPM, UFR MIM, 3 Rue Augustin Fresnel, F-57070 Metz, France; soulayma.sahraoui9@etu.univ-lorraine.fr (S.S.); christophe.sauvey@univ-lorraine.fr (C.S.); nathalie.sauer@univ-lorraine.fr (N.S.)

* Correspondence: sadok.turki@univ-lorraine.fr

Received: 15 April 2020; Accepted: 18 May 2020; Published: 20 May 2020

Abstract: Due to environmental concerns, firms are under increasing pressure to comply with legislations and to take up environmental strategies. This leads researchers and firms to develop new sustainable supply chains, where a new area has emerged for a manufacturing and reconditioning system. The originality of this work consists in simultaneously considering carbon emissions strategies, carbon tax and mandatory emission in a manufacturing-reconditioning system. The proposed system is composed of two parallel machines, a manufacturing stock, a reconditioning stock and a recovery inventory. In order to make the proposed green manufacturing system more realistic, it is assumed that manufactured (new products) and reconditioned products are distinguishable. The quantity of worn products (used products) depends on the sales in the previous periods, and the repair periods of the machines are stochastic and independent. The aim of this work is to determine the optimal capacities of manufacturing and reconditioning stocks that maximize the total profit, as well as the optimal value of worn products under two carbon emissions' limitations. An evolutionary algorithm is developed, along with an efficient improvement method, to find the optimal value of decision variables. Ultimately, numerical results are provided to show the impact of the period of carbon limit and the worn products (returned products) on decision variables.

Keywords: production planning; carbon regulation; reconditioning; green logistics; optimization

1. Introduction

Throughout the past few decades, massive carbon emissions have caused serious global environmental damage, such as thick haze and worsening greenhouse gas effects. The Intergovernmental Panel on Climate Change (IPCC), which is the international body for the assessment of climate change, has pointed out in its fifth assessment report that it is necessary to curb the global greenhouse gas (GHG) emissions by 40%–70% from the 2010 level before 2050, and to curb the global GHG emissions to the level of near zero by the end of the 21st century [1]. The IPCC has defined a complete method to standardize the computation of GHG emissions at the national level. Today, most countries are monitoring GHG emissions by using IPCC guidelines to perform annual inventories assessing the quantity of six main gases (CO_2: carbon dioxide, CH_4: methane, N_2O: nitrous oxide, HFCs: hydrofluorocarbons, PFCs: perfluorocarbon, SF_6: sulfur hexafluoride) [2]. It was found that CO_2 makes up the broad majority of GHG emissions—it represents about three-quarters of total GHG emissions. For instance, in the USA, the emissions from the industry occupies 24% of total emissions. Indeed, GHG emissions from industry primarily come from burning fossil fuels for energy, as well as GHG emissions from certain chemical reactions necessary to produce goods from raw materials [3,4]. To achieve the IPCC targets, the development of climate change mitigation technologies have played a pivotal role [5]. Therefore, the industrial companies have adopted new technologies and strategies to alleviate climate change impact associated with their activities. For example, to reduce the carbon emissions, several companies are devoted to recuperating and remanufacturing worn products instead of producing new ones. Indeed, carbon emissions from producing new parts is higher than that for remanufactured ones.

Nowadays, carbon emission control as well as maintaining sustainable economic development has become an increasing challenge, and many countries have attempted to curb carbon emission compulsively through enacting legislation [6]. Several governments promulgated some carbon emission laws, such as carbon taxes, mandatory carbon emissions capacity, and carbon emission cap and trade [7]. The most known regulations are carbon pricing and emissions trading [8]. Those policies can be classified into two categories: mandatory and non-mandatory. The mandatory policy are strict regulations where governments set a strict emissions cap on a firm for a given period. Non-mandatory policies allow companies to choose between reducing and paying for emissions [9], including carbon tax and emission trade (also known as cap and trade). According to the European Commission, European Union Emission Trading System (EU-ETS) is the first and largest emission trading scheme: 75% of international carbon trading covers more than 11,000 companies in 31 countries. Under this policy, companies are allocated a cap or quota on carbon emissions [10]. Indeed, when a company exceeds the allocated cap, it can purchase extra carbon allowance trough the carbon market trading. Conversely, when its carbon emission is under the cap, it can sell its surplus via trading [11]. Besides, carbon tax is a tax on carbon emissions. Companies under this policy are charged a tax proportional to the amount of carbon emitted [12]. Since the implementation of the Kyoto Protocol in 1997, several countries have enacted a variety of carbon tax schemes that have attracted wide attention in the manufacturing management [13]. According to the International Monetary Fund (IMF), carbon tax is the most effective policy to mitigate carbon emissions. With these environmental regulations, companies are constrained to adopt new production policies in order to curb their carbon emissions. From the beginning of the nineties until today, several published papers in the literature have dealt with production decision, taking into account carbon emissions regulations. Ingham and Ulph [14] have presented the case for using a carbon tax to control carbon emissions, and have illustrated the implications for the UK manufacturing. In addition, they have computed a total abatement cost curve corresponding to different levels of emissions' curbing to see whether there are critical points at which costs rise quickly. Fang et al., [15] have developed a mathematical programming model of a flow shop-scheduling problem that considers peak power load, energy consumption, and associated carbon emission in addition to cycle time. The objective is to determine the optimal scheduling that increases the energy consumption and the carbon emissions. Turki and Rezg [16] have proposed an optimal design for a manufacturing/remanufacturing system that sorts worn products into three quality levels. The authors have determined optimal production decisions regarding new and remanufactured products while considering carbon tax policy. As in a real case, the authors have considered that the carbon emissions from producing remanufactured products is different from those of new ones. Indeed, the benefit of the remanufacturing is that the worn products are recovered and reused. In addition, the carbon emission from producing remanufactured products is lower than that for producing new ones. More recently, Dou et al., [17] have considered a manufacturer who produces new parts in the first period and makes new and remanufactured parts in the second period under carbon tax policy, where the tax price differs over the two periods. The authors have determined optimal manufacturing and remanufactured plans that reduce the carbon emissions when the carbon tax varies. He et al., [18] have considered a supply chain network constrained by a stringent mandatory carbon cap, the purpose of which is to examine how stringent carbon regulations and operational decision modes jointly influence the profitability and emissions control of the system. The authors have shown that only when the mandatory cap is set in intermediate level rather than excessively mild or tight can it be effective to balance system profitability and emission control well. Indeed, the mandatory carbon policy is very efficient to curb the carbon emissions; however, due to its sternness, this policy drives away the investors. Some European countries are periodically applying the mandatory carbon regulation. For example, the French government applies the mandatory carbon policy on transport and manufacturing activities for a defined period when a high pollution rate is revealed. To the best of our knowledge, there is no work in the literature which deals with a periodic mandatory carbon policy. In this paper, we will combine the carbon tax and mandatory carbon regulation. Indeed, we will consider the carbon tax as a permanent policy to reduce the overall emissions, and the mandatory carbon as a periodic regulation when the emissions exceed a tolerated threshold.

One of the processes to curb carbon emissions in the industrial domain is remanufacturing. Remanufacturing, or refurbing, is considered as a great industrial process, within which worn-out products are restored to new products, providing benefits both environmentally and economically, while at the same time enhancing their image as environmentally responsible, since products are reused instead of being discarded, including curbing carbon emissions [19]. It is supposed that remanufactured products are profitable, at 45%–65% of the price of a new product. Furthermore, they are beneficial under emission regulation [20]. In general, the remanufacturing is a series of industrial processes: disassembly of the worn product on parts, restoring of the reusable parts, replacement of the parts if necessary, and reassembly of the parts [21]. Most of the published papers in the literature have assumed that remanufactured products have the same quality as new ones [22–25]. However, in practice, the quality of new products is mostly higher than for remanufactured ones. In this case, the remanufactured parts are called "reconditioned parts" and are sold with a lower price than new ones. Over the lasts years, the reconditioning production keeps increasing. In recent years, for example, in the automotive sector, reconditioning firms cover 35% of all production firms in the sector. That in the aerospace sector represents 25%. Concerning the sales, for example, in Asia, the sales of reconditioned smartphones represent 23% of the mobile phones market. In France and according to BACK MARKET and REMADE TECHNOLOGY companies, which produce and sell remanufactured products such as smart phones, laptops and home appliances, the market of remanufactured products posted a growth rate of 7% in 2018, unlike that of new ones, which fell by 6.5%. In the literature, few researchers distinguish between new and reconditioned parts. Gaur et al., [26] have conducted a real case of a battery manufacturer based in India. The authors have proposed a supply chain system, which considers that reconditioned batteries have a lower performance specification and more limited warranty relative to the equivalent new product. They have determined an optimal sales and production plan as well as configuration of the proposed supply chain system. Turki and Rezg [27] have determined the optimal storage, manufacturing, and reconditioning planning, while taking into consideration the difference between new and reconditioned parts. Indeed, they considered that the reconditioned and new parts are distinguishable, and the reconditioned parts are sold at lower price in a market different from that existing for new ones. Therefore, in order to conduce a real-world case of a manufacturing-reconditioning system, we will consider two distinguished markets: the first market for selling new parts and the second for reconditioned ones. Indeed, we will propose a reverse supply chain system that recovers the worn parts from the first market, and then reconditions them to then be sold in the second market. Furthermore, to satisfy markets demands, we will consider two machines: the first for producing new products and the second for reconditioning worn parts. Moshtagh and Taleizadeh [28] have considered a manufacturing-reconditioning system with two machines, one for manufacturing and one for reconditioning. The authors have taken into account inventory costs, production and reconditioning costs, ordering cost, and sales revenue. However, the authors have neglected some characteristics of manufacturing systems, such as machine breakdowns and time to repair [29]. In fact, in real-life manufacturing systems, the time to repair is a manufacturing dead time that cannot be neglected as it causes production delays and raises the risk of stock shortage. Therefore, when the time to repair is stochastic, as in practice, certainly, the control of the manufacturing, reconditioning, and parts' stocks becomes much more complicated. In this paper, in order to make our proposed system closer to the real-life case, we will consider that both machines are subject to random breakdowns and repairs. Furthermore, to control manufacturing and reconditioning processes, we will apply hedging point policy [30] that ensures that the number of parts does not exceed a defined stock threshold. Moreover, concerning the return of the worn parts, most works have assumed that the return of worn products is proportional to the demands, while this assumption is not true and causes a suboptimal in production policies. Therefore, in this work, we assume that the quantity of worn products depends on the sales in the previous periods. The configuration of the proposed system is inspired by the real-life firm, which produces new and reconditioned products such as smart phones or laptops. In addition, due to already existing environmental preoccupations and hard economic concurrence, we assume that the proposed system is operating in a competitive environment. Thus, in order to keep the system competitive, we aim to find the optimal stocking strategy and production planning that maximize the total profit. This consists to determine the optimal capacities of manufacturing and reconditioning stocks and the optimal production control, under carbon tax and periodic mandatory regulation, taking into account stochastic machine breakdowns and repairs. To perform the study and the simulation of the proposed system closer to reality, we adapt the discrete flow model [31] that is adopted in several works. The benefit of the discrete flow model is that it is faithful to describe discrete production systems. In addition, the simulation of the discrete flow model is clear and not painful. However, its optimization takes a huge amount of time to search the solutions. The efficient optimization method chosen was inspired from the evolutionary algorithm [32], and we developed an improvement method based on local search to give better solutions.

Compared with the existing works, this paper contributes mainly in two ways: the first is to examine a manufacturing-reconditioning system simultaneously considering the periodic mandatory carbon emission and carbon tax. The second is to develop a new approach to determine the quantity of new and reconditioned products to produce when the government imposes a limit carbon emission in a random period, the optimal levels of the manufactured-reconditioned stock, and the optimal percentage of returned worn products that maximizes the total profit. The present work is important for implementation of a new circular economy and green deal, which is now on top of the agenda for Europe.

After having stated our purpose, this paper is organized as follows: Section 2 introduces the manufacturing-reconditioning system and the mathematical models. Section 3 presents the developed optimization method inspired from the evolutionary algorithm to find the optimal value of decision variables. Section 4 analyzes the computational results. Finally, in Section 5, we conclude our work.

2. Materials and Methods

In this section, we present and explain the proposed system.

In the Table 1 we present parameters and decision variables that are used to formulate the models in this paper:

Table 1. Parameters and decision variables.

t	instant time.
Δt	time period length.
T	total simulation time.
S_A	stock capacity for new products.
$S_A{}^*$	optimal stock capacity for new products.
S_{Ar}	stock capacity for reconditioned products.
$S_{Ar}{}^*$	optimal stock capacity for reconditioned products.
$s_A(t)$	stock level of new products at time t.
$s_{Ar}(t)$	stock level of reconditioned products at time t.
$s_r(t)$	stock level of worn products at time t.
$u_A(t)$	production rate of the machine M_1 at time t.
U_A	maximum production rate of the machine M_1.
$u_{Ar}(t)$	production rate of the machine M_2 at time t.
U_{Ar}	maximum production rate of the machine M_2.
$d_A(t)$	demand for new products at time t.

Table 1. *Cont.*

$d_{Ar}(t)$	demand for reconditioned products at time t.
$P_A(t)$	quantity of unmet demand for new products at time t.
$P_{Ar}(t)$	quantity of unmet demand for reconditioned products at time t.
$V_A(t)$	quantity of new products sold at time t.
$V_{Ar}(t)$	quantity of reconditioned products sold at time t.
$R(t)$	quantity of worn products that are collected and returned from market 1 at time t.
p	percentage of the returned products (worn products).
p^*	optimal percentage of the returned products.
α	lifetime of products.
$\theta(t)$	state of the machine M_1 at time t.
$\beta(t)$	state of the machine M_2 at time t.
cv_A	unit selling price for new products.
cv_{Ar}	unit selling price for reconditioned products.
cu_A	unit production cost of new products.
cu_{Ar}	unit production cost of reconditioned products.
cs_A	unit storage cost for new products.
cs_{Ar}	unit storage cost for reconditioned products.
cr	unit storage cost for returned product.
cp_A	unit lost sales cost for new products.
cp_{Ar}	unit lost sales cost for reconditioned products.
ct	unit carbon emission cost.
q_l	quantity limit of carbon emission.
q_{pm}	carbon emission quantity of new products.
q_{pr}	carbon emission quantity of reconditioned products.
W	length of the mandatory carbon period.
$F(t)$	total profit function.
$F(t)^*$	optimal value of total profit.

The aim of our work is to find the optimal stock capacity for new products, $S_A{}^*$, for reconditioned products, $S_{Ar}{}^*$, and the percentage of the returned products, p^*, that maximizes the total profit, $F(t)$.

2.1. Description of the System

Figure 1 presents the model system, that is composed of two machines denoted M_1 for manufacturing and M_2 for reconditioning, which are subject to random repairs and failures. The machine M_1 produces new products from raw material, and we supposed that it never starves. This assumption defines the general case of the manufacturing system where the raw materials are usually acquired. The machine produces new products at variable rate $u_A(t)$ and they are stored in the stock S_A. The variable $u_A(t)$ represents the decided amount for producing new parts in the period t. In fact, $u_A(t)$ is determined at the period t-Δt according the machine state ($\theta(t)$) and the stock level of new parts $s_A(t)$. The production rate $u_A(t)$ takes a value between 0 and its maximum U_A. The stock S_A satisfies the demand of new products $d_A(t)$ coming from the Market 1. The sold new products will be used for a period α and then will be collected and returned to the manufacturer, then, after, they are stored in the recovery inventory R, to be reconditioned by the machine M_2, and finally are stored in S_{Ar}. The production rate of the reconditioning machine (M_2) is denoted $u_{Ar}(t)$, which represents the decided amount for producing reconditioned parts in the period t. Indeed, $u_{Ar}(t)$ is determined at the period t-Δt according the machine state ($\beta(t)$), the stock level of reconditioned parts $s_{Ar}(t)$ and the stock level of worn products $s_{Ar}(t)$. The production rate $u_{Ar}(t)$ takes a value between 0 and its maximum U_{Ar}. The stock S_{Ar} satisfies the demand of reconditioned products $d_{Ar}(t)$ coming from the Market 2. We supposed that the sold reconditioned products in Market 2 will be destroyed after their end of life. Indeed, in the real case, most reconditioned products are reused one time. In order to study the system under the carbon emission regulations used, we take into consideration the carbon emission emitted by producing the new/reconditioned products. In our model, the carbon emissions represent the quantity of carbon emitted from producing either new or reconditioned products. In other words, we do not consider the carbon footprint, but we directly calculate the carbon quantity generated by the production of new and reconditioned parts.

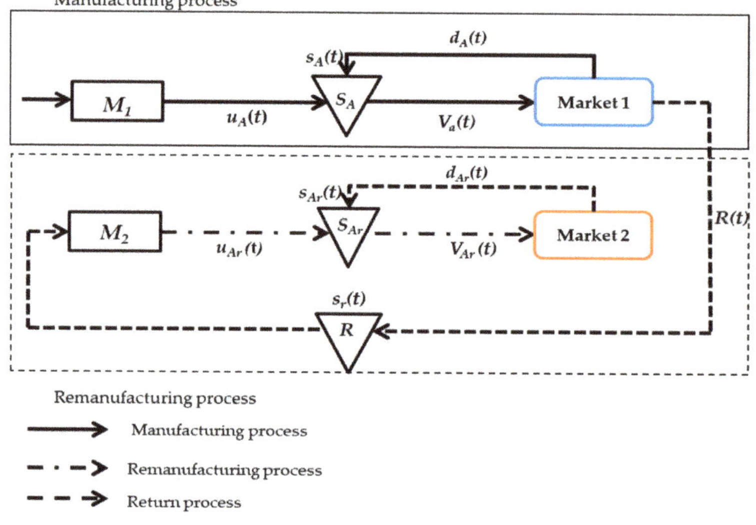

Figure 1. Manufacturing-Reconditioning system.

2.2. Mathematical Model

As assumed before, the machine M_1 is always supplied from raw materials, the state of this machine is either up or down. The machine states are presented in the Equation (1).

$$\theta(t) = \begin{cases} 1 & \text{if machine } M_1 \text{ is up} \\ 0 & \text{if machine } M_1 \text{ is down} \end{cases} \tag{1}$$

The state of machine M_2 is given by in the Equation (2).

$$\beta(t) = \begin{cases} 1 & \text{machine } M_2 \text{ is up} \\ 0 & \text{machine } M_2 \text{ is down} \end{cases} \tag{2}$$

We supposed that the machines M_1 and M_2 are subject to a random failure exponentially distributed. When M_1 is up, $u_A(t)$ is between 0 and its maximum U_A and when it is down, $u_A(t) = 0$. It is the same for the machine M_2. The production rates of the machines M_1 and M_2 are given by the Equations (3) and (4).

$$\begin{cases} 0 < u_A(t) \le U_A & \text{if } \theta(t) = 1 \\ u_A(t) = 0 & \text{if } \theta(t) = 0 \end{cases} \tag{3}$$

$$\begin{cases} 0 < u_{Ar}(t) \le U_{Ar} & \text{if } \beta(t) = 1 \\ u_{Ar}(t) = 0 & \text{if } \beta(t) = 0 \end{cases} \tag{4}$$

The stocks $s_A(t)$ and $s_{Ar}(t)$ at time t are equal to the level of the stock in the previous period plus the number of products produced in the previous period, minus the number of products outgoing from the stock. In order to simplify the calculations, it assumed that $\Delta t = 1$. Thus, the stock levels at time t are given by the Equations (5) and (6).

$$s_A(t) = s_A(t - \Delta t) + u_A(t - \Delta t) - V_A(t) \tag{5}$$

$$s_{Ar}(t) = s_{Ar}(t - \Delta t) + u_{Ar}(t - \Delta t) - V_{Ar}(t) \tag{6}$$

The demands $d_A(t)/d_{Ar}(t)$ are satisfied respectively by the stock level in the previous time $s_A(t - \Delta t)/s_{Ar}(t - \Delta t)$ if stock levels are higher or equal to $d_A(t)/d_{Ar}(t)$), otherwise the number of the sold products is equal to the stock levels at $t - \Delta t$. The numbers of new/reconditioned products sold at time t are expressed by by the Equations (7) and (8).

$$V_A(t) = \begin{cases} d_A(t) & if \; s_A(t - \Delta t) \geq d_A(t) \\ s_A(t - \Delta t) & otherwise \end{cases} \tag{7}$$

$$V_{Ar}(t) = \begin{cases} d_A(t) & if \; s_A(t - \Delta t) \geq d_A(t) \\ s_A(t - \Delta t) & otherwise \end{cases} \tag{8}$$

The numbers of unmet demand $P_A(t)$ and $P_{Ar}(t)$ equal the demand at time t minus the number of sold products at time t. We assume that unmet demands generate a lost cost. This assumption is considered in practice. Indeed, when the customer is not satisfied, a lost cost equivalent to the loss of the business relevance is considered. The Equations (9) and (10) show the number of lost demands at time t:

$$P_A(t) = d_A(t) - V_A(t) \tag{9}$$

$$P_{Ar}(t) = d_{Ar}(t) - V_{Ar}(t) \tag{10}$$

The number of worn products that are returned at time t is proportional to the quantity of satisfied demand after their lifetime. The number of used products equals a percentage of the number of the products sold at $t - \alpha$, where α is the lifetime of the products and p represents the percentage of returned products $(0 < p < 1)$. Thus, when $t - \alpha < 0$, there is no return. Therefore, the quantities of returned products are given by the Equation (11).

$$R(t) = \begin{cases} p \cdot V_A(t - \alpha) & if \, (t - \alpha) \geq 0 \\ 0 & otherwise \end{cases} \tag{11}$$

The Equation (12) represents the level of the recovery inventory at time t, which equals the number of the recovery inventory at $t - \Delta t$ plus the number of worn products that are returned at time t, minus the number of products reconditioned by the machine at $t - \Delta t$.

$$s_r(t) = s_r(t - \Delta t) + R(t) - u_r(t - \Delta t) \tag{12}$$

For the machine M_1, the production rate at time t (see Equation (13)) depends on the machine state $\theta(t)$ and on the stock level of new products at time t. We assumed that the machine is never starved from raw materials. The production rate follows these constraints:

- Equal to the maximum rate production when the state of the machine is up and the stock level capacity is lower than its maximum level.
- Equal to difference between the stock level capacity and its maximum level, when the state of the machine is up and the stock level at time t is between zero and its maximum level.
- Equal to the demand when the state of the machine is up or the stock level at time t has reached its maximum.

- Is null when the state of the machine is down.

$$u_A(t) = \begin{cases} U_A & if\ \theta(t) = 1\ and\ s_A(t) + U_A \le S_A \\ S_A - s_A(t) & if\ \theta(t) = 1\ and\ S_A - U_A < s_A(t) < S_A \\ d_A(t) & if\ \theta(t) = 1\ or\ s_A(t) = S_A \\ 0 & if\ \theta(t) = 0 \end{cases} \tag{13}$$

For the reconditioned products, the machine M_2 is supplied by the recovery inventory. Then, when it is empty or very low, the machine starves. The production rate for reconditioned items looks like the previous one, but also depends on the recovery inventory $R(t)$. Indeed, the production rate for reconditioned products at time t is given the Equation (14).

$$u_{Ar}(t) = \begin{cases} U_{Ar} & if\,\beta(t) = 1\ and\ s_{AR}(t) + U_{Ar} \le S_{Ar}\ and\ s_r(t) \ge U_{Ar} \\ S_{Ar} - s_{AR}(t) & if\,\beta(t) = 1\ and\ S_{AR} - U_{Ar} \le s_{Ar}(t) < S_{AR}\ and\ s_r(t) \ge S_{AR} - s_{Ar}(t) \\ d_{Ar}(t) & if\,\beta(t) = 1\ and\ s_{AR}(t) = S_{Ar}\ and\ s_r(t) \ge d_{Ar}(t) \\ s_r(t) & \begin{cases} if\,\beta(t) = 1\ and\ s_{AR}(t) + U_{Ar} \le S_{Ar}\ and\ s_r(t) < S_{Ar} \\ or\ if\,\beta(t) = 1\ and\ S_{AR} - U_{Ar} \le s_{Ar}(t) < S_{AR}\ and\ s_r(t) < S_{AR} - s_{Ar}(t) \\ or\ if\,\beta(t) = 1\ and\ s_{Ar}(t) = S_{AR}\,and\,s_r(t) < d_{AR}(t) \end{cases} \\ 0 & if\,\beta(t) = 0\ and\ s_r(t) = 0 \end{cases} \tag{14}$$

In this section, we describe the methodology used to plan the production for manufactured and reconditioned products, in the period where the government sets a limitation for carbon emission on a random period (periodic mandatory carbon emission). First, we present the time horizon discretization for the studied problem (see Figure 2).

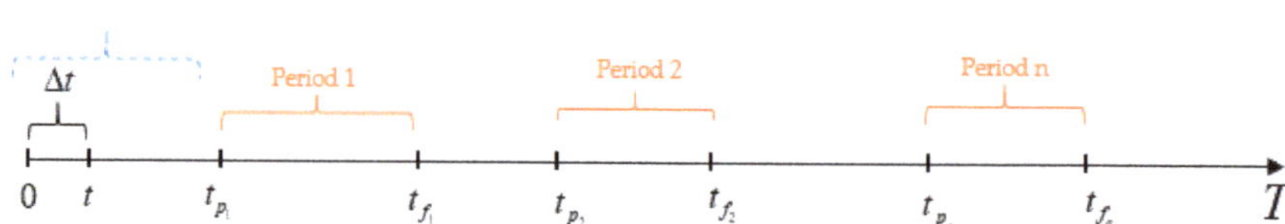

Figure 2. Studied time horizon.

The horizon $[0, T]$ is the finite horizon discretized into periods of time Δt, where Δt is the simulation time step. The occurrence of a period under mandatory carbon emission in the system and the length of the period are generated by truncated normal distribution.

- In order to determine the production quantity for manufacturing and reconditioning products under carbon regulation, we have provided a method to determine the quantity that respects the limited carbon emission and answers to the demand according to the determined production quantity under hedging point policy. We recall q_{pm} and q_{pr} present the carbon emissions of manufacturing and reconditioning respectively, and q_l is the limit quantity of carbon emission to not exceed. We have developed two algorithms (please see Algorithms 1 and 2 that are given in Appendices A and B) that build a function system capable to calculate the proportion of the production rate of the nearest manufactured and reconditioned products under the constraint to not exceed the limit of carbon emissions. The steps are as follows: We calculate the proportion of the production for new and reconditioned products:

$$P_{SC} = \frac{U_A(t)}{U_A(t) + U_{Ar}(t)} \tag{15}$$

$$P_{AC} = \frac{U_{Ar}(t)}{U_A(t) + U_{Ar}(t)} \tag{16}$$

- We search, with two counters i and j running from 1 to the upper bound ($u_A(t)$ and $u_{Ar}(t)$), to respect the constraint of not exceeding the limit of carbon emission:

$$i.q_{pm} + j.q_{pr} <= q_l \tag{17}$$

- We calculate the proportion:

$$P_{m2co} = \frac{i}{i+j} \tag{18}$$

$$P_{R2CO} = \frac{j}{i+j} \tag{19}$$

Then, we developed a function that gives the nearest values corresponding to the sum value P_{m2co} and P_{R2CO} closest to P_{SC} and P_{AC}. The total profit function is determined by the difference between the total revenue and the total costs. The total revenue includes the total cost from the new products $\sum_{t=0}^{T} V_A(t) \cdot cv_A$, and the total cost for reconditioned products $\sum_{t=0}^{T} V_{Ar}(t) \cdot cv_{Ar}$. The total cost includes: the total production costs for new products $\sum_{T=0}^{T} u_A(t) \cdot cu_A$ and the total production costs for reconditioned products $\sum_{T=0}^{T} u_{Ar}(t) \cdot cu_{Ar}$, the total storage cost for new products $\sum_{T=0}^{T} s_A(t) \cdot cs_A$, the total storage cost for reconditioned products $\sum_{T=0}^{T} s_{Ar}(t) \cdot cs_{Ar}$, and the total storage cost for returned products $\sum_{T=0}^{T} s_r(t) \cdot cs_r$. The total return cost for worn products $\sum_{T=0}^{T} R(t) \cdot cr$, the total lost sales for new products $\sum_{T=0}^{T} P_A(t) \cdot cp_A$, and the total lost sales for reconditioned products $\sum_{T=0}^{T} P_{Ar}(t) \cdot cp_{Ar}$. The total amount of carbon emitted from producing new products over the horizon $[0, T]$ is $\sum_{T=0}^{T} u_A(t) \cdot ct \cdot q_{pm}$ and from reconditioned products is $\sum_{T=0}^{T} u_{Ar}(t) \cdot ct \cdot q_{pr}$. We try to maximize the total profit function given by:

$$F(t) = \sum_{t=0}^{t=T} \begin{bmatrix} (v_A(t) \cdot cv_A + v_{Ar}(t) \cdot cv_{Ar}) - (s_A(t) \cdot cs_A + s_{Ar}(t) \cdot cs_{Ar} + \\ s_r(t) \cdot cs_r + P_A(t) \cdot cp_A + P_{Ar}(t) \cdot cp_{Ar} + R(t) \cdot cr + u_A(t) \cdot cu_A + \\ u_{Ar}(t) \cdot cu_{Ar} + u_{Ar}(t) \cdot ct \cdot q_{pr} + u_A(t) \cdot ct \cdot q_{pm}) \end{bmatrix} \tag{20}$$

3. Optimization Method

The mathematical model presented above is clearly based on simulation. Machines states are simulated, random failures are also simulated according to an exponential distribution law, and time flow is also simulated. From all of this, values of stock levels, number of new and reconditioned products, lost demands, return products quantities, production rates of both machines, limit of carbon emission constraint, production plan, and total profit function are calculated, according to Equations (5) to (20). It is obviously not an integer linear mathematical model, but we have developed an optimization method to cope with the non-linearity and random variables challenges, and managed to obtain a total profit function maximization.

The total profit function was programmed in the language C++ with the free software Dev-C++. The optimization method was also developed in this language. In order to facilitate the method handling, we have put the decision variables in a vector. Each value of the vector can be between a minimum and a maximum value. The optimization process is illustrated in Figure 3.

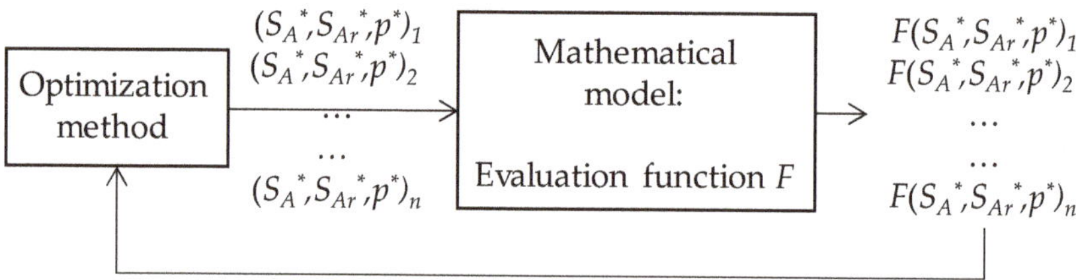

Figure 3. Optimization process of the decision variables $(S_A{}^*, S_{Ar}{}^*, p^*)$

The model is stochastic by construction. Therefore, we calculate the average of three simulations to obtain a more precise value of the objective function, in order to be able to have the best surface possible to perform optimization. The simulation time horizon is fixed to 10^7 time units. The optimization method is inspired from evolutionary algorithms and returns the optimal values of the vector which maximizes the total profit function. The quality of the results obtained by this method depends on the number of tests. The larger the number of tests, the better the results. However, the simulation time can also increase too much. Thus, we developed an improvement method based on local search that can give better solutions by testing the function the least amount possible. The steps are the following:

Initialization: random individuals are created, with each vector of values respecting the boundaries of each of its parameters.

Evaluation: In addition to the pre-determined constraints on variables, the following constraint is written: $S_A > S_{Ar}$. This equation imposes that the storage of new products should be higher than that of reconditioned ones in order to ensure the return of worn products. Consequently, unadapted individuals are rejected.

Selection: We retain the best adapted individuals, those who give satisfying total profit function results.

Mutation: Among the best adapted individuals, we try some neighborhood tests, thanks to a deviation of some values of the vector describing an individual, to see if its characteristics are better adapted.

Evaluation: If a new generated individual meets the constraints, it is evaluated with its total profit value.

Replacement: If the new total profit function is better than the previous, it is retained.

In the end, the final solution is represented by the following vector (S_A, S_{Ar}, p).

The second part of the proposed optimization method is composed of local improvement procedures, which are very useful to perform a better optimization [32]. We developed three main procedures. The first one changes the value of a parameter, chosen at random, in one sense or in the other, here again chosen at random (positive or negative). It can go up until the appropriate bound, but the variation is also at random (it may not "hit" the bound). The set of parameters is then inside each variable bounds, by construction, but the constraints may be violated anyway. Then, after a constraint checking, its objective function is evaluated. If it improves the current solution, it is kept, and if not, the counter of tested solutions is incremented. We allow to test up to 100 valid neighborhoods.

The second neighborhood iteratively repeats the procedure described below, until there is no improvement between two successive iterations. Positive and negative alterations are tested on each of the chromosomes of an individual. If any best improves the objective function, it is kept. Since this neighborhood is iterative, it is only launched once.

The third neighborhood we have implemented is also iterative. For each couple of values of the decision variables vector, we test an alteration of each possible direction's combinations. Since the vector we have to optimize is composed of 3 values ($S_A{}^*$, $S_{Ar}{}^*$, p^*), at each time this neighborhood method is called, it evaluates ($c_3^2 \times 4 = 12$) combinations of values around a current vector. A Pseudo-code algorithm of this iterative neighborhood is given in Appendix C (Algorithm 3). The optimization method we have developed combines both the necessity to consider a stochastic function, with the uncertainty and weakness associated, and the need of efficiency with a relatively long computing, time-consuming objective function. As well as the objective function, this method is also stochastic, and its results cannot be considered as optimal. Nevertheless, they allow a quick convergence towards the optimal region by testing an infinitesimal part of the solution space, thanks to the power of the neighborhoods developed. Moreover, in a stochastic solution space, all but convex, and where quasi-optimal solutions have values in a relatively narrow range, the method we have proposed, designed, and developed, optimizes correctly. This allows us first to work on this exhaustive model and to exploit its richness to come to interesting scientific results. This also proves the relevance of this method to treat optimization problems with high computing, time-consuming objective functions.

4. Numerical Results

This section conducts experimental computations to illustrate the proposed model and explore different results to investigate the impact of interesting parameters on the optimal values of the decision variables. In the beginning of this section, we introduce numerical data, then we present four subsections. In the first, we investigate the impact of the returned products percentage, p, on the optimal stock capacity for new products, $S_A{}^*$, and reconditioned products, $S_{Ar}{}^*$. This study analyzes the impact of the quantity of returned worn products on the manufacturing, reconditioning, and stocking decisions. In the second, we examine the influence of the period of mandatory carbon regulation (W) on the optimal stock capacity for new products, $S_A{}^*$, reconditioned products, $S_{Ar}{}^*$, and the percentage of the returned products, p^*. This study allows a firm leader to determine the optimal production and stocking planning when the mandatory carbon period changes. In the third, we analyze the influence of the limited quantity of carbon emission imposed by the government on optimal values $S_A{}^*$, $S_{Ar}{}^*$, and p^*. This study proposes the optimal stocking management when the government imposes a quantity limit of carbon emissions. In the fourth, we analyze the impact of the carbon emission cost imposed by the government on optimal values $S_A{}^*$, $S_{Ar}{}^*$, and p^*.

The used input data are:

- The time simulation, $T = 10^7$ periods
- The life cycle of new product, $\alpha = 100$ periods The carbon emission quantity for producing new products, $q_{pm} = 100$ carbon units
- The carbon emission quantity for producing reconditioned products, $q_{pr} = 10$ carbon units
- The quantity limit of carbon emission, $q_l = 1000$ carbon units
- The unit carbon emission cost, $ct = 0.01$ monetary unit (for example, dollars or euros)
- The maximum production rate, $U_A = 2500$ products/period
- The maximum production rate, $U_{Ar} = 2300$ products/period
- The unit selling price for new products, $cv_A = 400$ monetary units
- The unit selling price for reconditioned products, $cv_{Ar} = 180$ monetary units
- The unit storage cost for new products, $cs_A = 0.0005$ monetary units
- The unit storage cost for reconditioned products, $cs_{Ar} = 0.0005$ monetary units
- The unit lost sales cost for new products, $cp_A = 1200$ monetary unit
- The unit lost sales cost for reconditioned products, $cp_{Ar} = 875$ monetary units
- The unit production cost of new products, $cu_A = 50$ monetary unit

- The unit production cost of reconditioned products, $cu_{Ar} = 20$ monetary unit
- The unit storage cost for returned product $cr = 0.0003$ monetary unit.

The demand $d_A(t)$ is generated by truncated normal distribution, in which the average = 15 and the standard deviation = 7. For $d_{Ar}(t)$, the average = 7 and the standard deviation = 4. The generation of time to repair and time between failures are exponentially distributed. The occurrence of the period of carbon emission limit and length of the period denoted (W) are generated by truncated normal distribution. For length of the period W, the lower truncated is 40 and the upper truncated is 60, with the average = 50 and standard deviation = 10, and for the occurrence of the period, the upper truncated is 4100 and the lower truncated is 3900, and the standard deviation = 100.

4.1. Impact of the p on $S_A{}^*$, $S_{Ar}{}^*$, and F(T)

In this subsection, the we investigate the impact of the returned products percentage, p, on the optimal stock capacity for new products, $S_A{}^*$, reconditioned products, $S_{Ar}{}^*$, and the total profit, $F(T)$. Thus, we vary the value of p, and by using the optimization algorithm, we determine the corresponding $S_A{}^*$, $S_{Ar}{}^*$, and $F(T)$. The simulations' results are illustrated in Table 2. In order to improve the visualization of the obtained results in the analyzed part, we add two Figures, Figures 4 and 5, to the presented Table.

Table 2. Optimal decision variables in function of p.

p	$S_A{}^*$	$S_{Ar}{}^*$	$F(T)$
10%	1026	5	2.0804929×10^8
20%	1019	167	1.57228×10^{10}
30%	1032	188	2.9439×10^{10}
40%	1025	253	4.5063×10^{10}
50%	**1029**	**411**	**6.0738×10^{10}**
60%	1021	468	4.07471×10^{10}
70%	1028	517	2.06977×10^{10}
80%	1033	532	$-1.8739222 \times 10^{10}$

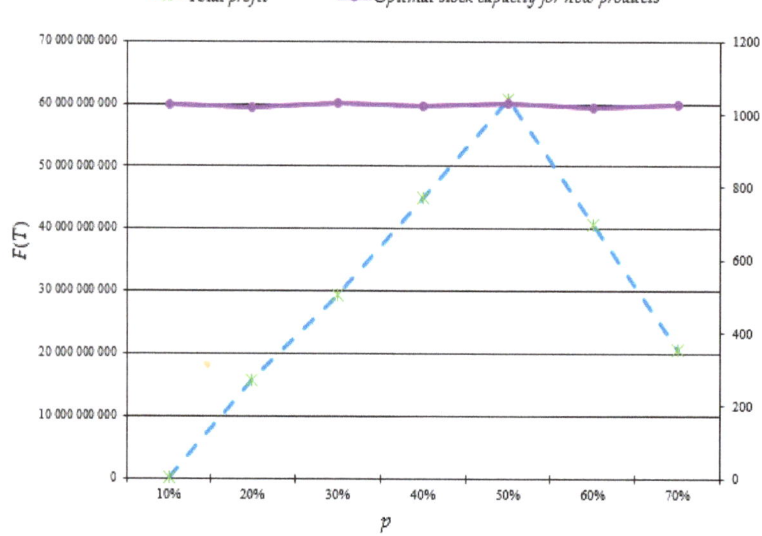

Figure 4. $S_A{}^*$ and $F(T)$ in function of p.

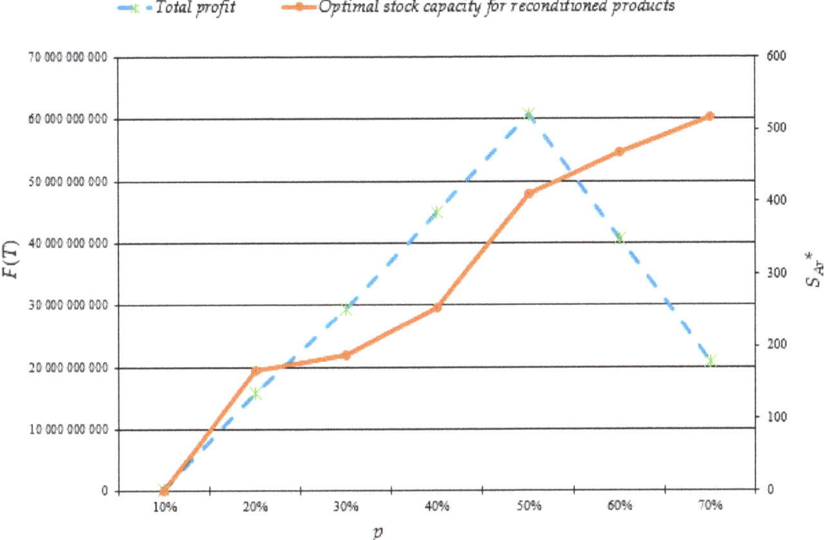

Figure 5. $S_{Ar}*$ and $F(T)$ in function of p.

As shown in Table 2, the total profit is, at the first time, proportional to the percentage of returned products. Then, as the percentage of returned products increases, the profit again decreases and leads to losses. Indeed, it is observed that we have an optimum when p is equal to 50%. This is explained by the fact that when p increases, the number of worn products returned, $R(t)$, increases and replenishes the recovery inventory. Thus, when p is below 50%, the value of profit function increases, which is normal, and the quantity of returned products increases and replenishes the recovery inventory, then the reconditioned products are satisfied. Nevertheless, when p it is above 50%, many of the returned products will be stored in the recovery inventory, which leads to overstocking and generates a high storage cost, and thus explains the losses. The optimal manufacturing stock, S_A*, remains constant, which means that it does not depend on p (see Figure 4). In fact, the production of new parts is independent from the production of reconditioned parts. The optimal reconditioning stock, $S_{Ar}*$, increases when p increases (see Figure 5), and this is explained by the need of the worn parts for supplying the demand of reconditioned parts in Market 2. If more worn parts are returned to the production system, more reconditioned parts will be available in the stock, S_{Ar}, to satisfy the demand, $d_{Ar}(t)$. This study allows a firm leader to manage, in an optimal way, the manufacturing, reconditioning, and stocking when the quantity of returned worn products varies. Furthermore, when the leader has the option to set the quantity of the returned worn products, he can improve the system management by proposing the optimal quantity of the returned worn products that maximize the profit. Therefore, in the next studies, we add p to the decision variables.

4.2. Impact of the W on S_A*, $S_{Ar}*$, $p*$, and $F(T)$

In this subsection, we investigate the impact of the length of the mandatory carbon period, W, on the optimal stock capacity for new products, S_A*, reconditioned products, $S_{Ar}*$, returned products percentage, p, and total profit, $F(T)$. Thus, we vary the value of W, and by using the optimization algorithm, we determine the corresponding S_A*, $S_{Ar}*$, $p*$, and $F(T)$. The simulations' results are illustrated in Table 3. In order to improve the visualization of the obtained results in the analyzed part, we add the Figure 6 to the presented Table.

Table 3. Impact of length of the mandatory carbon period W on $S_A{}^*$, $S_{Ar}{}^*$, p^*, and $F(T)$.

W	Lower	Upper	*Standard Deviation*	$S_A{}^*$	$S_{Ar}{}^*$	p^*	$F(T)$
50	**25**	**75**	**25**	**1029**	**411**	**50%**	**6.0735×10^{10}**
100	50	150	50	1241	617	50%	6.07324×10^{10}
500	250	750	250	1641	747	50%	5.11676×10^{10}
750	375	1025	375	2394	783	50%	4.57892×10^{10}

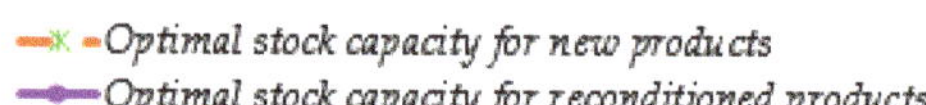

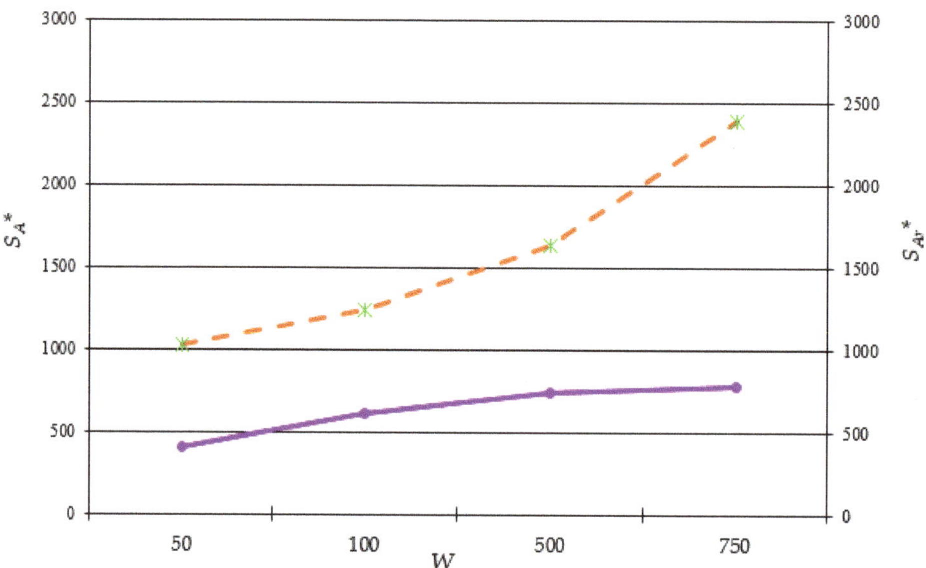

Figure 6. $S_{Ar}{}^*$ and $S_A{}^*$ in function of W.

As it is shown in Table 3, the larger the mandatory carbon period is, the more the total profit decreases, which is explained by lower production in a large period. Also, the optimal of the length of the period of carbon is given (the bold line). Indeed, when the mandatory carbon period is low, the limitation for producing either for new or reconditioned products is low, thus the production is less limited, and the number of satisfying products increases, and then the lost sales costs decreases (see Figure 6). Consequently, to hold the produced products, the optimal storage capacities for both types of products increase. Thus, the demands will be satisfied more, and of course, the profit increases. This study allows a company manager to propose optimal decisions on the manufacturing, reconditioning, stocking, and quantity of returned worn products when a mandatory carbon period is imposed. Indeed, the government imposes a mandatory carbon period when the pollution reaches a level that is considered harmful to humans. In this period, the manager has to respect a fixed limit of carbon emissions. Consequently, the production is limited, and the manager has to find the optimal planning of production, stocking, and returning of worn products.

4.3. Impact of the q_l on $S_A{}^*$, $S_{Ar}{}^*$, p^*, and $F(T)$

In this subsection, the we investigate the impact of the limited quantity of carbon emission, q_l, on the optimal stock capacity for new products, $S_A{}^*$, reconditioned products, $S_{Ar}{}^*$, returned products percentage, p, and the total profit, $F(T)$. Thus, we vary the value of q_l, and by using the optimization algorithm, we determine the corresponding $S_A{}^*$, $S_{Ar}{}^*$, p^*, and $F(T)$. The simulations' results are illustrated in Table 4. In order to improve the visualization of the obtained results in the analyzed part, we add the Figure 7 to the presented Table.

Table 4. Impact of the limited quantity of carbon emission, q_l, on $S_A{}^*$, $S_{Ar}{}^*$, p^*, and $F(T)$.

q_l	$S_A{}^*$	$S_{Ar}{}^*$	p^*	$F(T)$
100	1578	517	50%	6.06718×10^{10}
250	1401	483	50%	6.06987×10^{10}
500	1292	458	50%	6.07126×10^{10}
750	1173	431	50%	6.07297×10^{10}
1000	**1029**	**411**	**50%**	$\mathbf{6.07350 \times 10^{10}}$

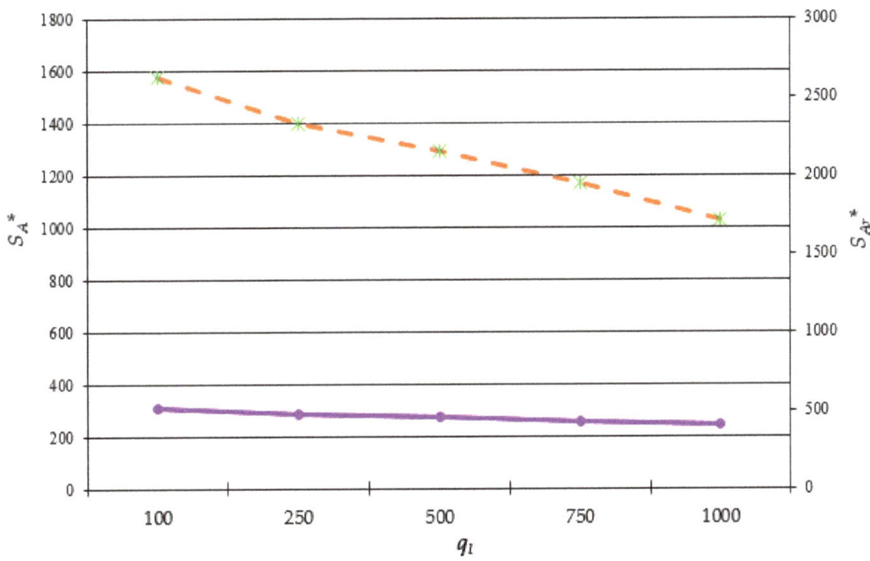

Figure 7. $S_{Ar}{}^*$ and $S_A{}^*$ in function of q_l.

As it observed, when the limited quantity of carbon increases, the profit slightly increases, but both $S_A{}^*$ and $S_{Ar}{}^*$ decrease (see Figure 7). In fact, when the limited quantity of carbon emission is high, the production either for the new or reconditioned products are less limited, and then both types of products are more available, thus, both demands $d_A(t)$ and $d_{Ar}(t)$ are better satisfied. Moreover, when both types of products are available, the stocking decreases as the demands are satisfied. However, when the limited quantity of carbon emission is low, the production either for the new or reconditioned products are very limited. Thus, both types of products are less available and then it shall increase the storing to satisfy the demands. This study allows a firm manager to propose optimal decisions on the manufacturing, reconditioning, stocking, and quantity of returned worn products when the government imposes a limited quantity of carbon emission.

4.4. Impact of the ct on $S_A{}^*$, $S_{Ar}{}^*$, p^*, and $F(T)$

In this subsection, we investigate the impact of the cost of carbon emission, ct, on the optimal stock capacity for new products, $S_A{}^*$, reconditioned products, $S_{Ar}{}^*$, returned products percentage,

p, and the total profit, $F(T)$. Thus, we vary the value of ct, and by using the optimization algorithm, we determine the corresponding S_A^*, S_{Ar}^*, p^*, and $F(T)$. To study the influence of the cost of carbon emission, we changed the parameters' values for $cs_A = 0.00002$, $cs_{Ar} = 0.00002$, and $cr = 0.0000001$ monetary units. The simulations' results are illustrated in Table 5. In order to improve the visualization of the obtained results in the analyzed part, we add the Figure 8 to the presented Table.

Table 5. Impact of the cost of carbon emission on S_A^*, S_{Ar}^*, p^*, and $F(T)$.

ct	S_A^*	S_{Ar}^*	p^*	$F(T)$
0.01	1135	434	50%	6.15698×10^{10}
1	922	471	50%	4.65295×10^{10}
4	883	502	50%	9.42485×10^{8}

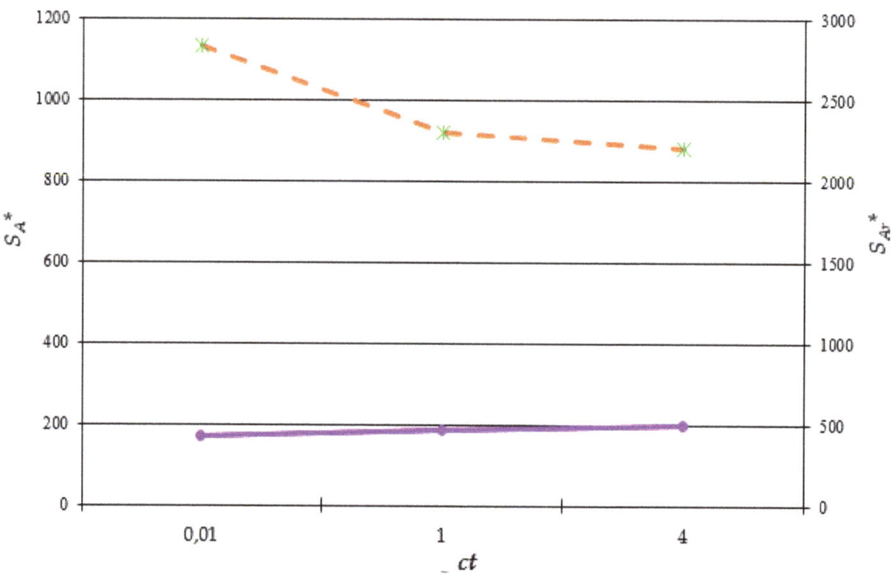

Figure 8. S_{Ar}^* and S_A^* in function of ct.

The first observation in Table 5 is that the total profit obviously decreases when the cost of carbon increases. Indeed, when the cost of carbon emission increases, the system produces less new parts in order to decrease the cost of emissions, and then, optimal stock, S_A^*, decreases (see Figure 8). On the other hand, in order to decrease the cost of emissions, the system produces more reconditioned parts instead of new ones. That explain the increase of optimal stock, S_{Ar}^*, when the cost of carbon emission increases (see Figure 8). This study allows a firm manager to propose optimal decisions on the manufacturing, reconditioning, stocking, and quantity of returned worn products when the government increases the carbon emission cost.

5. Conclusions

This paper has proposed a new design of a manufacturing and reconditioning system, taking into account two carbon regulations: carbon tax and periodic mandatory emission. New and reconditioned products are distinguished and sold in different markets. Our work provides, in the framework of this system, a new and quick strategy to quantify the production of parts under the periods of carbon limitation. To make this green manufacturing more realistic, we have considered the failure and repair of the machines, and the amount of returned products (worn products) depends on the previous sales. The demands are stochastic. To simulate the proposed system and to faithfully describe the system behavior, we have developed a mathematical model based on a discrete flow model. Then, an optimization method based on an evolutionary algorithm was developed, to determine the optimal quantity of the returned worn products, and the stock capacity for new and reconditioned products. Four numerical results were obtained by using the optimization algorithm, which help a firm manager to make decisions on the production and storing of new and reconditioned parts. The results have provided optimal strategies of production, storing, and recovery of worn parts in function of the mandatory carbon period length, emission limit, and carbon cost.

In future research, we will consider the mentioned limitations and the carbon emissions exceeded by transportation from stores to markets.

6. Discussion

This work concerned the consideration of a periodic mandatory carbon policy that is combined with carbon tax. A new approach was developed to determine the quantity of new and reconditioned products to produce when the government imposes a limited carbon emission in a random period, the optimal levels of the manufactured-reconditioned stock, and the optimal percentage of returned worn products that maximizes the total profit. Furthermore, the proposed approach combined different options that make the study more realistic, such as: distinction between new and reconditioned parts, the dependence of worn products on the sales in the previous periods, and the variability of the machines repair periods. An efficient optimization method based on evolutionary algorithms was developed that returns the optimal values of the vector, which maximizes the total profit function. The studies provided in this work allow a company manager to propose optimal decisions on the manufacturing, reconditioning, stocking, and quantity of returned worn products when a mandatory carbon period is imposed or when the government increases the carbon emission cost. This work is not exempt from limitations. First, it is considered that the reconditioning process is assured by one machine, but in practice, usually it consists of several machines that are subject to random failures. Second, only one type of product was considered, whereas in practice, almost all companies are producing multiple products. Therefore, in our future research, we will extend the model by assuming that the manufacturing process as well as the reconditioning process consists of several activities carried out at several workplaces. Also, we will consider a multi-product system of production and reconditioning.

Author Contributions: S.T. is the responsible for conducting this research. S.T. proposed the main paper contribution. S.S. developed the mathematical model. C.S. developed the optimization method and algorithms. N.S. supervised this work and contributed to revise this paper. All authors have read and agreed to the published version of the manuscript.

Funding: This research received no external funding.

Conflicts of Interest: The authors declare no conflict of interest.

Appendix A

Algorithm 1: Function near.

Near $(M[n], x, S[n], y)$
01: $M[n], x, S[n], y$
02: **Int** i, result; **Double** D;
03: $D = |TT [0] - x| + |SS [0] - y|$
04: $Result = 0$;
05: **For** $(i = 1; i < $ n$; i++)${
06:    **If** $(|TT[i] - x| + |SS[i] - y|; <= D)$
07:      $D= |TT[i] - x|+|SS[i] - y|$
08:      $Result = i$;
09:    **End if**
10: **End for**
11: **Return** $Result$

Appendix B

Algorithm 2: Production planning under mandatory carbon emission.

Production plan $(u_A(t), u_{Ar}(t))$
01: $k = 0$;
02: **If** $(u_A(t)! \ 0 \text{ and } u_{Ar}(t)! = 0)$
03:    $P_{SC} = u_A(t)/(u_A(t) + u_{Ar}(t))$
04:    $P_{AC} = u_{Ar}(t)/(u_A(t) + u_{Ar}(t))$
05:    **For** $(i = 1 \text{ to } u_A(t))$
06:      **For** $(j = 1 \text{ to } u_{Ar}(t))$
07:        **If** $(i{\cdot}q_{pm} + j{\cdot}q_{pr} < = q_L)$
08:          $P_{m2co} = i/(i + j)$
09          $P_{R2CO} = j/(i + j)$
10:          $h[k] = P_{m2co}$
11:          $B[k] = P_{R2CO}$
12:          $v[k] = i$
13:          $c[k] = j$
14:          $k++$
15:        **End if**
16:      **End for**
17:    **End for**
18:    $lignenear = $ **Near** (h, P_{SC}, B, P_{AC})
19:    $i = v[lignenear], j = c[lignenear]$
20: **End if**
21: $u_A(t) = i$
22: $u_{Ar}(t) = j$

Appendix C

Algorithm 3: Pseudo-code of couple of coordinates, iterative neighborhood.

```
Iterative_Neighborhood (Vres; Delta)
01: Do
02:    Vtest = Vres
03:    C_Min_Temp = Cost_Function (Vtest)
04:    For (i = 1 to nb_var-1)
05:       For (j = i + 1; j < nb_var; j++)
06:          Switch (l = 0; l < 4; l++)
07:             l = 0:
08:                Positive_Variation (Vtest [i])
09:                Positive_Variation (Vtest [j])
10:             l = 1:
11:                Positive_Variation (Vtest [i])
12:                Negative_Variation (Vtest [j])
13:             l = 2:
14:                Negative_Variation (Vtest [i])
15:                Positive_Variation (Vtest [j])
16:             l = 3:
17:                Negative_Variation (Vtest [i])
18:                Negative_Variation (Vtest [j])
19:          End Switch//on l
20:          If (Constraints_Respected (V))
21:             C_Temp = Cost_Function (V)
22:             If (C_Temp < C_Min_Temp)
23:                Vres = V
24:             End if
25:          End if
26:       End for//j
27:    End for//i
28: While (Vres! = Vtest)
```

References

1. Wang, D.D. Do United States manufacturing companies benefit from climate change mitigation technologies? *J. Clean. Prod.* **2017**, *161*, 821–830. [CrossRef]
2. Jorgensen, S.E.; Fath, B.D. *Encyclopedia of Ecology*; Newnes: Oxford, UK; Boston, MA, USA, 2014.
3. Poizot, P.; Dolhem, F. Clean energy new deal for a sustainable world: From non-CO_2 generating energy sources to greener electrochemical storage devices. *Energy Environ. Sci.* **2011**, *4*, 2003–2019. [CrossRef]
4. Benhelal, E.; Zahedi, G.; Shamsaei, E.; Bahadori, A. Global strategies and potentials to curb CO_2 emissions in cement industry. *J. Clean. Prod.* **2013**, *51*, 142–161. [CrossRef]
5. Grafton, R.Q. Intergovernmental Panel on Climate Change (IPCC). In *A Dictionary of Climate Change and the Environment*; Edward Elgar Publishing Limited: Cheltenham, UK, 2012.
6. Cao, K.; Xu, X.; Wu, Q.; Zhang, Q. Optimal production and carbon emission reduction level under cap-and-trade and low carbon subsidy policies. *J. Clean. Prod.* **2017**, *167*, 505–513. [CrossRef]
7. Wang, X.; Zhu, Y.; Sun, H.; Jia, F. Production decisions of new and remanufactured products: Implications for low carbon emission economy. *J. Clean. Prod.* **2018**, *171*, 1225–1243. [CrossRef]
8. Wang, X.Y. Effect of Carbon Pricing on Optimal Mix Design of Sustainable High-Strength Concrete. *Sustainability* **2019**, *11*, 5827. [CrossRef]
9. Zhou, P.; Wen, W. Carbon-constrained firm decisions: From business strategies to operations modeling. *Eur. J. Oper. Res.* **2020**, *281*, 1–15. [CrossRef]
10. Dong, C.; Shen, B.; Chow, P.S.; Yang, L.; Ng, C.T. Sustainability investment under cap-and-trade regulation. *Ann. Oper. Res.* **2016**, *240*, 509–531. [CrossRef]

11. Du, S.; Ma, F.; Fu, Z.; Zhu, L.; Zhang, J. Game-theoretic analysis for an emission-dependent supply chain in a 'cap-and-trade'system. *Ann. Oper. Res.* **2015**, *228*, 135–149. [CrossRef]

12. Zakeri, A.; Dehghanian, F.; Fahimnia, B.; Sarkis, J. Carbon pricing versus emissions trading: A supply chain planning perspective. *Int. J. Prod. Econ.* **2015**, *164*, 197–205. [CrossRef]

13. Poterba, J.M. Tax policy to combat global warming: On designing a carbon tax (No. w3649). *Natl. Bur. Econ. Res.* **1991**. [CrossRef]

14. Ingham, A.; Ulph, A. Market-based instruments for reducing CO_2 emissions: The case of UK manufacturing. *Energy Policy* **1991**, *19*, 138–148. [CrossRef]

15. Fang, K.; Uhan, N.; Zhao, F.; Sutherland, J.W. A new approach to scheduling in manufacturing for power consumption and carbon footprint reduction. *J. Manuf. Syst.* **2011**, *30*, 234–240. [CrossRef]

16. Turki, S.; Rezg, N. Impact of the quality of returned-used products on the optimal design of a manufacturing/remanufacturing system under carbon emissions constraints. *Sustainability* **2018**, *10*, 3197. [CrossRef]

17. Dou, G.; Guo, H.; Zhang, Q.; Li, X. A two-period carbon tax regulation for manufacturing and remanufacturing production planning. *Comput. Ind. Eng.* **2019**, *128*, 502–513. [CrossRef]

18. He, L.; Mao, J.; Hu, C.; Xiao, Z. Carbon emission regulation and operations in the supply chain supernetwork under stringent carbon policy. *J. Clean. Prod.* **2019**, *238*, 117652. [CrossRef]

19. Sundin, E.; Lee, H.M. In what way is remanufacturing good for the environment? In *Design for Innovative Value towards a Sustainable Society*; Springer: Dordrecht, The Netherlands, 2012; pp. 552–557.

20. Yenipazarli, A. Managing new and remanufactured products to mitigate environmental damage under emissions regulation. *Eur. J. Oper. Res.* **2016**, *249*, 117–130. [CrossRef]

21. Tighazoui, A.; Turki, S.; Sauvey, C.; Sauer, N. Optimal design of a manufacturing-remanufacturing-transport system within a reverse logistics chain. *Int. J. Adv. Manuf. Technol.* **2019**, *101*, 1773–1791. [CrossRef]

22. Assid, M.; Gharbi, A.; Hajji, A. Production planning of an unreliable hybrid manufacturing–remanufacturing system under uncertainties and supply constraints. *Comput. Ind. Eng.* **2019**, *136*, 31–45. [CrossRef]

23. Turki, S.; Rezg, N. Unreliable manufacturing supply chain optimisation based on an infinitesimal perturbation analysis. *Int. J. Syst. Sci. Oper. Logist.* **2018**, *5*, 25–44. [CrossRef]

24. Liu, B.; Holmbom, M.; Segerstedt, A.; Chen, W. Effects of carbon emission regulations on remanufacturing decisions with limited information of demand distribution. *Int. J. Prod. Res.* **2015**, *53*, 532–548. [CrossRef]

25. Turki, S.; Didukh, S.; Sauvey, C.; Rezg, N. Optimization and analysis of a manufacturing–remanufacturing–transport–warehousing system within a closed-loop supply chain. *Sustainability* **2017**, *9*, 561. [CrossRef]

26. Gaur, J.; Amini, M.; Rao, A.K. Closed-loop supply chain configuration for new and reconditioned products: An integrated optimization model. *Omega* **2017**, *66*, 212–223. [CrossRef]

27. Turki, S.; Sauvey, C.; Rezg, N. Modelling and optimization of a manufacturing/remanufacturing system with storage facility under carbon cap and trade policy. *J. Clean. Prod.* **2018**, *193*, 441–458. [CrossRef]

28. Moshtagh, M.S.; Taleizadeh, A.A. Stochastic integrated manufacturing and remanufacturing model with shortage, rework and quality based return rate in a closed loop supply chain. *J. Clean. Prod.* **2017**, *141*, 1548–1573. [CrossRef]

29. Guiras, Z.; Turki, S.; Rezg, N.; Dolgui, A. Optimal maintenance plan for two-level assembly system and risk study of machine failure. *Int. J. Prod. Res.* **2019**, *57*, 2446–2463. [CrossRef]

30. Turki, S.; Rezg, N. Impact of the Transport Activities within a Closed-loop Supply Chain: Study of theLost Profit Risk. In Proceedings of the 2019 IEEE 6th International Conference on Industrial Engineering and Applications (ICIEA), Tokyo, Japan, 12–15 April 2019; pp. 800–804.

31. Turki, S.; Hennequin, S.; Sauer, N. Perturbation analysis for continuous and discrete flow models: A study of the delivery time impact on the optimal buffer level. *Int. J. Prod. Res.* **2013**, *51*, 4011–4044. [CrossRef]

32. Trabelsi, W.; Sauvey, C.; Sauer, N. Heuristics and metaheuristics for mixed blocking constraints flowshop scheduling problems. *Comput. Oper. Res.* **2012**, *39*, 2520–2527. [CrossRef]

© 2020 by the authors. Licensee MDPI, Basel, Switzerland. This article is an open access article distributed under the terms and conditions of the Creative Commons Attribution (CC BY) license (http://creativecommons.org/licenses/by/4.0/).

Article

Spray and Aerosolised pH-Neutral Electrochemically Activated Solution Reduces *Salmonella* Enteritidis and Total Bacterial Load on Egg Surface

Sangay Tenzin [1], Sergio Ferro [2], Samiullah Khan [3], Permal Deo [4,*] and Darren J. Trott [1,*]

[1] Australian Centre for Antimicrobial Resistance Ecology, School of Animal and Veterinary Sciences, The University of Adelaide, Mudla Wirra Rd, Roseworthy 5371, Australia; sangay.tenzin@adelaide.edu.au

[2] Ecas4 Australia Pty. Ltd., Unit 8/1 London Road, Mile End South 5031, Australia; sergio@ecas4.com.au

[3] School of Animal and Veterinary Sciences, The University of Adelaide, Roseworthy 5371, Australia; samiullah.khan@adelaide.edu.au

[4] Health and Biomedical Innovation, UniSA Clinical and Health Sciences, University of South Australia, Adelaide 5000, Australia

* Correspondence: permal.deo@unisa.edu.au (P.D.); darren.trott@adelaide.edu.au (D.J.T.)

Abstract: The effectiveness of sprayed and aerosolised pH-neutral electrochemically activated solutions (ECAS) containing 150 mg/L of free available chlorine in reducing total bacteria load and artificially inoculated *Salmonella enterica* serotype Enteritidis 11RX on eggs surfaces was investigated. Treatment groups included untreated control, sodium hypochlorite (positive control), sprayed and aerosolised water and sprayed and aerosolised ECAS. Sprayed ECAS (150 mg/L, 45 s) showed a significant reduction in total bacterial load (2.2 log reduction, $p < 0.0001$) and *S.* Enteritidis (5.4 log reduction, $p < 0.0001$) when compared with the untreated control. Aerosolised ECAS (120 s) was effective in reducing both the total bacterial load (1.4 log reduction, $p < 0.01$) and *S.* Enteritidis (4.2 log reduction, $p = 0.0022$). However, aerosolised ECAS (60 s) only significantly reduced *S.* Enteritidis counts (2.8 log reduction, $p < 0.0008$), indicating that a longer time for bacterial reduction during fogging sanitisation is needed. Tests performed with one egg per oscillating tray were more effective in reducing both the total bacterial load and the *S.* Enteritidis counts than those with three eggs per oscillating tray. Sprayed ECAS (45 s) and aerosolised ECAS (120 s) did not deteriorate the egg cuticle integrity (ΔE_{ab}^*), which was evaluated using Cuticle Blue dye solution and colour intensity measurement. Overall, both the reduction in total bacteria counts and *S.* Enteritidis from the egg surface and retention of cuticle integrity suggest that sprayed and aerosolised ECAS could be used as alternative sanitising approaches to improve the food safety aspect of table eggs.

Keywords: pH-neutral electrochemically activated solution; total bacterial count; *Salmonella* Enteritidis; egg cuticle integrity

Citation: Tenzin, S.; Ferro, S.; Khan, S.; Deo, P.; Trott, D.J. Spray and Aerosolised pH-Neutral Electrochemically Activated Solution Reduces *Salmonella* Enteritidis and Total Bacterial Load on Egg Surface. *Appl. Sci.* **2021**, *11*, 732. https://doi.org/10.3390/app11020732

Received: 21 December 2020
Accepted: 11 January 2021
Published: 13 January 2021

Publisher's Note: MDPI stays neutral with regard to jurisdictional claims in published maps and institutional affiliations.

Copyright: © 2021 by the authors. Licensee MDPI, Basel, Switzerland. This article is an open access article distributed under the terms and conditions of the Creative Commons Attribution (CC BY) license (https://creativecommons.org/licenses/by/4.0/).

1. Introduction

Pathogenic serotypes of *Salmonella* are a major cause of foodborne diseases worldwide. The annual proportion of food origin salmonellosis in Australia is about 40,000 out of an estimated total of about 4.1 million foodborne gastroenteritis cases [1]. *Salmonella* associated foodborne illnesses have risen during the past 20 years and the rate of salmonellosis in Australia is much higher compared to economically similar countries [2]. It has been estimated that foodborne illnesses due to *Salmonella* spp. have caused up to 35% of hospitalisations and 28% of mortalities [1], and the hospitalisation and death cases were higher in comparison to other foodborne illnesses [1]. Among the salmonellosis cases of foodborne origin, raw eggs and egg-based products have the highest frequency [3–5]. For example, between 2001 and 2016, 50% of all foodborne *Salmonella* illnesses in Australia were attributed to the consumption of contaminated eggs [6]; 84% of these cases were caused by *Salmonella enterica* subspecies *enterica* serotype Typhimurium (*S.* Typhimurium) and three cases were caused by *Salmonella enterica* subspecies *enterica* serotype Enteritidis (*S.* Enteritidis) [6]. In other countries, salmonellosis is caused predominantly by the serotype Enteritidis [7–9]. As egg consumption in Australia is approx. 245 eggs per capita and growing [10], the industry is continuously exploring alternative means to address *Salmonella* contamination.

Unlike the one-off input costs for the establishment of farm infrastructure and human resources, farms and industries incur ongoing costs for egg hygiene and egg safety management. Washing and disinfection of egg surface are the key steps involved in egg production to reduce the risk of egg-related foodborne illnesses and to maintain consumer confidence on the microbiological safety of eggs. Elimination of pathogenic bacteria from the egg surface is achieved using several techniques and many eggshell sanitisation methods are also employed to reduce the contamination of eggs by *Salmonella* in commercial egg production premises.

Protocols employed for the reduction of *Salmonella* can be broadly classified into thermal and non-thermal disinfection procedures. Thermal disinfection, such as egg pasteurisation, is a highly effective method, but negatively affects egg proteins and rheological properties [11]. The most common non-thermal processes employ quaternary ammonium compounds (QAC) and chlorine-based chemicals [12] to sanitise eggs after washing with a high pH (11.0) detergent solution at a temperature above 40 °C. Unfortunately, bacteria may develop resistance to QAC [13], which in turn induces selection of genes for co-resistance to other antimicrobials [14], thus limiting its use. In the case of chlorine-based sanitisers, besides the development of bacterial resistance due to its persistent usage [15], the accumulation of organic load from dirt, manure and broken eggs reduces the chlorine concentration, affecting the efficacy of chlorine-based sanitisation. Moreover, due to the environmental impact caused by chlorine-based by-products and the problems with wastewater disposal, its usage in the food industry is limited. Other decontamination methods used are ultraviolet (UV) irradiation of eggs after washing and formaldehyde fumigation. However, the antibacterial activity of UV irradiation protocol is limited to the egg surface directly exposed to UV rays [16], while formaldehyde is a known occupational health hazard and a carcinogenic product [17].

Since occupational health safety and environmental regulations continually push towards safer eggshell sanitisers, electrochemically activated solutions (ECAS) (also called electrolysed oxidizing (EO) water) could be a potential alternative for eggshell cleaning and disinfection. The three forms of ECAS (acidic, slightly acidic and neutral) have been previously assessed for the sanitisation of table eggs in safe food production [18–21] and fertile eggs for quality production of chicks [22]. In most of the available research, a two-step process for ECAS disinfection of eggs was followed. In the initial washing step, dirt and debris are washed off the egg surface with water or alkaline detergent, followed by the ECAS disinfection. Bialka et al. [19] reported that immersion washing of eggs with acidic ECAS significantly reduced *S.* Enteritidis and *Escherichia coli* from the egg surface but also damaged the egg cuticle layer. On the other hand, the spray washing of eggs with slightly acidic EO water reduced total aerobic bacteria without negatively affecting the cuticle layer [22]. In other studies, an immersion washing with pH-neutral ECAS was not effective in reducing the total bacterial load [23], while a spray wash significantly decreased the level of *Listeria monocytogenes* without affecting the egg cuticle layer [20]. More recently, Medina-Gudino et al. [24] reported that pH-neutral EO spray treatment for 30 s significantly reduced *S.* Enteritidis and *E. coli* loads on the egg surface without adversely affecting egg cuticle integrity.

In this study, we explored the potential of pH-neutral ECAS (150 mg/L of free available chlorine (FAC)) as spray and aerosol fog for the sanitisation of unwashed, visibly clean eggs, assessing the reduction in total bacterial counts and *Salmonella* Enteritidis, and its effects on the cuticle layer. ECAS with neutral pH still contains hypochlorous acid (HOCl) as the main oxidising component (active chlorine compounds also include hypochlorite ions and dissolved gaseous chlorine) [25–27] but is less corrosive and more durable than the acidic and slightly acidic forms.

2. Materials and Methods

2.1. Chicken Egg Source and Selection

Freshly laid eggs were sourced from Hy-Line Brown hens (aged between 36 and 40 weeks) raised in conventional battery cages housing individual hens at the School of Animal and Veterinary Sciences, The University of Adelaide. Eggs stored for 24 h at room temperature were visually screened for thermal cracks and dirt; relatively clean, intact and uniformly sized eggs were selected and randomly divided into groups to determine the effectiveness of sanitisation treatments on total microbial load and artificially inoculated *S.* Enteritidis load.

2.2. ECAS Spray and Fog Disinfection Generation

The electrochemically activated solution was sourced from Ecas4 Australia Pty Ltd. (Adelaide, Australia) and its physicochemical properties such as temperature, pH, and oxidation/reduction potential (ORP) were measured using a handheld meter (Model MC-80, TPS Pty Ltd., Brisbane, Australia). Free and total available chlorine was measured using a Free Chlorine Checker® HC-HI701 and a Total Chlorine Checker® HC-HI711 (Hanna Instruments, Melbourne, Australia), respectively. ECAS was stored at $4 \pm 1\ °C$ and used within one week of preparation.

The working concentration of ECAS (pH $\approx$ 6.8–7.0, 150 mg/L of free available chlorine (FAC)) was freshly prepared each time prior to sanitisation experiment. We have previously optimised the aerosolised ECAS at 150 mg/L of FAC to significantly reduce the total microbial load in an animal farm environment, and for this reason we have chosen this concentration [28]. For the spray wash, ECAS was sprayed for 45 s using a handheld bottle sprayer. For the fog sanitisation, ECAS fog was generated using an ultrasonic humidifier that generates droplets sized between 1 and 3 μm in diameter (Ultrasonic Humidifier HU-85, Contronics Engineering B.V., Sint-Oedenrode, The Netherlands) as previously optimised [28].

2.3. Effectiveness of Spraying and Fogging on Total Bacterial Load Reduction on Eggshell Surface

For total bacterial counts, intact and visibly clean eggs were selected for each of the treatment groups (three eggs per treatment): unwashed control, NaOCl spray (~200 mg/L of FAC, positive control), water spray (45 s), ECAS spray (150 mg/L of FAC; 45 s), water fog (60 and 120 s) and ECAS fog (150 mg/L of FAC; 60 and 120 s). The disinfection procedures were performed in a biosafety cabinet (BSC) and eggs were placed in separate compartments on an oscillating tray for uniform exposure during spray-wash and fogging treatment. Each treatment was independently repeated for at least two times. In addition, ECAS (150 mg/L of FAC) fog treatment was compared for one and three eggs per oscillating tray, for 60 s and 120 s, respectively.

After treatment, individual eggs were immediately placed into a sterile Whirl-Pak bags (Nasco, Fort Atkinson, WI, USA) containing 5 mL of buffered peptone water (BPW), massaged gently (without breaking the egg) for 1 min, then the broth was transferred into 10 mL sterile tubes (SARSTEDT Australia Pty Ltd., Adelaide, Australia). Samples were centrifuged at $5444 \times g$ (MPW-351e Centrifuge, Med Instruments, Adelab Scientific, Adelaide, Australia) for 10 min, supernatant discarded, and the pellet was resuspended in 200 μL of $1 \times$ PBS. Aliquots (100 μL) of 10-fold dilutions (up to 10^{-5}) of the samples were spread plated on plate count agar (PCA; CM0325, Oxoid, Melbourne, Australia) in duplicates and incubated overnight at 37 °C for the enumeration of colonies. Plates with 25 to 300 colonies were used for colony forming unit (CFU) calculation and data presented as $\log_{10}$ CFU/egg.

2.4. Salmonella Enteritidis Seeding on Outer Eggshell Surface

2.4.1. Pre-Wash of Eggs for S. Enteritidis Seeding

To understand the efficacy of ECAS on *Salmonella* load reduction, eggs selected for *Salmonella* seeding were washed as per wash steps specified in Gole et al. [29] before being inoculated with *S.* Enteritidis. Briefly, eggs were placed on an oscillating tray, which helped in exposing the entire eggshell surface, in a BSC and initially spray-washed with a 0.45% (*v/v*) solution of NaOCl (ThermoFisher, Melbourne, Australia; pH $\approx$ 12) at 40 °C for 45 s. Then, spray-sanitised with a 0.16% (*v/v*) solution of NaOCl at 32 °C for 22 s and left on the sterilised BSC bench to dry for 60 min. The eggs were sanitised to ensure the complete removal of the microbiota of the egg surface and to achieve a uniform *S.* Enteritidis colonisation of the egg surface.

2.4.2. *S.* Enteritidis Preparation for Inoculation of Eggshells

Salmonella Enteritidis 11RX was used for this experiment. *S.* Enteritidis stored at -80 °C in 80% glycerol was plated on xylose lysine deoxycholate (XLD) agar (Oxoid CM0469) and incubated overnight at 37 °C to obtain isolated colonies. An inoculum was prepared by suspending colonies in phosphate buffered saline (1 $\times$ PBS) to obtain an absorbance ($OD_{600\,nm}$) value of 0.45. Viable *Salmonella* was enumerated by plating 10-fold serial dilutions of the inoculum on XLD agar and incubating overnight at 37 °C. After enumeration, a 200 mL inoculum containing $\sim 10^5$ CFU per mL was prepared in 1 $\times$ PBS.

For eggshell seeding, eggs were immersed for 90 s either in 1 $\times$ PBS (control) or in 1 $\times$ PBS containing $\sim 10^5$ CFU/mL of *S.* Enteritidis. Eggs were then placed into sterile zip lock bags and incubated at 37 °C. After 18–24 h post-inoculation, three eggs from each treatment were placed in separate Whirl-Pak bags containing 5 mL BPW and massaged for 1 min. Aliquots (100 µL) from a 10-fold serial dilution were spread plated on XLD and PCA media (in duplicate) and incubated overnight at 37 °C for enumeration of counts.

2.4.3. Effectiveness of Spraying and Fogging on *S.* Enteritidis on Eggshell Surface

The *S.* Enteritidis-inoculated eggs were subjected to the same treatment as above (Section 2.3). A 10-fold serial dilutions were prepared for each treatment as above in 1 $\times$ PBS and aliquots were spread plated both on PCA and XLD agar media (in duplicates).

2.5. Eggshell Cuticle Assessment

Twelve eggs per each treatment (water and ECAS spray washing for 45 s and water and ECAS fogging for 2 min) were screened and selected based on colour intensity measured using a MiniScan EZ colourimeter (4500 L Spectrophotometer, Hunter Associates Laboratory, Inc., Reston, VA, USA). The selected eggs were treated as in Section 2.3 and dried in biosafety cabinet for 60 min. The eggs were stained with MST Cuticle Blue dye (MS Technologies, Kettering, UK) as described by Khan et al. [30] and the cuticle coverage was assessed using the $\Delta E_{ab}{}^{*}$ method. The average of four readings of the L^* (lightness), a^* (red/green value) and b^* (yellow/blue value) values, before and after staining, were used for the calculation of $\Delta E_{ab}{}^{*}$ with Equation (1). A higher value of $\Delta E_{ab}{}^{*}$ denotes a higher cuticle staining:

$$\Delta E_{ab}{}^{*} = \sqrt{[(\Delta L^*)^2 + (\Delta a^*)^2 + (\Delta b^*)^2]} \tag{1}$$

2.6. Statistical Analysis

Total bacterial and *S.* Enteritidis counts were expressed as $\log_{10}$/egg and the data were analysed using ANOVA in GraphPad Prism v.8 (GraphPad Software, San Diego, CA, USA). Since the bacterial counts were log transformed (subjected to normal distribution) and each experiment had an equal number of samples, a one-way ANOVA was performed to compare differences of means among untreated control versus different sanitising treatments followed by a Tukey's multiple comparison test. A *p* value of < 0.05 was considered statistically significant.

3. Results

3.1. ECAS Spray and Aerosol Treatment Reduced Total Bacterial Count on Egg Surface

The mean bacterial load ($\log_{10}$ CFU/egg) for untreated and after sanitising treatments are presented in Table 1 and Figure 1. Eggs treated with sprayed water (45 s) (2.4 ± 0.1 $\log_{10}$

CFU/egg) showed no significant reduction ($p = 0.3662$) in total bacterial load when compared with the untreated control (2.2 ± 0.2 $\log_{10}$ CFU/egg; Table 1). Significant reduction in total bacterial load was observed when the eggs were treated with sprayed ECAS (45 s) (2.2 log reduction, $p < 0.0001$) and aerosolised ECAS (120 s) (1.4 log reduction, $p = 0.01$) compared to untreated control (Table 1). Aerosolised water (60 s and 120 s) and aerosolised ECAS (60 s) showed no significant reduction in total bacterial load when compared with sprayed water (45 s) or the untreated control (Table 1). In addition, the spray versus aerosol techniques were compared for their effectiveness in reducing the total bacterial load (Figure 1A). Sprayed ECAS (45 s) showed a significant reduction ($p < 0.0001$) in total bacterial load when compared with sprayed water (45 s) (Figure 1A). Aerosolised ECAS (120 s) also showed a significant reduction ($p < 0.01$) in total bacterial load when compared with aerosolised water (120 s); however, no significant reduction was shown for the 60 s treatment group (Figure 1A).

Table 1. Effect of sanitisation treatments on reduction of total bacterial load on egg surface.

Treatment	Total Bacterial Count ($\mathbf{Log_{10}}$ CFU/Egg)	Log Reduction [&]	Significance (*p*-Value)
Untreated control	2.2 ± 0.2	—	—
NaOCl (45 s spray)	0	2.2	<0.0001
Water (45 s spray)	2.4 ± 0.1	(+0.2)	0.3662
ECAS (45 s spray)	0	2.2	<0.0001
Water fog (60 s)	2.5 ± 0.2	(+0.3)	0.1201
ECAS fog (60 s)	2.1 ± 0.1	0.1	0.9206
Water fog (120 s)	2.3 ± 0.2	(+0.1)	0.9895
ECAS fog (120 s)	0.8 ± 0.6	1.4	0.01

Total bacterial counts were calculated as $\log_{10}$ CFU/egg and presented as mean $\pm$ standard deviation (SD); [&] $\log_{10}$ reduction = ($\log_{10}$ counts of untreated control)—($\log_{10}$ counts after sanitising treatment); a '+' sign means log increase in counts.

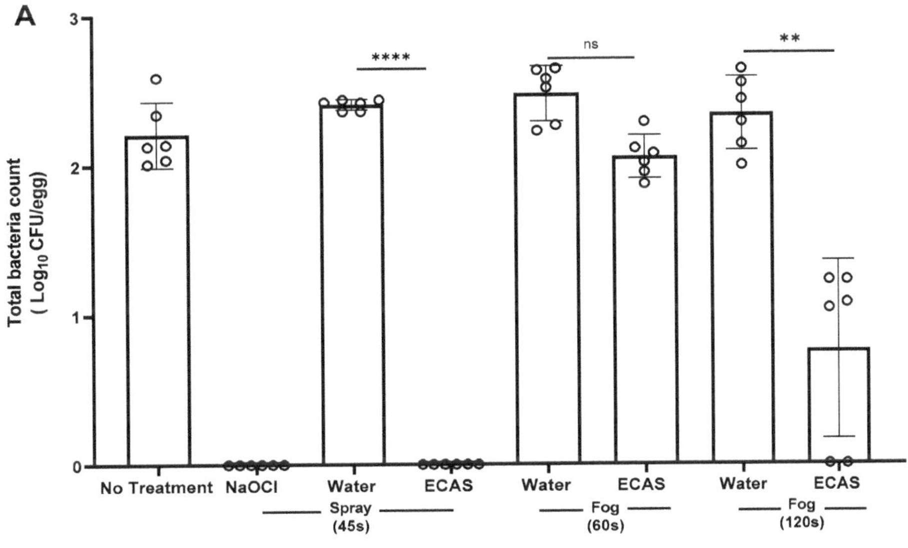

Figure 1. *Cont.*

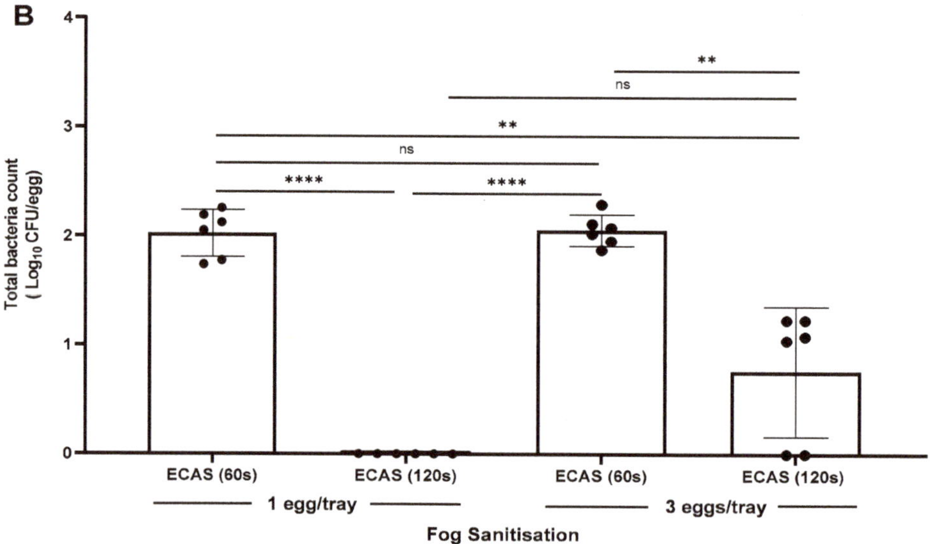

Figure 1. Effectiveness of sprayed and aerosolised ECAS on reduction of total bacterial load on egg surface. (**A**) Total bacterial load after the various sanitisation treatments. (**B**) Aerosolised ECAS sanitisation of individual (one egg/tray) and simultaneous (three eggs/tray) treated eggs for 60 s and 120 s. NaOCl—sodium hypochlorite (~200 mg/L FAC); ECAS—electrochemically activated solution (150 mg/L of FAC); ns—not significant; ** $p < 0.01$, **** $p < 0.0001$.

The effectiveness of aerosolised ECAS was further assessed when the eggs were sanitised simultaneously (three eggs/tray) or individually (one egg/tray) for 60 s and 120 s, respectively. Aerosolised ECAS (120 s) significantly reduced the total bacteria load for both one egg/tray ($p < 0.0001$) and three eggs/tray ($p < 0.01$; Figure 1B). The treatment of three eggs/tray with aerosolised ECAS (120 s) did not eliminate the total bacterial load completely; however, the level was not significantly different from the treatment of one egg/tray aerosolised ECAS (120 s) (Figure 1B).

3.2. ECAS Spray and Fog Treatment Reduced S. Enteritidis on the Egg Surface

S. Enteritidis counts (log_{10} CFU/egg) are presented in Table 2 and Figure 2. Sprayed water (45 s) showed a significant reduction (1.0 log reduction, $p = 0.0005$; Table 2) when compared with the untreated control. All the ECAS treatments significantly reduced the inoculated *S.* Enteritidis counts on the egg surfaces compared to the untreated control: sprayed ECAS (45 s) (5.4 log reduction, $p < 0.0001$; Table 2), aerosolised ECAS (60 s) (2.8 log reduction, $p = 0.0008$; Table 2) and aerosolised ECAS (120 s) (4.2 log reduction, $p = 0.0022$; Table 2). A significant reduction in *S.* Enteritidis counts (1.1 log reduction, $p < 0.0001$) was also achieved with aerosolised water (120 s), whereas no significant reduction was observed with aerosolised water (60 s) when compared with the untreated control (Table 2). For the spray versus aerosol techniques comparison, sprayed ECAS (45 s) showed a significant reduction ($p < 0.0001$) in *S.* Enteritidis load when compared with the sprayed water (45 s) (Figure 2A). Significant reduction in *S.* Enteritidis load was observed for aerosolised ECAS (60 s, $p < 0.01$) and aerosolised ECAS (120 s, $p < 0.01$) when compared with respective aerosolised water treatments (Figure 2A). *S.* Enteritidis level was significantly reduced in spray water (45 s, $p = 0.0005$) and aerosolised water treatment groups (120 s, $p < 0.0001$), respectively.

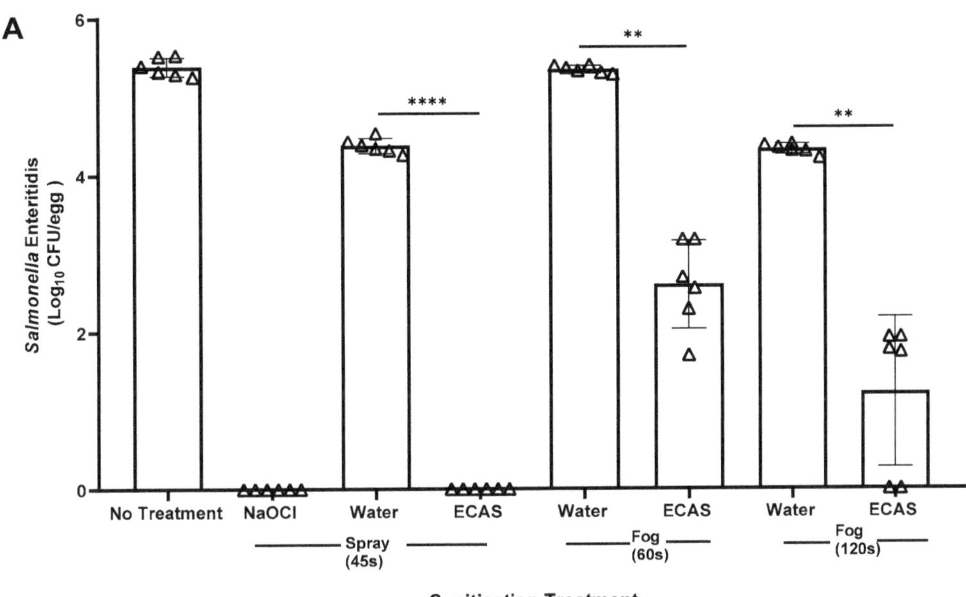

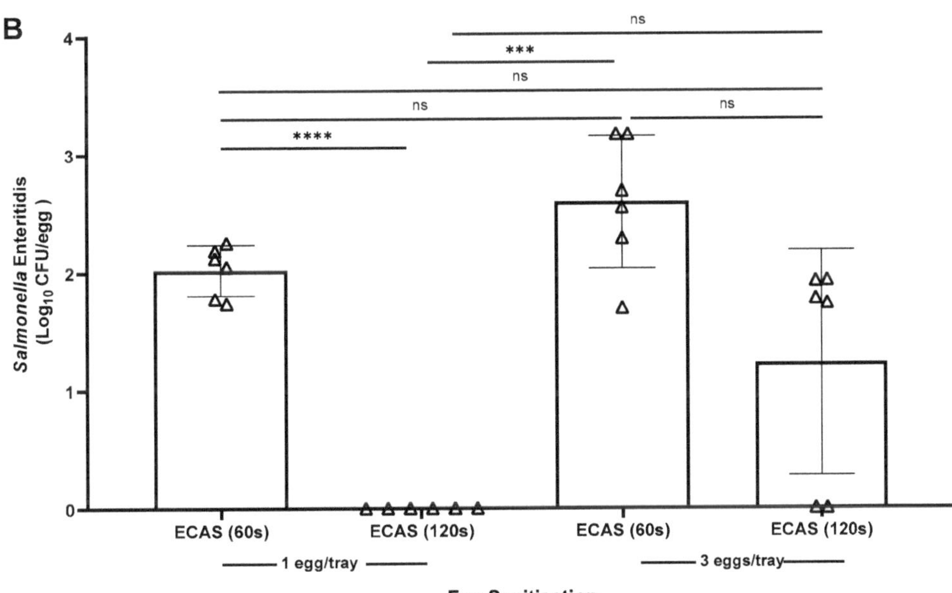

Figure 2. Effectiveness of sprayed and aerosolised ECAS on reduction of *S.* Enteritidis counts. (**A**) *S.* Enteritidis counts after the various sanitisation treatment. (**B**) Aerosolised ECAS sanitisation of individual (one egg/tray) and simultaneous (three eggs/tray) treated eggs for 60 s and 120 s. NaOCl—sodium hypochlorite (~200 mg/L FAC); ECAS—electrochemically activated solution (150 mg/L of FAC); ns—not significant; ** $p < 0.01$, *** $p < 0.001$, **** $p < 0.0001$.

Table 2. Effect of sanitisation treatments on the reduction level of *S.* Enteritidis counts on egg surface.

Treatment	S. Enteritidis Counts (log_{10} CFU/Egg)	Log Reduction [&]	Significance (*p*-Value)
Untreated control	5.4 ± 0.1	—	—
NaOCl spray (45 s)	0	5.4	<0.0001
Water spray (45 s)	4.4 ± 0.1	1.0	0.0005
ECAS spray (45 s)	0	5.4	<0.0001
Water fog (60 s)	5.3 ± 0.1	0.1	0.9926
ECAS fog (60 s)	2.6 ± 0.6	2.8	0.0008
Water fog (120 s)	4.3 ± 0.7	1.1	<0.0001
ECAS fog (120 s)	1.2 ± 1.0	4.2	0.0022

S. Enteritidis counts calculated as log_{10} CFU/egg and presented as mean $\pm$ standard deviation (SD); [&] log_{10} reduction = (log_{10} counts of untreated control)—(log_{10} counts after sanitising treatment).

Aerosolised ECAS (120 s) significantly reduced *S.* Enteritidis counts for one egg/tray ($p < 0.0001$); however, no significant reduction trend was observed in the 3 eggs/tray treatment group (Figure 1B).

3.3. ECAS Spray and Aerosol Treatments Did Not Affect Egg Cuticle Integrity

The effect of sprayed and aerosolised ECAS treatments on cuticle coverage (ΔE_{ab}^{*}) was measured (Figure 3). Sprayed ECAS (45 s) and aerosolised ECAS (120 s) did not show any significant difference when compared with the respective water controls.

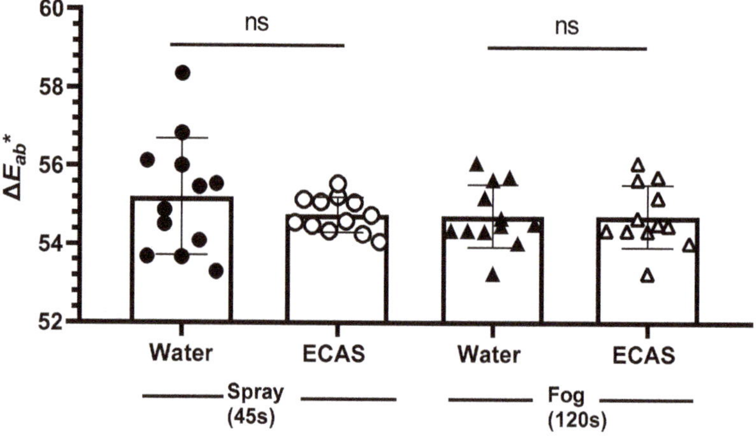

Figure 3. Effect of sprayed and aerosolised ECAS treatments on cuticle integrity (ΔE_{ab}^{*}). Data presented as mean $\pm$ SD, n = 12; ns—not significant.

4. Discussion

This study assessed the effectiveness of sprayed and aerosolised ECAS in reducing total bacterial load and inoculated *S.* Enteritidis on egg surface and their effects on the cuticle layer. The bacterial load on eggshells is usually acquired through contamination from the farm environment; therefore, farm type and poultry housing system influence the egg surface total bacterial count [31–34]. Eggshells from conventional-caged hens usually harbour lower total bacterial counts compared to other housing systems [35]. The total bacterial load of 2.2 ± 0.2 log_{10} CFU/egg found in this study was lower than those observed by Wall et al. [36] (2.7 log_{10}) and Alvarez-Fernandez et al. [31] (2.3 log_{10}) on eggshells from conventional-caged hens, probably because the eggs used in this study were laid by hens housed individually with a low density of hens in the shed (49 hens/shed).

ECAS spray-wash of eggs for 45 s and aerosolisation of individual eggs with ECAS for 120 s completely reduced the native total bacterial load and *S.* Enteritidis on eggshell surface. These two sanitisation approaches showed an effectiveness like that of a sodium hypochlorite (~200 mg/L of FAC) spray-wash for 45 s, which was used as a positive control. Although relevant literature is not available on the sanitisation of eggshell surface with ECAS fog, previous research has shown a significant reduction in total bacterial counts from the eggshell surface when the latter is spray-washed with pH-neutral EO water or acidic EO water [22,37]. In the present study, the effectiveness of aerosolised ECAS with one and three eggs per oscillating tray for 60 s and 120 s was also tested. ECAS fogging of multiple eggs (three simultaneously) as well as of a single egg for 120 s showed a significant reduction in total bacterial load when compared with the respective 60 s treatments, confirming that a longer treatment time may be more appropriate for total bacteria load reductions. It is noteworthy that a higher variation was observed in treating three eggs simultaneously, which could be attributed to an uneven distribution of the fog, as fogging was performed in a biological safety cabinet that quickly sucked out the surrounding air including the fog.

Previous studies have assessed the effectiveness of various forms of electrolysed water (acidic, slightly acidic, neutral, alkaline) by immersion or spray-washing of eggshell surface inoculated with *S.* Enteritidis [24,38–40]. In this study, we assessed a pH-neutral EO water in the form of fog in addition to spray-washing. The significant microbial reduction observed in our study for sprayed water and aerosolised water (120 s) could be due to the dislodgment of the inoculated *S.* Enteritidis from the egg surface; however, these techniques were ineffective in the reduction of total bacterial load. Sprayed ECAS (45 s) and aerosolised ECAS (60 s and 120 s) showed significant reduction in *S.* Enteritidis counts. Medina-Gudino et al. [24] reported that pH-neutral EO spray treatment (30 s) significantly reduced *S.* Enteritidis (>1.45 $\log_{10}$ CFU/egg) on egg surface. In another study [39], a slightly acidic (pH 6.53) EO water (15 mg/L of FAC) used at 20 and 45 °C showed a 4.2 $\log_{10}$ CFU/mL reduction of *S.* Enteritidis. Aerosolised ECAS (120 s) showed a significant log reduction of *S.* Enteritidis for one egg/tray treatment but no significant reduction was observed in the three eggs/tray treatment group. As previously mentioned, the higher variation in the three eggs/tray treatment could be due to the fog being quickly sucked by the BSC and not moistening the egg surface with enough FAC to reduce the bacterial cells.

The main purpose of egg washing is to reduce the bacterial load on the egg surface; however, one of the major concerns in the egg washing process is the likely damage of the cuticle layer. In the present study, no difference was observed in $\Delta E_{ab}{}^*$ values in eggs sprayed and aerosolised with ECAS, indicating that cuticle integrity was not altered. In a recent study, Medina-Gudino et al. [24] reported no differences in the overall $\Delta E_{ab}{}^*$ values when eggs were treated with pH-neutral EO. In contrast, for eggs washed with alkaline and acidic EO, changes in the *a** and *b** values were observed, hence the reduction in overall $\Delta E_{ab}{}^*$ values, indicating damage to the egg cuticle [19].

Although electrolysed water is an environmentally friendly, non-hazardous sanitiser with proven antibacterial efficacy against foodborne pathogens such as *E. coli* O157: H7, *L. monocytogenes* [41] and *S.* Enteritidis on shell eggs [38,39], it is not currently being considered in commercial decontamination settings to obtain pathogen-free eggs. The reasons for this are probably related to the corrosiveness of acidic formulations towards steel surfaces and the lack of knowledge of consumers on the impact of chemicals for disinfection of environments.

The current scenario shows a shift in table egg production towards free-range systems [42]. For instance, in Australia, free-range production has increased from 39% in 2015 to 45% in 2018 [10], driven by consumer demand for bird welfare and access to a range area. However, this approach poses risks to public health as the eggshell bacterial load, including counts of *Campylobacter* and *Salmonella* spp. [42], in free range production systems is considerably higher than in conventional battery cage systems [42,43]. The problem is further aggravated by the increase in the consumption of meals consisting of raw egg products [44]. Therefore, egg producers should consider eggshell cleaning and disinfection of table eggs as a priority to produce safe eggs and maintain consumer confidence. Besides issues with wastewater disposal, commonly used chlorine-based egg washing requires intensive monitoring of water temperature, pH, and chlorine concentration to retain its optimal efficacy and eggshell cuticle integrity. This suggests that electrolysed water could fill an important niche if washing and disinfection systems can be developed on a commercial level.

5. Conclusions

pH-neutral ECAS (150 mg/L of FAC), when used in the form of a spray (45 s) or as an aerosol (120 s), allows significant reductions in total bacterial load and *S.* Enteritidis counts from contaminated egg surface while retaining cuticle integrity. However, aerosolised ECAS (60 s) did not show significant reduction in either the total bacterial load or *S.* Enteritidis counts, suggesting a longer fogging time is needed to sanitise the eggs. The aerosolised ECAS sanitisation technique could be incorporated into cleaning and disinfection protocol to improve egg safety without the use of hazardous biocidal agents. Moreover, this disinfection protocol is easily implementable as ECAS can be easily generated on site with automated controls for FAC concentration and pH measurements. For this process to lead to large-scale tests and industrial implementation, additional testing needs to be performed, including the elimination of other pathogenic bacteria from eggshells, interior egg quality and consumer sensory evaluation, as per the regulatory requirements.

Author Contributions: Conceptualization, S.T., P.D., D.J.T.; Methodology, S.T., S.K., P.D.; Formal analysis, S.T., P.D.; Resources, S.F., S.K., P.D., D.J.T.; Data curation, S.T.; Writing—original draft preparation, S.T., P.D.; Writing—review and editing, S.T., S.F., S.K., P.D., D.J.T.; Visualization, S.T.; Supervision, P.D., D.J.T.; Funding acquisition, D.J.T. All authors have read and agreed to the published version of the manuscript.

Funding: S.T. was supported by the Endeavour Postgraduate Scholarship.

Conflicts of Interest: The Authors declare no conflict of interest.

References

1. Kirk, M.; Ford, L.; Glass, K.; Hall, G. Foodborne Illness, Australia, circa 2000 and circa 2010. *Emerg. Infect. Dis.* **2014**, *20*, 1857–1864. [CrossRef] [PubMed]
2. NNDSS, National Notifiable Diseases. Australia's notifiable diseases status 2015: Annual report of the National Notifiable Diseases Surveillance System. *Commun. Dis. Intell.* **2019**, *43*. [CrossRef]
3. Painter, J.A.; Hoekstra, R.; Ayers, T.; Tauxe, R.V.; Braden, C.R.; Angulo, F.J.; Griffin, P.M. Attribution of foodborne illnesses, hospitalizations, and deaths to food commodities by using outbreak data, United States, 1998–2008. *Emerg. Infect Dis.* **2013**, *19*, 407–415. [CrossRef] [PubMed]
4. Braden, C.R. *Salmonella enterica* serotype Enteritidis and eggs: A national epidemic in the United States. *Clin. Infect Dis.* **2006**, *43*, 512–517. [CrossRef] [PubMed]
5. Moffatt, C.R.; Musto, J.; Pingault, N.; Miller, M.; Stafford, R.; Gregory, J.; Polkinghorne, B.G.; Kirk, M.D. *Salmonella* Typhimurium and outbreaks of egg-associated disease in Australia, 2001 to 2011. *Foodborne Pathog. Dis.* **2016**, *13*, 379–385. [CrossRef] [PubMed]
6. Ford, L.; Moffatt, C.R.M.; Fearnley, E.; Miller, M.; Gregory, J.; Sloan-Gardner, T.S.; Polkinghorene, B.G.; Bell, R.; Franklin, N.; Williamson, D.A.; et al. The epidemiology of *Salmonella enterica* outbreaks in Australia, 2001–2016. *Front. Sustain. Food Syst.* **2018**, *2*, 86. [CrossRef]
7. Bélanger, P.; Tanguay, F.; Hamel, M.; Phypers, M. Foodborne Illness: An overview of foodborne outbreaks in Canada reported through outbreak summaries: 2008–2014. *Can. Commun. Dis. Rep.* **2015**, *41*, 254–262. [CrossRef]
8. Jackson, B.R.; Griffin, P.M.; Cole, D.; Walsh, K.A.; Chai, S.J. Outbreak-associated *Salmonella enterica* serotypes and food commodities, United States, 1998–2008. *Emerg. Infect. Dis.* **2013**, *19*, 1239–1244. [CrossRef]
9. Sasaki, Y.; Tsujiyama, Y.; Asai, T.; Noda, Y.; Katayama, S.; Yamada, Y. Salmonella prevalence in commercial raw shell eggs in Japan: A survey. *Epidemiol. Infect.* **2011**, *139*, 1060–1064. [CrossRef]
10. AECL. Annual Report 2017–2018. In *Australian Egg Corporation Limited Annual Report 2017-18*; Australian Egg Corporation: North Sydney, Australia, 2018.
11. Perry, J.J.; Rodriguez-Romo, L.A.; Yousef, A.E. Inactivation of *Salmonella enterica* serovar enteritidis in shell eggs by sequential application of heat and ozone. *Lett. Appl. Microbiol.* **2008**, *46*, 620–625. [CrossRef]

12. Al-Ajeeli, M.N.; Taylor, T.M.; Alvarado, C.Z.; Coufal, C.D. Comparison of eggshell surface sanitization technologies and impacts on consumer acceptability. *Poult. Sci.* **2016**, *95*, 1191–1917. [CrossRef] [PubMed]

13. Langsrud, S.; Sundheim, G.; Borgmann-Strahsen, R. Intrinsic and acquired resistance to quaternary ammonium compounds in food-related *Pseudomonas* spp. *J. Appl. Microbiol.* **2003**, *95*, 874–882. [CrossRef]

14. Fernandez Marquez, M.L.; Burgos, M.J.; Pulido, R.P.; Gálvez, A.; López, R.L. Biocide tolerance and antibiotic resistance in Salmonella isolates from hen eggshells. *Foodborne Pathog. Dis.* **2017**, *14*, 89–95. [CrossRef]

15. Ridgway, H.F.; Olson, B.H. Chlorine resistance patterns of bacteria from two drinking water distribution systems. *Appl. Environ. Microbiol.* **1982**, *44*, 972–987. [CrossRef]

16. De Reu, K.; Grijspeerdt, K.; Herman, L.; Heyndrickx, M.; Uyttendaele, M.; Debevere, J.; Putirulan, F.F.; Bolder, N.M. The effect of a commercial UV disinfection system on the bacterial load of shell eggs. *Lett. Appl. Microbiol.* **2006**, *42*, 144–148. [CrossRef] [PubMed]

17. IARC. Formaldehyde, 2-Butoxyethanol and 1-tert-Butoxypropan-2-ol, in formaldehyde, 2-butoxyethanol and 1-tert-butoxypropan-2-ol. *IARC Monogr. Eval. Carcinog. Risks Hum.* **2006**, *88*, 1–478.

18. Achiwa, N.; Nishio, T. The use of electrolyzed water for sanitation control of eggshells and GP center. *Food Sci. Technol. Res.* **2003**, *9*, 100–103. [CrossRef]

19. Bialka, K.L.; Demirci, A.; Knabel, S.J.; Patterson, P.H.; Puri, V.M. Efficacy of electrolyzed oxidizing water for the microbial safety and quality of eggs. *Poult. Sci.* **2004**, *83*, 2071–2078. [CrossRef]

20. Rivera-Garcia, A.; Santos-Ferro, L.; Ramirez-Orejel, J.C.; Agredano-Moreno, L.T.; Jimenez-Garcia, L.F.; Paez-Esquiliano, D.; Andrade-Esquivel, E.; Cano-Buendia, J.A. The effect of neutral electrolyzed water as a disinfectant of eggshells artificially contaminated with *Listeria monocytogenes*. *Food Sci. Nutr.* **2019**, *7*, 2252–2260. [CrossRef] [PubMed]

21. Zang, Y.T.; Bing, S.; Li, Y.J.; Shu, D.Q.; Huang, A.M.; Wu, H.X.; Lan, L.T.; Wu, H.D. Efficacy of slightly acidic electrolyzed water on the microbial safety and shelf life of shelled eggs. *Poult. Sci.* **2019**, *98*, 5932–5939. [CrossRef] [PubMed]

22. Fasenko, G.M.; O'Dea Christopher, E.E.; McMullen, L.M. Spraying hatching eggs with electrolyzed oxidizing water reduces eggshell microbial load without compromising broiler production parameters. *Poult. Sci.* **2009**, *88*, 1121–1127. [CrossRef] [PubMed]

23. Surdu, I.; Vatuiu, D.; Jurcoane, S.; Olteanu, M.; Vatuiu, I. The Antimicrobial Activity of Neutral Electrolyzed Water against Germs and Fungi from Feedstuffs, Eggshells and Laying Hen House. *Rom. Biotechnol. Lett.* **2018**, *23*, 13607–13614.

24. Medina-Gudiño, J.; Rivera-Garcia, A.; Santos-Ferro, L.; Ramirez-Orejel, J.C.; Agredano-Moreno, L.T.; Jimenez-Garcia, L.F.; Paez-Esquiliano, D.; Martinez-Vidal, S.; Andrade-Esquivel, E.; Cano-Buendia, J.A. Analysis of neutral electrolyzed water anti-bacterial activity on contaminated eggshells with *Salmonella enterica* or *Escherichia coli*. *Int. J. Food Microbiol.* **2020**, *320*, 108538. [CrossRef] [PubMed]

25. Cheng, K.-C.; Dev, S.R.S.; Bialka, K.L.; Demirci, A. Electrolyzed oxidizing water for microbial decontamination of food. In *Microbial Decontamination in the Food Industry*; Elsevier: Amsterdam, The Netherlands, 2012; pp. 563–591. [CrossRef]

26. Guentzel, J.L.; Liang Lam, K.; Callan, M.A.; Emmons, S.A.; Dunham, V.L. Reduction of bacteria on spinach, lettuce, and surfaces in food service areas using neutral electrolyzed oxidizing water. *Food Microbiol.* **2008**, *25*, 36–41. [CrossRef] [PubMed]

27. Liao, L.B.; Chen, W.M.; Xiao, X.M. The generation and inactivation mechanism of oxidation–reduction potential of electrolyzed oxidizing water. *J. Food Engineer.* **2007**, *78*, 1326–1332. [CrossRef]

28. Tenzin, S.; Ogunniyi, A.D.; Khazandi, M.; Ferro, S.; Bartsch, J.; Crabb, S.; Abraham, S.; Deo, P.; Trott, D.J. Decontamination of aerosolised bacteria from a pig farm environment using a pH neutral electrochemically activated solution (Ecas4 anolyte). *PLoS ONE* **2019**, *14*, e0222765. [CrossRef] [PubMed]

29. Gole, V.C.; Chousalkar, K.K.; Roberts, J.R.; Sexton, M.; May, D.; Tan, J.; Kiermeier, A. Effect of egg washing and correlation between eggshell characteristics and egg penetration by various *Salmonella* Typhimurium strains. *PLoS ONE* **2014**, *9*, e90987. [CrossRef]

30. Chousalkar, K.K.; Roberts, J.R.; Sexton, M.; May, D.; Kiermeier, A. Effects of egg shell quality and washing on *Salmonella* Infantis penetration. *Int. J. Food Microbiol.* **2013**, *165*, 77–83. [CrossRef]

31. Alvarez-Fernandez, E.; Domínguez-Rodríguez, J.; Capita, R.; Alonso-Calleja, C. Influence of housing systems on microbial load and antimicrobial resistance patterns of *Escherichia coli* isolates from eggs produced for human consumption. *J. Food Prot.* **2012**, *75*, 847–853. [CrossRef]

32. De Reu, K.; Grijspeerdt, K.; Heyndrickx, M.; Zoons, J.; De Baere, K.; Uyttendaele, M.; Debevere, J.; Herman, L. Bacterial eggshell contamination in conventional cages, furnished cages and aviary housing systems for laying hens. *Br. Poult. Sci.* **2005**, *46*, 149–155. [CrossRef]

33. De Reu, K.; Grijspeerdt, K.; Messens, W.; Heyndrickx, M.; Uyttendaele, M.; Debevere, J.; Herman, L. Eggshell factors influencing eggshell penetration and whole egg contamination by different bacteria, including *Salmonella enteritidis*. *Int. J. Food Microbiol.* **2006**, *112*, 253–260. [CrossRef] [PubMed]

34. De Reu, K.; Rodenburg, T.B.; Grijspeerdt, K.; Messens, W.; Heyndrickx, M.; Tuyttens, F.A.; Sonck, B.; Zoons, J.; Herman, L. Bacteriological contamination, dirt, and cracks of eggshells in furnished cages and noncage systems for laying hens: An international on-farm comparison. *Poult. Sci.* **2009**, *88*, 2442–2448. [CrossRef]

35. Dukic-Stojcic, M.; Peric, L.; Bjedov, S.; Milosevic, N. The quality of table eggs produced in different housing systems. *Biotechnol. Anim. Husbandry.* **2009**, *25*, 1103–1108.

36. Wall, H.; Tauson, R.; Sorgjerd, S. Bacterial contamination of eggshells in furnished and conventional cages. *J. App. Poult.Res.* **2008**, *17*, 11–16. [CrossRef]
37. Ni, L.; Cao, W.; Zheng, W.-C.; Chen, H.; Li, B.-M. Efficacy of slightly acidic electrolyzed water for reduction of foodborne pathogens and natural microflora on shell eggs. *Food Sci. Technol. Res.* **2014**, *20*, 93–100. [CrossRef]
38. Park, C.-M.; Hung, Y.C.; Lin, C.S.; Brackett, R.E. Efficacy of electrolyzed water in inactivating *Salmonella enteritidis* and *Listeria monocytogenes* on shell eggs. *J. Food Prot.* **2005**, *68*, 986–990. [CrossRef] [PubMed]
39. Cao, W.; Zhu, Z.W.; Shi, Z.X.; Wang, C.Y.; Li, B.M. Efficiency of slightly acidic electrolyzed water for inactivation of *Salmonella enteritidis* and its contaminated shell eggs. *Int. J. Food Microbiol.* **2009**, *130*, 88–93. [CrossRef]
40. Venkitanarayanan, K.S.; Ezeike, G.O.; Hung, Y.C.; Doyle, M.P. Efficacy of electrolyzed oxidizing water for inactivating *Escherichia coli* O157:H7, *Salmonella enteritidis*, and *Listeria monocytogenes*. *Appl. Environ. Microbiol.* **1999**, *65*, 4276–4279. [CrossRef]
41. Park, H.; Hung, Y.C.; Chung, D. Effects of chlorine and pH on efficacy of electrolyzed water for inactivating *Escherichia coli* O157:H7 and *Listeria monocytogenes*. *Int. J. Food Microbiol.* **2004**, *91*, 13–18. [CrossRef]
42. Parisi, M.A.; Northcutt, J.K.; Smith, D.P.; Steinberg, E.L.; Dawson, P.L. Microbiological contamination of shell eggs produced in conventional and free-range housing systems. *Food Control.* **2015**, *47*, 161–165. [CrossRef]
43. De Reu, K.; Messens, W.; Heyndrickx, M.; Rodenburg, T.B.; Uyttendaele, M.; Herman, L. Bacterial contamination of table eggs and the influence of housing systems. *World's Poult. Sci. J.* **2008**, *64*, 5–19. [CrossRef]
44. Kretser, A.; Dunn, C.; DeVirgilis, R.; Perry, K.L. Utility of a new food value analysis application to evaluate trade-offs when making food selections. *Nutr. Today.* **2014**, *49*, 185–195. [CrossRef]

Article

Automated Chlorine Dosage in a Simulated Drinking Water Treatment Plant: A Real Case Study

Javier Gámiz [1,2], **Antoni Grau** [1,*], **Herminio Martínez** [3] and **Yolanda Bolea** [1]

[1] Automatic Control Department, Technical University of Catalonia, 08034 Barcelona, Spain; jgamiz@agbar.es (J.G.); yolanda.bolea@upc.edu (Y.B.)

[2] Industrial Control Systems Department, Aigües de Barcelona, 08038 Barcelona, Spain

[3] Department of Electronics Engineering, Technical University of Catalonia, 08034 Barcelona, Spain; herminio.martinez@upc.edu

* Correspondence: antoni.grau@upc.edu

Received: 21 May 2020; Accepted: 8 June 2020; Published: 11 June 2020

Featured Application: This research was applied in the Sant Joan Despí drinking water treatment plant in Barcelona city, providing drinking water to 1,900,000 inhabitants. The tool was validated with real data obtained in the plant in different conditions of effluents.

Abstract: In this paper, we present a simulator of a drinking water treatment plant. The model of the plant was based in hydraulic and matter transportation models. In order to not introduce more inaccuracies in the simulation, the control system was based in the real equipment deployed in the plant. This fact was the challenging part of the simulator, and an accurate design is presented in this research, wherein the sampling time had to be limited to interchange data between the SCADA in the plant and the simulator in real time. Due to the impossibility to stop the plant when testing the new control strategy, a simulator implemented the plant behavior under different extreme conditions. The validation of the simulator was performed with real data obtained from the plant.

Keywords: drinking water treatment plant; advection–diffusion–reaction; chlorine dosage; simulator validation; industrial application

---

1. Introduction

Drinking water is a necessary but scarce good. Despite the fact that water is the most abundant and common substance on our planet—as it covers 70% of its surface—96.5% of it is contained in the oceans. Of the remaining 3.5%, approximately 2.5% is found in the polar ice caps and glaciers and only 0.61% is liquid fresh water. Of the latter, around 0.98% is found in underground aquifers, which are difficult to access, while only 0.009% constitutes fresh surface water (rivers and lakes). Furthermore, only 0.003% of the total is fresh water available to be used for residential purposes. That is, if the Earth's total water were a 100 L container, only half a teaspoon of water would be suitable for human consumption [1]. The sources of water pollution can be natural (rain, decomposing vegetable matter, soil erosion, etc.) or anthropogenic (activity livestock, by-products of industrial activity, home waters, etc.), but both give rise to water that does not meet the necessary requirements to ensure its potability. The basic water treatment processes include several stages: coagulation, flocculation, particle separation (sedimentation/flotation), filtration, and disinfection (chlorination/ozonation) [2]. In many of these stages, the addition of chemicals into the flow of water to be treated is performed, and it is at this very point that this research provides a contribution with respect to correct dosage and control. Of all the treatments mentioned above, this paper focuses on disinfection. This process attempts to destroy or inactivate pathogenic organisms present in water, mainly bacteria, viruses, and protozoa [3].

Disinfection treatments can be physical (gamma radiation, X-rays, ultraviolet radiation, thermal sterilization, etc.) or chemical (heavy metals, acids or bases, halogens, ozone, permanganate, etc.), with the latter being the most common form of treatment. Among the chemical reagents, chlorine and its derived compounds are the most widely used disinfecting agents worldwide. Many organisms regulate the residual chlorine values and they depend on the end use of the water. Thus, for drinking water, it is recommended that the residual free chlorine be between 0.2 and 1 ppm (parts per million), while in the case of swimming pools and spas, it should be kept between 1.5–3.0 ppm. However, these values are general and each competent body has determined its own thresholds.

The use of chlorine as a disinfecting agent began in the early 20th century and went on to complete the filtration process, which was already widely used. The most common chlorine family products for water disinfection are chlorine gas, chloramines, sodium hypochlorite, and calcium hypochlorite. Chlorine (Cl_2) is a yellowish green, denser than air, toxic gas. It is a very oxidizing product that reacts with many compounds [4]. In fact, the most widely used chemical in the world is chlorine.

In the presence of humidity, chlorine is extremely corrosive and therefore the conduits and the materials in contact with it must be made of special alloys. Chlorine vapor is irritating by inhalation and can cause serious injury if exposed to high concentrations. Chlorine management must therefore be carried out by specialized staff, and very effective control and alarm systems are necessary. For these reasons, the use of hypochlorites in solution or in solid form is preferable. Sodium hypochlorite (NaClO) in solution is a disinfectant that has been used since the 18th century and is popularly known as bleach. At an industrial level, it is obtained by reacting chlorine gas with a sodium hydroxide solution. After the reaction, greenish yellow aqueous solutions are obtained, which have a determined concentration of active chlorine per liter [5]. It is marketed in solutions with concentrations between 3% and 15% by weight. Sodium hypochlorite is a very powerful and unstable oxidant, and thus a solution of 100 g of active chlorine per liter, after being stored for 3 months, can contain 90 g or even less. Calcium hypochlorite ($Ca(ClO)_2$) is a white solid with a content of between 20% and 70% active chlorine. It is highly corrosive and can ignite on contact with certain acidic materials. However, it has two advantages over sodium hypochlorite: its higher chlorine content and its greater stability. To be used, it is diluted with water to obtain a more manageable concentration solution, for example 2%. When Cl_2 dissolves in water, it rapidly hydrolyses to generate hypochlorous acid and hydrochloric acid.

$$Cl_2 + H_2O \quad \leftrightarrow \quad HClO + HCl$$

In the case of hypochlorites, the dissociation of both salts occurs according to the following equations:

$$NaClO + H_2O \quad \leftrightarrow \quad NaOH + HClO$$
$$Ca(ClO)_2 + 2H_2O \quad \leftrightarrow \quad Ca(OH)_2 + 2HClO$$

Therefore, hypochlorous acid, which is actually the disinfectant variety, ends up forming in either case chlorine, sodium hypochlorite and calcium hypochlorite.One of the disadvantages of using chlorine and its derivatives is that it reacts with a large amount of organic matter and gives rise to trihalomethanes (THM), many of which have been shown to be toxic or carcinogenic [6]. Another drawback is the formation of chlorophenols in waters containing phenols, which would lead to bad odors. Chlorine also reacts with ammonia dissolved in water to form chloramines. These products also have some disinfecting power, although they are approximately 25 times less effective than free chlorine. However, their residence time in the water is long and, for this reason, they have sometimes been used as a reserve for residual chlorine. They present two major drawbacks: they can give rise to odors and flavors and are potentially toxic in a chronic way. Figure 1 shows an interesting representation of the risks associated with disinfection with chlorine. We adapted this figure from [7], indicating the optimal point for chlorination; due to the dynamics of chemical and microbiological risks, the optimal point of chlorination is at the intersection, in order to avoid both risks. It is clear that not dosing can lead to a very high risk of infection (microbiological risk), but in any case, overdosing is not a valid solution, as it does not guarantee the elimination of health risks, since it favors the increase of chemical risks. An insufficient dose can cause vomiting or diarrhea (microbiological risk) and an overdose has carcinogenic effects (chemical risk); thus, the objective is an optimal dose. At very high levels of chlorine, the microbial risk increases, as taste and odor may cause the use of unsafe supplies.

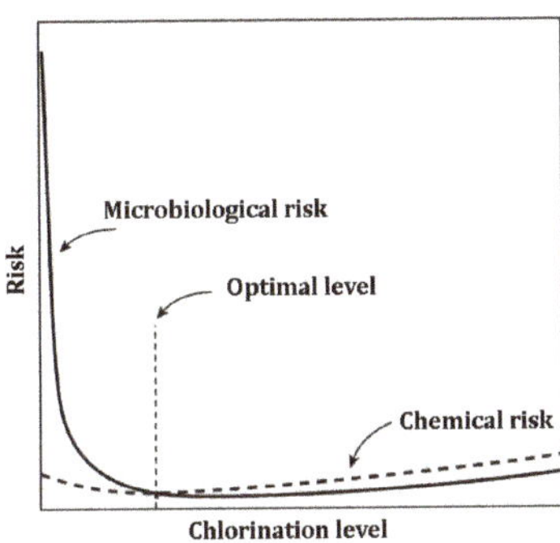

Figure 1. Relationship between chlorination level and associated risk, adapted from [7].

On the basis of the foregoing, it can be deduced that chlorine (and derivatives), in addition to reacting with microorganisms, also reacts with other matter dissolved in the medium: organic matter, metals (iron, manganese), and mainly plastic derivatives. For this reason, to achieve a certain level of residual chlorine, the necessary amount to be added is much higher than the residual obtained. Therefore, before deciding on the dose of chlorine to be used to disinfect, the demand for chlorine must be determined, that is, the amount of chlorine consumed within the tank until the residual appears at the end of the plant, prior to entering the water distribution step.

Regarding the legislation of drinking water, there are some regulations at different levels. The higher level in Europe is dictated by the European Parliament and of the Council in its Directive 2008/105/EC on environmental quality standards in the field of water policy [8]. This legislation guarantees that water intended for human consumption can be consumed safely and without danger to health. Specifically, in Spain (valid in the current drinking water treatment plant (DWTP) location), the regulation is articulated through royal decrees RD 140/2003, RD 314/2016, and RD 902/2018 [9], specifying that the residual chlorine value must be between 0.2 and 1.0 ppm at all points in the supply network. In short, this legislation will be replaced by the future European Directive on Drinking Waters, which is at the moment in the final revision phase, with the date of possible approval at the end of 2020 and with mandatory entry in 2022. This new standard requires a series of new requirements and is mainly aimed at avoiding unnecessary water loss and helping to reduce the carbon footprint of the EU member states. The new water directive emerges to achieve the 2030 sustainable development goals (Goal 6) and the Paris agreement on climate change. Information will be key to increasing consumer confidence in tap water, and thus information on the quality and supply of drinking water in each area should be provided on the Internet. The idea is that the more information there is, the more confidence there will be about this resource that comes out of our taps, and therefore there will be a lower purchase of plastic bottled water, reducing the waste from this material.

The main goal of this research was to develop a simulator of a drinking water treatment plant (DWTP). New technologies deployed in this kind of plant allow for the automatization of the chlorination process, and the use of this novel simulator will allow for the testing of any chlorination strategy without involving such a critical infrastructure. The plant cannot stop delivering drinking water to a large portion of the population and the tests cannot be done with the plant at full capacity. The simulator will have the ability to be connected in real-time with automatic chlorination system (specific hardware in the plant), and all this hardware equipment does not need to be simulated, thus enhancing the reliability of the whole process.

2. Materials and Methods

2.1. Simulator Design

This is one of the most important contributions of the presented research. The design of the simulator starts with defining the requirement, and proposing a novel architecture that will be able to fulfill them. This proposal involves the software implementation of different mathematical models (hydraulic and transportation models) and a hardware interconnection between plant computers, that is, programmable logic controller (PLC) and the main computer executing the simulator.

The chlorination process in the plant has been subjected to different phases of improvement at the automation level in recent years. The problems inherent in the automatic control have been diverse: variability in the chemistry of chlorine, punctual existence in the water to deal with significant amounts of ammonium, large dead times due to the hydraulics of its facilities, or disturbances caused by sudden changes in flow. The use of a simulator that reflects in the most realistic way the behavior of the chlorination process greatly facilitates the design of the control system in the effluent and allows decisions to be made on the optimization of the actual installation. Simulation has been seen over time as a modelling tool that has a very broad development and does not require sophisticated mathematics or statistics to develop a model and its usage [10]. On the other hand, the realization of experiments or tests on the real plant was expensive because of the need to have additional electronics and infrastructure, and moreover it mainly is especially dangerous because this is the last process of the DWTP and there is no correction mechanism in case of error (overdosage or underdosage of chlorine in tap water). The DWTP cannot be closed for testing new chlorination strategies, and the large duration of tests cannot be afforded by all the citizens and small industries that need drinking water uninterruptedly. Therefore, we decided to use a simulator that would allow for the study and analysis of the most appropriate online controller to later transfer the simulation results to the real control system of the plant.

When selecting a simulator to implement the desired model, we defined four strong requirements due to the complexity of the plant and the implications of stopping it:

1. The simulation tool has to be able to simulate a hydraulic model considering the turbulent flow rate.
2. The tool has to simulate a model of transport of diluted species where the reaction of chlorine with compounds of different sources would take place.
3. The tool needs a connection to the plant control system in real time through the OPC (object linking and embedding (OLE) for process control) platform that allows testing.
4. The tool has to allow importing Supervisory Control and Data Acquisition (SCADA) data and be able to compare it, offline, with those produced by the model implemented in the simulator.

In recent years, there is a growing interest by large automation companies to integrate computational fluid dynamics (CFD) software into online or real-time control scenarios. Through analyzing different software or platforms existing in the market to be used as a simulator in the chlorine automatic dosage, we finally decided to implement our own simulator, mainly for the simplicity of equations that we developed in this research. With our proposal, it is possible to reduce the sampling time to near 1 second. This fact is very important in order to have a good description and knowledge of water behavior and chlorine diffusion into tanks. After accurate research among many commercial simulators (18 software packages in total), we did not find any commercial simulator fulfilling the aforementioned conditions with such a reduced sampling time that was open-source and internally customized. Next, the architecture of the proposed simulator is presented, which would allow testing chlorination control algorithms for treatment plants (DWTP), fulfilling the following features:

- Large contact tanks.
- Variability in the demand for chlorine by water quality.
- Considerable disturbances due to the appearance of ammonium.
- Disturbances in the water inflow.

One of the main advantages of working with a model implemented in an open source language is the possibility to start with a simple approximation of the process and then gradually refine (programming) the model as the understanding of the process improves. The continuous process of refining leads to achieving a good (accurate enough) approximation to solve the problem of the chlorination tank simulator fulfilling the previously mentioned requirements.

The contribution of this paper is the design and implementation of such a simulator. The architecture depicted in Figure 2 is the key of the contribution, and the starting point for the implementation. With the proposal of this architecture, we will fulfil the four requirements described above. The simulator integrates the equations to describe a hydraulic model and a transport model of diluted species—in this case, the chorine reaction.

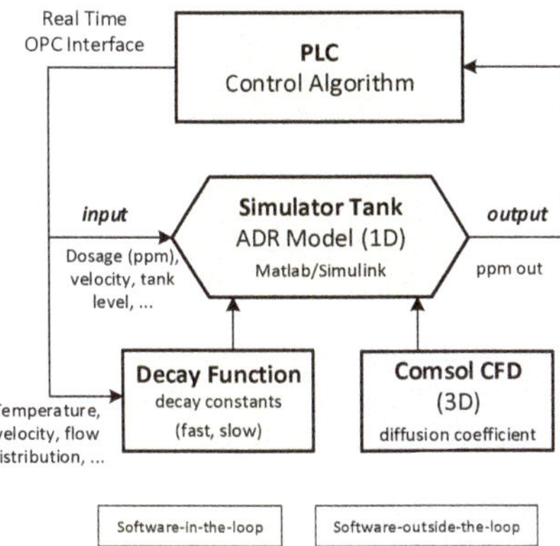

Figure 2. General architecture of the simulation system for tank chlorination.

The tank behavior (*Simulator Tank* block) was implemented by means of a Simulink S-function, which works in real time, communicating the simulator with the control algorithm implemented in the PLC. The communication of the control loop is the OPC platform. This communication will allow for the importing of SCADA real data from the plant to compare and validate the simulator behavior. In addition to the control variable, the output of the controller transfers the water speed and the level of the tank to the simulator. Conversely, the simulator provides two outputs: (1) the response of the simulated plant to the input of the control algorithm, and (2) the error between the measured chlorine dose and the necessary dose. On the other hand, the output of the controller provides a set of variables such as temperature, speed, or the distribution of flows between osmosis and carbon filters, among others, which allow the chlorine decay to be estimated using the *Decay Function* block. The *Comsol CFD (3D)* block provides a table of diffusion coefficients for locations at different distances of the chlorine output meter of the treated tank in an offline process. In the next sections, we present how this design was implemented and validated, showing relevant results compared with real data.

2.2. Description of Sant Joan Despí DWTP

Since its inception in 1955, the DWTP of Sant Joan Despí (Barcelona, Spain) has aimed to purify the surface waters captured from the Llobregat River and groundwater from the delta aquifer of the river, today supplying drinking water to more than half of the inhabitants (1,900,000) of the city of Barcelona, as well as the southwest conurbation.

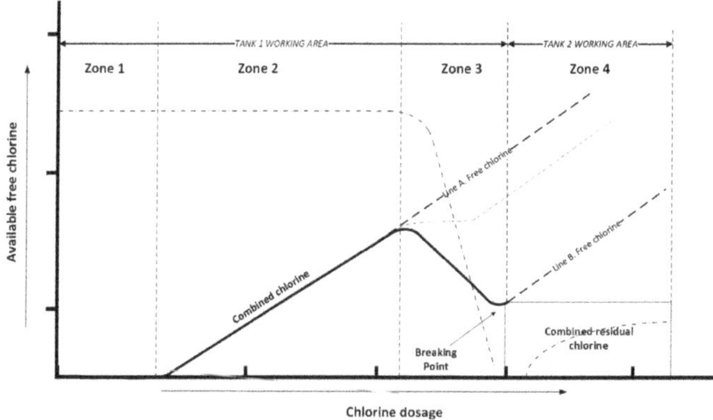

Figure 4. Transition in the dosing areas of the chlorination in tank 1 and inlet of tank 2.

2.4. Tanks and Equipment

As previously mentioned, the disinfection system has two contact tanks, tank 1 and 2, with a capacity of 10,000 m^3 each. These tanks are hydraulically coupled and, in case of an anomaly in any of them, the DWTP can treat water only with the available tank (only for a limited time of operation). The water flow to be treated in tank 1 (Figure 5) arrives at a small chamber where water treated with a traditional treatment by granulated active carbon filters (GACF) is mixed with water treated with a reverse osmosis (RO) treatment. At the exit of both treatment processes, there are flow meters that allow sensing the flow coming from both the osmosis and the carbon filters. The sum of these two flows will be the variable of total flow to be treated, as well as the input for the control system and the simulator presented in this article. The actual flow range treated by the plant is between 5.5 m^3/s at maximum performance and a minimum flow in certain operating situations of 1.0 m^3/s. In normal regime, the distribution by GACF and by RO is 50%, however, this distribution may change in a few hours due to the operating conditions of the DWTP. The theoretical residence time of water in the one tank can vary between 23 min and 55 s for a flow rate of 5.5 m^3/s up to 2 h and 11 min for a flow rate of 1.0 m^3/s.

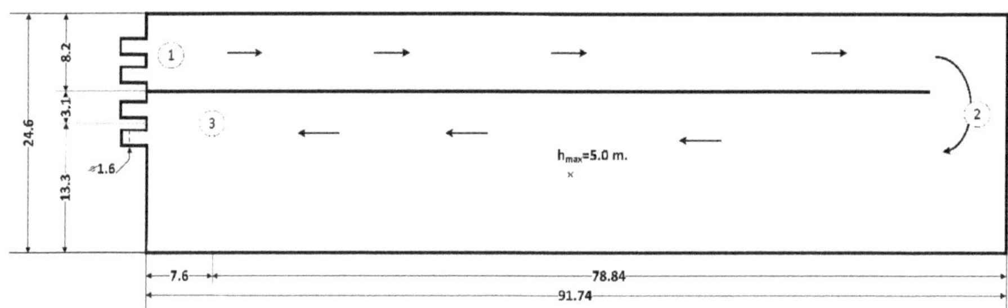

Figure 5. Plant view of chlorination tank 1 (measured in meters).

In the inlet of tank 1, the mixing chamber, we installed an ammonium meter (An1), allowing the control system to acquire the amount of ammonium contained in the water to be treated, with the first chlorinated water dosage point (1) being located a few meters apart downstream of the entrance.

It is essential for the control system to know the delay times between the production of chlorinated water and the dosage in point 1 since they are dead times that must be compensated and considered in the calculation of the dose. At a distance of 87 m from the dosing point 1, there is the first residual chlorine meter (2). This distance between the dosing point and the measurement of free chlorine is considered sufficient in most cases for the chlorine to react with organic matter and other compounds. Depending on the velocity of the water and the amount of ammonium, sometimes the measurement is no longer representative for errors in the meter and for this reason there is a second meter (3) located at the exit of the tank. As shown in Figure 5, the tank is divided into two parts and forces the treated water to travel twice the distance so that the reaction is complete and the contact time is sufficient for an effective disinfection. In the center of the tank, there is an ultrasonic level meter with a height range of 0–4 m.

The analytical instrumentation provided by the control system is composed of a colorimetric ammonium-type analyzer at the entrance of the tank and two amperometric meters that measure the residual chlorine at the entrance and exit of the two tanks. The response of the amiometric residual chlorine meters for combined chlorine samples has been quite satisfactory once the reaction time of the ammonium with the free chlorine is guaranteed. To measure the flow at the output of each of the 20 GACF and the 10 RO frames, we used electromagnetic flow meters. It is important to parameterize these flowmeters at the level of integration of the measure to dampen possible sudden changes in the reading of the control system. Finally, an ultrasonic level meter is available in both tanks that allow us to know the variations of the volume stored in each tank on the basis of the knowledge of their dimensions.

3. Mathematical Process Modeling

In this section, the graybox model implemented is presented. This model allows for a simulation of the chlorination process that describes the behavior of the fluid by the advection and diffusion equations, as well as a chlorine-demand estimation function to model the reaction.

The chlorine dosing system in contact tanks 1 and 2 can be characterized as a distributed parameter system that takes place in both coupled reactors. The chlorine dosed at the beginning of each tank (reactor) is subject to a first-order decay which allows the mass balance shown in Equation (1). This partial differential equation (PDE) represents the model of a solute in a turbulent flow, in 3D:

$$\frac{\partial c}{\partial t} = D_{t-x}\frac{\partial^2 c}{\partial x^2} + D_{t-y}\frac{\partial^2 c}{\partial y^2} + D_{t-z}\frac{\partial^2 c}{\partial z^2} - U\frac{\partial c}{\partial x} - V\frac{\partial c}{\partial y} - W\frac{\partial c}{\partial z} \tag{1}$$

where c is the chlorine concentration; $D_{t-x,y,z}$ is the dispersion; and U, V, and W are the water velocities in the coordinate axes x, y, and z, respectively. Simplifying it to a 1D equation and adding the reactive term, we obtained the following equation:

$$\frac{\partial c}{\partial t} = D\frac{\partial^2 c}{\partial x^2} - U\frac{\partial c}{\partial x} - kc \tag{2}$$

where U is the water velocity in the tank (m/s). Therefore, Equation (2) is the advection–diffusion–reaction equation (ADR) and belongs to the group of partial differential equations of parabolic type.

The simulator developed and presented in this article is based on the discretization of Equation (2) [13], treating the chlorine decay part (kc). The diffusion coefficient values were obtained from the Comsol CFD Software on the basis of the tank and flow characteristics, treated as a RANS k-ε turbulence model.

The proposed simulator, despite working in 1D, collects all the necessary information for the characterization of the process and allows the control system to be tested as if it were actually being dosed in the tank. In the following sections, we detail the more relevant aspects and considerations that were considered at the time of the design, as well as the implementation of the simulator related to advection transport, flow characteristics and the parameterization of the chlorine reaction, and diffusion of the chlorine through the tank.

In the following sections, the turbulent flow regime (Reynolds number) is verified, and the Péclet number is estimated in order to assess the influence of the convective versus diffusive effect. Finally, the Schmidt number chosen for this turbulent flow model is presented.

3.1. Flow Characteristics

The flow in the tank under study corresponds to a three-dimensional speed distribution. This is a non-stationary flow, considering the fluctuation of the flow in the tank and not its uniformity due to its geometry considering the characteristic velocity profile of the flow in open channels.

The consideration of constant velocity throughout the domain would be a relatively correct simplification if the chlorine were evenly distributed at all times. However, in principle, this hypothesis is not acceptable due to the characteristic velocity profile for this type of geometry. In addition, the geometry of the studied domain should be considered where changes occur in the direction of flow. This situation even drives away the actual behavior of the fluid from the idealized one-dimensional flow situation.

On the other hand, we faced a turbulent flow, according to the Reynolds number range, Equation (3), for the featured width of the tank $L_{(width)}$ = 8.2 m and according to the properties of the water at normal working temperature, with a velocity for maximum and minimum operations points $[0.052 \ldots 0.128]$m/s being

$$Re = \frac{\rho v L}{\mu} = \frac{1000 \times v \times 8.2}{0.001} \tag{3}$$

where ρ is the density of the fluid (gr/L), v the velocity, L the tank width, and μ the dynamic viscosity of the fluid. The water velocity range was obtained on the basis of the flow and the elapsed time to cross the tank of known dimensions. The minimum and maximum values for the Reynolds number were 4.29×10^5 and 1.06×10^6, respectively, defining a range of values within the turbulent regime. In [14], a comparative study is presented between different configurations of contact tanks for a disinfection process and the calculation of the Reynolds number, using as a parameter, in this case, the kinematic viscosity of the water as opposed to the dynamic viscosity such as in this research.

3.2. Diffusion Coefficient

The diffusion coefficient depends largely on the nature of the particles, the solvent, the temperature, and the viscosity of the solvent. In this case, neither the type of particles nor the solvent is modified during the process. On the other hand, temperature variations are not important with respect to diffusion and, as a consequence, there are also no significant variations in the viscosity of the solution. Considering all these conditions, we can affirm that the diffusion coefficient will remain constant.

In general, if there is an active flow, as presented in this paper, the diffusion effect is negligible. However, to determine the relationship between the advective and diffusive terms, the Péclet number Pe, Equation (4), was estimated:

$$Pe = \frac{L \times v}{D} \tag{4}$$

where L is the distance between concentration measurement points, v the fluid velocity, and D (m/s^2) the diffusion coefficient.

In this study case, the Comsol CFD software was used to obtain the value of the diffusion coefficient of the particles from the characterization of the tank and different flow rates giving a range of 0.0055 to 0.013 m^2/s for water flow in the range between 1.5 m^3/s and 3.7 m^3/s, respectively. Similar to this strategy, in [15], CFD software was used to obtain the dissemination number of the treated tank.

For the calculation of the minimum and maximum values of the number of Péclet, according to the extreme values of the velocity, we took the most unfavorable value of the diffusion coefficient (D = 0.013 m^2/s). Considering these conditions, the Péclet number oscillated between 71.92 and 1831.0, both higher than 1. These values suggest that the diffusive term was negligible compared to the effects of advection, and since the diffusion can be neglected for *Pe* >> 1 under these conditions, the effects of advection exceeded those of diffusion in determining the overall mass flow.

3.3. Turbulence Model

The turbulence model chosen for the CFD study is the R-based k-epsilon (k-ε) model, which is a two-equation model that provides a general description of the turbulence through turbulent kinetic energy (k) and dissipation (ε). The k-ε turbulence model is widely used to model flow behavior in chlorination tanks of these characteristics [16].

The Schmidt number (Sc), Equation (5), is a relevant parameter in the configuration of the model used in this study, which is used in RANS k-ε to avoid the resolution of the boundary layer. The k-ε model predicts turbulent viscosity thanks to the Schmidt number:

$$S_c = \frac{\mu}{\rho D_t} \tag{5}$$

where μ is the dynamic viscosity of the fluid, ρ the density of the fluid, and D_t is mass diffusivity of the fluid—a value that depends on the fluid, water plus chlorine in this case.

The Schmidt number is an empirical constant with typical values between 0.1 and 1; in the present study giving a value of 0.7, following the criteria of many other documented works on water treatment and contact tanks of similar characteristics [17–20]. This parameter is relatively insensitive to the properties of the molecular fluid (the particular value obtained from other experiments can be used despite the differences between the simulation domain). It represents a significant parameter for fully developed turbulent flows, and it is considered that, for the present case, the average Reynolds number is in the transition to the turbulence zone with moderate values, and thus it was not a key parameter for the present case. A sensitivity analysis of the Sc was carried out in the present study, showing the little relevance of the changes in its value close to 0.7; between 0.5 and 0.9 the concentration variations obtained in the effluent were around ±0.01 ppm. Finally, with respect to Sc, it should be borne in mind that the experimental determination of this parameter was not within the scope of this project.

3.4. Chlorine Reaction through the Tank

The expression in Equation (2) is a generalization that must be developed with more precision for the present case. In [21,22], different models are presented, describing a chlorine decay more adjusted to the reality of the studied plant. Considering the characteristics of the water that reaches the exit of the DWTP, a significant part of the chlorine (fraction f in Equation (6)) reacts quickly with the existing matter (organic and inorganic) and the rest (fraction $1-f$) of the matter reacts with the remaining chlorine fraction. Therefore, a combined first-order model according to [21], plus a combination of first and second order presented in [22], yields Equation (6):

$$\frac{\partial c}{\partial t} = D\frac{\partial^2 c}{\partial x^2} - U\frac{\partial c}{\partial x} - k_R c(f) - k_r c^2(1-f) - k_s c(1-f) \tag{6}$$

where k_R is the decay coefficient for the rapid reaction, and k_r and k_s would be the decay constants of rapid and slow reaction for the remaining chlorine fraction with different characteristics to those that react following k_R. Let f be the fraction of chlorine that reacts quickly in response to the decay constant k_R.

It has been proven by laboratory analysis that, for treated waters without ammonia, the demand for chlorine varies between 0.2 ppm and 0.5 ppm depending on whether its source is from the traditional treatment of granular active carbon filters (GACF) or osmosis (RO). These reactions take place between 5 to 10 min after the chlorine is exposed to the water and follows a decay according to the constant k_R. The demand for chlorine is very high, and k_R varies significantly when the existence of ammonium occurs, with values between 7 and 10 times the amount of NH_4-N detected by the ammonium meter (An1). Constants k_r and k_s were estimated in [23], with values between 3.3934×10^{-5} (s^{-1}) and 5.4259×10^{-7} (s^{-1}), respectively. These values (k_r and k_s) are significant because of the importance in the parameterization of the proposed simulator. With those values of k_r and k_s, and knowing that the demand for chlorine in its rapid reaction stage is between 0.2 ppm and 1.0 ppm and that the fraction that reacts with the chlorine is around 40%, then k_R is between 3.046×10^{-4} (s^{-1}) y 14.05×10^{-4} (s^{-1}), according to experimental data.

Figure 6 shows the simulator response to a step in the dosage of 1 ppm for a flow of 1.5 m^3/s parameterized with k_r and k_s, and different values of k_R with a reaction fraction of $f = 0.4$ (40%).

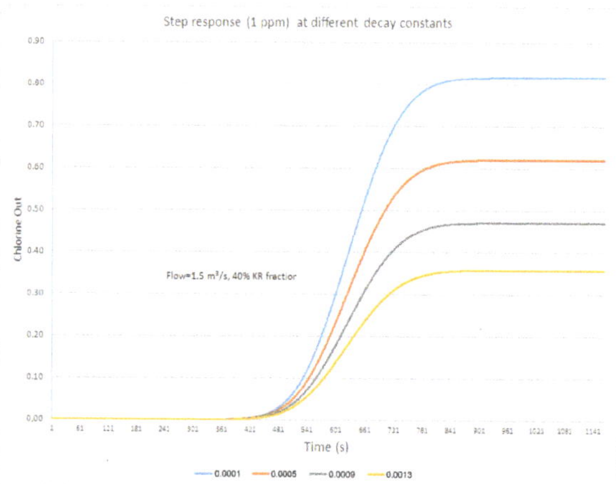

Figure 6. Step response for a dose of 1 ppm of chlorine for different values of k_R (0.0001, 0.0005, 0.0009, 0.0013 s^{-1}).

Figure 7 shows a view of one of the studies carried out in steady state with the Comsol CFD on the diffusion of chlorine along the contact tank. Specifically, on the basis of a color scale, the diffusion of chlorine along the tank is represented for a turbulent flow rate RANS k-ε with a flow rate of 2.7 m^3/s, with a decay constant of $9.0 \times 10{-4}$ (s^{-1}), a Schmidt number of 0.7, and an initial dosing step of 2 ppm of chlorine.

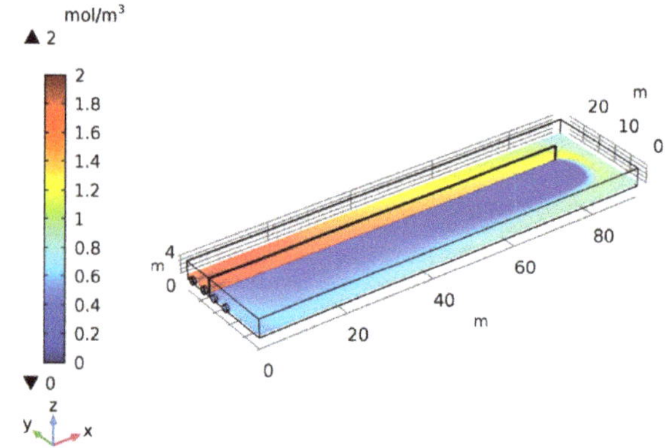

Figure 7. Comsol simulation for decay constant of $k_R = 9.0 \times 10^{-4}$ (s^{-1}).

4. Discretization

The partial derivative equation (PDE) presented in Equation (6) in continuous space was reduced to a second-order ordinary differential equation (ODE), Equation (7), in the discretization of the simulator. It was considered that the type of simulator flow belongs to the "Plug Flow" pattern where the fluid circulates through the tank evenly and along parallel paths from the entrance to the exit of the tank.

$$0 = D\frac{d^2C}{dx^2} - U\frac{dC}{dx} - k_R c(f) - k_r c^2(1-f) - k_s c(1-f) \tag{7}$$

Among the different methods of discretization for the type of equations that govern the behavior of the simulated system process (Quick, Upwind, etc.), we chose a central finite differences equation, which was adapted in an acceptable way for the values of flow, diffusive characteristics, and Péclet numbers of the present problem [24]. In [25], different discretization methods were presented and discussed to address advection, diffusion, and reaction problems in contact tanks; in [26], a discretization scheme was presented for storage tanks in transport processes and distribution of the same type of water treated by the DWTP of this article.

When applying central finite differences by replacing the first and second derivative in Equation (7), then Equation (8) is obtained:

$$0 = D\frac{C_{i+1} - 2C_i + C_{i-1}}{x^2} - U\frac{C_{i+1} - C_{i-1}}{2x} - k_R c_i(f) - k_r c_i^2(1-f) - k_s c_i(1-f) \tag{8}$$

for a diffusion coefficient D and a chlorine concentration c at different instants of time.

Figure 8 shows a diagram of the diffusion effect in chlorine concentration. It supposes a tank of length L with constant increments of space on the mesh (Δx) at different periods of time. An increment of $\Delta x = L/(n-1)$ was considered, with n being the number of mesh divisions. To avoid the effects of dynamic instability, the increment values of x fulfilling $x \leq \frac{2D}{v}$ were restricted, and the stability criterium $t \leq \frac{(x)^2}{2D+k_R x^2}$ was applied according to [27]. Applying the stability restriction on the spacing Δx in the most unfavourable case, a value of $\Delta x \leq 0.2$ m was obtained and $\Delta t \leq 2.16$ s. In the simulation, knowing that the actual dosage analyzer 2 (An2) is located at 83 m from the tank 1 entrance, a value $\Delta x = 0.127$ m and $\Delta t = 1$ s with a mesh of $n = 685$ slots was used, in accordance with a trade-off to minimize the error with a computing time that would allow the simulation in real time.

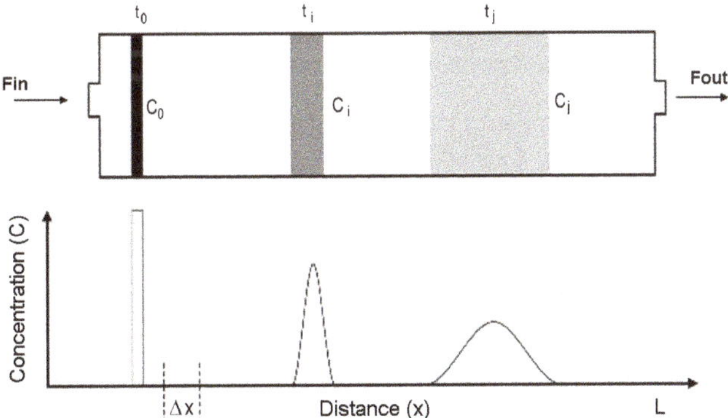

Figure 8. Diagram representing the diffusion effect respect time and space (not at scale) in the tank for a water inflow *Fin*. At different instant times (t_0, t_i, t_j ...), the concentration of chlorine is different in the tank (c_0, c_i, c_j ...) depending on the distance to the tank inlet (function of x).

5. Simulator Implementation

For the implementation of the simulator, the Matlab/Simulink tool was used using the C language to encode the functions that describe the behavior of the model. The simulation code was programmed and encapsulated in an S-Function (Level 2) in Simulink environment and the interface was adapted to be able to contrast data from the output of the supervision and control software (SCADA) with the output of the simulation. The proposed simulator could be configured for any programmable logic controller (PLC) and SCADA controller with an OPC interface (OLE for process control, object linking, and embedding). One of the advantages of using a S-Function is that it allows for the creation of a general-purpose block that can be used in all the iterations of the model, varying the parameters with each instance of the block and thus interacting in real time with the controller (PLC). The simulator implements the central finite differences for the discretized model in Equation (8) [28].

Simulation main screen (in Simulink environment) for the interaction scheme between the simulator and the PLC controller according to the simulator architecture (Figure 2) is shown in Figure 9. The data from the controller are evaluated by the *estimateCd* function that calculates the chlorine demand and, thus, the value of the rapid decay coefficient (k_R) for each iteration. Depending on the velocity of the flow, we determined the diffusion coefficients that are part of the input variables of the simulator, together with the dose, k_R, and velocity. The designed interface allows for the setting of the distance to the reading point, the rapid reaction fraction for k_R, and constant values k_r and k_s.

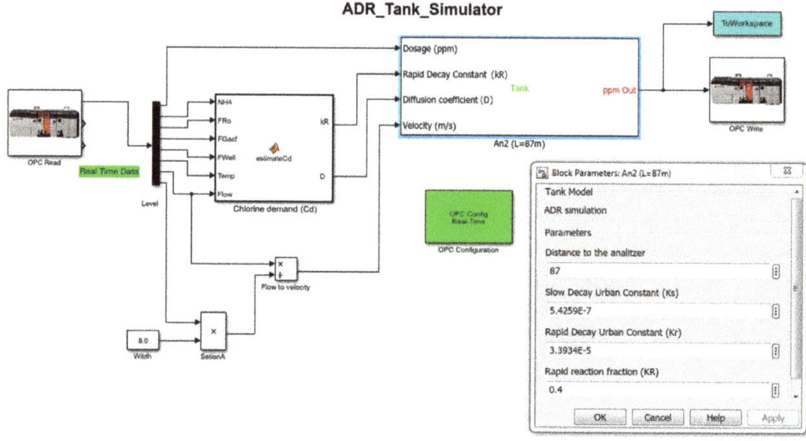

Figure 9. Connection between the simulator and a programmable logic controller (PLC) in the OPC (object linking and embedding (OLE) for process control) platform.

The simulator comprises two working zones. If there is less than 0.02 ppm of ammonium in the treated water (sensed with analyzer An1), the chlorine demand is calculated on the basis of the temperature and the source of treated water. We developed a classification of the chlorine demand on the basis of the water temperature (Temp) and the percentage of water passing through the carbon filter treatment (TPGacf) with respect to the total treated flow. In the tree, each child node represents a chlorine demand value based on the classification by temperature and source and percentage of treated water. This study was generated through taking real data from the last 5 years in the plant. Figure 10 (left) shows the decision tree for classified data. The study was implemented with RStudio. For values greater than 0.02 ppm in the ammonium concentration, the meter yields accurate and reliable results, having proven experimentally that there is a linear correlation (greater than 0.93) between the ammonium concentration and the chlorine demand for treated water (see Figure 10, right).

The validation of the simulator is an important part of this work, because it allows for the measurement of the goodness of such a simulator in its use in forecasting different situations in the real plant in front unknown events that can happen in the future. Results are shown in the next section, but here the implementation of how the simulation is validated is presented.

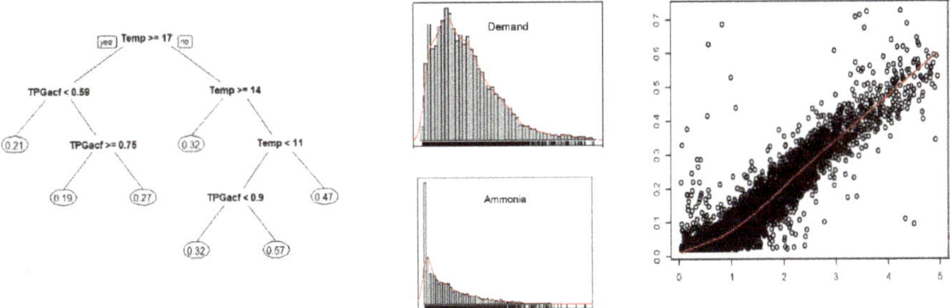

Figure 10. Left: decision tree for chlorine demand; **right**: correlation between ammonium concentration and chlorine demand (about $R = 0.93$).

The procedure to compare real data obtained with the SCADA and simulated data is shown in Figure 11. Two blocks (*Tank_secA* and *Tank_secB*) are instantiated in the test, one for each installed analyzer, An2 and An3, at 87 and 163 m away from the dosing point, respectively. The level of the tank (*Level*) and the flow rate (*TotalFlow*) are used to calculate the velocities in sections A and B of the tank. The function blocks (*An2 Sample* and *An3 Sample*) simulate the measurement behavior of chlorine analyzers with periodic samples every 5 min. The *Chlorine Dosage System* function block simulates the process of injecting chlorine gas into water from its generation point and its delay in transport to the contact tank, in order to be more realistic in the simulation procedure.

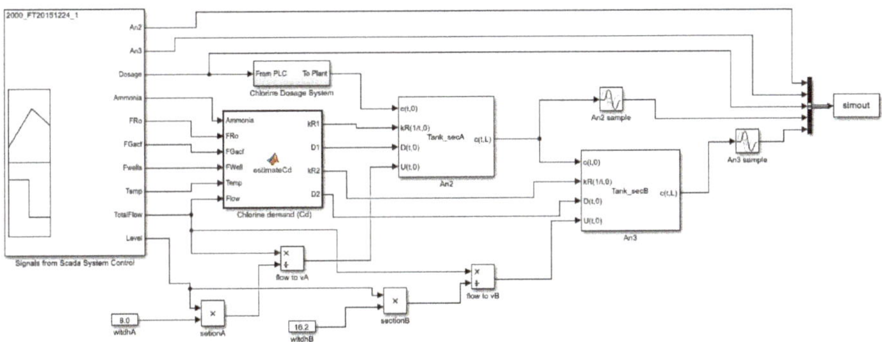

Figure 11. Simulator block diagram to compare simulation data and real data from SCADA.

At present, the chlorination stage is performed in a semi-automatic regime. From the plant control room, the operators check the level of chlorine concentration on the basis of the sensors' readings, and they order PLC (programmable logic controllers) to open the valves of chlorine bottles, which adds a specific dosage to reach the demand at the control point (located as indicated in Figure 5).

6. Results: Validation of the Simulator

The validation of the simulator was divided into two stages with specific objectives. In a first stage, data from the Comsol CFD, considered as a reference computational fluid dynamics software, were contrasted with respect to the output of the simulator for conditions of advective flow, diffusion, and decay of chlorine. The objective of this phase was to verify the correct implementation and validation of the advection–diffusion–reaction model in the Simulink S-Function. In a second stage, the *estimateCD* function was parameterized and implemented for the characteristics of the water treated by the DWTP, and data from the SCADA database were validated on the basis of this parameterization.

Table 2 shows a comparison of values obtained from Comsol CFD with respect to those provided by the simulator for different decay constants and a flow rate of 3.7 m^3/s. In any case, the relative error in the sample is larger than 2% considering that chlorine meters have an absolute error of 0.04ppm. This fact implies that the error made by the simulator would widely meet the requirements in terms of accuracy.

Table 2. Comparison of chlorine decay between the simulator and the Comsol computational fluid dynamics (CFD).

| k_R (s^{-1}) | Chlorine Concentration (ppm) for a Flow Rate of 3.7 m^3/s | | | | | |
| | At 35 m from Tank 1 Inlet | | | At 70 m from Tank 1 Inlet | | |
	Comsol CFD	Proposed Simulator	Relative Error at 35 m	Comsol CFD	Proposed Simulator	Relative Error at 70 m
0.0001	0.9180	0.9028	−1.69%	0.8869	0.8802	−0.76%
0.0003	0.8620	0.8554	−0.77%	0.7859	0.7859	−0.01%
0.0005	0.8107	0.8105	−0.02%	0.6978	0.7016	0.55%
0.0007	0.7634	0.7680	0.60%	0.6206	0.6264	0.93%
0.0009	0.7198	0.7277	1.09%	0.5529	0.5593	1.15%
0.0011	0.6794	0.6895	1.48%	0.4934	0.4994	1.21%
0.0013	0.6419	0.6534	1.76%	0.4409	0.4459	1.13%
0.0015	0.6071	0.6192	1.95%	0.3946	0.3982	0.91%

Figure 12 shows graphically a comparison of data obtained with Comsol CFD and the simulator, with different water flows between 1.50 and 3.70 m^3/s. On the basis of these results, the simulator gave excellent results.

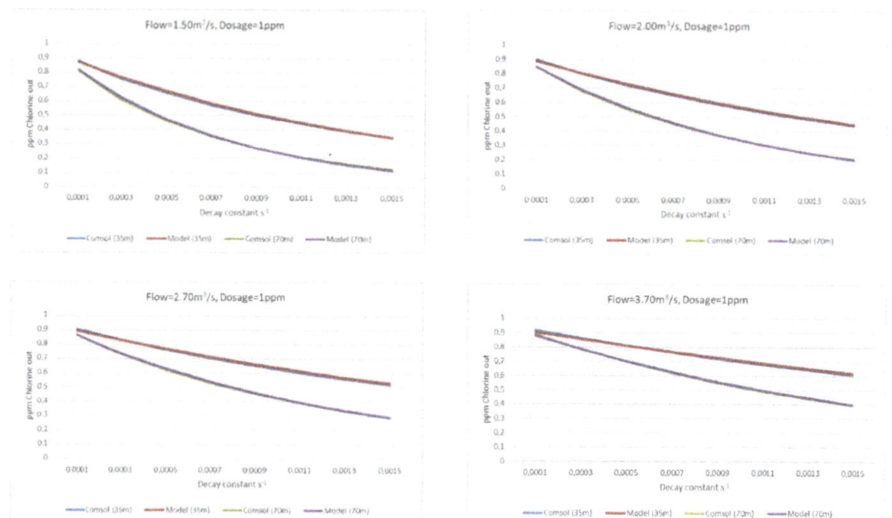

Figure 12. Results comparison between CFD Comsol data and simulator data for different flow rates (1.50, 2.00, 2.70, and 3.70 m^3/s).

However, the most important validation of simulator was carried out with real data; this is perhaps the most effective way to demonstrate the power of a simulator. In this case, the experiments were performed with real data obtained from the plant in real situations. Most of the time, the plant has neither variation on the input flow nor the effluent where the water comes from, and thus the chlorine dosage is constant. Such situations are easy to model. However, when weather conditions are varying, such as in episodes of flooding or big storms, or even in large episodes of draught when the river flow in very low, the situation in the plant becomes exceptional. In such periods, the chlorine dosage has to abruptly change due the bad conditions of input flow to the plant. We focused our attention upon such days, and obviously to the associated data recorder during those extreme episodes. Together with the flow, the dosage is already recorded as in a log file. Therefore, it is easy to reproduce the conditions that generated a specific behavior in the plant, and those conditions were used as input in the simulator to recreate the real situation that occurred in the plant. The output of the simulator was compared with the chlorine concentration in the measurement point (output of the plant) as a consequence of the dosage that was injected in the inflow water. This comparison study was carried out in all the operating points of the plant, that is, at the different input flows of 1.5, 2.0, 2.7, and 3.7 m^3/s.

As can be seen in Figure 13, the plant behavior refers to the chlorine concentration at the end of the tank, that is, the concentration of chlorine in parts per million ready to enter the pipes for distribution. The model is the execution result of the simulator as the chlorine concentration at the same point of the plant, and therefore plots in Figure 13 compare both concentration in the same location of the tank. As it can be observed, the simulator behavior worked exceptionally well following the reality in front of different input flow situations. This qualitative assessment aside, the variations between real data and simulated data are quantified in Figure 14.

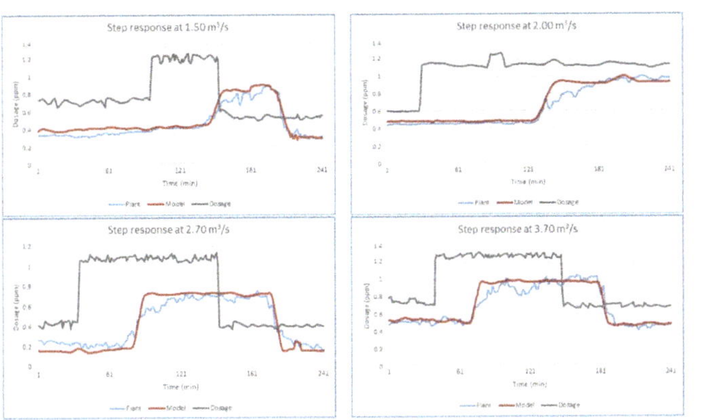

Figure 13. Data comparison between real data from SCADA and simulator data for different flow rates (1.5, 2.0, 2.7, and 3.7 m^3/s).

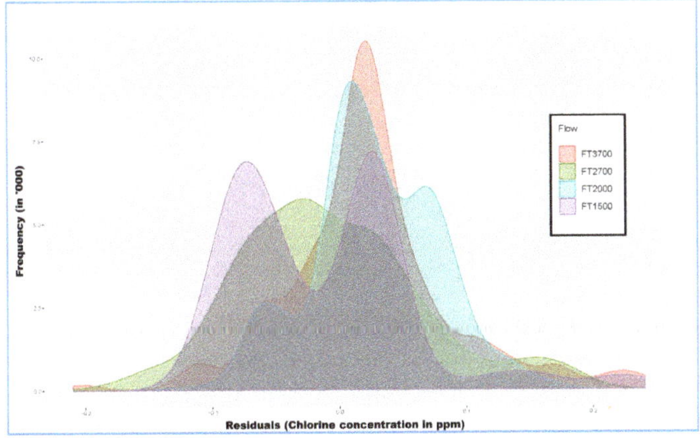

Figure 14. Plot of residuals for different flow rate operating points (1.5, 2.0, 2.7, and 3.7 m^3/s).

Defining a residual value (or residuals with sign) as the difference between the observed values and the values predicted by the model (estimated values), Figure 14 shows a graph of the residual values obtained in the different operating points. For all flow values, a distribution of bell-shaped residuals can be seen, except for flows close to 1.5 m³/s, which is multimodal. At large flow rates with high velocity (above from 1.5 m³/s), it is easier to predict errors and residuals closer to zero and with less dispersion, because the dispersion, either multimodal or unimodal, increases when the flow drops. One explanation is that the simulation treats the problem as a case of one-dimensional flow (and Plug Flow), and it does not consider that the velocity is not uniform throughout the domain. Not all the dosed chlorine moves at the same velocity in the tanks and it is also possible that there are fluid recirculations in the studied domain. As a consequence, at the measurement point, the concentration is detected with a time lag with respect to the simulation, and at low flow rates this fact is greatly accentuated. Finally, the chlorine analyzer error of ± 5% over 5.00 ppm must be considered and, therefore, those simulation errors can be perfectly accepted. The abscissa axis indicates the error in concentration of chlorine between real and simulated data. The ordinate axis represents the frequency of such residuals appears, that is, a kind of histogram of number of appearances. Note that the frequency is in thousands because we studied large amounts of data, searching for exceptional situations in the plant, and then for every particular study the sequence is for 241 min, sampled at 1 s, the sampling time of the designed simulator.

As it can be seen in Table 3, the mean square error (MSE) are collected as a measure of the quality of the simulator. In this case, the difference between the real data and the simulated data is measured as:

$$MSE = \frac{1}{N} \sum_{t=1}^{N} \left(c_{real\_data} - c_{simulated\_data} \right)^2 \tag{9}$$

that is, the sum of the square differences between the chlorine concentration in the real plant and the result of the simulator in front the same dosage and flow. It is worthwhile to note that variable t, the time, has a step of 1 s.

Table 3. Mean square error (MSE) between the observed values (real data) and the estimated values (simulated data).

Operating Point (Input Flow)	MSE Error
1.5 m³/s	0.071307
2.0 m³/s	0.057033
2.7 m³/s	0.072119
3.7 m³/s	0.070618

7. Conclusions

In this article, we proposed the design and implementation of a simulator that allows for the verification of control methodologies in a simple and visual way before its implementation directly in a real drinking water treatment plant DWTP. The proposed simulator is specific for processes that involve a hydraulic model where transport of diluted species can be simulated. The simulator provides simplicity, easy connection to plant control equipment (with OPC platforms), and reliability. The connection between the specific hardware of the plant and the simulator operated in a satisfactory manner, allowing a data interchange at a sample rate of 1 s, time enough in this kind of phenomenon. The simulator was validated with a more complex simulator (Comsol CFD) unable to operate with the plant equipment. The result of such a comparison was very satisfactory, giving an error less than 2% in the worst case. Moreover, and a crucial test, the simulator was validated with real data acquired under extreme circumstances in tough periods in terms of water effluents arriving to the DWTP. The results were completely satisfactory, with a good performance. Those good results confirmed that the reduction of the model from partial derivatives equations to ordinary differential equations was correctly performed, and thus the model is reliable enough to be confident in its functioning in the plant. After these results, we are confident in the use of the simulator as a forecast tool in a real plant.

Author Contributions: Conceptualization, J.G. and A.G.; methodology, J.G. and Y.B.; software, J.G. and H.M.; validation, H.M. and Y.B.; formal analysis, A.G. and H.M.; investigation, J.G. and Y.B.; resources, A.G. and H.M.; data curation, J.G. and Y.B.; writing—original draft preparation, J.G.; writing—review and editing, A.G.; visualization, H.M. and Y.B.; supervision, A.G. All authors have read and agreed to the published version of the manuscript.

Acknowledgments: We would like to thank Francisco Luque Montilla; David Ibarra from Aquated, Inc.; Aigües de Barcelona analytic laboratory staff; Ricardo Torres, UPC; and Mercedes Garcia, Education Consortium Barcelona.

Conflicts of Interest: The authors declare no conflict of interest.

References

1. World Health Organization WHO. Drinking-water. June 14th 2019 report. Available online: https://www.who.int/es/news-room/fact-sheets/detail/drinking-water (accessed on 5 June 2020).
2. Gargas, J. Combined Ozonation and Electrolytic Chlorination Water Purification System. U.S. Patent No. 6,277,288B1, 21 August 2001.
3. Jia, S.; Shi, P.; Hu, Q.; Li, B.; Zhang, T.; Zhang, X.-X. Bacterial Community Shift Drives Antibiotic Resistance Promotion during Drinking Water Chlorination. *Environ. Sci. Technol.* **2015**, *49*, 12271–12279. [CrossRef] [PubMed]
4. Sconce, J.S. *Chlorine: Its Manufacture, Properties and Uses*; Reinhold Publishing Corporation: New York, NY, USA, 1962; pp. 1–45.
5. National Academy of Sciences. *Drinking Water and Health: Disinfectants and Disinfectant By-Products, Volume 7*; National Academy Press: Washington, DC, USA, 1987.
6. Gopal, K.; Tripathy, S.S.; Bersillon, J.L.; Gopal, K. Chlorination byproducts, their toxicodynamics and removal from drinking water. *J. Hazard. Mater.* **2007**, *140*, 1–6. [CrossRef] [PubMed]
7. Morris, J.C.; Baum, B. Precursors and mechanisms of haloformation in the chlorination of water supplies. In *Water Chlorination: Environmental Impact and Health Effects, Volume 2*; Ann Arbor Science Publishers: Ann Arbor, MI, USA, 1978.
8. Directive 2008/105/EC of the European Parliament. Official Journal of the European Union 24/12/2008. Available online: https://eur-lex.europa.eu/eli/dir/2008/105/oj (accessed on 5 June 2020).
9. Spanish Royal Decree RD 902/2018 on health criteria for the quality of water for human consumption. Available online: https://www.boe.es/eli/es/rd/2018/07/20/902 (accessed on 5 June 2020).
10. Roberts, S.D.; Pegden, D. The history of simulation modeling. In Proceedings of the 2017 Winter Simulation Conference (WSC), Las Vegas, NV, USA, 3–6 December 2017; Volume 18, pp. 308–323.
11. Garcia, D.; Puig, V.; Quevedo, J. Prognosis of Water Quality Sensors Using Advanced Data Analytics: Application to the Barcelona Drinking Water Network. *Sensors* **2020**, *20*, 1342. [CrossRef] [PubMed]
12. Kruger, E. *Water Quality Deterioration in Potable Water Reservoirs Relative to Chlorine Decay*; Water Research Commission WRC: Pretoria, South Africa, 2001.
13. Hoffman, J.D. *Numerical Methods for Engineers and Scientists*, 3rd ed.; CRC Press: London, UK, 2014.
14. Rauen, W.; Lin, B.; Falconer, R.A.; Teixeira, E. CFD and experimental model studies for water disinfection tanks with low Reynolds number flows. *Chem. Eng. J.* **2008**, *137*, 550–560. [CrossRef]
15. Kingham, T.J.; Hoggart, T.M. Chlorination control in a large water treatment works. In Proceedings of the IEE Colloquium on Application of Advanced PLC (Programmable Logic Controller) Systems with Specific Experiences from Water Treatment (Digest No.1995/112), London, UK, 29 June 1995.
16. Gualtieri, C.; Angeloudis, A.; Bombardelli, F.A.; Jha, S.; Stoesser, T. On the Values for the Turbulent Schmidt Number in Environmental Flows. *Fluids* **2017**, *2*, 17. [CrossRef]
17. Martínez-Solano, F.J.; Iglesias, P.; Gualtieri, C.; López-Jiménez, P.A. Modelling flow and concentration field in rectangular water tanks. In Proceedings of the 5th International Congress on Environmental Modelling and Software, Ottawa, ON, Canada, 5–8 July 2010; pp. 1–11.
18. Kim, D.; Stoesser, T.; Kim, J.-H. Modeling aspects of flow and solute transport simulations in water disinfection tanks. *Appl. Math. Model.* **2013**, *37*, 8039–8050. [CrossRef]
19. Demirel, E.; Aral, M.M. Unified Analysis of Multi-Chamber Contact Tanks and Mixing Efficiency Based on Vorticity Field. Part I: Hydrodynamic Analysis. *Water* **2016**, *8*, 495. [CrossRef]
20. Angeloudis, A.; Stoesser, T.; Falconer, R.A. Predicting the disinfection efficiency range in chlorine contact tanks through a CFD-based approach. *Water Res.* **2014**, *60*, 118–129. [CrossRef] [PubMed]

21. Gang, D. Modeling Chlorine Decay in Surface Water. *J. Environ. Inform.* **2003**, *1*, 21–27. [CrossRef]
22. Clark, R.M. Chlorine Demand and TTHM Formation Kinetics: A Second-Order Model. *J. Environ. Eng.* **1998**, *124*, 16–24. [CrossRef]
23. Nejjari, F.; Puig, V.; Pérez, R.; Quevedo, J.; Cugueró-Escofet, M.À.; Sanz, G.; Mirats, J. Chlorine Decay Model Calibration and Comparison: Application to a Real Water Network. *Procedia Eng.* **2014**, *70*, 1221–1230. [CrossRef]
24. Brasseur, G.P.; Jacob, D.J. Numerical Methods for Advection. In *Modeling of Atmospheric Chemistry*; Cambridge University Press: Cambridge, UK, 2017; Volume 7, pp. 275–341.
25. Wang, H.; Falconer, R.A. Simulating disinfection processes in chlorine contact tanks using various turbulence models and high-order accurate difference schemes. *Water Res.* **1998**, *32*, 1529–1543. [CrossRef]
26. Codina, R.; Principe, J.; Muñoz, C.; Baiges, J. Numerical modeling of chlorine concentration in water storage tanks. *Int. J. Numer. Methods Fluids* **2015**, *79*, 84–107. [CrossRef]
27. Chapra, S.C.; Canale, R.P. *Numerical Methods for Engineers*; McGraw-Hill Higher Education: Boston, MA, USA, 2010.
28. Rubio, A.; Zalts, A.; El Hasi, C. Numerical solution of the advection–reaction–diffusion equation at different scales. *Environ. Model. Softw.* **2008**, *23*, 90–95. [CrossRef]

© 2020 by the authors. Licensee MDPI, Basel, Switzerland. This article is an open access article distributed under the terms and conditions of the Creative Commons Attribution (CC BY) license (http://creativecommons.org/licenses/by/4.0/).

 *applied sciences*

Article

Effects of an Eco-Friendly Sanitizing Wash on Spinach Leaf Bacterial Community Structure and Diversity

Sangay Tenzin [1,*], Abiodun D. Ogunniyi [1], Sergio Ferro [2], Permal Deo [3] and Darren J. Trott [1,*]

1   Australian Centre for Antimicrobial Resistance Ecology, School of Animal and Veterinary Sciences, The University of Adelaide, Mudla Wirra Rd, Roseworthy 5371, Australia; david.ogunniyi@adelaide.edu.au
2   Ecas4 Australia Pty. Ltd., Unit 8/1 London Road, Mile End South SA 5031, Australia; sergio@ecas4.com.au
3   Health and Biomedical Innovation, UniSA Clinical and Health Sciences, University of South Australia, Adelaide 5000, Australia; permal.deo@unisa.edu.au
*   Correspondence: sangay.tenzin@adelaide.edu.au (S.T.); darren.trott@adelaide.edu.au (D.J.T.)

Received: 16 February 2020; Accepted: 21 April 2020; Published: 24 April 2020

Abstract: Ready-to-eat (RTE) spinach is considered a high-risk food, susceptible to colonization by foodborne pathogens; however, other microbial populations present on the vegetable surface may interact with foodborne pathogens by inhibiting/inactivating their growth. In addition, sanitizers applied to minimally processed salad leaves should not disrupt this autochthonous barrier and should be maintained throughout the shelf life of the product. This investigation aimed at comparing the effects of a pH neutral electrochemically activated solution (ECAS), a peroxyacetic acid (PAA)-based commercial sanitizer (Ecolab Tsunami® 100), and tap water wash on the minimally processed spinach leaf microbiome profile for 10 days after washing. The bacterial microbiota composition on spinach samples was assessed by 16S rRNA pyrosequencing and downstream analyses. Predominant phyla observed in decreasing order of abundance were Proteobacteria, Bacteroidetes, Actinobacteria and Firmicutes corresponding with the dominant families *Micrococcaceae*, *Clostridiales Family XII*, *Flavobacteriaceae*, *Pseudomonadaceae*, and *Burkholderiaceae*. Bacterial species richness and evenness (alpha diversity) and bacterial community composition among all wash types were not significantly different. However, a significant difference was apparent between sampling days, corresponding to a loss of overall heterogeneity over time. Analysis of composition of microbiome (ANCOM) did not identify any amplicon sequence variants (ASVs) or families having significantly different abundance in wash types; however, differences (17 ASVs and five families) were found depending on sampling day. This was the first bacterial microbiome composition study focused on ECAS and PAA-based wash solutions. These wash alternatives do not significantly alter microbial community composition of RTE spinach leaves; however, storage at refrigerated temperature reduces bacterial species heterogeneity.

Keywords: *Spinacia oleracea* microbiota; electrochemically activated solution; peroxyacetic acid; sanitization; 16s rRNA pyrosequencing; amplicon sequence variants; alpha diversity; bacterial community composition

---

1. Introduction

A wide range of microbes, with distinct phylogenetic structure, is associated with the aerial organs (phyllosphere) of plants through parasitic or symbiotic interactions; in particular, bacteria are the most common microorganisms colonizing plant phyllosphere in comparison to fungi and archaea. The bacterial communities associated with edible leafy vegetables are less diversified than those of farm soil and coastal seawater habitats [1]. Actinobacteria, Bacteroidetes, Firmicutes, and Proteobacteria are the predominant bacterial phyla present in ready-to-eat (RTE) leafy vegetables (which are consumed raw, either treated or minimally processed) [2–6]. The core bacterial genera identified in most studies are *Pseudomonas*, *Sphingomonas*, *Methylobacterium*, *Bacillus*, *Massilia*, *Arthrobacter*, and *Pantoea* [2,3]. Human pathogens mostly associated with RTE leafy vegetables include *Escherichia coli* O157:H7, *Listeria monocytogenes*, and *Salmonella* spp. [7,8], but these are greatly affected by the vegetable type and bacterial community structure [9,10].

Lettuce and spinach are minimally processed RTE vegetables highly susceptible to colonization by foodborne pathogens [11]; therefore, various post-harvest sanitizing washing strategies are generally implemented to reduce spoilage and eliminate human pathogens. Today, the effectiveness of a post-harvest sanitizer is assessed based on its effect on the overall microbial populations, in addition to its propensity to reduce the microbial load and eliminate foodborne pathogens [12]. The composition of the microbiome community is assessed because the microbiome present on fresh produce is not only responsible for spoilage but rather acts as a natural biological barrier against spoilage organisms and pathogens, which constitute a smaller subset of the whole soil microbial population [13–15]. Furthermore, the bacterial microbiota on the surface of the plant inhibits or inactivates the growth of bacterial pathogen by producing acidic antimicrobial peptides and other secondary metabolites [16–18] that adversely affect the survival of the pathogen [19].

Bacterial population on RTE spinach is generally assessed using traditional culture-based techniques or specific polymerase chain reaction (PCR) to detect pathogens known for public health risk and quantify the population of indicator bacteria. Molecular techniques such as denaturing gradient gel electrophoresis and terminal restriction length polymorphism have been used for the analysis of 16S ribosomal RNA (rRNA) gene to understand the bacterial community of the phyllosphere on spinach leaves [20–22]. Contemporary next-generation sequencing techniques are now widely used for comprehensive analysis of the composition of bacterial community due to the increase in the depth of sequence readings and improved easier to use bioinformatics pipelines [23,24]. This method, in addition to providing information on the community structure, provides insights into the association of bacterial phyllosphere diversity with environmental factors [6,23], use of biocidal agents [6,23], and pesticides [6,25]. It also provides the interaction dynamics of the composition of the bacterial community with the various stages of plant growth, post-harvest, during processing and storage [3,23,26,27].

For leafy vegetable processing, chlorine- or peroxyacetic acid (PAA)-based sanitizers are commonly used. Chlorine is used for its effectiveness and low cost, whereas PAA for its activity over a wide pH range and limited reaction with organic matter. Electrochemically activated solution (ECAS) with an approximately neutral pH (6.5–7.5) has been suggested as a promising alternative washing solution with disinfection capability comparable to that of other commonly used disinfection chemicals such as chlorine and PAA [28–32]. Izumi [28] reported that neutral ECAS containing 50 mg/L of free available chlorine (FAC), completely inactivated the total bacteria on leaf surface. Guentzel et al. [31] reported a reduction of 4.0–5.0 Log$_{10}$ CFU/mL of *E. coli*, *S. typhimurium*, *S. aureus*, *L. monocytogenes*, and *E. faecalis* inoculated on spinach leaves, working with 100 mg/L and 200 mg/L of FAC.

The sanitizers used in washing RTE vegetables have a different influence on bacterial microbiota. Some sections of the bacteria composition of plants affect the survival of pathogens through competition for limited nutrients or production of growth inhibitors [16,19,33], and others facilitate the growth of pathogens through the metabolism of different carbon sources [24]. Chlorine-based washing has previously been reported to reduce the number of microbes that inhibit the growth of pathogens in lettuce and spinach [18]. Gu et al. [25] observed changes in the bacteria community in spinach leaves washed with chlorine. Tatsika et al. [34] reported a reduction in the richness of the bacterial community of RTE spinach without affecting bacterial diversity after washing the spinach leaves with vinegar. However, the effect of washing with ECAS on the composition of the microbiome of RTE spinach leaves compared to that of PAA sanitizer has not previously been assessed.

This study evaluated the effect of an ECAS at neutral pH with proven efficacy against foodborne pathogens and in reducing the overall bacterial load in RTE spinach [28,35–37] focusing on the structure of the bacterial community present on RTE spinach leaves. We compared the changes in the profile of the bacterial microbiome in minimally processed fresh spinach leaves washed with tap water, PAA (50 mg/L), and ECAS (50 mg/L and 85 mg/L of FAC) on days 0, 5, and 10 after the sanitizing wash and storage at 4 ± 1 °C. Furthermore, a comparative analysis of the bacterial composition was performed through an analysis of the composition of microbiomes among all the treatment types and sampling days.

2. Materials and Methods

2.1. Sanitizers Treatment of Spinach Leaves

Freshly cut Tasmanian baby spinach leaves, grown in soil, stored and shipped at 4 ± 1 °C, were used within 24–48 h of receipt. ECAS (produced by Ecas4 Australia Pty Ltd., Mile End South, Adelaide, South Australia, Australia) was also stored at 4 ± 1 °C and used within one week of production, diluted in Milli-Q water (Milli-Q academic A10 deionizer, Millipore Corporation, Molsheim, France) to 50 mg/L and 85 mg/L of FAC. Peroxyacetic acid (Ecolab Tsunami® 100, which nominally contains 30–60% acetic acid, 10–30% peroxyacetic acid and 10–30% H_2O_2), commonly used as a post-harvest sanitization of fresh agriculture produce, was used at 50 mg/L of PAA. The temperature, pH, and oxidation-reduction potential (ORP) of ECAS, Tsunami® 100, and tap water were measured using a portable MC-80 m (TPS Pty Ltd., Brendale, Queensland, Australia). The quantities of free and total chlorine in ECAS were measured using a Free Chlorine Checker® HC-HI701 and a Total Chlorine Checker HC-HI711, both from Hanna Instruments (Keysborough, Victoria, Australia). The amount of PAA in Tsunami® 100 was measured using specific test strips (Hydrion PAA160 Peroxyacetic Acid (PAA) Sanitizer Test Strips, Brooklyn, New York, USA).

Three samples of spinach leaves (200 g each) were washed with 800 mL of either tap water (control, pH 7.4 ± 0.1) or sanitizers (52 ± 2 mg/L of PAA, ORP of 492 ± 15 mV, pH 3.6 ± 0.1; ECAS with 48 ± 4 mg/L of FAC, ORP of 833 ± 13, pH 7.1 ± 0.2; and ECAS with 82 ± 4 mg/L of FAC, ORP of 864 ± 13, pH 7.0 ± 0.2) at 4 ± 1 °C for 60 s, and the excess liquid removed using a salad spinner at 70 rpm for 30 s. Samples (3×25 g) from each treatment were homogenized in 225 mL of sterile 0.1% peptone water for 60 s in a stomacher (BA 6021 Stomacher, Seward Ltd., Worthing, UK) immediately after treatment (day 0) and stored at -20 °C. Spinach samples from each treatment were stored at 4 ± 1 °C and further processed on day 5 and day 10, as described by Ogunniyi et al. [37].

2.2. Samples Preparation for Variable V3-V4 Region Sequencing

Samples stored at -20 °C were thawed in a shaking incubator kept at 20 °C for about 45 min. Samples from each type of treatment and for the various sampling days were centrifuged at $15,000 \times g$ for 15 min; the supernatants were discarded, and the pellets were frozen at -20 °C for DNA extraction. The DNA from the samples was then isolated and purified using the Qiagen QIAamp DNA Mini Kit (Cat. #51304, Germantown, MD, USA) as per the manufacturer's instructions. DNA concentrations were measured using the multi-mode microplate reader (CLARIOstar Plus).

The amplicon-sequence PCR was performed using the 16S DNA V3-V4 region primers from Klindworth et al. [38] and following the guidelines provided in "16S Metagenomic Sequencing Library Preparation" (Part #15044223 Rev. B) [39]. PCR products were confirmed to produce a single amplicon size of ~460 bp after electrophoresis on a 2.0% agarose gel. Aliquots (25 µL) of all samples were subjected to clean-up PCR, index PCR, second clean-up PCR and MiSeq 16S metagenomic sequencing at the South Australian Health and Medical Research Institute (SAHMRI), Adelaide, South Australia. The data analyzed were based on Illumina Miseq sequences of 300 bp paired amplicon sequences from the V3 and V4 region of 16S rRNA gene from baby spinach leaf samples with and without sanitizing treatments. The profile of the demultiplexed fastq paired-end reads was assessed using FastQC [40] for sequence quality scores and adapter contents. First, the forward reads were truncated at position 260 and the reverse at position 220 to remove low quality reads (<26 Phred). Trimming was set up for the first 20 nucleotides for forward reads and 10 nucleotides for reverse reads to remove primer sequences and low-quality reads. The trim and filter parameters were performed jointly on the paired-end read by setting a maximum of two errors expected per read [41], so that both paired-end reads passes the filter for the pair to pass. Downstream analysis to infer the amplicon sequence variants (ASVs) was performed in R version 3.5.3 [42] using the DADA2 workflow that resolves variants that differ by a single nucleotide [43]. Taxonomic assignments were made for the sequence variants data implementing the naïve Bayesian classifier method [44] using the SILVA reference data set (version 132) [45] formatted for DADA2 [46]. The DECIPHER R package [47] was used for the alignment of multiple sequences, and a phylogenetic tree was built using the phanghorn R package [48]. The phyloseq R package [49] was used to synthesize sample data, phylogeny and taxonomic assignment objects into a single phyloseq object. Further downstream analyses and graphical visualization of the microbiome data were performed in phyloseq [49] and Shiny-phyloseq [50] R packages.

2.3. Statistical Analysis

Calculations of alpha diversity indexes were performed in R versions 3.5.3 [42] with the phyloseq R package [49]. The Shannon and inverse Simpson indexes were compared among the variables since these indexes consider the richness and evenness that are powerful in providing insights into the structure of the microbial community [51,52]. In addition, the number of ASVs (species) was estimated using the observed richness and Chao1 richness estimator. The alpha diversities among the groups of samples were statistically tested using the analysis of variance (ANOVA) test to evaluate any differences in the microbial composition among treatment types and sampling days, as both variables (treatment type and sampling day) had more than two levels and the data distribution was normal according to the Shapiro–Wilk normality test. Tukey's honest significance test as a post hoc test was performed on the ANOVA results to compare within-group alpha diversity.

Measurements of samples similarity (beta diversity) with the R phyloseq and vegan packages [53] were also performed at ASV level based on non-metric multidimensional scaling (NMDS) Bray–Curtis dissimilarity [54] and Unifrac distances [55], which include abundance and phylogenetic information respectively, in addition to taxon counts. Statistical significance testing among the groups, such as the type of sanitization and the days post sanitizing treatment, was performed using permutational multivariate analysis of variance (PERMANOVA) [56] using the adonis function in the R package vegan. The community pattern of microbial composition among the groups using taxon dissimilarity information was visualized by NMDS Bray–Curtis and Unifrac ordination methods. In addition, microbiota heterogeneity, a measure of dissimilarity of the beta diversity (Bray–Curtis) of each sample with respect to the group, was compared between the various types of treatment (sanitizing and control washes) and days of sampling to evaluate the differences in homogeneity of each treatment group and homogeneity of sampling day using the R package microbiome [57]. Statistical tests for multiple variables within the type of treatment and sampling days were performed by the betadisper function on distance matrix (Bray–Curtis), and an ANOVA was performed to compare the variances between pairs of groups using the permutest function by setting the pairwise variable to true and the number of permutations to 1000 on R package vegan [56].

Analysis of differentially abundant taxa among the types of sanitization and days 0, 5, and 10 post treatment, at ASV and family level, were performed using analysis of composition of microbiomes (ANCOM) [58] plugin in QIIME 2 [59], at ASV and genus levels. For ANCOM analysis, ASVs present in less than three samples and ASV frequencies below fifty were removed before the analysis.

2.4. Data Submission

The access number for raw reads submitted to GenBank-SRA is PRJNA576552.

3. Results

We characterized the overall bacterial composition of minimally processed spinach leaves using high-throughput amplicon sequences from the V3–V4 region of the 16S rRNA gene. Moreover, changes in bacterial composition at phyla and families levels were compared for the washed samples and the control (unwashed) on day 0, day 5 and day 10 post sanitization, and between the types of washing (ECAS, Tsunami® 100, and Tap Water). In addition, the differences in bacterial diversity associated with the days post-treatment and the types of sanitizer were evaluated.

3.1. Composition of Spinach Bacterial Community

Overall, a total of 1,093,364 ASVs were observed, with a maximum of 113,737 and minimum of 39,474 reads. After removing uncharacterized phyla and contaminants and normalizing the data to the lowest number of reads (1000), the total number of ASVs reduced to 383,290 with a minimum of 1044 (observed for samples washed with ECAS at 85 mg/L of FAC on day 0) and a maximum of 50,871 (observed for samples washed with ECAS at 50 mg/L of FAC on day 5). The above reads were assigned to 12 distinct phyla, with the majority identified as Proteobacteria (2949 distinct ASVs), followed by Bacteroidetes (1876 ASVs), Actinobacteria (756 ASVs) and Firmicutes (396 ASVs). All other phyla had ≤8 ASVs (Table 1) and were excluded from further analysis [27] as the percentage abundance of these phyla were approximately 0.1% which would not affect the biological interpretation. All ASVs were assigned to one of 65 bacterial family identified and 84% of reads were further assigned to different bacterial genus with 158 genera identified. The five most abundant families identified were *Micrococcaceae* (28.2%), *Clostridiales Family XII* (19.7%), *Flavobacteriaceae* (17.9%), *Pseudomonadaceae* (12.8%), and *Burkholderiaceae* (10.1%). The five most abundant genera identified were *Exiguobacterium* (19.7%), *Flavobacterium* (17.7%), *Arthobacter* (15.4%), *Pseudomonas* (12.6%), and *Paeniglutamicibacter* (10.3%) (Table S1).

Table 1. Abundance and percentage abundances of phyla present in spinach leaf samples identified from 16S rRNA gene sequences analyzed using DADA2 package in R and taxonomic assignment performed according to the SILVA rRNA database.

Phylum	Phyla Abundance	Percentage Abundance
Actinobacteria	756	12.59
Bacteroidetes	1876	31.25
Deinococcus-Thermus	7	0.12
Firmicutes	396	6.60
Fusobacteria	4	0.07
Patescibacteria	8	0.13
Planctomycetes	7	0.12
Proteobacteria	2949	49.13

The overall relative abundances (RA) of phyla observed for all types of sanitization wash are presented in Figure 1a, and the relative abundances for the samples collected immediately after treatment (day 0) as well as on day 5 and day 10 after storage at 4 °C are presented in Figure 1b. On day 0, the phyla Proteobacteria had the highest RA (0.36 ± 0.07), while the phyla Actinobacteria had the lowest RA (0.18 ± 0.05). On day 5, Proteobacteria was still the most abundant phyla (0.35 ± 0.08), whereas phyla Bacteroidetes was the least abundant (0.09 ± 0.01). However, on day 10, Actinobacteria was the most abundant phyla (0.34 ± 0.05) and Firmicutes was the least abundant phyla (0.10 ± 0.05). The relative abundances of bacterial taxonomy at order level for sanitization wash types and sampling days are presented in Supplementary Figure S1.

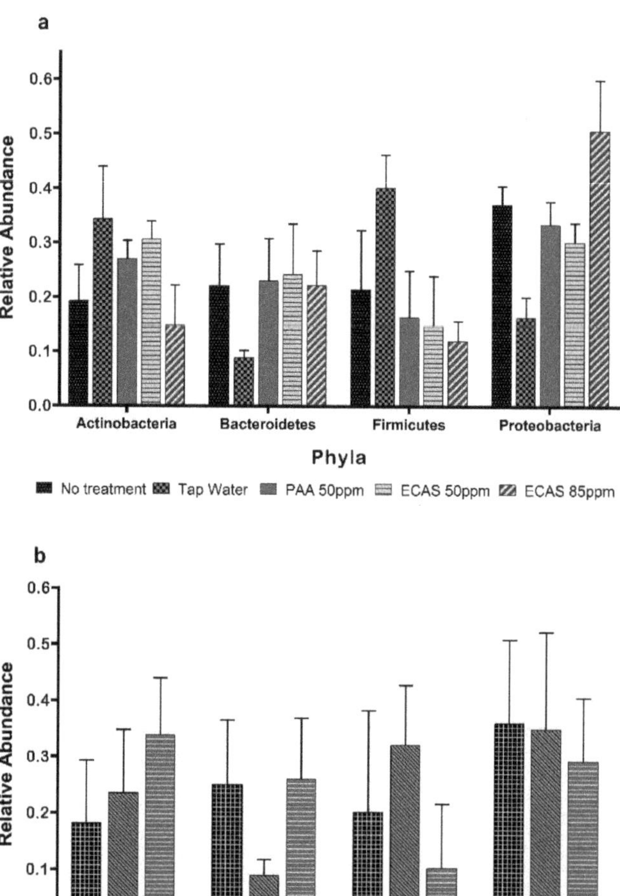

Figure 1. Relative abundance of phyla (Proteobacteria, Bacteroidetes, Firmicutes, Actinobacteria) for samples collected (**a**) after the sanitizing wash (no wash, tap water, peroxyacetic acid (PAA) at 50 mg/L, ECAS at 50 mg/L and 85 mg/L of free available chlorine (FAC)), and (**b**) immediately after the treatment (day 0) and on days 5 and 10 post sanitizing wash.

3.2. Alpha Diversity

The alpha diversity metrics (Figure S2) and the Shapiro–Wilk tests for normality showed that the data were normally distributed. The alpha diversity measures for all samples are presented in Table 2. The mean ratio between observed to expected (Chao1) richness was >0.99 for all samples. The lowest Shannon (3.4), Inverse Simpson (17.6), and richness (Chao1 = 60) indexes were recorded for samples that were not washed (control) on day 0. The highest Shannon (5.6) and richness (771) indexes were recorded for the samples that were washed in ECAS at 85 mg/L of FAC on day 0, while the highest Inverse Simpson index (134.5) was observed for the no-wash control on day 5. Species richness (Shannon diversity and Inverse Simpson indexes) and species evenness (Chao1 and abundance-based coverage estimator, ACE) measures of the bacterial community structure were assessed for the four types of treatment plus control and the three sampling days. For all samples washed with sanitizers, the Shannon and Inverse Simpson diversity measures were higher than those found for the no-wash and tap water wash, but these measures were not significantly different (ANOVA and Tukey's honestly significant difference (HSD)). Similar results on species richness (Chao1 and ACE) were observed, with no significant differences between all types of washing (Kruskal–Wallis and pairwise Wilcox (FDR corrected) (Table S2).

Table 2. Alpha diversity metrics of species richness (Shannon and Inverse Simpson & Fisher) and evenness (Chao1 and abundance-based coverage estimator, ACE) for all samples.

Treatment Type	Sampling Day	Chao1	ACE	Shannon	InvSimpson	Fisher
Tap water	0	128.00	128.29	3.86	26.38	22.58
	5	371.08	371.89	4.45	39.49	61.56
	10	337.42	338.75	4.38	41.90	51.58
ECAS 50 mg/L	0	771.06	771.86	5.59	125.72	128.92
	5	487.38	488.94	4.47	38.84	75.21
	10	542.10	542.79	5.42	118.15	96.19
ECAS 85 mg/L	0	60.00	60.00	3.42	17.66	13.74
	5	445.04	446.17	5.31	134.56	77.66
	10	630.76	633.55	5.42	113.40	133.90
PAA 50 mg/L	0	577.32	578.44	5.28	91.59	94.20
	5	368.25	369.16	4.69	50.61	58.87
	10	468.20	469.06	5.19	88.46	83.08
No treatment (control)	0	64.00	64.00	3.54	22.51	15.04
	5	342.00	342.00	4.50	35.73	53.66
	10	415.60	416.57	5.11	80.86	73.09

3.3. Bacterial Diversity Associated with Treatment Type and Sampling Day

The PERMANOVA analyses of microbial communities for the different types of treatment were not significantly different among all the variables tested ($p = 0.053$). PERMANOVA analysis of Bray–Curtis distances for sampling days determined that the microbial communities were significantly different on sampling days ($p = 0.006$) (Table 3a). Moreover, non-metric multidimensional scaling (NMDS) cluster analysis showed that the microbial communities for different treatment groups did not cluster into distinct treatment groups (Figure 2a); however, the bacterial communities on day 5 and day 10 assembled distinctly, with a divergent microbial community observed for day 0 (Figure 2b). Also the quantification of the group divergence between the treatment types (ECAS at 50 and 85 mg/L of FAC, tap water and PAA washing) plotted as a box and whisker diagram showed that the group homogeneity among treatment types did not differ (Figure 2c). However, it shows that the microbiota of the ECAS and PAA wash treatments were more homogenous, whereas the tap water wash and the no wash (control) samples were more divergent (Figure 2c). The group divergence measurement for the sampling days shows that samples on day 0 had a higher value (>0.7), indicating that the composition was more heterogeneous. On the contrary, samples on day 5 and day 10 had lower divergence values (>0.3 and >0.2, respectively), indicating homogenous microbiota (Figure 2d).

The statistical homogeneity test of the multivariate dispersion of microbial composition among the types of treatment and the sampling days showed that the variances between the different washing treatments were not significantly different. In the case of the sampling days, the composition changes between day 5 and day 10 were not significantly different, but changes between day 0 and day 5 and between day 0 and day 10 were significantly different ($p < 0.05$) (Table 3b).

Table 3. (**a**) Permutational multivariate analysis of variance (PERMANOVA) results based on Bray-Curtis dissimilarities using abundance data for treatment types and sampling days. (**b**) Analysis of variance (ANOVA) pairwise comparison tests of dispersion of microbial composition among sampling days (significant if p value < 0.05).

	Df	Sum Sq	F.Model	R^2	P
Treatment Type	4	0.455	1.813	0.330	0.053
Sampling Day	2	0.421	3.359	0.305	0.006
Residual	8	0.502			
Total	14	1.378			

(**a**) Df—degrees of freedom; Sum Sq—sum of squares; F.Model—F value by permutation. R^2—the effect size. Boldface indicates statistical significance with $p < 0.05$ based on 1000 permutations.

Sampling Day	p-Value (Observed)	p-Value (Permutated)
Day 0–Day 5	0.022	0.029
Day 0–Day 10	0.031	0.027
Day 5–Day 10	0.856	0.854

(**b**) Boldface indicates statistical significance with $p < 0.05$ based on 1000 permutations.

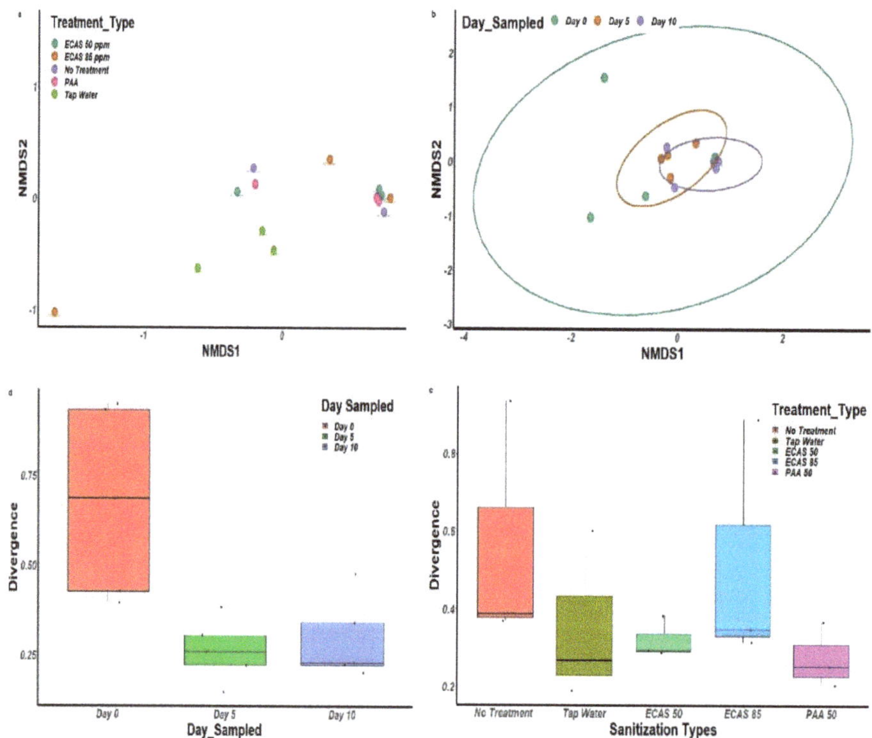

Figure 2. Microbial community cluster analysis of assembled sequence variants (ASV) non-metric multidimensional scaling (NMDS) based on Bray-Curtis dissimilarity index for (**a**) electrochemically activated solution (ECAS), tap water, and PAA washing, and (**b**) sampling days. Dispersion of the beta diversity group based on Bray–Curtis dissimilarity index for (**c**) ECAS, tap water, and PAA washing, and (**d**) sampling days.

3.4. Taxa Differences Among the Different Sampling Days

ANCOM performed with a false discovery rate (FDR) of 0.05 identified 17 ASVs and four families with significantly different abundances (ANCOM W ≥ 5) among the different sampling days (Table 4). No significantly different ASVs and families were identified for the various washing treatments. The relative abundance ratios of ASVs and taxa rank family were calculated using day 0 as the basis for displaying the relative abundance in Figure 3a,b, respectively. Out of 17 ASVs identified as significantly different, 4 ASVs on day 5, and 8 ASVs on day 10 had an increase in relative abundance. An ASV identified as belonging to the *Pseudomonadaceae* family (unclassified genus) had the highest relative abundance (RA ratio of 8.82), followed by an ASV belonging to the *Moraxellaceae* family (*Alkanindiges illinoisensis*—RA ratio of 6.41) on day 5. ASVs identified as belonging to the *Flavobacteriaceae* (unclassified genus) and *Pseudomonadaceae* (unclassified genus) families had RAs of 4.3 and 3.2, respectively. ANCOM family-level analysis revealed that *Pseudomonadaceae* had the highest relative abundance (2.9) on day 5. The relative abundance of three additional families (*Spingobacteriaceae*, *Flavobacteriaceae* and *Xanthomonadaceae*) on day 5 and day 10 were lower than on day 0 (Table 4).

Table 4. Taxa (17 ASVs) and genera (5 genera) identified as significantly different in abundance on sampling days 0, 5 and 10 by analysis of composition of microbiomes (ANCOM) analysis at a false discovery rate (FDR) of 0.05. The higher the W value, the more significant are the differences in abundance levels between the sampling days.

ASVs *		Relative Abundance (RA)			Reject Hypothesis
Taxon [a]	W	Day 0	Day 5	Day 10	
f *Pseudomonadaceae*; g unclassified (ASV1)	37	0.0010	0.0087	0.0032	TRUE **
g *Alkanindiges*; s *illinoisensis* (ASV2)	14	0.0030	0.0079	0.0017	TRUE **
f *Flavobacteriaceae*; g unclassified (ASV3)	13	0.0016	0.0000	0.0019	TRUE **
f *Flavobacteriaceae*; g unclassified (ASV4)	12	0.0002	0.0000	0.0009	TRUE **
g *Herminiimonas*; s aquatilis (ASV5)	10	0.0011	0.0000	0.0012	TRUE **
f *Micrococcaceae*; g *Arthrobacter* (ASV6)	10	0.0107	0.0076	0.0162	TRUE**
f *Oxalobacteraceae*; g unclassified (ASV7)	10	0.0142	0.0099	0.0187	TRUE **
f Flavobacteriaceae; g *Persicivirga*(ASV8)	10	0.0028	0.0000	0.0001	TRUE **
o Bacillales; f unclassified (ASV9)	10	0.0010	0.0017	0.0000	TRUE **
g *Achromobacter*; s *xylosoxidans* (ASV10)	7	0.0009	0.0000	0.0002	TRUE **
g *Flavobacterium*; s *frigidarium* (ASV11)	7	0.0010	0.0000	0.0001	TRUE **
g *Alkanindiges*; s *illinoisensis* (ASV12)	6	0.0060	0.0388	0.0066	TRUE **
f *Flavobacteriaceae*; g unclassified (ASV13)	6	0.0013	0.0000	0.0002	TRUE **
f *Pseudomonadaceae*; g unclassified (ASV14)	6	0.0022	0.0005	0.0000	TRUE **
g *Flavobacterium*; s *frigidarium* (ASV15)	5	0.0013	0.0002	0.0000	TRUE **
f *Micrococcaceae*; g *Arthrobacter* (ASV16)	5	0.1405	0.0667	0.1766	TRUE **
f *Flavobacteriaceae*; g unclassified (ASV17)	5	0.0010	0.0002	0.0007	TRUE **
Family					
Taxon	W				
o Sphingobacteriales; f *Sphingobacteriaceae*	6	0.5810	0.1827	0.2363	TRUE **
o Pseudomonadales; f *Pseudomonadaceae*	5	0.2025	0.5894	0.2082	TRUE **
o Xanthomonadales; f *Xanthomonadaceae*	2	0.4953	0.1604	0.3443	TRUE **
o Flavobacteriales; f *Flavobacteriaceae*	2	0.4343	0.1861	0.3796	TRUE **

* Amplicon sequence variants, [a] Taxa are identified from Greengenes database. ** Indicate rejected null hypothesis. o-order, f-family, g-genus, s-species.

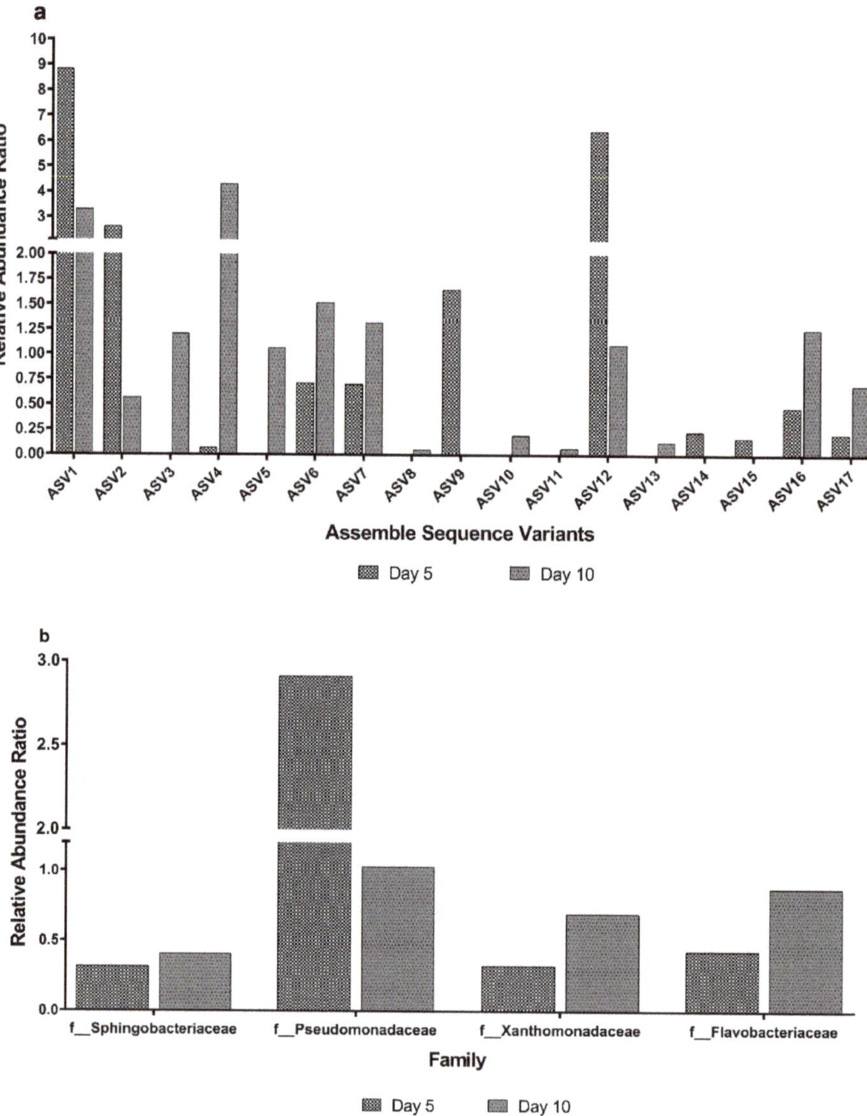

Figure 3. The relative abundance (RA) ratio of (**a**) ASVs (17 ASVs with taxa identification in Table 4) and (**b**) taxa rank family (4) identified as significantly different in abundance on sampling days 5 and 10 by ANCOM analysis at a false discovery rate (FDR) of 0.05. The RA ratio is calculated as RA on day 5 or day 10 divided by RA on day 0. f-family.

4. Discussion

This study investigated the microbiome profiles of RTE spinach leaves washed with different sanitizers (ECAS, PAA) and compared with leaves washed with tap water and not washed at all (control), at three time points over 10 days (day 0, day 5, and day 10). Although a higher proportion of ASVs was found compared to previous studies [26,27], their richness and evenness (alpha diversity) did not significantly differ among the types of sanitizer and the sampling points. We also found that the types of sanitizing washing, apart from a reduced heterogeneity over time, did not significantly influence the community structure of the bacteria (beta diversity). ANCOM analyses identified that the composition of ASVs and families changed significantly over the sampling days.

The number of ASVs identified (>2000) in the present study was much higher than that observed in the spinach leaf microbiome profiling studies by Gu et al. [26] and Söderqvist et al. [27], who identified 673 and 190 operational taxonomic units respectively. In addition, 12 phyla were identified in this study compared to the four phyla observed by Söderqvist et al. [27] and 14 phyla detected by Gu et al. [26]; however, the number of predominant phyla ($n = 4$) and their relative proportions are similar in all three studies. In agreement with previous observations, the phylum Proteobacteria showed the highest total abundance on day 0, followed by phyla Bacteroidetes, Firmicutes, and Actinobacteria [3,6,26,34]. The basal bacterial microbiome of RTE spinach leaves is therefore very similar to that of other minimally processed fruits and vegetables [3,6,27,34,60].

Our analyses also showed that the Shannon and Inverse Simpson diversity indexes and richness (ACE and Chao1) measures did not differ significantly among all spinach samples. Furthermore, the community composition of bacteria (beta diversity) for all types of washing did not differ significantly, indicating that ECAS treatments did not affect bacterial microbial diversity. This could be seen as a good outcome, since it has been suggested that the microbiome on fresh produce is not responsible for spoilage but acts as a natural biological barrier against spoilage organisms and pathogens [13–15]. On the other hand, a significant grouping of spinach microbial community structures was observed for sampling days and reduction over time of the heterogeneity of bacterial composition. The reduction in heterogeneity can be attributed to the reduction in the relative abundance of phylum Proteobacteria on day 5 and day 10, in accordance with the reduction observed by Gu et al. [26] in RTE spinach leaves washed with chlorine and stored at 4 °C for a week. Moreover, the microbiome community on day 10 clustered distinctly due to a significant increase in the relative abundance of Bacteroidetes, similar to that observed by Gu et al. [26] when the spinach leaves were stored at 4 °C for a week.

ANCOM is a method based on compositional log-ratios to detect differences in relative abundance and has been used to detect taxa abundance in the spinach microbiome at ASV and family level. Taxa at ASV and family level for the different types of treatment were not significantly different, but differences in ASVs and family-related abundances were identified at different sampling days. ASVs identified as *Pseudomonadaceae* and *Moraxellaceae* families, and the order Bacillales (unclassified family) had a high relative abundance on day 5. The increase in the relative abundance of these families of bacteria (*Pseudomonadaceae* and *Moraxellaceae*) has been correlated strongly with the spoilage of leafy vegetables at cold storage temperatures [34]. Increases in the relative abundance of the order Bacillales were also observed by Söderqvist et al. [27] and have been positively correlated with the increase in the viable counts of bacteria causing food safety concerns (*Yersinia enterocolitica*, *Listeria monocytogenes* and *E. coli* O157:H7) [27]. A significant increase was observed for four ASVs (classified as *Pseudomonadaceae*, *Flavobacteriaceae*, *Micrococcaceae* and *Oxalobacteraceae*) on day 10 and it is interesting to note that the relative abundance of the order Flavobacteriales was negatively correlated with foodborne pathogens in a previous study [27]. The predominance of *Micrococcaceae* and *Oxalobacteraceae* may be explained by their ability to grow at extremely low temperatures [61]; they are considered putative protectors against *Rhizoctonia* (fungal) rot of root crops [62]. The abundance of the family *Xanthomonadaceae* was significantly reduced on day 10, as observed by Lopez-Velasco et al. [3] and Schwartz et al. [60]. Similarly, the abundance of *Spingobacteriaceae* was significantly reduced, and the order Sphingobacteriales was correlated positively to *Escherichia coli* O157:H7 counts and negatively to *L. monocytogenes* and *Y. enterocolitica* counts [27].

5. Conclusions

To our knowledge, this study represents the first documented profile of the bacterial microbiome present on minimally processed RTE Australian spinach treated with ECAS. We have shown that washing with a neutral ECAS did not significantly change the composition of the bacterial communities compared to washing with PAA (Tsunami® 100) and tap water. In addition, complete changes over time in the community composition of bacterial species have been documented during storage at refrigeration temperature (4 ± 1 °C) on day 5 and day 10 after washing treatments, compared to day 0. The information that ECAS does not change the structure of the bacterial community could help select an environmentally friendly biocidal agent capable of meeting the aesthetic needs of current consumers and production industries.

Supplementary Materials: The following are available online at http://www.mdpi.com/2076-3417/10/8/2986/s1, Figure S1. Relative Abundance ratio of bacterial taxonomy level order for sanitization wash types and sampling days.; Figure S2. Visualization of Shannon and Inverse–Simpson diversity (alpha-diversity) and Chao and ACE richness metrics of all samples.; Table S1. Total abundances and percentage abundances of most abundant taxa at family and genus level.; Table S2. Probability values of analysis of variance (ANOVA), Tukey's HSD test on ANOVA of Shannon diversity index, Kruskal–Wallis H test, and Wilcoxon pairwise rank-sum test of Chao1 richness comparing alpha diversity metrics among the types of sanitizing treatment (Treatment Types) and day post-sanitation treatment (Day Sampled). Alpha diversity was not significantly different among the types of treatment and the sampling days, as determined by ANOVA and Tukey's HSD test.

Author Contributions: Conceptualization, S.T. and A.D.O.; Data curation, S.T.; Formal analysis, S.T.; Methodology, S.T. and A.D.O.; Supervision, D.J.T. and P.D.; Visualization, S.T.; Writing, original draft, S.T.; Writing, review and editing, S.T., A.D.O, S.F., P.D., and D.J.T. All authors have read and agreed to the published version of the manuscript.

Funding: Ecas4 Australia Pty Ltd. funded the study, including the 16s rRNA pyrosequencing.

Acknowledgments: S.T. was supported by the Endeavour Postgraduate Scholarship. ANCOM analysis was performed with supercomputer resources provided by the Phoenix HPC service of the University of Adelaide. We would like to acknowledge Mark Van der Hoek (David Gunn Genomics Facility, South Australian Health and Medical Research Institute, Australia) for his generous technical guidance on preparing samples for genomic analysis.

Conflicts of Interest: Page: 12, Ecas4 Australia Pty Ltd. played no role in the study design, data collection and interpretation, and decision to submit the article for publication. The authors declare no conflict of interest.

References

1. Lindow, S.E.; Brandl, M.T. Microbiology of the Phyllosphere. *Appl. Environ. Microbiol.* **2003**, *69*, 1875–1883. [CrossRef] [PubMed]

2. Delmotte, N.; Knief, C.; Chaffron, S.; Innerebner, G.; Roschitzki, B.; Schlapbach, R.; Von Mering, C.; Vorholt, J.A. Community proteogenomics reveals insights into the physiology of phyllosphere bacteria. *Proc. Natl. Acad. Sci. USA* **2009**, *106*, 16428–16433. [CrossRef] [PubMed]

3. Lopez-Velasco, G.; Welbaum, G.; Boyer, R.; Mane, S.; Ponder, M.A. Changes in spinach phylloepiphytic bacteria communities following minimal processing and refrigerated storage described using pyrosequencing of 16S rRNA amplicons. *J. Appl. Microbiol.* **2011**, *110*, 1203–1214. [CrossRef] [PubMed]

4. Rastogi, G.; Sbodio, A.; Tech, J.J.; Suslow, T.V.; Coaker, G.L.; Leveau, J.H.J. Leaf microbiota in an agroecosystem: Spatiotemporal variation in bacterial community composition on field-grown lettuce. *ISME J.* **2012**, *6*, 1812–1822. [CrossRef]

5. Whipps, J.; Hand, P.; Pink, D.; Bending, G. Phyllosphere microbiology with special reference to diversity and plant genotype. *J. Appl. Microbiol.* **2008**, *105*, 1744–1755. [CrossRef] [PubMed]

6. Leff, J.W.; Fierer, N. Bacterial Communities Associated with the Surfaces of Fresh Fruits and Vegetables. *PLoS ONE* **2013**, *8*, e59310. [CrossRef]

7. Cordano, A. Occurrence of Listeria monocytogenes in food in Chile. *Int. J. Food Microbiol.* **2001**, *70*, 175–178. [CrossRef]

8. De Oliveira, M.A.; De Souza, V.M.; Bergamini, A.M.M.; De Martinis, E.C.P. Microbiological quality of ready-to-eat minimally processed vegetables consumed in Brazil. *Food Control.* **2011**, *22*, 1400–1403. [CrossRef]

9. Klerks, M.M.; Franz, E.; Van Gent-Pelzer, M.; Zijlstra, C.; Van Bruggen, A.H.C. Differential interaction of Salmonella enterica serovars with lettuce cultivars and plant-microbe factors influencing the colonization efficiency. *ISME J.* **2007**, *1*, 620–631. [CrossRef]

10. Mitra, R.; Cuesta-Alonso, E.; Wayadande, A.; Talley, J.; Gilliland, S.; Fletcher, J. Effect of Route of Introduction and Host Cultivar on the Colonization, Internalization, and Movement of the Human Pathogen Escherichia coli O157:H7 in Spinach. *J. Food Prot.* **2009**, *72*, 1521–1530. [CrossRef]

11. Hackl, E.; Hölzl, C.; Konlechner, C.; Sessitsch, A. Food of plant origin: Production methods and microbiological hazards linked to food-borne disease. Reference: CFT/EFSA/BIOHAZ/2012/01 Lot 1 (Food of plant origin with high water content such as fruits, vegetables, juices and herbs). *EFSA Support. Publ.* **2013**, *10*, 402E. [CrossRef]

12. Gaglio, R.; Miceli, A.; Sardina, M.T.; Francesca, N.; Moschetti, G.; Settanni, L. Evaluation of microbiological and physico-chemical parameters of retail ready-to-eat mono-varietal salads. *J. Food Process. Preserv.* **2019**, *43*, e13955. [CrossRef]

13. Andrews, J.H.; Harris, R.F. The Ecology and Biogeography of Microorganisms on Plant Surfaces. *Annu. Rev. Phytopathol.* **2000**, *38*, 145–180. [CrossRef] [PubMed]

14. Janisiewicz, W.J.; Korsten, L. Biological Control of Postharvest Diseases of Fruits. *Annu. Rev. Phytopathol.* **2002**, *40*, 411–441. [CrossRef]

15. Barth, M.; Hankinson, T.R.; Hong, Z.; Breidt, F. Microbiological Spoilage of Fruits and Vegetables. In *Compendium of the Microbiological Spoilage of Foods and Beverages*; Springer: Berlin, Germany, 2009; pp. 135–183.

16. Schuenzel, K.M.; Harrison, M.A. Microbial Antagonists of Foodborne Pathogens on Fresh, Minimally Processed Vegetables. *J. Food Prot.* **2002**, *65*, 1909–1915. [CrossRef]

17. Scolari, G.; Vescovo, M. Microbial antagonism of Lactobacillus Casei added to fresh vegetables. *Ital. J. Food Sci.* **2004**, *16*, 465–475.

18. Johnston, M.A.; Harrison, M.A.; Morrow, R.A. Microbial Antagonists of Escherichia coli O157:H7 on Fresh-Cut Lettuce and Spinach. *J. Food Prot.* **2009**, *72*, 1569–1575. [CrossRef]

19. Cooley, M.B.; Miller, W.G.; Mandrell, R.E. Colonization of Arabidopsis thaliana with Salmonella enterica and Enterohemorrhagic Escherichia coli O157:H7 and Competition by Enterobacter asburiae. *Appl. Environ. Microbiol.* **2003**, *69*, 4915–4926. [CrossRef] [PubMed]

20. Jackson, C.R.; Randolph, K.C.; Osborn, S.L.; Tyler, H.L. Culture dependent and independent analysis of bacterial communities associated with commercial salad leaf vegetables. *BMC Microbiol.* **2013**, *13*, 274. [CrossRef] [PubMed]

21. Rudi, K.; Flateland, S.L.; Hanssen, J.F.; Bengtsson, G.; Nissen, H. Development and Evaluation of a 16S Ribosomal DNA Array-Based Approach for Describing Complex Microbial Communities in Ready-To-Eat Vegetable Salads Packed in a Modified Atmosphere. *Appl. Environ. Microbiol.* **2002**, *68*, 1146–1156. [CrossRef] [PubMed]

22. Handschur, M.; Piñar, G.; Gallist, B.; Lubitz, W.; Haslberger, A.G. Culture free DGGE and cloning based monitoring of changes in bacterial communities of salad due to processing. *Food Chem. Toxicol.* **2005**, *43*, 1595–1605. [CrossRef] [PubMed]

23. Truchado, P.; Gil, M.I.; Suslow, T.; Allende, A. Impact of chlorine dioxide disinfection of irrigation water on the epiphytic bacterial community of baby spinach and underlying soil. *PLoS ONE* **2018**, *13*, e0199291. [CrossRef] [PubMed]

24. Lopez-Velasco, G. Molecular Characterization of Spinach (Spinacia Oleracea) Microbial Community Structure and Its Interaction with Escherichia Coli O157:H7 in Modified Atmosphere Conditions. Ph.D. Thesis, Virginia Polytechnic Institute and State University, Blacksburg, VA, USA, 2010. Available online: https: //vtechworks.lib.vt.edu/handle/10919/37601 (accessed on 19 September 2019).

25. Gu, L.; Bai, Z.; Jin, B.; Hu, Q.; Wang, H.; Zhuang, G.; Zhang, H. Assessing the impact of fungicide enostroburin application on bacterial community in wheat phyllosphere. *J. Environ. Sci.* **2010**, *22*, 134–141. [CrossRef]

26. Gu, G.; Ottesen, A.; Bolten, S.; Ramachandran, P.; Reed, E.; Rideout, S.; Luo, Y.; Patel, J.; Brown, E.; Nou, X. Shifts in spinach microbial communities after chlorine washing and storage at compliant and abusive temperatures. *Food Microbiol.* **2018**, *73*, 73–84. [CrossRef]

27. Söderqvist, K.; Osman, O.A.; Wolff, C.; Bertilsson, S.; Vågsholm, I.; Boqvist, S. Emerging microbiota during cold storage and temperature abuse of ready-to-eat salad. *Infect. Ecol. Epidemiol.* **2017**, *7*, 1328963. [CrossRef]

28. Izumi, H.; Kiba, T.; Hashimoto, S. Efficacy of electrolyzed water as a disinfectant for fresh-cut spinach. *ACIAR Proc.* **2000**, *100*, 216–221.

29. Rahman, S.; Ding, T.; Oh, D.-H. Inactivation effect of newly developed low concentration electrolyzed water and other sanitizers against microorganisms on spinach. *Food Control.* **2010**, *21*, 1383–1387. [CrossRef]

30. Gómez-López, V.M.; Marín, A.; Medina-Martínez, M.S.; Gil, M.I.; Allende, A. Generation of trihalomethanes with chlorine-based sanitizers and impact on microbial, nutritional and sensory quality of baby spinach. *Postharvest Boil. Technol.* **2013**, *85*, 210–217. [CrossRef]

31. Guentzel, J.L.; Lam, K.L.; Callan, M.A.; Emmons, S.A.; Dunham, V.L. Reduction of bacteria on spinach, lettuce, and surfaces in food service areas using neutral electrolyzed oxidizing water. *Food Microbiol.* **2008**, *25*, 36–41. [CrossRef]

32. Park, E.-J.; Alexander, E.; Taylor, G.; Costa, R.; Kang, D.-H. Effects of organic matter on acidic electrolysed water for reduction of foodborne pathogens on lettuce and spinach. *J. Appl. Microbiol.* **2008**, *105*, 1802–1809. [CrossRef]

33. Babic, I.; Watada, A.E.; Buta, J.G. Growth of Listeria monocytogenes Restricted by Native Microorganisms and Other Properties of Fresh-Cut Spinach. *J. Food Prot.* **1997**, *60*, 912–917. [CrossRef] [PubMed]

34. Tatsika, S.; Karamanoli, K.; Karayanni, H.; Genitsaris, S. Metagenomic Characterization of Bacterial Communities on Ready-to-Eat Vegetables and Effects of Household Washing on their Diversity and Composition. *Pathogens* **2019**, *8*, 37. [CrossRef] [PubMed]

35. Rivera-Garcia, A.; Santos-Ferro, L.; Ramirez-Orejel, J.C.; Agredano-Moreno, L.T.; Jimenez-Garcia, L.F.; Paez-Esquiliano, D.; Andrade-Esquivel, E.; Cano-Buendia, J.A. The effect of neutral electrolyzed water as a disinfectant of eggshells artificially contaminated with Listeria monocytogenes. *Food Sci. Nutr.* **2019**, *7*, 2252–2260. [CrossRef] [PubMed]

36. Premier, R. Evaluation of Vegetable Washing Chemicals. In Australia: Horticulture Australia Ltd. 2013. Available online: https://www.ausvegvic.com.au/pdf/r%26d_VG09086_vegetable_wash_water_extract.pdf (accessed on 12 September 2019).

37. Ogunniyi, A.D.; Tenzin, S.; Khazandi, M.; Ferro, S.; Venter, H.; Pi, H.; Amorico, T.; Deo, P.; Trott, D.J. A pH-neutral electrolyzed oxidizing water significantly reduces microbial contamination of fresh spinach leaves. *Food Microbiol.* **2020**. under review (Accessed on 25/02/2020).

38. Klindworth, A.; Pruesse, E.; Schweer, T.; Peplies, J.; Quast, C.; Horn, M.; Glöckner, F.O. Evaluation of general 16S ribosomal RNA gene PCR primers for classical and next-generation sequencing-based diversity studies. *Nucleic Acids Res.* **2012**, *41*, e1. [CrossRef]

39. Amplicon, P.; Clean-Up, P.; Index, P. 16s Metagenomic Sequencing Library Preparation. 2013. Available online: https://support.illumina.com/documents/documentation/chemistry_documentation/16s/16s-metagenomic-library-prep-guide-15044223-b.pdf (accessed on 17 November 2018).

40. Andrews, S. FastQC a Quality Control Tool for High Throughput Sequence Data. Available online: https://www.bioinformatics.babraham.ac.uk/projects/fastqc/ (accessed on 22 January 2019).

41. Edgar, R.C.; Flyvbjerg, H. Error filtering, pair assembly and error correction for next-generation sequencing reads. *Bioinformatics* **2015**, *31*, 3476–3482. [CrossRef]

42. Core Team, R. R: A Language and Environment for Statistical Computing. R Foundation for Statistical Computing, Vienna. 2019. Available online: https://www.R-project.org/ (accessed on 14 February 2019).

43. Callahan, B.J.; McMurdie, P.; Rosen, M.J.; Han, A.W.; Johnson, A.J.; Holmes, S. DADA2: High-resolution sample inference from Illumina amplicon data. *Nat. Methods* **2016**, *13*, 581–583. [CrossRef]

44. Wang, Q.; Garrity, G.M.; Tiedje, J.M.; Cole, J.R. Naïve Bayesian Classifier for Rapid Assignment of rRNA Sequences into the New Bacterial Taxonomy. *Appl. Environ. Microbiol.* **2007**, *73*, 5261–5267. [CrossRef]

45. Quast, C.; Pruesse, E.; Yilmaz, P.; Gerken, J.; Schweer, T.; Yarza, P.; Peplies, J.; Glöckner, F.O. The SILVA ribosomal RNA gene database project: Improved data processing and web-based tools. *Nucleic Acids Res.* **2012**, *41*, D590–D596. [CrossRef]

46. Callahan, B. Silva taxonomic training data formatted for DADA2 (Silva version 132). *Zenodo* **2018**. [CrossRef]

47. Wright, E.S. Using DECIPHER v2.0 to analyze big biological sequence data in R. *R Journal* **2016**, *8*, 352–359. Available online: https://journal.r-project.org/archive/2016-1/wright.pdf (accessed on 3 March 2019). [CrossRef]

48. Schliep, K. phangorn: Phylogenetic analysis in R. *Bioinformatics* **2010**, *27*, 592–593. [CrossRef] [PubMed]

49. McMurdie, P.; Holmes, S. phyloseq: An R Package for Reproducible Interactive Analysis and Graphics of Microbiome Census Data. *PLoS ONE* **2013**, *8*, e61217. [CrossRef] [PubMed]

50. McMurdie, P.; Holmes, S. Shiny-phyloseq: Web application for interactive microbiome analysis with provenance tracking. *Bioinformatics* **2014**, *31*, 282–283. [CrossRef]

51. Lozupone, C.A.; Knight, R. Species divergence and the measurement of microbial diversity. *FEMS Microbiol. Rev.* **2008**, *32*, 557–578. [CrossRef]
52. Knight, R.; Vrbanac, A.; Taylor, B.C.; Aksenov, A.; Callewaert, C.; Debelius, J.; González, A.; Kosciolek, T.; McCall, L.-I.; McDonald, D.; et al. Best practices for analysing microbiomes. *Nat. Rev. Genet.* **2018**, *16*, 410–422. [CrossRef]
53. Oksanen, J.; Blanchet, F.; Friendly, M.; Kindt, R.; Legendre, P.; McGlinn, D.; Minchin, P.; O'Hara, R.; Simpson, G.; Solymos, P. Ordination methods, diversity analysis and other functions for community and vegetation ecologists. In Vegan: Community Ecology Package. 2019. Available online: https://www.researchgate.net/publication/319525938_vegan_Community_Ecology_Package_Ordination_methods_diversity_analysis_and_other_functions_for_community_and_vegetation_ecologists_Version_22-4_URL_httpsCRANR-projectorgpackagevegan/link/59b0e308a6fdcc3f888bf5da/download (accessed on 3 March 2019).
54. Bray, J.R.; Curtis, J.T. An Ordination of the Upland Forest Communities of Southern Wisconsin. *Ecol. Monogr.* **1957**, *27*, 325–349. [CrossRef]
55. Lozupone, C.; Knight, R. UniFrac: A New Phylogenetic Method for Comparing Microbial Communities. *Appl. Environ. Microbiol.* **2005**, *71*, 8228–8235. [CrossRef]
56. Anderson, M.J. Permutational Multivariate Analysis of Variance (PERMANOVA). In *Wiley StatsRef: Statistics Reference Online*; John Wiley & Sons, Ltd.: Hoboken, NJ, USA, 2014; pp. 1–15.
57. Lahti, L.; Shetty, S.; Blake, T.; Salojarvi, J. Microbiome R package. 2012–2019. [CrossRef]
58. Mandal, S.; Van Treuren, W.; White, R.A.; Eggesbø, M.; Knight, R.; Peddada, S.D. Analysis of composition of microbiomes: A novel method for studying microbial composition. *Microb. Ecol. Health Dis.* **2015**, *26*, 27663. [CrossRef]
59. Bolyen, E.; Rideout, J.R.; Dillon, M.R.; Bokulich, N.A.; Abnet, C.C.; Al-Ghalith, G.A.; Alexander, H.; Alm, E.J.; Arumugam, M.; Asnicar, F.; et al. Reproducible, interactive, scalable and extensible microbiome data science using QIIME. *Nat. Biotechnol.* **2019**, *37*, 852–857. [CrossRef]
60. Jarvis, K.; Daquigan, N.; White, J.R.; Morin, P.M.; Howard, L.M.; Manetas, J.E.; Ottesen, A.; Ramachandran, P.; Grim, C.J. Microbiomes Associated With Foods From Plant and Animal Sources. *Front. Microbiol.* **2018**, *9*, 2540. [CrossRef] [PubMed]
61. Schwartz, E.; Van Horn, D.J.; Buelow, H.N.; Okie, J.G.; Gooseff, M.; Barrett, J.E.; Takacs-Vesbach, C. Characterization of growing bacterial populations in McMurdo Dry Valley soils through stable isotope probing with (18) O-water. *FEMS Microbiol. Ecol.* **2014**, *89*, 415–425. [CrossRef] [PubMed]
62. Araujo, R.; Dunlap, C.; Barnett, S.; Franco, C. Decoding Wheat Endosphere-Rhizosphere Microbiomes in Rhizoctonia solani-Infested Soils Challenged by Streptomyces Biocontrol Agents. *Front. Plant Sci.* **2019**, *10*, 1038. [CrossRef] [PubMed]

© 2020 by the authors. Licensee MDPI, Basel, Switzerland. This article is an open access article distributed under the terms and conditions of the Creative Commons Attribution (CC BY) license (http://creativecommons.org/licenses/by/4.0/).

 *applied sciences*

Article

Minimizing the Environmental Impact of Industrial Production: Evidence from South Korean Waste Treatment Investment Projects

Olga A. Shvetsova  and Jang Hee Lee *

School of Industrial Management, Korea University of Technology and Education, 1600, Chungjeol-ro, Byeongcheon-myeon, Dongnam-gu, Cheonan-si, Chungcheongnam-do, Cheonan City 31253, Korea; shvetsova@koreatech.ac.kr
* Correspondence: janghlee@koreatech.ac.kr

Received: 4 April 2020; Accepted: 14 May 2020; Published: 18 May 2020

Abstract: This research deals with the theoretical and practical issues of investment support activities for industrial waste management in developed countries, based on the example of South Korea. The main goal of this research is the evaluation of waste treatment investment projects and understanding their impact on the development of environmental policies. The problems of forming the sustainable systems for controlling the disposal of industrial wastes are being studied. The authors discuss the practical application of environmental policies and modern technologies of South Korean companies in the field of industrial waste processing. The approaches of waste investment project's evaluation are applied and multi-criteria decision making (MCDM) methods were discussed for various cases and applications. Using MCDM methods, the authors study the effectiveness of investment projects in waste treatment activities in Korea. The analyses of MCDM methods are implemented in this research to provide some instructions on how to effectively apply these methods in waste treatment investment project analyses. Furthermore, the authors propose a combination of multi-criterial selection and interval preferences to evaluate waste treatment projects. The proposed approach improves the method of calculating economic efficiency based on a one-dimensional criterion and sensitivity analysis. The main results of this research perform the investment impact and risk-analysis on the environmental policies development.

Keywords: environmental impact; waste management; industrial production; investment project; risk evaluation; multicriterial approach

---

1. Introduction

The rapid industrialization of South Korea has three characteristics: authoritarian state control over the industrial sphere, high economic growth rates due to export heavy industry, and rapid capital accumulation [1,2]. In particular, heavy industry has had a significant impact on the environment. The problems of forming sustainable systems for controlling the disposal of industrial wastes are become more and more apparent in South Korea. Business and government organizations discuss the practical application of environmental policies and modern technologies of Korean companies in the field of industrial waste processing. That is why it is necessary to discover all of the industrial factors which have a strong impact on the environment, and to develop effective models of waste treatment processing. Large industrial complexes involved in heavy industry have strongly polluted the earth, water and air [3]. In the areas that neighbor industrial complexes, sulfur was found on rice leaves, and some elementary school students fainted on the way to school due to toxic gas emitted from factory stacks. The economic development of South Korea was achieved at the expense of the environment. The same process has been followed by most European countries [4].

Solid waste was not considered an environmental issue in Korea for a long time. Bekun et al. in 2019, and Paramati in 2018 [5–7], discuss that there was no concern about how much solid waste was generated, and any waste collected from households was dumped in open landfills without regard for environmental hazards. At the same time, the government charged only a fixed amount for household waste disposal services, regardless of how much waste was disposed of. The Korean economic boom over the past few decades has led to a significant increase in solid waste generation. The Korea Institute of Environmental Technology Development in 1996 announced that, in just 20 years, the total amount of municipal solid waste generated per day increased from about 12,000 tons in 1970 to 84,000 tons in 1990 [8]. As the amount of waste increased rapidly, several problems related to waste disposal developed in Korea.

The Republic of Korea officially recycles more than 85 percent of all waste, as stated in a 2017 government study [8]. On the other hand, illegally dumped waste can be found in rural areas of the country. Despite this criticism, the Korean waste management system is quite effective.

According to the 2018 Korea Environmental Review (ECOREA), 10.3 percent of the country's waste (including household and business waste) was disposed of, 6.3 percent was burnt, 82.4 percent was recycled, and 1.0 percent was dumped into the sea. As of 2012, 97.3 percent of construction waste and 76.5 percent of commercial waste was recycled, 14.9 percent of them was buried in landfills, 6.5 percent was burnt, and 2.1 percent dumped into the sea. Moreover, 54.4 percent of designated waste (the term refers to industrial waste) was processed, 16.4 percent was burnt, 23.0 percent was buried and 6.2 percent was treated with other measures (storage) [9]. Even though the percentage of waste processed during incineration or processing is increasing annually, the rate of processing waste generated in landfills and discharges into the ocean is reduced. The reduction in waste discharged into the sea is the result of a ban on the discharge of wastewater, wastewater from food waste, and cattle wastewater into the ocean in 2012–2013; treatment methods for these types of waste have been replaced by incineration or recycling (Figure 1).

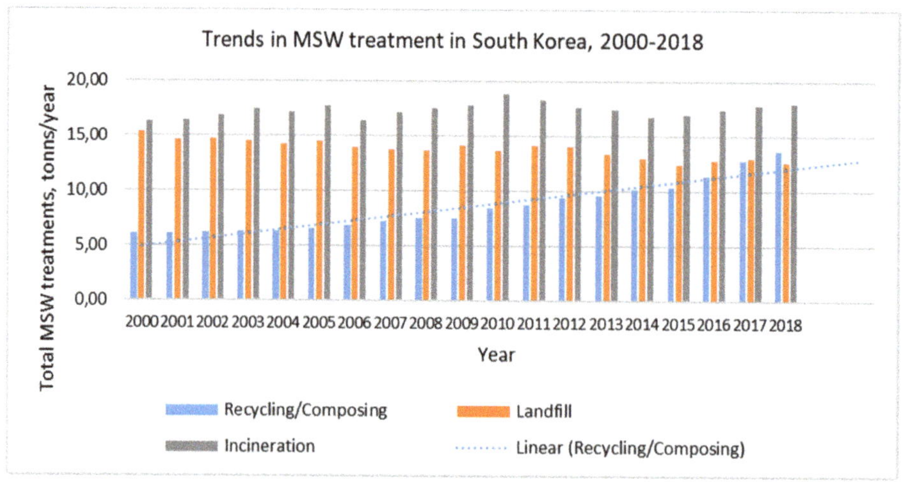

Figure 1. Trends in Municipal Solid Waste treatment in South Korea, 2000–2018. Source: Korea's Environmental Review, 2018, Ministry of Environment, ECOREA, http://eng.me.go.kr [9].

Korea has been one of the fastest growing OECD economies over the past decades, driven by a large export-oriented manufacturing sector [10]. However, growth has come with high pollution and resource consumption. With increasing energy demand, greenhouse gas (GHG) emissions have risen significantly and air pollution remains a major health concern. Despite significant improvements in wastewater treatment, diffuse pollution increasingly affects scarce water resources. Urbanization and industrialization also put considerable pressure on biodiversity. The environmental challenges are exacerbated by Korea's population density—the highest in the OECD [11]. Access to environmental goods and services, and exposure to environmental risk vary significantly by region.

To tackle these challenges, Korea has invested considerable effort into improving environmental management, for example, by introducing strategic environmental assessment, reforming the environmental permitting system and strengthening air and water quality standards. Korea introduced the world's second largest emission trading scheme, and remains one of the most innovative countries in terms of climate change mitigation technology. However, coal is set to remain integral to energy production, and road transport continues to be supported as the dominant form of mobility. Current energy prices and taxes do not reflect the environmental costs of energy production and use. The OECD Review Report in 2017 emphasizes that Korea needs to align its energy and climate policies to reduce GHG emissions to 37% below business-as-usual levels by 2030 [12].

There are many investment projects currently being supported in Korea. Korea's transition to a low-carbon economy is vital for its future prosperity. This is a core message of the OECD report in 2018, which provides 45 recommendations to help Korea pursue the implementation of green growth and strengthen environmental performance [13]. To attain this end, the Korean Government incorporates foreign investment projects in its development of a National Green Strategy. The problem refers to considerations about risk identification for such investment projects, and the further development of waste treatment technologies and methods.

2. Environmental Policies and Waste Management Methods: Case Study of South Korea

Starting from 1978, the Korean Government actively enacted comprehensive environmental protection legislation and policies. Since that time, approximately every 10 years, the Korean Environmental Ministry and other government institutions—together with companies—work together to improve environmental policies and develop new technologies for waste treatment. In this section, we discuss the successful waste management methods and regulations in Korea.

2.1. Waste Management Methods and Environmental Policy in Industrial Production

Industrialized countries face the challenge of quickly and safely disposing of waste. Non-biodegradable and toxic wastes, such as radioactive residues, can cause irreparable damage to the environment and human health if they are not disposed of strategically.

Although waste management has been a concern for several decades, the main problem has been the enormous proportions of generated waste due to population growth and industrialization—the two main factors contributing to waste generation. Although some successes have been achieved in waste disposal methods, they are still insufficient. The challenge is to discover new and dangerous methods of waste disposal and use them.

Several global companies and local Korean companies (Samsung, CMG, Ovarro, Presona AB, Amandus Kahl GmbH & Co, Ion Science Ltd., Tana Oy, AMETEK Brookfield and others)—together with leading scientists—have developed a unique technology for processing solid waste [14,15]. The latest international developments in the field were combined into a single complex, including the unique technology of catalytic pyrolysis. The uniqueness of this technology is that the garbage is not burned whatsoever. Newly patented in 2013, catalytic pyrolysis makes it possible to completely recycle solid waste without careful sorting measures. It turns waste into valuable commercial products of a high quality and secondary raw materials. Based around the method, a recycling complex of a new type was created. However, the problems of risk-management and cost-analysis of these methods' implementation and operation quickly arose. Nowadays, these companies try to invent new risk-analysis methods within the waste management process to reduce operational costs. The managers understand that project-management concepts help them understand certain basic requirements and cooperative management strategies.

Innovative technology deserves attention. The complex processing of solid industrial waste in a vortex air-mineral flow—which is based on in-depth study of the chemical, mineral composition and technological properties of the secondary raw materials, and is an individual solution for each type of industrial solid waste with maximum processing efficiency—eliminates the use of water and chemical reagents.

Industrial waste, as it turned out, can be used as a secondary resource in various industries, especially in the production of building materials, products and structures. The main tasks in the field of processing industrial waste should be aimed at improving existing processes, creating and implementing the latest technology. It is necessary to develop and implement measures aimed at implementing waste directly in the manufacturing process.

There are six effective waste disposal methods in South Korea which foreign and domestic companies prefer to venture into:

1. The prevention or reduction of waste: the widespread use of new or unnecessary products is the main reason for the generation of unverified waste. Rapid population growth necessitates reusing products or the judicious use of existing products, as, otherwise, there is a potential risk that people may suffer from the harmful effects of toxic waste. Waste disposal should also take a formidable form. On a personal and professional level, a conscious decision must be made to curb the dangerous increase in waste.

2. Recycling: serves to turn waste into products of its genre through industrial processing. Paper, glass, aluminum and plastic are typically recycled. The environmentally friendly reuse of waste instead of adding it to nature. However, the processing technology is quite expensive.

3. Incineration: incineration involves the incineration of waste, turning it into basic components, and the heat generated is then used to generate energy. Various gases and inert ash are common byproducts. Pollution is caused in varying degrees, dependent on the nature of the waste burned and the design of the incinerator. In 2019, Claborn said that using filters can check for pollution. Waste incineration is relatively inexpensive, and waste is reduced by about 90% [16]. Nutrient-rich ash obtained from burning organic waste can contribute to hydroponic solutions. Using this method, you can easily get rid of hazardous and toxic waste. The extracted energy can be used for cooking, heating and supplying energy to the turbines. However, strict vigilance and due diligence should be observed to check for the accidental leakage of micro-level contaminants—such as dioxins—from waste incineration lines.

4. Composting: involves the decomposition of organic waste by microbes, allowing waste to accumulate in the pit for a long period of time. In 2018, Freeman suggested that nutrient-rich compost can be used as plant manure [17]. However, the process is slow and takes up a significant amount of land. Biological processing significantly improves soil fertility.

5. Sanitary landfill: This is a landfill for waste. The base is made of a protective lining, which serves as a barrier between the waste and groundwater, and prevents the leaking of toxic chemicals into the water zone. The layers of waste are compacted and then covered with a layer of earth. Non-porous soil is preferred to reduce vulnerability to the accidental leakage of toxic chemicals. In 2017, Wackernagel M. et al. said that landfills should be established in areas with low groundwater levels and away from flood sources. However, a sufficient amount of skilled labor is required to maintain sanitary landfills [18].

6. Disposal in the ocean/sea: waste, usually of a radioactive nature, is discharged into the oceans away from the active human habitat. Nevertheless, environmentalists dispute this method, since it is believed that such an action means death for aquatic life, depriving ocean waters of its natural nutrients.

There is a three-tier approach to assessing the risk associated with air and water emissions from waste management facilities. With this approach, an acceptable level of protection is provided at all levels, but with each progressive level, the level of uncertainty in the risk analysis decreases [19–22].

Reducing the level of uncertainty in risk analysis can reduce the level of control required by the waste management unit (if necessary for the site), while maintaining an acceptable level of protection. An enterprise performing a risk assessment incurs higher costs associated with a more complex risk assessment in exchange for greater confidence and potentially lower construction and operating costs. The advantages and relative costs of each tier are outlined below in Table 1.

Table 1. Three-tiered approach for assessing the risk of waste treatment projects.

Tier	Characteristics (Advantages and Disadvantages)
Tier 1 Evaluation	– Allows for a quick but conservative assessment. – Lower cost. – Minimum site data required. – Assumptions about the fate and transfer of pollutants and exposure are developed using conservative assumptions that are not specific to the sites provided by the EPA. Values are presented in "look-up tables," which provide a quick and easy means of assessing risk. These values are designed for protection in a wide range of conditions and situations and are very conservative in design.
Tier 2 Evaluation	– Represents a higher level of difficulty. – Moderate cost. – Provides the ability to enter some site-specific data into a risk assessment and, thus, provide a more accurate picture of the site's risk. – Uses relatively simplified fate-and-transport models.
Tier 3 Evaluation	– Provides a sophisticated risk assessment. – Higher cost. – Allows for the maximum use of specific site data and, thus, provides the most accurate picture of the risk of the site. – Uses more sophisticated fate-and transport models and analysis.

Source: adopted from Tonmoy F. N., Rissik D. and Palutikof J. P. A three-tier risk assessment process for climate change adaptation at a local scale. Climatic Change, 153, 2019, pp. 539–557 [22].

We try to solve these practical problems associated with conducting an assessment of the risk of waste management in an organization, and offer a multi-level structure that supports the training path necessary for adaptation, allowing organizations to optimize their resources for adaptation, systematically increase the knowledge base on waste management, and develop targeted interaction strategies.

2.2. Waste Management Regulations and Environmental Policies in South Korea

Currently, the regulatory regime for environmental protection in South Korea consists of legislative acts, executive decrees, ministerial decrees and regulatory acts related to the general environment, including:

1. Nature conservation.
2. Preservation of air quality.

3.   Preservation of water quality.
4.   Water supply/Sewer management system.
5.   Recycling/Waste recycling.
6.   Green growth.

The basic framework is laid in the Framework Act on Environmental Policy (FAEP), which contains the main objectives of environmental policy, including pollution prevention and natural resource management for sustainable use. To create such a sustainable waste management system, Korea went through many regulatory improvements and discoveries, which are presented below [23–27].

In 1995, a volume-based waste disposal system was introduced, which is a waste containment system that uses the principle of payment for emissions. This system, which is representative of market incentives in Korea, represents a shift from the old system in which a fixed fee was charged regardless of the volume of waste disposed, to a system in which a proportional charge is applied to the volume discharged. As reported, this provides an incentive to reduce discharge and increase processing. In addition, a Promotion of Installation of Waste Facilities and Assistance Law was issued in 1995 to related areas in order to prevent the NIMBY phenomenon (not in my backyard) due to the installation of incinerators, and also to help resolve social conflicts through means such as relief projects for affected communities [28].

In the 2000s, the foundation was created for a society of resource recycling. According to this master plan, the waste was not just recycled, but recycled as a resource. Korea is currently pursuing a "Zero Waste" policy that seeks to use waste as a source of resource, in addition to minimizing waste generation, which was emphasized by the authors of [29].

Since the late 2000s, a reduction in greenhouse gases has been necessary due to the rapid increase in prices for resources and energy, as well as global warming. Therefore, the government emphasized the need to restore resources and energy from waste. In September 2011, the Korean Ministry of the Environment developed the First Framework Plan (2011–2015) for the management of resources, to form the basis for waste management and thereby promote green growth aimed to at encouraging the effective societal dealing with resources (without waste) [30]. In addition, the measure to facilitate the transition to a resource circulation society (2013) facilitates the collection and transportation of recyclable resources through the free collection of large household electronic waste, the consolidation of the sorting system, and the expansion of equipment installation. Waste energy utilization facilities and other similar complexes are the foundations of a waste recycling society [31]. The creation of a market for processed products and support for their industries was also announced.

Currently, intensive advertising and educational programs are beginning to educate the population on how to classify recyclables and use designated garbage bags, and, more importantly, to educate the public on the purpose and significance of waste reduction [31–33]. Due to difficulties in monitoring the levels of flies in rural areas, residents are not required to take out the trash in specially designed trash bags. Instead, public trash and recycle bins were installed in rural areas, and the fees are shared among all residents of the community. Local officials noted that low-income people were offered appropriate assistance to alleviate the burden of paying for MSW. It is interesting to compare waste quantities in different Asian countries; many of these countries employ similar approaches to waste treatment (Table 2).

The Framework Act on Resource Circulation (FARC) was adopted in 2016 to form the basis for implementing this policy; the Korean government has been applying it since 2018 [33]. The country intends to transform the economic structure, focused on mass production and mass waste, into a much more stable and efficient, resource-oriented structure at a fundamental level. The provisions of this framework can be divided into three categories, each of which creates a framework for the circulation of resources, stimulates the circulation of resources and supports the processing industry. It introduces new waste management related programs such as the Recyclable Resource Recognition Program (RRRP), Resource Circulation Performance Management Program (RCPMP), Waste Disposal Tariffs, etc. [33,34]. The government expects to obtain economic, environmental and social benefits by preventing environmental pollution, but is also aware of the fact that the country needs to make further efforts to change the paradigm of its waste management policy.

Table 2. Comparison of waste quantities among countries of Asia.

Area (year)	Hong Kong (2015)	Taiwan (2015)	Taipei (2015)	South Korea (2017)	Seoul (2017)	Hong Kong (2016)	Singapore (2011)
Type of Waste	Domestic Waste	General Waste	General Waste	Municipal/ Domestic Waste	Municipal/ Domestic Waste	MSW	Total Solid Waste Excluding Construction Debris, Sludge and Used Slag
Disposal (tones per day)	5.973	9.893	1.037	19.78	3.428	8.996	7.397
Disposal rate per capita (Kg/day)	0.84	0.43	0.39	0.41	0.32	1.27	1.43
Generation rate per capita (Kg/day)	1.36	0.88	1.00	1.04	0.95	2.44	2.75
Recycling rate	38%	52%	61%	61%	66%	48%	48%

Source: Environmental Protection Department of Hong Kong; Environmental protection authority of Taiwan; Ministry of Environment of South Korea; National Environment Agency of Singapore [32].

2.3. Waste Management System in Samsung Electronics Corporation

To protect the environment in times of crisis and make more efficient use of resources, Samsung Electronics is working to focus on a circular economy. Going beyond the normal practice of single-use resources and throwing them away, the Samsung Corporation is working to ensure that resources are reused through recovery, reuse, and disposal at the end of the product's life. By minimizing the type of materials used and optimizing the assembly method, the company has developed production methods that minimize the use of resources [35]. By collecting products that have expired, they acquire valuable materials. Through this circular economy, the Samsung Corporation is reducing the amount of natural resources it requires, reducing greenhouse gas and pollutant emissions from waste incineration, and preventing soil and groundwater pollution from landfills.

Samsung responsibly has to collect and recycle old, unwanted or nonworking electronic products in the U.S.; therefore, the company has to develop new and adaptive programs to manage the recycling needs of its business partners. This is not always easy to manage—but the reasons behind it are twofold:

- Firstly, to provide a stable service for customers (no matter where they are located in the U.S.), so there is the task of using such a project to provide an easy way for customers to responsibly recycle their products.
- Secondly, to abide by the Samsung Corporation's commitment to responsible recycling.
- The problem is that all these projects require critical analysis, risk-management efforts and cost-analysis. Unfortunately, these were not initially conducted, and the company now has severe restrictions to its waste-management strategies.

The best way to conserve resources is to make quality, durable products. The Samsung Corporation extends the life of products by making an additional contribution to the circular economy and resources by increasing the longevity of its products prior to release, carrying out a series of rigorous reliability tests, and providing convenient repair services through their global service centers, including ongoing software updates.

The amount of waste generated during the product development and manufacturing processes is significant. For complex electronic devices with numerous components, even the packaging is thrown away for every part. At the end of the project, many of the pilot products, which are used to improve the product, are lost. If all of them are burned or buried, the environment will be polluted and resources depleted. By processing waste through an environmentally responsible recycling company, the Samsung Corporation has increased the amount of recyclable materials it utilizes, and achieved its overall waste recycling goal of 95% in 2016—four years earlier than was planned (Figure 2).

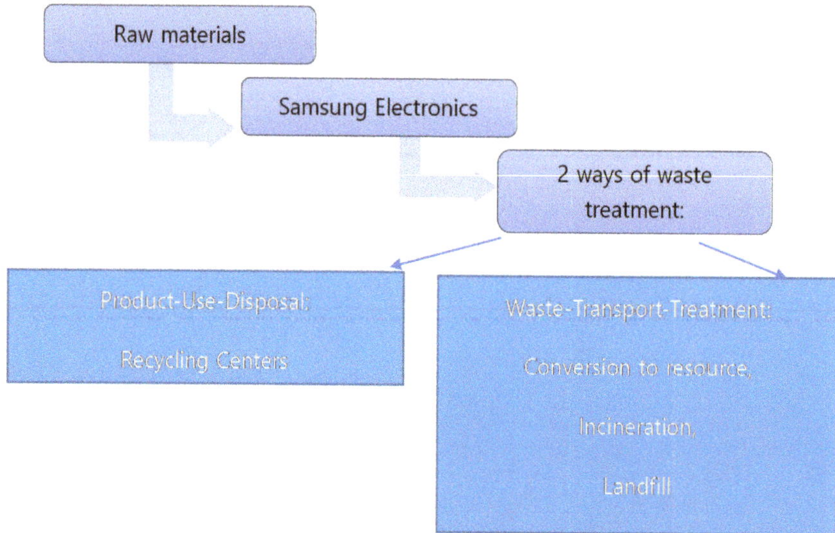

Figure 2. Waste treatment process in Samsung Electronics [36].

Since 2016, Samsung Electronics has been continuously implementing the Galaxy Upcycling Project, which turns old Galaxy smartphones that are no longer in use into new IoT devices. Using such IoT devices, for example, pet feeders and doorbells, etc., the company was able to improve the quality of life of its consumers and, at the same time, protect the environment.

In 2018, the company completed a project to develop a low-level ophthalmoscope using Galaxy Upcycling technology as an appropriate technology [36]. In collaboration with the Yonsei University Health System Design Specification—which Samsung Electronics supports as part of Tomorrow's Decision Program (based on competition events)—ophthalmoscopes have been developed that can be used in developing countries where people cannot receive proper medical treatment because of the difficulty in distributing expensive diagnostic devices. As a result, many people in developing countries are expected to develop blindness. In addition, the company has a plan to expand the use of technology for cervical diagnostic devices—among other things—and promote better health in developing countries. The Galaxy Upcycling plans offer a variety of concepts that go beyond the health sector to reduce resource waste in collaboration with institutions that seek to achieve sustainable development with limited resources.

So, the Samsung Corporation has put forward many proposals to invest in the Korean system of waste management. However, these proposals should have business planning directions and risk-management support. Even if Samsung Engineering offers a full range of EPC services for a variety of industries, from hydrocarbon offshore facilities to wastewater treatment, from initial financing through operation and management, many consulting companies will try to develop an effective financial and risk-preventive model for it. EPC stands for "Engineering, Procurement, Construction", and is a prominent form of contracting agreement in the construction industry. In this case, we want to present the waste treatment investment project's evaluation approach.

3. Methods

All national and foreign enterprises in Korea, to one degree or another, are engaged in investment activities in waste processing; in addition, the decision-making associated with such investment in waste processing involves various complex factors, including limited financial resources, the type of investment, and the possible losses that an enterprise may suffer if the project subsequently turns out to be less profitable or if it fails completely for unforeseen circumstances [37–40]. Thus, risk management allows us to confirm the viability of decisions for the project and reduce the likelihood of adopting an ineffective or unprofitable project.

In light of the discussions held so far, we consider the following research questions (RQ) in our study:

RQ1: What is the relevance of investment projects in South Korea for reducing the environmental impact of industrial production?

RQ2: What are the positive effects and shortcuts of the application of multi-criteria approaches for the evaluation of waste processing investment projects with uncertainty consideration?

RQ3: What are the risks and disadvantages of multi-criteria approaches for the assessment of waste treatment investment projects?

The remainder of this paper is organized as follows: The current trend of waste management in South Korea is discussed in Introduction Section 1. Environmental policies and waste management methods are discussed in Section 2. Section 3 describes the method, within which we consider the analysis of the feasibility and management an optimal investment project in a risk environment based on the Pareto model. Section 4 presents the results and discussion, which present various possible calculations within the framework of the discussed model. These calculations prove various possible situations for applying this method, as well as the possibility of using other performance indicators with our method. In addition, various approbations of this method in future research projects are discussed. In conclusion, Section 5 presents the main results of testing the methodology for evaluating investment projects in the field of waste management, describes our findings and predicts possible future discussions in this area.

In this research, we continuously develop this approach and take into account the risk of multivariate estimation. Vedernikov and Mogilenko in 2011 [41] suggest that uncertainty has a number of factors that affect the results of actions. Actions in this case cannot be clearly defined, and questions about how to determine the degree of possible influence of these factors on the results are accumulated. Therefore, when determining methods for industrial waste management, and methods of investing in the development of processing technologies are considered, it is important to identify the possible risks and damages from the effects of poorly predicted external factors in detail. In doing this, it is possible to carry out a scenario analysis of market risks, as well as assess the possible effects of new approaches in industrial waste management systems. This approach allows us to take the assessment of the aggregated scenario of factors into account, which enables us to represent various types of risks as components of the analysis [42–45]. This is shown in Figure 3.

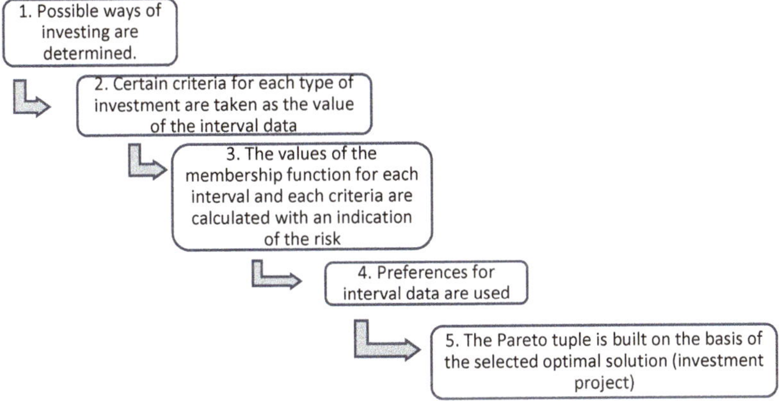

Figure 3. Research methodology.

In 2013–2017, Rodionova et al. began to analyze an integrated approach to making investment decisions, including the calculation of NPV, DPP and IRR for each alternative [42–44]. In addition, this approach is specific, since it takes the uncertainty of the external environment into account [45–48]. For this, expert assessments of the probability of damage from the implementation of the project and the intervals of fluctuations of the three criteria are used to assess the effectiveness of the investment project.

IRR measures the effectiveness of capital investments; thus, this indicator partially allows a comparison between investment projects with different capital investments and terms of implementation. The typical methodological recommendations for calculating the effectiveness of investment projects solve the problem of selecting from alternative projects by using the NPV indicator for risk evaluation. This method is useful for certain cases, such as efficiency comparison within existing external circumstances. This recommendation helps avoid conflict of interests in terms of which indicators to use. We suggest to include them all, as they each reflect different aspects (e.g., uncertainty, market situation, project capacity, etc.) of the economic system. Each of these aspects is important for the formation of criteria in the economic system.

There are four primary reasons that justify the use of multicriterial (MCA) methods; these are listed as follows:

(i)   MCA methods allow the investigation and integration of the interests and objectives of multiple actors, because both quantitative and qualitative information from every actor is considered in forming the criteria and weight factors.
(ii)   These methods address the complexity of a multi-actor setting by providing output information.
(iii)   These are well-known and commonly used methods for the assessment of investment alternatives. Moreover, different versions of these methods are developed for specific contexts.
(iv)   MCA methods allow for the objectivity and inclusiveness of the different perceptions and interests of actors.

Because Cost Benefit Analysis (CBA) is dependent on the time at which it is being performed, it is more appropriate as an ex-ante instrument; in contrast, the multi-criteria approach can be adopted for both ex-ante and ex-post assessments, which is an advantage of the MCA. Considering the dimensions of the project or the policy to be evaluated, the characteristics (evaluative standpoint, decision-relevance, comparability, verifiability, accountability, and scientific progression) of CBA and MCA render the dimensions of the project useful. In particular, on a large scale—i.e., when public and private costs are consistent—the CBA approach is necessary, whereas MCA appears to be useful on a small-scale, where all the stakeholders can be considered individually and can be consulted or express informed opinions on their priorities.

Based on the discussion thus far, it is necessary to use methods for the evaluation of the effectiveness of alternative investment projects that are based on multi-criteria selection. However, the known methods for multi-criteria selection are still not considered in the commonly used methods that can solve the problem of selecting the optimal investment solution (Roy, 1976). In particular, the selection of an effective investment project involves the best combination of values based on the analysis of disparate indicators characterizing the investment project.

We assess the variety of values for all components of the model, taking into account the risk associated with alternative waste treatment investment projects. Intervals are determined by both the absolute values of indicators and by estimates [49–52].

To evaluate the effectiveness of alternatives and to select the most preferable, our method is based on the built-in interval preference ratio (IPR).

Let us suggest that $I = \{I_\alpha, \alpha = 1 \ldots n\}$ is a pool of types of waste treatment investment projects; $K_i(I_\alpha) = [A_i(I_\alpha); B_i(I_\alpha)]$ represents the standards for assessing the effectiveness of every investment project within the interval type; $i = 1 \ldots r$, r is the total range of evaluation criteria; $A_i(I_\alpha)$ and $B_i(I_\alpha)$ are the area units of the lower and higher bounds of the interval analysis; $K(I_\alpha) = \{K_1(I_\alpha), K_2(I_\alpha), \ldots K_r(I_\alpha)\} = \{[A_1(I_\alpha); B_1(I_\alpha)], \text{ and } [A_2(I_\alpha); B_2(I_\alpha)], \ldots [A_r(I_\alpha); B_r(I_\alpha)]\}$ are the direction indicators of every waste treatment investment project's effective results. We tend to introduce the notation *II* for the set of Pareto-optimal information processing *IP* (*II⊂I*), with the amount of parts $\gamma \leq n$ meeting the main condition $II_{m1} > II_{m2} > \ldots II_{my}$, $m_j = 1 \ldots y$. Then, the matter may be developed as follows to construct the economic expert tuple of thought of the variants of waste treatment investment flows, the parts of which satisfy one in all the conditions: $K_i(I_{yj}) = min[K_i(I_\alpha)]$, $I_{yj} \in II$ or $K_i(I_{yj}) = max[K_i(I_\alpha)]$, $I_{yj} \in II$.

We also consider that if the exponent is a scalar amount, it may be mentioned as a degenerate interval with coincident ends $A_i(I_\alpha) = B_i(I_\alpha)$. This concept was presented by Orlovsky in 1981, Serguieva and Olson in 2014, and Stoyanova in 2006 [53–56].

There is an ambiguity in the choice of standards and the form of factors that the unit considers, due to the quality of the question of assessing the effectiveness of waste treatment investments. It must be assumed that the person responsible for making decisions (usually the project manager) does not have a transparent opinion about the preferences of the analyzed variants. Within the scope of the indicators of victimization, the values of the intervals and the qualitative difference of the measured values—which are expressed in the fact that the difference in the units of assessment—are built, it is convenient to check the options supported by the IPR. This concept was presented by Minakova L. V. and Anikanov P. V. in 2013 [57].

Let m_i be the breadth of the estimates' interval for the *i*-th criterion. Consistent with fuzzy methods—which were discussed by Parrino et al. in 2014, Roy in 1976, and Saaty in 1990—the interval relation of preference R on the set I_α is the set of the Cartesian product $I_k \times I_l$, $(k = 1, \ldots n, l = 1,\ldots n, k \neq l)$ [58–60]. For the characteristic of the set of the Cartesian product, we should take the interval membership operation $\mu K_i (I_k, I_l)$: $I_k \times I_l \rightarrow [-1;1]$ into account.

$$\mu^i K_i(I_k, I_l) = m_i^{-1}(K_i(I_k) - K_i(I_l)) \tag{1}$$

Each valuable measure of the membership function $\mu K_i (I_k, I_l)$ estimates the degree of injury and gain in recognizing position of I_k as the dominant variant I_l supported by the criteria K_i.

The degree of privilege of the choice I_k over the choice I_l, supported by the interval criterion K_i, is diagrammatically presented by the membership function $\mu_D K_i (I_k, P_l)$, which determines the quantitative relation of the strict interval preference.

$$\mu_D^u K_i(I_k, I_l) = \mu_u K_i(I_k, I_l) - \mu_u K_i(I_l, I_k) \tag{2}$$

For comparison, it is vital to establish that the alternative I_k is not undermined compared with the I_l alternative, which would be a mistreatment of the membership operation.

$$\mu_{ND} K_i(I_k, I_l) = 1 - x, x \geq 0; x = \mu_D^u(K_i(I_K, I_l)) \tag{3}$$

In this case, for the criterion of the *i*-th interval criteria, the approximation of the alternative I_k to the Pareto optimal variant is determined by the index of the membership function for the set of non-privileged alternatives [61,62].

$$\mu_D^* K_i(I_k) = min\mu_{ND} K_i(I_k, I_l) \tag{4}$$

Wang et al. in 2009, and Zare et al. in 2016 [63,64] suggested that the indicator NPV is based on the quantity of cash flows at a certain time and the discount rate r:

$$NPV = C_1(1 + r)^{-t1} + \ldots + C_n(1 + r)^{-tn} \tag{5}$$

The discount rate usually uses the risk-free interest rate or interest rate for investment projects with a similar degree of risk, as well as the market and industry coefficient of efficiency for capital investments. This criterion underlies the choice of an environmental management project with a maximum value, or with the same value of r. It is known that NPV is entirely dependent on the discount rate; therefore, a poor-quality and unverified forecast of the discount rate definitely leads to risky management decisions. For example, a good project with high-quality technologies, but with high costs, can be rejected, and a project with lower costs but low-quality technologies can be accepted for discussion and subsequent implementation. The refinement of the values of the NPV interval allows us to determine that the maximum possible determining factor for the NPV criterion is the maximum value.

In addition, DPP is represented as a time interval; the optimal condition for this criterion should correspond to its minimum value. Furthermore, IRR is presented as a percentage and is defined as the value of the interval; in accordance with this criterion, the waste treatment investment project that matches the maximum value is selected.

4. Results

Risk assessment is predicated on the interval values in the estimates. The presumption is that the rate of interest r may be a variable, and for that the likelihood of a random event may be found, $NPV\,(r, t) > 0, P\,(NPV\,(r, t) > 0) = P\,(r < IRR) = F\,(IRR)$. Here, $F\,(x) = P\,(r < x)$ is the distribution operation of r; IRR is the internal rate of pay back, that is obtained as an answer of the equation $NPV\,(t, r) = 0$. For various r, it is possible to ascertain the likelihood that the waste treatment project will not pay off at time t; then the estimates of victimization are obtained in the analysis procedure. Here, we tend to conduct a risk assessment for the project supported by the explained methodology for three doable and inevitable market conditions; those conditions are then evaluated by consultants, and an evaluation of the probability of every of them is enforced. It ought to be noted that the chance assessment criterion for a waste processing investment project needs to select the most effective possibility, supported by the minimum worth of the standards.

Given the well-known theoretical ideas, the values of m_i are selected as the most allowable values for the considered option (standard). The initial knowledge necessary for the calculations and investment analysis are presented in Table 3. Three different investment projects are presented (I_1—investment in landfill production; I_2—investment in industrial waste recycling R&D; I_3—investment in the implementation of green technologies).

Table 3. Data implication for different projects.

Projects/Indicators	I_1	I_2	I_3	m_i
$K_1(I\alpha)$-NPV (USD)	[50;60]	[70;120]	[80;100]	200
$K_2(I\alpha)$-DPP (annual)	[3;8]	[4;6]	[5;9]	10
$K_3(I\alpha)$-IRR (%)	[16;17]	[10;20]	[14;18]	30
$K_4(I\alpha)$-risk evaluation (points)—pessimistic forecast	[6;8]	[3;9]	[5;9]	10
$K_5(I\alpha)$-risk evaluation (points)—realistic forecast	[4.5;7]	[5;8.5]	[4;7]	10
$K_6(I\alpha)$-risk evaluation (points)—optimistic forecast	[4;5]	[4;6]	[3;5.5]	10

The risk evaluation process is presented using interval values in grade system. Assuming that the rate of interest r could be a variable quantity for which the chance of a random event is found, $NPV\,(r, t) > 0, P\,(NPV\,(r, t) > 0) = P\,(r < IRR) = F\,(IRR)$. Here, $F\,(x) = P\,(r < x)$ is the disseminative operation of r, IRR is the internal rate of pay back, which is suggested as an explanation of the formula $NPV\,(t, r) = 0$. For variety of r, it is important to determine the likelihood that the investment project will not be profitable at time t. Then the results will be determined through the evaluation analysis procedure. This study assesses the risk of an industrial waste management project using the aforementioned methodology for three defined and predicted market conditions, and an expert assessment of the likelihood of each of market condition is introduced. It is important to indicate that the risk assessment criterion for a waste processing investment project requires the selection of the best option based on the minimum value of the criteria.

Based on known theoretical models, m_i values are defined as the maximum allowable indicators for the criteria under consideration. The initial data necessary for calculations on the analysis of investment projects are presented in Table 3.

Taking in consideration Equation (1), we achieve the appraisal of the membership operation $\mu K_i (I_k, I_l)$ for each pair of variants for each criterion and calculate their estimated matrices. Thus, Equation (1) can be expanded as:

$$\mu^u K_i(I_k, I_l) = ([min\{A_i(I_k) - A_i(I_l); B_i(I_k) - B_i(I_l)\};$$

$$max\ \{A_i(I_k) - A_i(I_l); B_i(I_k) - B_i(I_l)\}])/m_i$$

and be denoted by

$$C_i^{kl} = min\{A_i(I_k) - A_i(I_l);\ B_i(I_k) - B_i(I_l)\}/m_i,$$

$$D_i^{kl} = max\{A_i(I_k) - A_i(I_l); B_i(I_k) - B_i(I_l)\}/m_i$$

Then,

$$\mu^i K_i(I_k, I_l) = [C_i^{kl}; D^{kl}_i] \tag{6}$$

Furthermore, the interval membership function for the I_l, I_k takes the following form:

$$\mu^u K_i(I_k, I_l) = [-D_i^{kl}; -C_i^{kl}] \tag{7}$$

Hence, if the relation $|C^{kl}_i| = D^{kl}_i$ is true, then the values $\mu K_i (I_l, I_k)$ $\mu K_i (I_k, I_l)$ coincide.

Based on Equation (2), we take into account the preference frequency for each pair of options for each indicator using the value of the membership operation $\mu_D K_i (I_k, I_l)$, and place them in the evaluation matrices. Using Equations (6) and (7), we move the calculations into a simple method.

Thus, we evidently have

$$M_D{}^u K_i (I_k, I_l) = [C_i^{kl}; D_i^{kl}] - [-D_i^{kl}; -C_i^{kl}] = [C_i^{kl} + D_i^{kl}; C_i^{kl} + D_i^{kl}]$$

Thus,

$\mu_D{}^u K_1(I\alpha) =$

-	−0.4	−0.35
0.4	-	0.05
0.35	−0.05	-

$\mu^D{}_u K_4(I\alpha) =$

-	0.2	0
−0.2	-	−0.2
0	0.2	-

$\mu^D{}_u K_2(I\alpha) =$

-	−0.1	−0.3
−0.1	-	−0.4
0.3	0.4	-

$\mu^D{}_u K_5(I\alpha) =$

-	−0.2	0.05
0.2	-	0.25
−0.05	−0.25	-

$\mu^D{}_u K_3(I\alpha) =$

-	0.1	0.03
−0.1	-	−0.06
−0.03	−0.06	-

$\mu^D{}_u K_6(I\alpha) =$

-	−0.1	0.05
0.1	-	0.25
−0.05	−0.25	-

From Equations (3) and (4), we achieve the valuable measures of the membership function $\mu_{ND} K_i(I_k, I_l)$ for each pair of options for each criterion, and assemble the membership function valuable measures for the set of non-privilege options $\mu_D K_i(I_k)$:

$$\mu_D{}^* K_1(I_k) = \{0.6, 1, 0.95\};$$

$$\mu_D{}^* K_2(I_k) = \{0.9, 1, 0.6\};$$

$\mu_D{}^*K_3(I_k) = \{1, 0.9, 0.93\};$

$\mu_D{}^*K_4(I_k) = \{1, 0.8, 1\};$

$\mu_D{}^*K_5(I_k) = \{0.9, 0.75, 1\};$

$\mu_D{}^*K_6(I_k) = \{0.95, 0.75, 1\}.$

After analyzing the values of $\mu_D{}^*K_i(I_k)$, we can conclude that the investment project I_2 is the best option based on the criteria $K_1(I_a)$ and $K_2(I_a)$, the investment project I_1 is the best option based on the criteria $K_3(I_a)$ while it is possible to perform risk management in a pessimistic scenario of market development, and investment project I_3 is optimal in terms of risk, based on the considered set of options for waste processing investment projects [65–67].

Savchuk in 2007, and Syroezhin in 1980 [68,69] suggested that in order to determine the best preference in this set of waste management investment projects, it is necessary to determine the vector preference using some previous studies in this area (membership functions $\mu_D{}^*K_i(I_k)$ determine the degree of proximity of variant I_k to the Pareto-optimal variant of an investment project using the K_i criterion; this justifies the use of special criteria instead of traditional factors, indicating the importance of the indicator. The next step is to compare the variants I_k and I_l in pairs, calculate the values of $\mu_D{}^*K_i(I_k)$, and introduce the subsets $I_{kl}{}^+$, $I_{kl}{}^-$, and $I_{kl}{}^=$ for optimal, pessimistic and realistic values of $\mu_D{}^*K_i(I_k)$ and $\mu_D{}^*K_i(I_k)$, (where $i = 1...4$; $k, l = 1,...$ 3, $k \neq l$) of these variants, respectively. The next step determines the elements of the estimation matrix $\mathbf{C} = \|C_{kl}^\mu\|$ based on these conditions; this is shown in Table 4.

$$C_{kl}^\mu = \left(\sum_{i=1}^{3} \mu_D^*K_i(I_k)\right)\left(\sum_{i=1}^{3} \mu_D^*K_i(I_l)\right)^{-1} \tag{8}$$

Table 4. Evaluation matrix.

I_{kl}^+	I_{kl}^-	$I_{kl}^=$	C_{kl}^μ	C_{lk}^μ	Notes
$\varnothing$	$\varnothing$	$\{1..3\}$	1	1	-
$\{1..3\}$	$\varnothing$	$\varnothing$	N_2	0	-
$\varnothing$	$\{1..3\}$	$\varnothing$	0	N_2	-
$\neq.$	$\varnothing$	$\neq 0$	N_3	0	$1 << N_3 < N_2$
$\varnothing$	$\neq <$	$\neq 0$	0	N_3	-
$\neq <$	$\neq <$	$\left|S_{kl}^=\right| \geq 0$	Formula (8)	$C_{lk}^\mu = 1/C_{kl}^\mu$	-

Source: Prisyach E., 2018 [70].

The matrix is created considering the risk criteria; therefore, it is necessary to pay attention to the possibility of various risk conditions being weighted evaluated options of the matrix component.

$$C_{kl}{}^\mu = \left(\sum{}^*_{i=1} a_i \mu_D{}^*k_i(I_k)\right)\left(\sum{}^*_{i=1} a_i \mu_D{}^*k_i(I_i)\right)^{-1},$$

$$\left\{ \begin{array}{l} a_i = 1,\ i = 1,2,3 \\ p_i,\ l = 4,5,6 \end{array} \right\}$$

Then, we get the following matrix of preferences

$$\mathbf{C} = \begin{array}{|c|c|c|} \hline - & 0.66 & 5.01 \\ \hline 1.51 & - & 0.94 \\ \hline 0.19 & 1.05 & - \\ \hline \end{array}$$

Using well-known theoretical methods and developing a methodology for evaluating investment projects, we introduce the indicators $G^\mu{}_l$ and $H^\mu{}_l$, which denote the set of elements of the l-th column in C, the value of which is less than one, but greater than zero and more than one, respectively, and the exponent $C^\mu{}_{kl\,max}$, which is equal to the indicator of the maximum value of the l-th column. It can be argued that $H^\mu{}_l$ represents the number of investment project options that dominate the l-th column. Furthermore, $G^\mu{}_l$ shows the number of investment project options that dominate the l-th column, and $C^{\mu}_{kl\,max}$ represents the maximum level of dominance of the k-th version of the investment project over the l-th column.

We include these indicators in the matrix, as shown in Table 5.

Table 5. Matrix of variations.

Investment Projects, Variants Indicators	I_1	I_2	I_3
G^{μ}_l	1	1	1
H^{μ}_l	1	1	1
C^{μ}_{klmax}	1.51	1.05	5.01

Source: made by authors.

According to the results of Table 5, investment project I_2 turned out to be the best variable alternative with a minimum value of $C^{\mu}_{kl\,max}$. In this regard, the second version of investment design should be included in the Pareto tuple, but is excluded from the subsequent analysis. For this exception procedure, we delete the corresponding row and column in the preference matrix.

At the next stage, we analyze the other (above mentioned) options for investment design, and analyze them using the new matrix of indicators in a similar way.

As a result, the Pareto preference tuple can be denoted as $II = \{I_2, I_1, I_3\}$. In this regard, the best alternative for the vector of the heterogeneous performance index is $K\,(I\alpha) = \{K1\,(I\alpha), K2\,(I\alpha), K3\,(I\alpha), K4\,(I\alpha), K5\,(I\alpha), K6\,(I\alpha)\}$. In the Pareto tuple of the three options considered, those criteria that characterize NPV—discounting for calculating DPP in the vector efficiency index—became preferable.

Based on Table 5, the best alternative to an investment project with a minimum value $_{max}$ is option I_2. Therefore, the second version of the investment project is included in the Pareto tuple and excluded from further analyses by deleting the corresponding row and the column in the preference matrix.

The remaining options are analyzed using the new matrix of indicators in a similar manner.

Finally, the tuple of Pareto preferences can be obtained as $II = \{I_2, I_1, I_3\}$. Therefore, the best alternative for the vector inhomogeneous efficiency index $K(I_\alpha) = \{K_1(I_\alpha), K_2(I_\alpha), K_3(I_\alpha), K_4\,(I_\alpha), K_5(I_\alpha), K_6(I_\alpha)\}$ should be recognized as the second variant. In the Pareto tuple of the considered variants, preference was expressed for the criteria characterizing the NPV and discounting the calculation of the DPP in the vector efficiency index.

So, to answer the research questions which were submitted in Section 3, we consider these research points and describe them above to show the significance of this research:

(i). The proposed methodology for using interval values and assessing the effectiveness of investment projects in the field of industrial waste management gives experts new opportunities for the simpler and more accurate analysis of proposed opinions. In addition, uncertainties can be considered without additional statistic indicators.

(ii). This allows managers to evaluate various criteria in different environmental conditions and the influence of factors; furthermore, it allows reflecting on various aspects of the measured phenomena (in particular, the effectiveness of the waste processing investment project).

(iii). The proposed set of indicators may include an extended list of influential factors beyond those that are part of the project environment.

5. Discussion

Since Korea's volume of waste is evidence of changing lifestyles in the midst of a trend toward a convenience-oriented life (single-use products, convenient goods, instant food, etc.) and an abundant capitalist socioeconomic environment (mass consumption/mass production), the current state of waste management in Korea is at a turning point, where a paradigm shift from a convenience-oriented society (single-use product society) to a society oriented toward resource conservation (resource-circulating society) is taking place [71,72]. Wastes have a close relationship with each country's life and cultural patterns, as well as with changes in society, patterns of waste generation and treatment, thereof, change.

A summary of the evolution of Korea's system of legislations for waste management shows that this evolution has been taking place alongside the flow of developmental processes in Korea, and each developing country needs to introduce waste policies suitable for its current economic and social conditions. These policies should be supported with investment perspectives and based upon the development of technology and innovation [73–75].

According to the well-developed support policies of waste treatment investment projects and new environmental regulations, the Korean Government expects to observe some effects in terms of the economic, environmental and social impacts of the Framework Act on Resource Circulation (Table 6).

Table 6. Expected results from investment activities in Korean waste management systems.

Impact	Current Situation	After Investment and Introduction of New Regulations and Technology Performance
Economic	Increasing dependence on foreign resources due to the depletion of natural resources (97% of energy, 90% of mineral resources are imported).	Increased use of circulated resources (yearly quantity of recycling increases by about 10 million tons) → Recycling market of about 1 trillion and 70 million won; about 10,000 jobs created.
Environmental	Concerns of a trash crisis due to decreasing capacity of remaining landfill sites.	Decreased quantity of landfill disposal due to the levying of landfill disposal and incineration fees → increased lifespan of remaining landfill sites.
Social	Intensifying opposition of residents in surrounding areas due to worsening conditions in areas surrounding landfill sites.	Improvement of surrounding environments, support for local residents → co-existence of dischargers and residents living in areas surrounding landfill sites.

Source: adopted by authors from Joint work of related departments and agencies, September 2011, the 1st Framework Plan for Resource Circulation (2011–2015) [76].

Waste treatment investment projects are needed to be classified according to the type of waste treatment and operational technology. Each investment project requires risk analysis, which should be evaluated. For further research, it is necessary to identify the correlation between risk type and used technology in waste treatment operations.

Risk-evaluation procedures are the importing starting points for waste management. A well-designed risk evaluation system presents a structured mechanism for searching for potential problems and creating judgements on the consequences. Assessing the risks of waste processing investment underpins the "suitable for use" approach adopted by the Korean regulatory mechanism and supports planning policy [77]. The main goal is to decide whether there are any unacceptable risks to people or the wider environment—including industries. The risk evaluation process can be very detailed, particularly where risks are diversified. For the discussion of investment in waste recycling, there are a range of specific technical approaches for different contaminants and circumstances.

However, these all broadly fit within a general process that can be seen as a tiered approach. Each tier is applied to the case circumstances, with an increasing level of detail information required by the assessor in progressing through each tier [78].

There are three tiers, or steps:

1. Basic risk evaluation.
2. Risk assessment using generic criteria and assumptions—where a contamination concentration is compared against a generic Soil Guideline Value (SGV).
3. Risk assessment using specific criteria and assumptions—where a detailed, site-specific approach is used.

During the management of investment projects in the field of industrial waste management, it is important to determine the possible influence of various external factors on the result (market factors, production factors, technological factors, social factors, etc.). In this case, the degree of occurrence of a particular risk is determined and an initial assessment of risk is carried out. Depending on the likelihood of a negative event in the external environment of the investment project and the degree of negative influence of factors on its successful implementation, managers can use only one approach to assess risks, or conduct a multi-stage risk analysis. Usually it depends on the properties (complexity and degree of influence) of pollutants—some of them can be evaluated using common criteria; others may need to develop specific indicators for a more detailed and comprehensive risk assessment.

6. Conclusions

Our proposed investment project selection algorithm discusses the associated risks and allows us to determine the degree of their influence on the result of improving the environmental safety systems of production facilities. Do not forget that this examines the various conditions and components of the economic and environmental systems, as well as identifies all possible risks in each of them. This is achieved by describing risk situations and introducing a multi-component presentation of the risk component as one of the decision criteria. In subsequent studies, it will be necessary to identify a correlation effect between environmental factors and elements of the ecological and economic systems.

The investment project appraisal approach proposed in this study opens up new possibilities for applying the multi-criteria selection method to the conditions of economic and environmental activities in the field of waste management.

For the successful selection and implementation of an investment project in the field of waste management, it is extremely important to carry out the following management activities:

(i)   Determine the form of investment policy of the enterprise.
(ii)   Develop the structure of the investment project.
(iii)   Study the multi-factor impact of local or global economic and environmental environment on the investment project management.
(iv)   Calculate the volume of risk conditions and classify all possible risks.
(v)   Develop a risk management scenario and process calculations for their management and dissemination.

This paper is a combination of three elements: (i) presentation of the South Korean policy for waste, (ii) case study of the Samsung Corporation waste management system, and (iii) a multidimensional approach to risk-analysis in investment projects. Let us summarize the correlation between these three parts:

(i)   Presentation of the South Korean policy for waste—this refers to the different programs and advanced technologies which are presented in Korean waste management practice. The authors show how Korean companies and legal organizations develop systems of waste management.
(ii)   The story of the Samsung Corporation is presented in case study format. The authors present some successful examples of waste management decisions by the company, with both risks and open questions, including how to successfully manage waste treatment projects in local and global markets, and how to install advance technologies within a cost-leadership strategy.
(iii)   The authors discuss four primary reasons justifying the use of multicriterial (MCA) methods in waste management practice; they are mentioned in Section 3.

Thus, our proposed method allows us to use multidimensional and specific information for the process of making comprehensive economic and environmental (managerial) decisions in a market system. Moreover, our algorithm can be used to make long-term strategic decisions in the field of investment risk management in environmental and economic systems.

Author Contributions: J.H.L. and O.A.S. conceptualized the study; J.H.L. designed the methodology; O.A.S. carried out the investigation; J.H.L. and O.A.S. did the analyses, validation, and data curation; O.A.S. wrote the paper, and J.H.L. contributed to reviewing and editing all sections; J.H.L. supervised the work. All authors have read and agreed to the published version of the manuscript.

Funding: This research was done as a part of Research project 2020-0108 and was funded in 2020 by Korea University of Technology and Education (KOREATECH).

Conflicts of Interest: The authors declare no conflict of interest.

References

1. Knowledge Sharing Program Development Research and Learning Network. Available online: http://www.ksp.go.kr/english/index (accessed on 12 March 2020).
2. Kim, K.; Kim, Y. *Volume-based Waste Fee System in Korea*; Ministry of Strategy and Financy: Seoul, Korea, 2012; pp. 45–78.
3. Hou, L. South Korea's Food Waste Solution: You Waste, You Pay. Commonwealth Magazine, 2013. Available online: http://english.cw.com.tw/article.do?action=show&id=14067 (accessed on 12 March 2020).
4. Bekun, F.V.; Alola, A.A.; Sarkodie, S.A. Toward a sustainable environment: Nexus between CO2 emissions, resource rent, renewable and nonrenewable energy in 16-EU countries. *Sci. Total Environ.* **2019**, *657*, 1023–1029. [CrossRef]
5. Shvetsova, O.A. Development of Environmental Management in South Korea: Practice of Industrial Waste Processing. In Proceedings of the 2018 IEEE International Conference in Management of Municipal Waste as an Important Factor of Sustainable Urban Development, WASTE, St. Petersburg, Russia, 4–6 October 2018; pp. 93–97.
6. Paramati, S.R.; Apergis, N.; Ummalla, M. Dynamics of renewable energy consumption and economic activities across the agriculture, industry, and service sectors: Evidence in the perspective of sustainable development. *Environ. Sci. Pollut. Res.* **2018**, *25*, 1375–1387. [CrossRef]
7. Kim, K. *Performance of Waste Management Policy in Korea*; Ministry of Environment: Seoul, Korea, 2008.
8. Waste Management Review. South Korea Legislates Towards a Zero Waste Society. Waste Management Review, 2015. Available online: http://wastemanagementreview.com.au/south-korea-legislates-towards-a-zero-waste-society/ (accessed on 12 March 2020).
9. Korea's Environmental Review, 2018, Ministry of Environment, ECOREA. Available online: http://eng.me.go.kr (accessed on 12 March 2020).
10. OECD Economic Report 2017. Available online: https://www.oecd-ilibrary.org/economics/oecd-economic-surveys-korea_19990707 (accessed on 12 March 2020).
11. OECD Waste Statistics. 2015. Available online: Data.oecd.org/waste/municipal-waste.htm (accessed on 20 November 2017).
12. Jones, R.; Yoo, B. *Korea's Green Growth Strategy: Mitigating Climate Change and Developing New Growth Engines*; OECD Economics Department Working Papers, No. 798; OECD Publishing: Paris, France, 2011. [CrossRef]
13. OECD. *OECD Economic Surveys: Korea 2018*; OECD Publishing: Paris, France, 2018. [CrossRef]
14. Daly, H.E. Some overlaps between the first and second thirty years of ecological economics. *Ecol. Econ.* **2019**, *164*, 106–372. [CrossRef]
15. CMG recycling solutions.-[on-line access]. Available online: http://www.thplastics.co.uk/tag/cmg-recycling-solutions/ (accessed on 12 March 2020).
16. Claborn, K.A.; Brooks, J.S. Can we consume less and gain more? Environmental efficiency of well-being at the individual level. *Ecol. Econ.* **2019**, *156*, 110–120. [CrossRef]
17. Freeman, R. A theory on the future of the rebound effect in a resource-constrained world. *Front. Energy Res.* **2018**, *6*, 81. [CrossRef]
18. Wackernagel, M.; Hanscom, L.; Lin, D. Making the Sustainable Development Goals consistent with sustainability. *Front. Energy Res.* **2017**, *5*, 18. [CrossRef]
19. Hogan, B. Technology Trumps Food Waste in South Korea. Food Waste Focus, 2015. Available online: http://blog.leanpath.com/technology-trumps-food-waste-south-korea (accessed on 12 March 2020).
20. Mancini, M.S.; Galli, A.; Coscieme, L. Exploring ecosystem services assessment through Ecological Footprint accounting. *Ecosyst. Serv.* **2018**, *30*, 228–235. [CrossRef]
21. Haasnoot, M.; Kwakkel, J.H.; Walker, W.E.; Ter Maat, J. Dynamic adaptive policy pathways: A method for crafting robust decisions for a deeply uncertain world. *Glob Environ. Chang.* **2013**, *23*, 485–498. [CrossRef]

22. Tonmoy, F.N.; Rissik, D.; Palutikof, J.P. A three-tier risk assessment process for climate change adaptation at a local scale. *Clim. Chang.* **2019**, *153*, 539–557. [CrossRef]

23. Ayers, J.; Anderson, S.; Pradhan, S.; Rossing, T. *Participatory Monitoring, Evaluation, Reflection and Learning for Community-Based Adaptation: A Manual for Local Practitioners*; CARE International: Atlanta, GA, USA, 2012.

24. *Research into Modifying/Supplementing Basic Guidelines for Investigations of Preliminary Feasibility (Report)*; Korea Development Research Institute: Seoul, Korea, 2014; pp. 89–90.

25. GiYoung, R. *Zero Waste City, Seoul's New Project*; SDI Policy Report No. 61; Seoul Research Institute: Seoul, Korea, 2010; pp. 56–59.

26. Hyun, S. *Analysis of the Sorting System under the 'Wastes Control Act'*; Legal Research Vol. 41; Korea Legal Research Institute: Seoul, Korea, 2013; pp. 34–45.

27. *Transitioning to a Resource Circulating Society, Choices for the Future of Society*; Korea Ministry of Environment: Seoul, Korea, 2015; pp. 67–71.

28. Tonmoy, F.N.; El-Zein, A. *Vulnerability of Infrastructure to Sea Level Rise: A Combined Outranking and System-Dynamics Approach*; European Safety and Reliability (ESREL-2013); CRC Press: Cleveland, OH, USA, 2013; pp. 2407–2414.

29. Pini, B.; River, S.W.; Mckenzie, F.M.H. Factors inhibiting local government engagement in environmental sustainability: Case studies from rural Australia. *Aust. Geogr.* **2007**, *38*, 161–175. [CrossRef]

30. JungIm, Y. *Methods of Establishing the Foundations of a Zero Waste City*; GRI Policy Research No. 63; Gyeonggi Development Research Institute: Gyeonggi, Korea, 2013; pp. 23–27.

31. Park, J.W. *3R Policies of Korea*; Ministry of Environment: Seoul, Korea, 2009. Available online: https://www.kdi.re.kr/kdi_eng/publications/reports.jsp (accessed on 15 November 2019).

32. *Annual Report of Environmental Protection Department of Hong Kong*; Ministry of Environment of South Korea: Seoul, Korea, 2018.

33. Ministry of Environment, Government of the Republic of Korea. 1997 to 2015. Current Status of Waste Generation and Treatment (in Korean). Available online: http://eng.me.go.kr/eng/web/index.do?menuId=394 (accessed on 15 November 2019).

34. Chen, C.; Hellmann, J.; Berrang-Ford, L. A global assessment of adaptation investment from the perspectives of equity and efficiency. *Mitig. Adapt. Strat. Glob Chang.* **2018**, *23*, 101–122. [CrossRef]

35. Samsung's, LG's Marketing Costs Rise in 2016 Amid Heightened Competition. Yonhap News Agency, 2017. Available online: http://english.yonhapnews.co.kr/news/2017/04/05/0200000000AEN20170405001800320.html?input=rss (accessed on 2 March 2020).

36. Samsung Corporation Website. Available online: https://www.samsung.com/ca/aboutsamsung/sustainability/environment/resource-efficiency/ (accessed on 17 February 2020).

37. Beinat, E.; Nijkamp, P. *Multi-Criteria Evaluation in Land Use Management*; Kluwer Academic Publishers: Dordrecht, The Netherlands, 1998; pp. 56–68.

38. Brav, A.; Graham, J.R.; Harvey, C.R.; Michaely, R. Payout policy in the 21st century. *J. Financ. Econ.* **2005**, *77*, 483–527. [CrossRef]

39. Shvetsova, O.A.; Rodionova, H.A.; Epstein, M.Z. International Evaluation of Investment Projects under Uncertainty: Multi-Criteria Approach Using Interval Data. *Int. J. Entrep. Sustain. Issues* **2018**, *7*, 46–68.

40. Brigham, E.F.; Ehrhardt, M.C. *Financial Management: Theory and Practice*; South-Western College Publishing: Cincinnati, OH, USA, 2015; pp. 23–35.

41. Vedernikov, Y.V.; Mogilenko, V.V. Scientific and methodical apparatus of vector preference for complex technical systems characterized by quality indicators specified in a limited-indefinite form, Issues of Modern Science and Practice, System Analysis, *Autom. Manag.* **2011**, *32*, 81–96.

42. Rodionova, E.A.; Trifonova, N.V.; Epstein, M.Z.; Shvetsova, O.A. Multicriterial approach to estimation of economic efficiency based on regional innovative cluster. In Proceedings of the 20th IEEE International Conference on Soft Computing and Measurements, SCM, St. Petersburg, Russia, 24–26 May 2017; pp. 89–97.

43. Rodionova, E.A.; Epshtein, M.Z.; Petukhov, L.V. Multivariate evaluation of investment projects based on interval preferences, Scientific and Technical Sheets of the Saint-Petersburg Polytechnic University. Information. *Telecommun. Manag.* **2013**, *169*, 141–148.

44. Rodionova, E.A.; Shvetsova, O.A.; Michael, Z.E. Multicriterial Approach to Investment Projects: Estimation under Risk Conditions. *Revista Espacios* **2018**, *39*, 28–44.

45. Bukhvalov, A.V.; Bukhvalov, V.V.; Idelson, A.V. *Financial Calculations for Professionals*; BHV-Petersburg: St. Petersburg, Russia, 2011; pp. 78–90.

46. Grierson, D.E. Pareto multi-criteria decision making. *Adv. Eng. Inform.* **2008**, *22*, 371–384. [CrossRef]

47. Gurumurthy, A.; Kodali, R. Multi-criteria decision-making model for the justification of lean manufacturing systems. *Int. J. Manag. Sci. Eng. Manag.* **2013**, *3*, 100–118. Available online: http://www.worldacademicunion.com/journal/MSEM/msemVol03No02paper02 (accessed on 12 December 2019). [CrossRef]

48. Hayashi, K. Multicriteria analysis for agricultural resource management: A critical survey and future perspectives. *Eur. J. Oper. Res.* **2000**, *122*, 486–500. [CrossRef]

49. Jelnova, C.V. Analysis of the Practice of Decision-Making in the Field of Investment Policy, Contemporary Economic Issues 4, 2013. Available online: http://economic-journal.net/index.php/CEI/article/view/83/70 (accessed on 11 March 2020). [CrossRef]

50. Mardani, A.; Jusoh, A.; Zavadskas, E.K.; Khalifah, Z.; Nor, K. Application of multiple-criteria decision-making techniques and approaches to evaluating of service quality: A systematic review of the literature. *J. Bus. Econ. Manag.* **2015**, *16*, 1034–1068. [CrossRef]

51. Mazur, I.I.; Shapiro, V.D.; Olderogge, N.G. *Project Management: Practical Allowance for Universities*; Omega-L: Moscow, Russia, 2014; pp. 12–17.

52. Kangas, J.; Kamgas, A. Multiple criteria decision support methods in forest management. An overview and comparative analyses. In *Multi-Objective Forest Planning*; Pukkala, T., Ed.; Kluwer Academic Publishers: Dordrecht, The Netherlands, 2002; pp. 37–70. [CrossRef]

53. Orlovsky, S.A. *Problems of Decision Making with Fuzzy Source Information*; Nauka: Moscow, Russia, 1981; pp. 78–89.

54. Serguieva, A.; Hunter, J. Fuzzy interval methods in the investment risk appraisal. *Fuzzy Sets Syst.* **2014**, *142*, 443–466. [CrossRef]

55. Stoyanova, E.S.; Krylova, T.B. *Financial Management: Theory and Practice*; Perspectiva: Moscow, Russia, 2006; pp. 56–78.

56. Olson, D. *Decision Aids for Selection Problems*; Springer: New York, NY, USA, 1995. [CrossRef]

57. Minakova, L.V.; Anikanov, P.V. Modelling of area of possible results of the innovative investment project. *Contemp. Econ. Issues* **2013**, *1*, 23–28.

58. Parrino, R.; Kidwell, D.; Bates, T. *Essentials of Corporate Finance*; Wiley: New York, NY, USA, 2014; pp. 55–57.

59. Roy, B. *Problems and Methods of Solutions in Problems with Many Objective Functions, in Analysis Questions and Decision-Making Procedures*; MIR: Moscow, Russia, 1976; pp. 20–58.

60. Saaty, T.L. *Multi-Criteria Decision Making: The Analytic Hierarchy Process*; RWS Publications: Pittsburgh, PA, USA, 1990; pp. 12–18.

61. Keshavarz Ghorabaee, M.; Zavadskas, E.K.; Olfat, L.; Turskis, Z. Multi-criteria inventory classification using a new method of evaluation based on distance from average solution (EDAS). *Informatica* **2015**, *26*, 435–451. [CrossRef]

62. Opricovic, S.; Tzeng, G.H. Compromise solution by MCDM methods: A comparative analysis of VIKOR and TOPSIS. *Eur. J. Oper. Res.* **2004**, *156*, 445–455. [CrossRef]

63. Wang, J.J.; Jing, Y.Y.; Zhang, C.F.; Zhao, J.H. Review on Multi-Criteria Decision Analysis Aid in Sustainable Energy Decision-Making. *Renew. Sustain. Energy Rev.* **2009**, *13*, 2263–2278. [CrossRef]

64. Zare, M.; Pahl, C.; Rahnama, H.; Nilashi, M.; Mardani, A.; Ibrahim, O.; Ahmadi, H. Multi-criteria decision-making approach in e-learning: A systematic review and classification. *Appl. Soft Comput.* **2016**, *45*, 108–128. [CrossRef]

65. Tarp, S. Theory-based lexicographical methods in a functional perspective. An overview. *Lexicographica* **2014**, *30*, 58–76. [CrossRef]

66. Tsoutos, T.; Drandaki, M.; Frantzeskaki, N.; Iosifidis, E.; Kiosses, I. Sustainable energy planning by using multi-criteria analysis application in the Island of Crete. *Energy Policy* **2009**, *37*, 1587–1600. [CrossRef]

67. Seitz, N.E.; Ellison, M. *Capital Budgeting and Long-Term Financing Decisions*; Harcourt Brace College Publishers: Fort Worth, TX, USA, 1999; pp. 45–56.

68. Savchuk, V.P. *Evaluation of the Investment Projects' Effectiveness*; Phoenix: Moscow, Russia, 2007; pp. 66–78.

69. Syroezhin, I.M. *Perfection of the System of Efficiency and Quality Indicators*; The Economy: Moscow, Russia, 1980; pp. 78–98.

70. Prisyach, E.Y.; Shvetsova, O.A. Elements of Innovative Scenario's Development of Waste Management System in Russia. In Proceedings of the 2018 IEEE International Conference in Management of Municipal Waste as an Important Factor of Sustainable Urban Development, WASTE, St. Petersburg, Russia, 4–6 October 2018; pp. 89–93.

71. Shvetsova, O.A.; Suthar, B. Business Trends and Opportunities of South Korea and India Cooperation. In Proceedings of the 2018 International Conference "Quality Management, Transport and Information Security, Information Technologies", IT and QM and IS, St. Petersburg, Russia, 24–28 September 2018; pp. 53–59.

72. Khokhlov, N.V. *Risk Management: Practical Allowance for Universities*; UNITY-DANA: Moscow, Russia, 2011.

73. Laufman, G. *To Have and Have Not*; CFO Publishing Corporation: New York, NY, USA, 1998; pp. 78–79.

74. Lukicheva, L.I.; Egorychev, D.N. *Decision Making Process in Management*; Omega-M: Moscow, Russia, 2016; pp. 99–109.

75. Epstein, M.Z.; Shvetsova, O.A. Development of Innovative Strategies: Comparative Analysis of External and Internal Environments. *Revista Espacios* **2018**, *40*, 45–56.

76. *National Disaster Recovery Framework Strengthening Disaster Recovery for the Nation*; FEMA: Washington, DC, USA, 2016; pp. 78–99.

77. Shvetsova, O.A. Practical and theoretical issues of innovation management in South Korea. *Int. J. Innov. Technol.* **2017**, *4*, 78–89.

78. Shvetsova, O.A. Management of small and medium enterprises in global environment. In Proceedings of the IEEE Proceedings of 2nd International Conference on Control in Technical Systems, CTS, St. Petersburg, Russia, 25–27 October 2017; Volume 2017, pp. 45–56.

© 2020 by the authors. Licensee MDPI, Basel, Switzerland. This article is an open access article distributed under the terms and conditions of the Creative Commons Attribution (CC BY) license (http://creativecommons.org/licenses/by/4.0/).

MDPI
St. Alban-Anlage 66
4052 Basel
Switzerland
Tel. +41 61 683 77 34
Fax +41 61 302 89 18
www.mdpi.com

Applied Sciences Editorial Office
E-mail: applsci@mdpi.com
www.mdpi.com/journal/applsci